Data Analysis for the Geosciences

Advanced Textbook Series

1. **Unconventional Hydrocarbon Resources: Techniques for Reservoir Engineering Analysis**
   Reza Barati and Mustafa M. Alhubail
2. **Geomorphology and Natural Hazards: Understanding Landscape Change for Disaster Mitigation**
   Tim R. Davies, Oliver Korup, and John J. Clague
3. **Remote Sensing Physics: An Introduction to Observing Earth from Space**
   Rick Chapman and Richard Gasparovic
4. **Geology and Mineralogy of Gemstones**
   David Turner and Lee A. Groat
5. **Data Analysis for the Geosciences: Essentials of Uncertainty, Comparison, and Visualization**
   Michael W. Liemohn

Advanced Textbook 5

Data Analysis for the Geosciences

Essentials of Uncertainty, Comparison, and Visualization

Michael W. Liemohn
University of Michigan, USA

This work is a co-publication of the American Geophysical Union and John Wiley and Sons, Inc.

This edition first published 2024
© 2024 American Geophysical Union

All rights reserved. No part of this publication may be reproduced, stored in a retrieval system, or transmitted, in any form or by any means, electronic, mechanical, photocopying, recording, or otherwise, except as permitted by law. Advice on how to obtain permission to reuse material from this title is available at http://www.wiley.com/go/permissions.

Published under the aegis of the AGU Publications Committee

Matthew Giampoala, Vice President, Publications
Carol Frost, Chair, Publications Committee
For details about the American Geophysical Union, visit us at www.agu.org.

The right of Michael W. Liemohn to be identified as the author of this work has been asserted in accordance with law.

Wiley Global Headquarters
111 River Street, Hoboken, NJ 07030, USA

For details of our global editorial offices, customer services, and more information about Wiley products, visit us at www.wiley.com.

Limit of Liability/Disclaimer of Warranty
While the publisher and authors have used their best efforts in preparing this work, they make no representations or warranties with respect to the accuracy or completeness of the contents of this work and specifically disclaim all warranties, including without limitation any implied warranties of merchantability or fitness for a particular purpose. No warranty may be created or extended by sales representatives, written sales materials, or promotional statements for this work. The fact that an organization, website, or product is referred to in this work as a citation and/or potential source of further information does not mean that the publisher and authors endorse the information or services the organization, website, or product may provide or recommendations it may make. This work is sold with the understanding that the publisher is not engaged in rendering professional services. The advice and strategies contained herein may not be suitable for your situation. You should consult with a specialist where appropriate. Further, readers should be aware that websites listed in this work may have changed or disappeared between when this work was written and when it is read. Neither the publisher nor authors shall be liable for any loss of profit or any other commercial damages, including but not limited to special, incidental, consequential, or other damages.

Library of Congress Cataloging-in-Publication Data applied for:
9781119747871 (Paperback); 9781119747888 (Adobe PDF); 9781119747895 (ePub)

Cover Design: Wiley
Cover Image: Aurora borealis, Iceland © Cavan Images/Getty Images

Set in 9.5/12.5pt STIXTwoText by Straive, Pondicherry, India

SKY10088349_101824

To Ginger, for inspiring me to take on a project like writing a book

Contents

Preface *xv*
Acknowledgments *xxi*
About the Companion Website *xxiii*

1 Assessment and Uncertainty: Examples and Introductory Concepts *1*
- 1.1 Chicken Little, Amateur Meteorologist — *2*
- 1.2 Uncertainty Ascribes Meaning to Values — *3*
- 1.3 Significant Figures — *3*
- 1.4 Types of Uncertainty — *7*
- 1.5 Example: Finding Saturn's Moons — *9*
- 1.6 Comparing Two Numbers: Are They Measuring the Same Value? — *11*
  - 1.6.1 Distributions of Number Sets — *12*
  - 1.6.2 The Gaussian Distribution — *13*
  - 1.6.3 Testing a Specific Value within a Data Set: The z Test — *14*
  - 1.6.4 Comparing Two Values Revisited — *18*
- 1.7 Use and Misuse of Statistics — *19*
- 1.8 Example: Solar Wind Density and Space Weather — *20*
- 1.9 Uncertainty and the Scientific Method — *22*
- 1.10 Further Reading — *24*
- 1.11 Exercises in the Geosciences — *26*

2 Plotting Data: Visualizing Sets of Numbers *27*
- 2.1 Plotting One-Dimensional Data — *27*
  - 2.1.1 What Makes a Good Plot? — *29*
  - 2.1.2 Exploratory Versus Explanatory Plot Styles — *31*
- 2.2 Example: Earth's Magnetic Field Strength — *33*
- 2.3 Probability Distributions—The Histogram — *35*
- 2.4 Plotting Two Data Sets Against Each Other — *39*
  - 2.4.1 Overlaid Histograms — *39*
  - 2.4.2 The Scatterplot — *40*
  - 2.4.3 The Box Plot — *42*
  - 2.4.4 The Box-and-Whisker Scatterplot — *43*
  - 2.4.5 The Running Average Plot — *44*

	2.5	Example: Temperature and Carbon Dioxide	48
	2.6	Scientific Visualization: A Sampling from the Literature	50
		2.6.1 A Very Brief History of Visualization	51
		2.6.2 Good Modern-Day Example Visualizations	53
	2.7	Visualization Best Practices	58
		2.7.1 Levels of Abstraction	58
		2.7.2 A Process for a Good Graphic	61
		2.7.3 Types of Colorblindness	63
		2.7.4 Color Scales	63
	2.8	Further Reading	65
	2.9	Exercises in the Geosciences	67

3 Uncertainty Analysis: Techniques for Propagating Uncertainty ... 69

3.1	Propagating Uncertainty	69
	3.1.1 Calculating Uncertainty with One Independent Variable	69
	3.1.2 Calculating Uncertainty with Two Independent Variables	70
	3.1.3 Calculating Uncertainty with Many Independent Variables	72
3.2	Example: Atmospheric Density	72
	3.2.1 The Hydrostatic Equilibrium Approximation	72
	3.2.2 One Independent Variable	73
	3.2.3 Two Independent Variables	74
	3.2.4 Many Independent Variables	74
3.3	Fractional and Percentage Uncertainties	75
3.4	Special Cases of Uncertainty Propagation	77
	3.4.1 Addition and Subtraction	77
	3.4.2 Multiplication and Division	78
	3.4.2.1 Multiplication of Two Parameters	78
	3.4.2.2 Uncertainty of Air Pressure	79
	3.4.2.3 Division with Correlated Variables	80
	3.4.2.4 Multiplication and Division with Independent Variables	81
	3.4.3 Power Laws	82
	3.4.4 Exponentials and Logarithms	82
	3.4.4.1 Exponential Functions	83
	3.4.4.2 Logarithmic Functions	84
	3.4.5 Trigonometric Functions	84
3.5	Stepwise Uncertainty Propagation	85
3.6	Example: Planetary Equilibrium Temperature	87
3.7	Multistep Processing	90
3.8	Final Advice on Uncertainty Propagation	91
3.9	Further Reading	93
3.10	Exercises in the Geosciences	93

4 Centroids and Spreads: Analyzing a Set of Numbers ... 95

4.1	Quantitatively Describing a Data Set: The Centroid	95
	4.1.1 Three Versions of Mean	96
	4.1.2 More Centroids: Median and Mode	98
	4.1.3 Histograms and the Arithmetic Mean	99

	4.2	Quantitatively Describing a Data Set: Spread	100
		4.2.1 Measures of Spread: Standard Deviation and Mean Absolute Difference	100
		4.2.2 Another Measure of Spread: Quantiles	102
		4.2.3 Spread Via Full Width at Half Maximum	106
		4.2.4 Spread as an L-p Norm	107
		4.2.5 Sample Versus Population	108
	4.3	Random and Systematic Error of a Data Set	109
	4.4	Which Centroid and Spread to Use and Other Tidbits of Advice	111
	4.5	Standard Deviation of the Mean	112
	4.6	Counting Statistics	113
	4.7	Example: Galactic Cosmic Rays	116
	4.8	Further Reading	119
	4.9	Exercises in the Geosciences	120

5 Assessing Normality: Tests for Assessing the Gaussian Nature of a Distribution ... 123

5.1	Histogram Check	124
5.2	Comparing Centroid and Spread Measures	126
5.3	Skew	128
5.4	Kurtosis	130
5.5	The Chi-Squared Test	132
5.6	The Kolmogorov–Smirnov Test	137
5.7	Example: pH in a Lake	139
5.8	Asymmetric Uncertainties	142
5.9	Outliers—Tests for a Single Data Value	144
5.10	Combining Centroid and Spread: The Weighted Average	146
5.11	Example: pH in a Lake Redux	148
5.12	Further Reading	149
5.13	Exercises in the Geosciences	150

6 Correlating Two Data Sets: Analyzing Two Sets of Numbers Together ... 153

6.1	Comparing Two Number Sets	153
	6.1.1 Chi-Squared and Kolmogorov–Smirnov Tests	154
	6.1.2 The Student's t Test	155
	6.1.3 The Welch's t Test	156
6.2	Linear Correlation	157
	6.2.1 Covariance of Two Data Sets	158
	6.2.2 Pearson Linear Correlation Coefficient	161
	6.2.3 Spearman Rank-Order Correlation	163
	6.2.4 Correlation with Logarithms	167
6.3	Example: Atmospheric Ozone and Temperature	168
6.4	Uncertainty of R	172
	6.4.1 The Jackknife Method	172
	6.4.2 The Bootstrap Method	173
	6.4.3 Uncertainty of R for the Ozone-Temperature Example	175
6.5	Correlation and Causation	177
6.6	Further Reading	178
6.7	Exercises in the Geosciences	179

7 Curve Fitting: Fitting a Line between Two Sets of Numbers ... 181
- 7.1 Linear Regression ... 181
  - 7.1.1 Obtaining A and B ... 181
  - 7.1.2 Uncertainties on A and B ... 185
  - 7.1.3 The Zero-Intercept Special Case ... 186
  - 7.1.4 Weighted Linear Fitting ... 187
- 7.2 Testing a Linear Fit ... 188
- 7.3 Example: Human-Induced Seismicity ... 191
- 7.4 Nonlinear Fitting ... 194
  - 7.4.1 Polynomial Fitting ... 194
  - 7.4.2 Generalized "Linear Coefficient" Fitting ... 196
  - 7.4.3 Exponential Fitting: Linearizing the Dependence on Coefficients ... 197
  - 7.4.4 Piecewise Linear Fitting ... 198
  - 7.4.5 Advice about Curve Fitting ... 199
- 7.5 Example: The Ozone Hole ... 200
- 7.6 Iterative Curve Fitting ... 203
  - 7.6.1 One-Dimensional Iterative Curve Fitting ... 203
  - 7.6.2 Multidimensional Iterative Curve Fitting ... 205
  - 7.6.3 Gradient Descent Curve Fitting ... 208
- 7.7 Final Thoughts on Curve Fitting ... 209
- 7.8 Further Reading ... 210
- 7.9 Exercises in the Geosciences ... 210

8 Data-Model Comparison Basics: Philosophies of Calculating and Categorizing Metrics ... 213
- 8.1 Example Model: River Flow Rate ... 213
- 8.2 What Is a Model? ... 214
- 8.3 Visualizing Observed and Modeled Values Together ... 217
  - 8.3.1 Scatterplots of Data and Model Values ... 217
  - 8.3.2 The 2D Histogram Plot ... 219
  - 8.3.3 Overlaid Histogram Plots ... 221
  - 8.3.4 Cumulative Probability Distribution Plots ... 222
  - 8.3.5 Quantile–Quantile Plots ... 224
- 8.4 Example: Total Solar Irradiance ... 226
- 8.5 A Diverse Zoo of Metrics ... 229
  - 8.5.1 The Primary Categories of Metrics ... 230
  - 8.5.2 Skill ... 231
  - 8.5.3 Metrics Categories Based on Subsetting ... 234
- 8.6 The Concept of Model "Goodness of Fit" ... 235
- 8.7 Application Usability Levels ... 236
- 8.8 Designing a Meaningful Data-Model Comparison ... 237
- 8.9 Further Reading ... 239
- 8.10 Exercises in the Geosciences ... 240

Contents | xi

9 Fit Performance Metrics: Data-Model Comparisons Based on Exact Observed and Modeled Values 243
- 9.1 What Is Fit Performance? — 244
- 9.2 Running Example: Dst and the O'Brien Model — 245
- 9.3 Accuracy — 250
  - 9.3.1 The Big Three of Accuracy: MSE, RMSE, and MAE — 251
  - 9.3.2 Neglecting Degrees of Freedom — 253
  - 9.3.3 Normalizing the Accuracy Measure — 256
  - 9.3.4 Percentage Accuracy Metrics — 257
  - 9.3.5 Choosing the Right Accuracy Metric — 261
- 9.4 Bias — 262
  - 9.4.1 Mean Error — 262
  - 9.4.2 Percentage Bias — 265
- 9.5 Precision — 266
  - 9.5.1 Modeling Yield — 266
  - 9.5.2 Definitions of Precision Using Standard Deviation — 268
- 9.6 Association — 268
  - 9.6.1 Correlation Coefficient — 269
  - 9.6.2 Nonlinear Association Metrics — 270
- 9.7 Extremes — 272
  - 9.7.1 Extremes of the Cumulative Probability Distribution — 272
  - 9.7.2 Using Skew and Kurtosis for an Extremes Assessment — 276
- 9.8 Skill — 278
  - 9.8.1 Prediction Efficiency — 278
  - 9.8.2 Other Options for Fit Performance Skill — 279
- 9.9 Discrimination — 281
- 9.10 Reliability — 283
- 9.11 Summarizing the Running Example — 286
- 9.12 Summary of Fit Performance Metrics — 287
- 9.13 Further Reading — 291
- 9.14 Exercises in the Geosciences — 292

10 Event Detection Metrics: Comparing Observed and Modeled Number Sets When Only Event Status Matters 295
- 10.1 Defining an Event — 296
- 10.2 Contingency Tables — 299
- 10.3 Data-Model Comparisons with Events — 301
- 10.4 Running Example: Will It Rain? — 303
- 10.5 Significance of a Contingency Table — 307
- 10.6 Accuracy — 310
- 10.7 Bias — 311
- 10.8 Precision — 313
- 10.9 Association — 314
  - 10.9.1 Odds Ratio — 315
  - 10.9.2 Odds Ratio Skill Score — 316
  - 10.9.3 Matthews Correlation Coefficient — 317
- 10.10 Extremes — 317

	10.11	Skill	321
		10.11.1 Heidke Skill Score	321
		10.11.2 Peirce and Clayton Skill Scores	323
		10.11.3 Gilbert Skill Score	324
	10.12	Discrimination	325
	10.13	Reliability	326
	10.14	Summarizing the Running Example	327
	10.15	Summary of Event Detection Metrics	328
	10.16	Further Reading	330
	10.17	Exercises in the Geosciences	331

11 Sliding Thresholds: Event Detection Metrics with a Variable Event Identification 333

	11.1	Sliding the Event Identification Thresholds	334
	11.2	Sweeping the Modeled Threshold	337
	11.3	Sweeping the Data Threshold	340
	11.4	Sweeping Both Thresholds Simultaneously	342
	11.5	Metric-Versus-Metric Curves	344
		11.5.1 ROC Curves	344
		11.5.2 Alt-ROC Curves	346
		11.5.3 STONE Curves	347
	11.6	Application of Sliding Thresholds to the Geophysical Running Examples	349
		11.6.1 Event Definitions for the Running Examples	349
		11.6.2 Metric-Versus-Modeled Threshold Curves for the Running Examples	352
		11.6.3 Metric-Versus-Observed Threshold Curves for the Running Examples	355
		11.6.4 Metric-Versus-Simultaneous Threshold Sweep Curves for the Running Examples	357
		11.6.5 Metric-Versus-Metric Analysis for the Running Examples	359
	11.7	The Power of Sliding Thresholds	362
	11.8	Further Reading	364
	11.9	Exercises in the Geosciences	365

12 Applications of Metrics and Uncertainty: Final Advice and Introductions to Advanced Topics 367

	12.1	Choosing the Right Set of Metrics	367
		12.1.1 Metrics for Fit Performance Assessment on Gaussian Distributions	368
		12.1.2 Metrics for Fit Performance Assessment on Non-Gaussian Distributions	369
		12.1.3 Metrics for Event Detection Assessment	372
	12.2	Combining Metrics for Robust Data-Model Comparisons	374
		12.2.1 The Accuracy–Bias–Precision Trifecta	374
		12.2.2 The Accuracy–Association Connection	376
		12.2.3 The Association–Extremes Linkage	377
		12.2.4 Expanding Our Understanding of Skill	378
		12.2.5 Using Discrimination and Reliability Together	379
	12.3	Uncertainty on Metrics	380
	12.4	Uncertainty on Fit Performance Metrics for the Dst Running Example	381

12.5	A Recipe for Robust Comparisons		*385*
12.6	Metrics and Decision-Making		*387*
	12.6.1	Choice Combination Statistics	*388*
	12.6.2	Example: Spacecraft-Charging Model	*390*
12.7	Additional Advanced Topics		*392*
	12.7.1	Periodicity Analysis	*392*
	12.7.2	Time-Lagged Analysis	*393*
	12.7.3	Additional Tests	*394*
	12.7.4	Multidimensional Data Analysis	*394*
	12.7.5	Multidimensional Data-Model Comparisons	*396*
	12.7.6	Uncertainty Quantification	*397*
	12.7.7	Design of Experiments	*398*
	12.7.8	Geographical Information System (GIS) Analysis	*398*
	12.7.9	Machine Learning	*399*
12.8	Uncertainty and the Scientist		*400*
12.9	Further Reading		*402*
12.10	Exercises in Geoscience		*406*

Index *407*

Preface

A critical element of a robust undergraduate education in science, technology, engineering, and mathematics (STEM) disciplines is an understanding of sets of numbers and how to process, plot, and compare them. This concept is usually first taught in introductory laboratory courses and reinforced in the advanced lab classes. These labs, however, are usually focused on the scientific concept being explored by the experiment as well as the methodologies of setting up the equipment and making the measurements. These classes often only give a brief glimpse into the many techniques for analyzing the obtained number sets, and these courses often devote even less attention to how one should visualize the number sets and compare number sets.

STEM students need to gain an appreciation of the uncertainties surrounding observations and model results. This uncertainty strongly governs the interpretation of the values and especially the comparison of several values. A related concept is uncertainty propagation, keeping track of the uncertainties as the values are processed (e.g., used as a value in an equation to yield a new number). Without a grasp on the uncertainty of a given number, its comparison with other numbers is meaningless.

Some STEM departments require their students to take a statistics course as part of the undergraduate degree program. While this is a highly worthwhile and useful topic for such students to understand, it typically does not cover the full range of issues concerning data analysis, visualization, and comparative metrics techniques that a practicing scientist or engineer should know. Applied statistics courses for STEM majors often cover the basic statistical processing for a single data set (calculating mean and standard deviation, for instance) and for two data sets (calculating a linear regression fit and correlation coefficient, for instance). It usually covers the basics of comparing those two data sets, including t tests, chi-squared tests, and F-statistic tests. It usually does not cover very much in the way of visualization of the data and almost never covers comparison metrics beyond the correlation coefficient.

Using numerous examples from the Earth, ocean, atmospheric, space, and planetary sciences, this volume presents a comprehensive introduction to data analysis, visualization, and data-model comparisons and metrics, within the framework of the uncertainty around the values. Currently, I teach an upper-level undergraduate course, "Data Analysis and Visualization for Geoscientists," at the University of Michigan, for which this textbook is written. The volume can be used as the text for a data analysis course, a supplement for an advanced laboratory series, or as a reference resource, for everyone from upper-level undergraduate students to experienced researchers in STEM fields.

While data-model comparisons have always been an essential component of scientific research, it is often a topic not rigorously introduced at the undergraduate level. This is no longer acceptable, especially with the advent of machine learning as a fast-growing field of analysis. A fundamental trait of machine learning is the optimization of the computer-developed model, fitting its result to the training data set. Undergraduate science students gain experience as data analysts, but for some reason, data-model comparisons are barely mentioned in most undergraduate curricula. Many of these students are going straight into the industrial and commercial sector at ever-increasing rates, often as data analytics experts. To be an effective data scientist and user of advanced statistical applications including machine learning, these students should have an understanding and appreciation of data-model comparison techniques.

How to Use This Book

This is a data analysis textbook for upper level undergraduate STEM students. It is designed to be their statistics course in the degree program, offering them a learning experience based on real geophysical observations. Data from geoscience examples are used throughout the book to actively engage the reader in the concept of uncertainty as a leading factor in the interpretation and usage of a set of numbers. Note that this book intentionally avoids many of the derivations of the formulas presented. Some are given, but only for context to understand the assumptions built into the formula so that students learn the limitations of that particular formula. No derivations are assigned in the homework problems at the end of the chapters. If an instructor wanted to add derivations to the assignments, then please feel free to do so, but I am omitting them intentionally because I want the focus to be on the application of the formulas for scientific investigations.

This book should be useful to students across all science disciplines, meant to serve as an initial course in scientific data analysis and hypothesis testing. This book provides the precursor knowledge to understanding machine learning techniques. While it does not explicitly cover machine learning, it provides a critical toolkit for students to fully understand, appreciate, and optimally use the latest machine learning advancements in data science. A key topic in geosciences that it does not cover is periodicity analysis. In my department at the University of Michigan, we have a separate course that deals explicitly with this subject, exploring Fourier transforms and other periodicity methodologies, and then teaching students how to interpret the resulting power spectral density graphs in a scientific context. Inclusion of this topic could occur in a future edition, when one merged book is used for both of these undergraduate geophysical data analysis courses, but that endeavor is reserved for the future.

While this book could be assigned as a reference text in conjunction with an advanced laboratory course, it is perhaps most effectively used as a separate course taken before, after, or in parallel to the lab class. The lab course is focused on the methods of data collection while this text is focused on the methods of data processing. These two topics are intimately related but the methods are completely different and each deserves its own focused learning experience.

The topic of data analysis and model metrics requires a computational approach. When teaching the course on which this book is based, half of the class sessions are held in a computer lab with experiential learning examples, walking the students through the usage of the concepts and equations presented in class. Specifically, these interactive analysis sessions are taught using the Python programming language via Jupyter notebooks, which include interleaved blocks of code and explanatory text. At least one version of these code files is already available online, as a supplement to the book content. Instructors should feel free to use this coding material in designing their own version of this type of course. The exercises at the end of the chapters assume some programming proficiency and most of them require some coding for completion.

I have not included any programming-language-specific content in this book; any necessary coding instruction should be provided in addition to the content of this book. For my class, I give them lots of code; this is not a class about the special tricks of opening a data set in a particular format but rather the usage of the analysis techniques to robustly assess one or more data sets. Note that some of the "Exercises and geosciences" problem sets at the ends of the chapters are quite long; instructors might want to think about whether to assign everything or a subset of it. For some of the chapters, I only assign half, switching back and forth each year.

Prerequisites for Using This Book

This is meant as an introductory statistics textbook for upper level undergraduates. It does not require any prior statistics coursework. If students have taken a statistics course already, then the some of the first half of the book (especially Chapters 4, 6, and 7) will be somewhat of a review. The content in these chapters, however, is different from a typical statistics approach to the material, so students should find it to be a new perspective on these concepts. There is a small bit of probability in the course, but again no prior knowledge of this topic is needed.

The math in Chapter 3 on uncertainty propagation includes differential calculus. It is assumed that students have this "Calculus 1" knowledge base, so proficiency to this level of math is essential for that part of the course. Chapter 3 includes partial derivatives, so higher level calculus is preferred, but this concept could be briefly introduced as part of this course (as I do, when I teach it).

Some programming experience is required to successfully navigate the homework problems in this book. I teach it in Python, via Jupyter Notebooks, going through coding technique and example code in class. I do not, however, teach the basic fundamentals of scientific programming, but rather go through a style guide of code format and documentation expectations. I allow students to work together on coding assignments, as I do not want the programming aspect to be a hurdle to understanding the statistical and data-model comparison concepts.

The book contains numerous examples in Earth, atmospheric, space, and planetary sciences. No prior knowledge of these topics is necessary to fully appreciate these examples; background context is provided and they are separated from the statistical concepts. They offer a real-world usage of the math that hopefully exposes readers to these geoscience topics and piques interest in them. From earthquakes to tornadoes to the auroral lights to the bright "wanderers" of the dark night sky, geoscience is a fascinating discipline that we experience every day. I hope that you enjoy these examples from this field.

Features of This Book

The content is organized into 12 chapters. A quick summary of the content is as follows:

- Chapter 1: Uncertainty around data, comparing a single number to a group, and the Gaussian "normal" distribution
- Chapter 2: Visualizing a data set, elements to consider when making a plot, and best practices for conveying a message with a figure
- Chapter 3: Calculating the uncertainty of a processed data set
- Chapter 4: Quantifying the centroid and spread of a number set, Poisson counting statistics in relation to centroids and spreads
- Chapter 5: Methods for determining if a number set follows a Gaussian distribution, including the chi-squared test and Kolmogorov-Smirnov test, rejecting single data points from a set, what to do if it isn't Gaussian
- Chapter 6: Comparing two data sets, t tests, covariance, correlation coefficients, and

calculating an uncertainty of a metric with the jackknife and bootstrap methods
- Chapter 7: Fitting a line between two paired data sets, including polynomial fitting and nonlinear curve fitting techniques
- Chapter 8: Visualization techniques for comparing a data set with a model trying to reproduce it, approaches to data-model comparison metrics, categories of metrics
- Chapter 9: Fit performance metrics (those that use the continuous nature of the two number sets)
- Chapter 10: event detection metrics (those that categorize the data and model values into event/non-event status)
- Chapter 11: techniques for sliding the threshold of event identification
- Chapter 12: summary of best options for data-model comparison metrics choices for certain applications, maximizing interpretive value with combinations of metrics, binomial distribution and decisions with metrics, and introductions to additional statistical topics

Each chapter includes similar content elements meant to make the material readily accessible and discoverable. At the end of each section or subsection is a light blue "Quick and Easy" box that recaps the primary point of that section. If you are skimming the book, reading only these should provide a concise, high-level summary of all of the main content. Most of the sections are focused on statistics, but interspersed within every chapter are sections providing example usage of those concepts in the Earth, atmospheric, space, and planetary sciences. These geoscience sections contain no new statistical concepts and could be skipped or replaced with different examples. This clear separation of the statistics content from the example geoscience content is slightly violated in Chapters 9, 10, and 11, which contain running examples. In these chapters, there is an early section introducing the running example and then an ending section that recaps the running example, and in the intervening sections that introduce the metrics for each category, there is one paragraph, usually near the end, that applies that metric to the running example.

Another accessibility feature of the book is the highlighting of key term definitions. At the first usage of a key statistics term, it is listed in bold font within the text and then a brief summary definition is given in red text at the bottom of that page. Similarly, key geoscience terms are also listed in bold font with a brief summary definition given in blue text at the bottom of the column. Again, for those skimming the book, looking for a particular concept, scanning these red-text or blue-text definitions should allow you to quickly find it.

Each chapter ends with the same two sections. The first of these is "Further Reading," which is an annotated reference list for the chapter. The ordering of entries in this section follows the sections of the book, providing key references, links to data sets or important websites, and additional commentary. The final section of every chapter is "Exercises in the Geosciences," which are a mixture of by-hand calculations, coding tasks, and short-answer interpretations about the results from other parts of the problem sets. All of the data sets are available at the companion website for this book, as well as Python code in Jupyter Notebook format for opening these files. As preferred programming languages shift, additional code files in these other languages could be made available at this website.

Final Thoughts

The long-range goal of writing this book is to improve the quality of methodology in scientific research. Many scientific studies conduct only cursory analysis of their selected data; approaches abound for conducting more rigorous assessments and this book serves as an introduction to many of those analysis techniques. Furthermore, many studies involving numerical models to assess and interpret observations implemented rather simplistic approaches to the data-model comparisons, often choosing only one or two metrics for the analysis. As presented in this book, there is a "zoo of metrics" available, each with their particular strengths for examining a particular aspect of the data-model relationship. Each metric, though, has its limitations, because it was designed for a particular purpose. With additional emphasis on the numerous options for data processing and data-model comparisons, including the role of uncertainty in these evaluations, the scientific discourse will be elevated.

Acknowledgments

I am a huge fan of John R. Taylor. He wrote An Introduction to Error Analysis: The Study of Uncertainties in Physical Measurements in 1982, which was assigned as the required text for my Advanced Physics Lab course series at Rose-Hulman Institute of Technology. This was my first full immersion into data analysis and processing, and the Taylor book was an outstanding resource for understanding these concepts. I thank John Taylor for providing a brilliant example of how to easily convey difficult concepts. If I could measure it, I am sure that the influence John Taylor has had on me through this book passes the threshold of statistical significance.

This book would not have happened without my appointment as the editor in chief of the Journal of Geophysical Research Space Physics. It was through my experiences reading and editing a few hundred manuscripts a year that I realized the need in the scientific community for a better awareness of data processing techniques and a stronger appreciation for conducting robust data-model comparisons using a wide array of metrics. Suggesting resources to people for additional information on these concepts led me to the realization that the community could really use a single resource for this collection of topics. Thanks to the editor selection committee that recommended me for the position and the many people at the American Geophysical Union (AGU) and across the space physics research community that helped make that experience a positive one for me.

AGU contracts with Wiley for the publication of their various journals and topic-specific monographs. Intending to reach the broad spectrum of disciplines included within the umbrella of AGU, I chose Wiley as the publisher for this. I am glad that they accepted my book proposal and that I took this path of writing this book. A special thanks goes out to Ritu Bose, Mandy Collison, and Layla Harden from Wiley and Jenny Lunn from AGU for encouraging and supporting me during this process. I would also like to thank the technical editor, Dr. Bea Gallardo-Lacort, for her time, effort, and encouragement through the long process of peer review.

I'd like to thank my colleagues in the Department of Climate and Space Sciences and Engineering at the University of Michigan. Our curriculum committee recognized a need for an upper-level undergraduate data processing and visualization course, and they looked around for someone to develop and teach it. This coincided with my editorship and my realization that the natural sciences could use a course like this, so I volunteered to do this task for the department. Thanks to the entire faculty that have contributed to my understanding of data analysis, data-model comparisons, data visualization, and, of course, for providing the opportunity for me to develop this course, from which I have now written this book.

I will be forever in awe of the intellect, audacity, and tenacity of Dr. Abigail Azari.

As my PhD student at the time when I was assigned the new course, she heard me pondering which programming language to use for the labs. After that group meeting, she came to my office and told me that the labs would be in Python, specifically Jupyter Notebooks, and that she would develop them. I wholeheartedly agreed, and while we developed that lab structure and outline together, she developed the content and actual code. In fact, she taught the labs for the first two years. Thank you, Abby; you will make an outstanding professor.

Many other coworkers had a direct impact on the content of this book. I am very grateful for my many conversations about data analysis, data visualization, and data-model comparisons with others in the space physics field, both those in my immediate group at the University of Michigan and those across the world. There are too many to name, but I would like to especially mention Dr. Natalia Ganushkina, Dr. Dan Welling, Dr. Janet Kozyra, Dr. Lutz Rastätter, Dr. Katherine Garcia-Sage, Dr. Steven Morley, Dr. Alexa Halford, Dr. Meghan Burleigh, Dr. Michael Balikhin, Dr. Aaron Ridley, Dr. Derek Posselt, Dr. Suzy Bingham, and Dr. Darren De Zeeuw. In addition, I would like to thank Prof. Ted Bergin from the Department of Astronomy at the University of Michigan who graciously provided an office during my sabbatical, a time when a part of this book was written. I would also like to thank my former and current PhD students that have greatly contributed to my thoughts on these topics (in addition to Abby); you have all profoundly and positively influenced me. In particular with respect to the subjects covered in this book, I extend an extra thanks to Roxanne Katus, Shaosui Xu, Alicia Petersen, Alexander Shane, Brian Swiger, and Agnit Mukhopadhyay.

I am indebted to the many students that took the class on which this is based. In particular, I would like to thank the students of class from winter term 2021, who were given draft—and often incomplete—versions of the chapters throughout the term. I would especially like to thank the robust comments and suggestions from Aiden Kingwell, John Delpizzo, and Neha Satish. I would also like to acknowledge Samuel Ephraim, whose class project provided the data set for the event detection metrics running example. I also thank the students who took the class from winter term 2022, the first to see the complete book, and for the comments provided throughout the term. I would especially like to appreciate the many comments from Jena Alidinaj, Kira Biener, Alanah Cardenas-OToole, Jessica Fisher, Akshay Gupta, Lunia Oriol, and Ollie Paulus.

I owe a big thanks to Asher/Anya Hurst, the artist that I contracted to create the Chicken Little illustrations used throughout the book. They are a very talented graphic artist with skills well beyond the simple figures I requested for this project. I hope that you enjoy this whimsical accent to the otherwise dry topic of statistics for the Earth and space sciences, and please consider commissioning Asher/Anya for your illustrative needs.

Finally, I would like to thank my family. Here is a message to my immediate family—my wife Ginger and children Annie and Derek—I love you very much and thank you for giving me the encouragement and feedback that I needed to keep going with this project. Additionally, I thank my parents and siblings for being there with me throughout, well, all the years of my life. I especially thank my dad, Dr. Harold Liemohn, who gave me the encouragement to pursue a technical career and advice along the way on how to navigate the scientific research landscape, and for graciously reading and editing several chapters of this manuscript.

About the Companion Website

This book is accompanied by a companion website.

www.wiley.com/go/liemohn/uncertaintyingeosciences

The website includes:

- All of the data files needed for the homework problems in the book
- Example code (in Python, in Jupyter Notebook format) for opening these data files

The instructor pages on this site include:

- Solution sets to the homework problems
- PDF and Powerpoint files of all figures from the book for downloading
- Latex and Powerpoint files containing all equations used in the text

1

Assessment and Uncertainty: Examples and Introductory Concepts

Analyzing data is a fundamental skill for someone pursuing a career in science, technology, engineering, and mathematics (STEM). That first word in the list, *science*, is defined broadly here and includes not only geosciences, the focus of the examples throughout the book, but also all natural sciences, physical sciences, social sciences, and the medical professions. Nearly every discipline uses "sets of numbers" in its standard analysis methods. Therefore, a firm grounding in how to approach a data set and understand its properties is a skill that allows you to confidently address a very wide variety of real-world problems. Similarly, being able to thoroughly assess and interpret how two data sets relate to one another (the core of the field of statistics) is fully transferrable from one discipline to another. *Knowing how to examine numbers is foundational for making good decisions*; those in managerial positions should also know these skills to properly distill large quantities of information into the essential elements needed for progress on the projects they lead.

A key element of data analysis is the use of models to decipher features of the measured values relative to physical processes. This topic of data-model comparisons is fundamental to the investigative methods across STEM disciplines. Yet, applying metrics to find the "goodness of fit" regarding a model output compared to an observational data set requires careful thought about how to conduct the analysis and how to interpret the values resulting from this "processing" of the number sets. The first half of that methodology, how to conduct the comparison analysis, could take many paths depending on the eventual use of the quantities to be calculated. Some equations are much better suited for some applications than others, or with certain types of data. Choosing which types of analysis to conduct is an important aspect of the process because there are no one-size-fits-all statistics. The second half of the methodology, interpreting the resulting metric values, is subjective and requires context about not only the origins of the numbers but also the final application.

All of these steps depend on the **uncertainty** connected with the values. Uncertainty here means "the range of other possible values for this particular number." That is, it is not about poor memory and being unsure of whether a procedure was conducted properly, but rather the spread of options that a value might reasonably take instead of the actual value reported. This is especially true for visually determined values, such as precipitation accumulated in a rain gauge. Uncertainty allows

---

Uncertainty: The range of other possible values for a particular number.

Data Analysis for the Geosciences: Essentials of Uncertainty, Comparison, and Visualization, Advanced Textbook 5, First Edition.
Michael W. Liemohn.
© 2024 American Geophysical Union. Published 2024 by John Wiley & Sons, Inc.
Companion website: www.wiley.com/go/liemohn/uncertaintyingeosciences

you to ascribe meaning to the measured values and properly compare it with another number. Uncertainty allows you to place two numbers in context with each other. Uncertainty is the key to applied statistics.

Unfortunately, it is often forgotten, unreported, or incorrectly included in the analysis and interpretation. To make uncertainty a little more tangible, let us consider a fresh take on a well-known story.

1.1 Chicken Little, Amateur Meteorologist

Chicken Little, being a methodical amateur meteorologist, went out every day at noon to measure atmospheric pressure. One day, they recorded a value of 100 kilopascals (abbreviated kPa) on the barometer. The next day, the reading was only 90 kPa. Figure 1.1 shows Chicken Little hard at work at their computer, analyzing the data. Did they take this drop in atmospheric pressure as a signal that the sky is falling and run off to tell the local authorities?

Before we answer, perhaps you want a bit of context. A pressure of 100 kPa is close to a normal atmospheric pressure here on Earth, while 90 kPa is a typical pressure in the eye of a *very* strong hurricane. At first glance, this looks like a big atmospheric pressure difference that should be reported.

The correct answer, to no one's satisfaction except a scientist, is that *it depends*. On what, you ask? It depends on the uncertainty in their two values. If Chicken Little is using a modern digital barometer with a pressure-sensing electronic transducer, then the uncertainty on the measurements is only a few kilopascals (approximately 2%). In this case, the drop is large compared to the uncertainty and they should run to report it. If, however, they were judging the atmospheric pressure based on the rise time of a balloon, then the uncertainty estimate on these numbers could be something like 30%. The 10 kPa **discrepancy**, that is, the difference between the two numbers, is quite a bit less than the 30 kPa uncertainty on either number, so they should not report it, as the difference is not meaningful. What if they made the first measurement of 100 kPa with a modern barometer and the second measurement of 90 kPa with the rising balloon? The combined uncertainty (simply adding the two uncertainties) is still greater than the discrepancy, so they should not report it. However, because atmospheric pressure only varies by about 10 kPa, if the measurement techniques were reversed and they measured 100 kPa with the rising balloon and then 90 kPa with the digital barometer, meaning that the range is between 88 and 92 kPa with the typical uncertainty of such devices, then Chicken Little should report the very low atmospheric pressure regardless of what it was the day before. The uncertainty on the numbers makes all the difference in whether their measurements warrant action.

Figure 1.1 Should Chicken Little be scared by a 10 kPa pressure drop? Should they declare that the sky is falling? It depends. Artwork by Asher/Anya Hurst.

Discrepancy: The difference between two numbers.

> **Quick and Easy for Section 1.1**
>
> The interpretation of numbers depends on their uncertainty.

1.2 Uncertainty Ascribes Meaning to Values

The story mentioned in the preceding text illustrates a main premise of science—the meaning of a particular number cannot be properly understood without a value for the uncertainty surrounding that number. This is especially important when, like in the above-mentioned example, two numbers are being compared. You can calculate the difference between the two numbers, but this is not enough; it is only with the uncertainty intervals around the two values that it becomes possible to ascribe a judgment on whether the two numbers are basically the same or fundamentally different.

Uncertainty is the amount that a measured or calculated number could be off from the true value. This can sometimes be inferred from the significant digits reported on the number, with the last nonzero digit indicating the degree of accuracy of the number. This would work except for two things: the first is that calculators, computers, and digital sensors make it very easy to report many digits beyond the true accuracy of the number; and the second is that many people have forgotten their middle school math unit on significant figures. Numbers are often reported with more nonzero digits than they should have, including in scientific experiments. If you trust that the source of the number took into account the sources of uncertainty when reporting the value, then you can proceed with this assumption of nonzero digits being the accuracy of the value; otherwise, caution is highly advised when using this technique to determine uncertainty.

Another issue here is that the final significant digit could be a zero. Reporting the number in scientific notation will avoid the confusion, but this can be awkward, especially if the number is only one or few places from the decimal point. In the above-mentioned example, we should report the first day's measurement from a digital barometer as 1.00×10^2 kPa, indicating that there are three significant digits and that the uncertainty is in the second zero. For a number like 100, most people do not usually make the extra effort to write out the number in scientific notation to ascribe meaning to the zeros and convey the proper uncertainty of the value. This means that zeros might or might not be significant and caution should be used with such numbers.

This book is about comparing numbers, or more specifically sets of numbers. While it is a guided tour through the essential statistics of working with data, in particular with application in the natural sciences, an underlying message throughout the chapters ahead is that the numbers calculated from those statistical methods are meaningless without context. A big part of that context is the uncertainty of the number. Uncertainty ascribes meaning to the many values that will be calculated later in this book.

> **Quick and Easy for Section 1.2**
>
> Learn to report the uncertainty and use the correct number of significant figures.

1.3 Significant Figures

Far too often in science, engineering, and mathematics, it is easy to get a value with a calculator that is reported on the screen to 10 digits. It is almost never the case, however, that the tenth digit in the list is important. The reason is because there is uncertainty in the values that were used to obtain that result, and the digits probably lost meaning much earlier in the lineup. This uncertainty of the original numbers propagates through the calculation to the newly obtained value; a topic to be discussed in detail in Chapter 3. For now, let us discuss the critical issue of **significant figures**, sometimes called **significant digits**. Specifically, how many significant digits should be reported when listing a number and its uncertainty?

The **basic rule of significant digits** is this—uncertainty values (denoted δx, for the measured value x) should only have one significant digit and the measured value should be reported to match the decimal place of the uncertainty. Round off the uncertainty and then round off the measured value. The second number in the uncertainty is not meaningful; because this number is an uncertainty, it is by definition not reliable past its first digit. Once the decimal place of the uncertainty is established, this defines the level of accuracy in the measured number. The uncertainty dictates the significant figures in the measured value, not the other way around.

Let us do a quick example. Given a length value of, say, 643.8 ± 26 m, the proper way to report this is to first examine the uncertainty value of 26 m. It is listed with two significant digits, but should not be and instead should be rounded up to 30 m. The measured value should then be reported with significant values only to the tens decimal place, removing the last two values originally listed, and reported as 640 m. The final style of listing this number, with its uncertainty, should be 640 ± 30 m.

The value of 30 m is called an **absolute uncertainty**. It has the same units as the measured value and a straightforward meaning—it provides a range around the measured value that most likely encompasses the true value you were trying to measure. The qualifier of "most likely" is used here because there could be outliers, and typically the uncertainty does not extend to include all possible options of where the true value could be, but rather a compromise range that has a high chance of containing the true value.

There is another commonly reported type of uncertainty—**percentage uncertainty**. This number, written δx_{perc}, is directly related to absolute uncertainty δx, and it is an easy conversion between the two: divide by the measured number and multiply by 100%. Without the 100% multiplier, it is called a **fractional uncertainty**, δx_{frac}, a unitless value that could also be reported as the uncertainty of the measured number. Because fractional uncertainty has no units, there is no clear way to designate it as a fractional uncertainty instead of an absolute uncertainty. That is, an absolute uncertainty could have its one significant digit to the right of the decimal place, making the listed uncertainty

---

Significant figures and significant digits: The number of decimal places that have meaning for that value.

Basic rule of significant digits: The uncertainty value has only one significant digit, and the measured value is rounded to the same decimal place as this one significant digit on the uncertainty value.

Absolute uncertainty: The uncertainty given in the same units as the measured value.

Percentage uncertainty: The uncertainty given as a percentage of the measured value.

Fractional uncertainty: The uncertainty given as a fraction of the measured value, it is used within the calculation of uncertainty propagation, but, to avoid confusion with absolute uncertainty, should always be converted to percentage uncertainty when being reported with a measured value.

highly confusing for a reader to interpret. Percentage uncertainty is better because it should be followed by a symbol that designates it as different from absolute uncertainty. So, while fractional uncertainty is sometimes convenient to use within a calculation, always multiply it by 100% and report it as a percentage uncertainty when listing it next to the measured value.

Let us continue our example. If you wanted to convert the absolute uncertainty into a percentage uncertainty, then we have this as the intermediate fractional uncertainty, 30/640 = 0.0469, which should be multiplied by 100% and rounded to one digit, yielding 5%. This should be reported as 640 m ± 5%, with different units on the measured value (meters) and the uncertainty value (percent). It could be argued to instead do the fractional uncertainty calculation with the original values, making it 26/643.8 = 0.0404, which would round down to a percent uncertainty 4%. Either 4% or 5% can be justified as the reported percent uncertainty; there is not a strict methodology to uncertainty reporting and examples of conducting the calculation both ways can be found in the scholarly literature across many disciplines. In this case, I prefer the larger value because it is consistent with the absolute uncertainty reported to only one significant digit.

An ambiguity of the reporting technique arises when the final significant digit of the measured value is zero. Let us consider the example of a wind speed measurement of 29.9 ± 2.1 m/s. In this case, the uncertainty should be rounded down to 2 m/s and the measured value rounded up to 30 m/s. The final value could be reported as 30 ± 2 m/s. It should be obvious that the zero at the end of the measured value is significant because the uncertainty is reported to the ones decimal place. This is sometimes confusing, though, and could be misinterpreted in later use. An alternative way to report this value is to add a period to the measured value to signify the last significant zero, in which case the speed would be reported as 30. ± 2 m/s. For this example, the period was easily added because it is a natural location for a period in a number (at least in the US number listing), separating the decimal places above and below one. If significant zeros occur elsewhere in the measured value, let us say that our example numbers were shifted by a factor of 10, and a period is desired to designate the significance of those zeros, then it could be reported as $30. \cdot 10^1 \pm 20$ m/s. This is also confusing, and a more natural way to report it would be in standard scientific notation as $3.0 \cdot 10^2 \pm 20$ m/s. The single zero after the decimal period indicates that this is a significant digit; it should only be listed as 3.0 and not 3 or 3.00, either of which would give the false impression of an incorrect number of significant figures on the measured value.

There is one exception that is useful to include as an **exception to the basic rule of significant digits**—when the first digit of the uncertainty is a one. In this case, rounding down significantly reduced the calculated uncertainty by tens of percent, giving the measured value extra precision that it should not have. This goes in both rounding directions; rounding up when the first uncertainty digit is a one would increase the uncertainty by tens of percent. The exception, therefore, is to leave the uncertainty with two significant digits when the first digit is a one. Take the example of a liquid volume of 36 ± 14 L. The basic rule would have us report it as 40 ± 10 L, yielding a percentage uncertainty of 10/40 = 0.25, which would be reported to one significant digit in uncertainty as 40 L ± 30%. If we used the original values to calculate the percentage uncertainty, it would be 14/36 = 39%. Similarly, if the value was 34 ± 16 L, the basic rule would

---

Exception to the basic rule of significant digits: If the first digit of an uncertainty value is a one, then it is acceptable to report two digits and then round the measured value to the same decimal place as this second place of the uncertainty value.

---

yield 30 ± 20 L, and two variations of percentage error would be 70% using the already-rounded values and 50% using the original values. The exception to the basic rule allows you to report these values with two significant digits on the uncertainty in the special case of a leading 1, greatly reducing the influence of the rounding process.

A similar rounding issue with wildly changing uncertainties can occur when the percentage uncertainty is just above 10%. Take the example of 36 ± 5 L. This is fine, as initially reported, according to the basic rule. When converted to a percentage uncertainty, it becomes 5/36 = 0.14, which would be rounded down to 10% uncertainty. That is, over a quarter of the uncertainty vanishes with the downward rounding. If the number had been 33 ± 5 L, then the division yields 5/33 = 0.151, and the basic rule would have us report it as 20%, an increase of nearly a quarter. The exception to the rule would allow these percentage uncertainties to be listed as 14% and 15%, respectively, rather than 10% and 20%.

The question arises of how many significant figures should be assigned to the measured value for a given percentage uncertainty. To determine this, let us consider the example of an uncertainty of 1 being applied to various measured values. A few examples of this relationship between the absolute uncertainty and the resulting percentage uncertainty are given in Table 1.1. If the measured value is also 1, then the percentage uncertainty is 100% and there is only one significant figure. For a measured value of 3 or 9, this is still the case; there is only one significant figure. The percentage uncertainty is decreasing however, and for 9 ± 1, the fractional uncertainty is just above 10%. At the change to 11 for the measured value, however, the significant figure changes to two. This is also the value at which the percentage uncertainty level decreases to a value below 10%. A very similar change occurs as the measured value crosses 100; the significant figure of the measured value changes from two to three as the measured value increases from 99 to 101

Table 1.1 Relationship of percentage uncertainties to given measured value significant figures.

Value: $x \pm \delta x$	Significant figures on the measured value	Exact δx_{perc}	Rounded δx_{perc} (with significant figures applied)
1.00 ± 1	1	100%	1 ± 100%
3.00 ± 1	1	33.3%	3 ± 30%
9.00 ± 1	1	11.1%	9 ± 10%
11.0 ± 1	2	9.09%	11 ± 9%
33.0 ± 1	2	3.03%	33 ± 3%
99.0 ± 1	2	1.01%	99 ± 1%
101.0 ± 1	3	0.990%	101 ± 1%
333.0 ± 1	3	0.300%	333 ± 0.3%
999.0 ± 1	3	0.100%	999 ± 0.1%
1001.0 ± 1	4	0.0999%	1001 ± 0.1%

with an absolute uncertainty of 1. The exact calculation of the percentage uncertainty also crosses a benchmark at this point, changing from above 1% to below 1%. Finally, another analogous change occurs as the measured value increases from 999 to 1001—as the significant figure of the measured value changes from 3 to 4, the exact percentage uncertainty crosses the 0.1% value.

From this pattern, a general rule for percentage uncertainties and significant figures can be written, given in Table 1.2. This is a rewriting of the examples shown in Table 1.1, but this time starting from the percentage uncertainty value with the resulting number of significant figures for the measured value listed in the third column. When the percentage uncertainty is exactly coincident with a boundary value listed in Table 1.2, then the larger number of measured value significant figures can be applied.

What about a value for which no uncertainty is given? If the only thing known is the number of significant figures in the measured value, then the final two columns of Table 1.2 can be used to determine a rough estimate of

Table 1.2 Significant figures for the measured value given a percentage uncertainty.

Upper δx_{perc} value	Lower δx_{perc} value	Significant figures on the measured value	Rough estimate δx_{perc} value for given significant figures
100%	10%	1	30%
10%	1%	2	3%
1%	0.1%	3	0.3%
0.1%	0.01%	4	0.03%

a percentage uncertainty. As seen, the suggested estimate is at the "three" level within the range of possible percentage uncertainties. This is a rounded value of the geometric mean using equal weighting across the full range of the percentage uncertainties for that number of significant figures (for example, $\sqrt{(10) \cdot (100)} = 31.6 \approx 30\%$). Some suggest a different value, such as the "five" or "six" level of each range, which is close to the linear average across the percentage uncertainty range for that number of significant figures in the measured value. Because each range spans a full order of magnitude, the geometric mean is more appropriate for the rough estimate, but this is only a suggestion. For any specific value, it is up to the user to apply whatever is deemed best for the estimated percentage uncertainty. Once a value is applied, this can then be used in other calculations and additional processing of the value. The estimated percentage uncertainty can even be converted into an absolute uncertainty, if needed. Even though this uncertainty might not be the correct value, it is within the right order of magnitude and, therefore, off by a factor of 3, at most, from the true uncertainty.

> **Quick and Easy for Section 1.3**
>
> Uncertainty should be reported with one significant digit, then round the measured value to match that decimal place.

1.4 Types of Uncertainty

There are two main causes of uncertainty in a measurement. The first is that the measurement technique is nearly always subject to small deviations. By using the same device to measure the same parameter from the same environment, the answer could be slightly different. A simple example of this is counting students in a classroom. There are many ways to do this. You could scan across the first row, then the second, and so on, cautiously noting each student in each place along each row. Repeating this process should give the same answer. But what if it doesn't? The counter might have inadvertently skipped a student or two, or double-counted some of them as they got to the back of the room and the heads are no longer in nicely aligned rows. Perhaps you only counted the first row of students, then counted the number of rows, and multiplied these two together. The counter could have chosen any row, not just the first, and so they might get a different answer. Or there could be empty seats sprinkled around the room. All of these counting methods result in uncertainty in the final count of students in the room. We will call this **measurement technique uncertainty**.

The second cause of uncertainty could be variability in the quantity being measured. Continuing with the students in the room example, let us say that you want the count of students coming to class on an average week. Each day in class, you count the students. It could be, however, that some students missed class on some days. Perhaps there was a visitor in the class some day, providing extra bodies to count beyond the official roster. At the end of the week, you have a spread of values. The variability in the values could be due to the

Measurement technique uncertainty: Uncertainty associated with the method of how the data were collected.

method and carefulness of the counting, as discussed in the earlier text, but let us say that your technique was the same and very thorough. In this case, the variation of the values is not due to the counting process but due to changes in the quantity being counted. This represents a different source of uncertainty than that is caused by the measurement technique itself. It is related to the variability of the parameter itself with respect to time or space. A collection of values that are supposedly from the same population might not be if the measurements are taken over a time span or across a spatial scale that is larger than the typical coherency of the parameter. This type of uncertainty will be called **parameter variability uncertainty**.

Either of these uncertainty sources could fall within the two primary types of uncertainty. The first of these categories is **random uncertainty**. This type of uncertainty results in a spread of values that could be either larger or smaller than the true value. For counting students, the measurement technique random uncertainty would be related to occasionally skipping over a student or double-counting a student. Regarding the weekly average in the student count, the parameter variability random uncertainty includes the coming and going of students that either increases or decreases the daily count.

The second category is **systematic uncertainty**. This type causes a shift in the measured values relative to what they should have been. Continuing the example, a possible measurement-technique-related systematic uncertainty could be if the classroom is long, making it hard to see all of the rows in the back of the room, resulting in students hidden behind others and therefore a regular undercounting of the number of students present. Or, similarly, it could arise by the counter regularly double-counting students at the back of the room, again because of ambiguity about which row the students are in. The result is not a spread of the values but an offset, to lower or higher values for the two examples, respectively. For the weekly average count, the parameter variability systematic uncertainty might be from a later influx of new students as more register for the course and start attending the class sessions. It could also be that a substantial fraction of the students regularly show up late to class, after the count is taken each day. In either case, the count during the chosen week is shifted with respect to the true student enrollment.

These two types of uncertainty exist in every data set, including throughout the natural science disciplines. When care has been taken to ensure reliability of the measurements, then they could be extremely small, allowing usage of several significant figures in the measured values. Other quantities are very difficult to measure, or they vary on such short temporal and spatial scales that it is difficult to reduce the spread of the values. In this case, the uncertainty might be relatively large, allowing for the use of only one significant figure of the measured value. There could also be times when there is an unknown systematic offset that shifts the observed values relative to what they actually should be.

Parameter variability uncertainty: Uncertainty associated with the physical quantity being observed.

Random uncertainty: Uncertainty associated with fluctuations that average to zero.

Systematic uncertainty: Uncertainty caused by a shift of the values from what they should be.

> **Quick and Easy for Section 1.4**
>
> Both systematic and random uncertainty exist in the quantity being measured as well as in the measurement technique.

1.5 Example: Finding Saturn's Moons

In late 2019, scientists using data from the Subaru Telescope in Hawaii announced that they had discovered 20 new moons orbiting Saturn. The newly seen moons are just a few kilometers across, but they bring the total of the "large" objects flying around Saturn up to 82. This puts Saturn in the lead, at least for now, with respect to the planet having the most moons, topping the recently expanded total of 79 for Jupiter.

How is it possible that there were undiscovered moons around Saturn that were just seen and identified for the first time? In a word: uncertainty.

The issue is pointing accuracy and pixel size. Before the invention of telescopes, only a few moons of Jupiter could be seen; none of Saturn's moons are visible with the naked eye. Your eye has a limit to its resolution and contrast in seeing faint objects next to a bright one (i.e., moons next to the planet). The dim glow from the moons of Saturn was lost, either because they are too close to Saturn and could not be resolved, or they looked the same as the background noise of the night sky. Human eyes could not detect the tiny signal from any of the moons. Whatever photons our eyes receive from these moons look like the random fluctuations of photons, at least on the scale of the resolution that our eyes could detect.

After Galileo's invention, however, those pointing this new device at Saturn quickly saw a few of the largest moons. The change was not that the moons were suddenly emitting more photons and appeared brighter at Earth; it was that the telescope increased the resolution to previously unreachable levels. What used to be combined into a single pixel was now, with the aid of this magnifier, distributed over many pixels. The light from the large moons were now distinguishable from the background noise, dominating the signal in one or even several pixels. We could track these newly seen light sources in their orbits around Saturn and deduced that they were objects in orbit around that planet—moons.

As telescopes keep getting better, not only does the resolution get better but also the photon-counting technology is improving and we are able to see ever smaller objects around the planet. Another innovation that greatly helped in our ability to detect moons around Saturn was sending satellites with cameras to this planet. This allowed another tremendous jump in resolution, separating the signal from the noise. Figure 1.2 shows two images—one

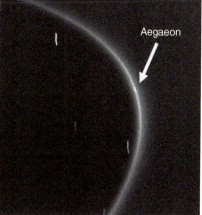

Figure 1.2 On the left, an image of Saturn from the Pioneer 11 satellite as it approached the planet, showing Titan, the planet's largest moon, and the rings. On the right, an image from the Cassini spacecraft showing the detection of the very small moon Aegaeon within one of the rings of Saturn. The short vertical streaks in the right image are star signals. NASA/Public Domain.

from the Pioneer 11 flyby of Saturn from 1973 and another from the Cassini spacecraft as it orbited Saturn in 2008. The ring in the right image is the G ring, located beyond the main rings and which is so faint it is not visible in the image on the left. Aegaeon, the moon seen as the bright spot along the ring in the right image, is only 240 m across (an average because it is not spherical).

To understand the resolution issue and how it relates to the detection of Saturn's moons, let us create an example to graphically show the process of revealing a signal from noise. Specifically, let us focus on a **histogram**, also known as a bar chart, in which counts within each bin are plotted as a function of the bin values. Several such plots are shown in Figure 1.3, illustratively representing photon counts from a telescope attempting to observe Titan from Earth. The upper panel (panel (a)) is an example collection of measurements made at low resolution. The uncertainty of the measurements is represented by the bin widths. The center bin has a higher photon count than the other two, but the difference is not large. It is very difficult to justify the declaration of a moon at this location of interest from these measurements. In the lower panels (panels (b)–(d)), the spatial resolution of the measurement increases, doubling the number of bins in each panel.

A photon count measurement with a telescope involves photons within a given area of the field of view. To see a moon around Saturn, the "area" under consideration needs to resolve that moon's cross-section within the viewing space. To get more photons, either you must count longer or expand the area. But when observing a moon around Saturn, the area is fixed (equal to the cross-sectional area of the moon within the telescope's field of view) and the time is fixed, as the moon moves in its orbit around the planet. Furthermore, you do not

---

Histogram: The plot of a number set showing the counts in each bin.

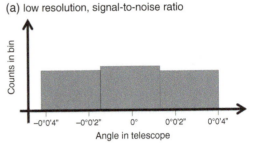

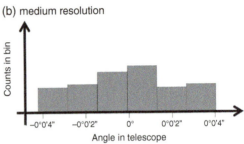

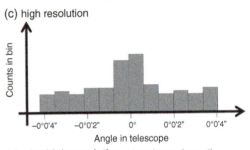

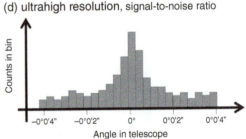

Figure 1.3 Illustrative observational sets representing a collection of telescope measurements and increasingly better resolution and signal-to-noise ratio, revealing the presence of an object at the location of interest (scaled for detecting Titan from Earth, which is 0.8" (arcseconds) across from here).

have many photons to count. The light source is the Sun, so the photons must travel from the Sun all the way to Saturn, reflect off of the surface of the moon (which are often rather dark and nonreflective), and then make their way back near the center of the solar system and directly into the telescope optical path.

The measurement has an uncertainty attached to it. The photon count from a particular pixel has a spread of possible values that it could have been. One source of 'uncertainty is the scatter of photons along their very long path from the Sun to Saturn to Earth. Another source of uncertainty is that the photons could be scattered within the telescope itself. The lenses or mirrors in the system might have imperfections, resulting in a spread from one pixel to another. The counts could be from light entering the telescope at an angle and bouncing off the walls within the device. These walls are usually very dark, coated to intentionally absorb this stray light, but some photons could still scatter off the walls and appear as signal. Finally, there could be issues with the sensor, which needs to record the photon striking it. This process is usually not 100% efficient, resulting in a loss of signal. Sometimes there is even digital noise within the sensor, resulting in a spread of an otherwise tightly confined signal.

There are several ways to improve the photon measurement. One is to build a bigger telescope in order to catch more photons. Another is to improve the optics within the telescope—the perfection of the lenses, the smoothness of the mirrors, the nonreflectivity of the sidewalls, or the sensitivity of the sensor recording the photons. Yet another method is to improve the processing of the images, perhaps combining images from different times, maybe following possible trajectories looking for a repeated signal in the image set.

The quality of the initial telescope observations of Saturn's moons was only good enough to resolve a few of them. That is, the uncertainty of the pixels was too large to be able to detect the signals from the smaller, less bright moons. As telescopes improved, the uncertainty of the measurements was reduced and more were discovered. These advancements in telescope technology and processing capabilities continue, and as the faint glow of smaller objects is revealed due to even smaller uncertainties on the photon counts, we catch our first glimpse of them; thus, the new discoveries.

> **Quick and Easy for Section 1.5**
>
> Reducing uncertainty allows for better scientific interpretation and discovery, as is the case with the continual discovery of new moons around Saturn.

1.6 Comparing Two Numbers: Are They Measuring the Same Value?

Given two numbers, it is reasonable to ask the question, "are they the same?" This is the question that Chicken Little asked themself in the story mentioned in the earlier text. As stated there, it depends on the method by which the values were calculated or measured. This depends on the uncertainty of each value. It depends on the assessment test used to judge the closeness of the two numbers. It depends on the confidence with which you want to accept or reject the "sameness" of the values. If making this assessment visually, then it depends on how the values are displayed and presented.

As a specific example, given a measurement of temperature outside of your residence and a collection of such measurements outside of your neighbors' homes, it is reasonable to ask, is your value the same as the neighborhood average? First, the location and time of the measurements should be compared. Were they close by to each other? Were they taken on the same day around the same time? Are the local environmental conditions around the thermometers similar? Are the thermometers identical, or at least both well calibrated? How were the temperatures recorded? Do they show the temperature by a column height of fluid, or by a dial, or a digital display? How carefully did the person observe the number? Was it a quick glance or did they take a long moment to digest the reading? All of these factors contribute to

12 | Data Analysis for the Geosciences

the uncertainty assigned to the numbers and the level of confidence that should be applied to the eventual similarity test results.

1.6.1 Distributions of Number Sets

To begin this analysis, it is useful to introduce the concept of a **distribution**. This will allow us to place percentages on how often a certain number should arise within a population, and therefore it is possible to test two such values against each other.

Distributions describe the "shape" of a population of numbers along the value axis. In its simplest form, it is a histogram, counting the values within equal-sized bins spanning the range of possible values within the population. If the x axis is the value range, then the y axis should be the count in that bin, or, dividing by the total number of values in the set, it is the frequency that a randomly selected value comes from that bin.

Let us take a simple function, like the cosine curve, shown in the top panel of Figure 1.4. We need to collect a **sample set** of values from this curve. Sampling this curve at various x axis values yields a collection of numbers ranging from −1 to +1. How the curve is sampled matters, though. The middle panel of Figure 1.4 shows a histogram made from 41 uniformly distributed samples of the cosine curve (that is, the x values were evenly spaced from 0 to 2π). For this evenly distributed x axis sampling, the counts in each bin follow a nice pattern. The slope of the curve is steepest when the cosine value is at zero, so there are fewest points in the bins closely surrounding this value and substantially more points in the bins near −1 and +1. Now consider the lower panel of Figure 1.4, for which the 41 samples were from randomly

---

Distribution: The shape of a number set when sorted by value and counted into bins.

Sample set: The number set collected from a larger possible population of numbers.

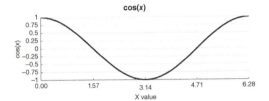
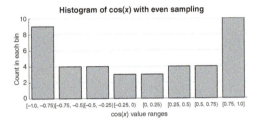
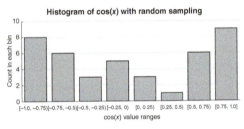

Figure 1.4 The cosine curve and two histograms of the distribution, the first sampled at 41 evenly spaced values from 0 to 2π and the second sampled at 41 randomly determined values from 0 to 2π.

distributed locations along the x axis. The pattern is somewhat similar to the middle panel, but not as well organized; most of the bins have slightly higher or lower counts than they did for the evenly spaced sampling histogram.

Both of the histograms in Figure 1.4 represent distributions of cos(x). They both reveal the bimodal nature of the cosine distribution, peaked at both −1 and +1. Both distributions are less than perfect, though. The finite number of samples does not reveal all of the features of the distribution; taking more samples and dividing into more bins would have been better to show the full curvature of the distribution. Both are pretty good, though, and inform the viewer of the nature of the cosine function.

Quick and Easy for Section 1.6.1

A histogram is a distribution of a number set.

1.6.2 The Gaussian Distribution

For most quantities in nature, the spread of values follows a particular pattern. Again, let us consider people around a neighborhood making temperature measurements. Rather than having two peaks like the cosine function, this number set of temperature measurements will most likely have a lot of the values that are fairly close to each other, designating a **centroid** to the eventual distribution of values. Values far away from this centroid become less frequent in the sample set than the values near the centroid value. That is, a distribution of values arises in which there is a peak of possible numbers at the centroid and dropping to lower probabilities in the two "wings" of the distribution. While there are many functional forms that this distribution could take, by far the overwhelming majority of this distribution of perturbations across the natural world takes the particular form known as the **Gaussian distribution**. It is also known as the normal distribution, not because it is so common—although it is common—but rather because of the other definition of normal as orthogonal, having to do with its original derivation. Yet another name for this is the "bell curve" because of its characteristic shape, shown in the upper panel (panel (a)) of Figure 1.5.

The Gaussian distribution of a variable x, $G(x)$, has a specific functional form based on the natural exponential, often written e or $\exp$, as well as a normalizing term out front:

$$G_{X,\sigma}(x) = \frac{1}{\sigma\sqrt{2\pi}} e^{-\frac{(x-X)^2}{2\sigma^2}} \quad (1.1)$$

Centroid: The value representing the middle of a number set's distribution.

Gaussian distribution: The distribution defined by a decaying exponential of the values squared.

(a) Gaussian distribution

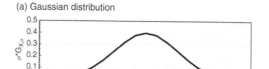

(b) Cumulative Gaussian

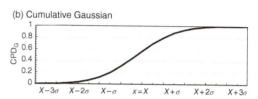

Figure 1.5 The upper panel (panel (a)) shows the Gaussian probability density function, along with the corresponding cumulative probability distribution of the Gaussian function in the lower panel (panel (b)). Both are plotted along an arbitrary x axis scale, with the peak at X and the spread normalized to σ. The y axis of the Gaussian distribution is not G but rather $\sigma*G$, making it an arbitrary scale.

In Equation (1.1), X is the centroid of the distribution and σ is the **spread**. Some alternative names for X are μ or m. These two parameters in Equation (1.1), X and σ, are the same quantities used as reference values along the x axis in Figure 1.5. It is seen that the argument of the exponential is negative as all terms in the fraction are squared or constants. At $x = X$, the argument of the exponential term is zero and so the exponential is one, leaving only the constant multipliers in front of the exponential term. As x deviates away from X, the argument increases, and therefore the value of the exponential term decreases. That is, the Gaussian peaks at $x = X$ and eventually decreases to vanishingly small values at lower and higher values of x far from X.

Note that the y axis of the upper panel (panel (a)) of Figure 1.5 is multiplied by σ. This makes the y axis values equal to the probabilities if the spread of the Gaussian were one, and use of another spread value would scale the curve.

Spread: The amount a distribution varies around its centroid.

That is, you should divide the y axis by σ to get the Gaussian probability for that specific σ value. A **probability** is the fractional chance of a specific outcome occurring. This is a word that will be used a lot in this book, especially with respect to uncertainty.

The extra terms in front of the exponential are **normalization factors**. Specifically, they set the integral of the function across all x to be equal to one:

$$I = \int_{x=-\infty}^{+\infty} G_{X,\sigma}(x)dx = 1 \quad (1.2)$$

This special characteristic of the Gaussian distribution—that the integral over all values of x yields an integral value of exactly one—indicates that is part of a family of curves known as **probability distribution functions (PDFs)**. Any distribution of a number set can be converted into a probability distribution by dividing by the total number of values in the number set.

This concept of an integral of the Gaussian PDF from negative infinity to positive infinity can be expanded. If the upper limit is not set to infinity but rather to a value along the x axis, a, then the integral yields the summed-up probability of the part of the Gaussian to the left of $x = a$, known as the **cumulative probability distribution** or CPD:

$$CPD_G(a) = \int_{x=-\infty}^{a} G_{X,\sigma}(x)dx \quad (1.3)$$

Equation (1.3) does not have an analytical form; it must be solved numerically. Luckily, there are several excellent approximations to this integral. This cumulative probability distribution can be plotted as a function of x (or, more specifically, a), shown in the lower panel (panel (b)) of Figure 1.5. The y axis gives the probability of getting a value from a Gaussian PDF at or below $x = a$. The lower panel (panel (b)) of Figure 1.5 is also useful for conducting the opposite calculation—obtaining an x value from a known Gaussian PDF above a particular percentage of the total possibility of values in the distribution.

Equation (1.3) can be translated into percentages for particular ranges within the Gaussian distribution. These are given in Figure 1.6, using the spread σ as the boundary marker for the regions. Again, the x axis in Figure 1.6 is listed in normalized units of σ, and the y axis is the Gaussian distribution multiplied by σ. It is seen that within the central peak of the Gaussian curve, a width of one standard deviation in each direction around the centroid contains over two-thirds of the area under the curve.

The concept of the Gaussian distribution is so fundamental to uncertainty calculations, data analysis, and data-model comparison metrics that it needs to be introduced here in this chapter. We will keep referring back to this throughout the rest of the book.

Probability: The fractional chance that a certain event will occur.

Normalization factor: A multiplier in front of an equation to scale the result to a certain range of values.

Probability distribution function (PDF): A distribution, defined by a mathematical function, for which the integral over all values of x is exactly one.

Cumulative probability distribution (CPD): The one-sided summation of a distribution.

Quick and Easy for Section 1.6.2

The Gaussian distribution is common in nature. The cumulative probability distribution is a one-sided sum of a distribution up to that point.

1.6.3 Testing a Specific Value within a Data Set: The z Test

Another fundamental calculation to be made throughout this book is a comparison of a

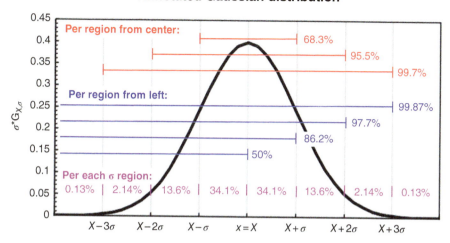

Figure 1.6 The Gaussian distribution, this time annotated with percentages in the sectors demarked by multiples of σ away from the centroid X, as an integral from the left, and per σ region within the graph. Note that the y axis is not simply G but rather $\sigma*G$.

single value against the **population** that it is from. This comparison is directly related to the Gaussian distribution defined and discussed in the previous section.

The comparison part is straightforward. First, we assume that the underlying population follows a Gaussian distribution and that the centroid and spread of this Gaussian distribution are known. Given a value, x, it can be converted into a nondimensional number relative to this centroid X and spread σ,

$$z = \frac{|x - X|}{\sigma} \quad (1.4)$$

This is called a **z score**, and we have already seen it as a term within the Gaussian distribution itself (compare Equation (1.4) with the term in the exponential of Equation (1.1)). It is

usually calculated with the absolute value operator included in the numerator, thus removing the directionality of the x value relative to the population centroid X. Equation (1.4) quantifies the distance of a measured value away from the centroid of the distribution. Given that the underlying distribution is a Gaussian, this z score carries with it a probability of values, $P(z)$, located below this value:

$$P(z) = \text{CPD}_G(z) \quad (1.5)$$

This probability can then be interpreted in a scientifically meaningful way.

It could be that you simply want to know its location with the full Gaussian distribution. In this case, you have your answer, and you are done. If x is below X and you want the percentage of values below x, then the value you want is actually the complement of that given by Equation (1.5):

$$P(z) = 1 - \text{CPD}_G(z) \quad (1.6)$$

This simple reversal of the probability is possible because the Gaussian distribution is symmetric.

It could be, however, that you are questioning whether this value belongs to that population. In this case, the probability can be used as a

---

Population: The very large number set from which the sample set, or individual value, was collected.

---

z score: The number of spread increments that a particular value is located away from the centroid of the population.

significance test; that is, as the probability that it belongs to the original population. This check of the probability is called a **z test**.

The z test can be either one-sided or two-sided. This is shown graphically in Figure 1.7. A **one-sided z test** assumes that we know that the value should be to the right of the population mean and beyond some reference z score (top panel of Figure 1.7) or the left of the population mean and therefore lower than some $-z_{ref}$ value. For a **two-sided z test**, the value could be either above $+z_{ref}$ or below $-z_{ref}$; that is, far from the mean in either direction.

For a one-sided z test, the probability to consider is just like Equation (1.6):

$$P_{1-sided}(z) = 1 - CPD_G(z) \qquad (1.7)$$

This is the percentage of values in the Gaussian distribution that are located beyond the location of x from X in the same direction of x relative to X. That is, it is only considering the part of the Gaussian beyond z from one side of the Gaussian distribution, above or below X. For a two-sided test, both ends of the Gaussian distribution should be considered, and the probability should be doubled and therefore changes to this:

$$P_{2-sided}(z) = 2 \cdot [1 - CPD_G(z)] \qquad (1.8)$$

---

z test: The conversion of a z score into a probability, based on a Gaussian distribution, and the comparison of this probability against an expected value.

One-sided z test: A z test that assumes the value under scrutiny is known to be located either above or below the mean, and the assessment should be made with probabilities calculated for only that tail of the Gaussian distribution.

Two-sided z test: A z test that conducts the assessment with probabilities calculated from both tails of the Gaussian distribution.

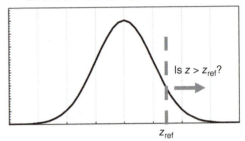
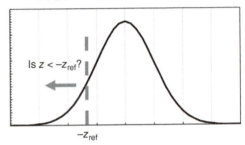
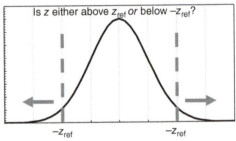

Figure 1.7 The regions of interest for one-sided and two-sided z tests. For a one-sided test, the region of interest could be either above a $+z_{ref}$ threshold or below a $-z_{ref}$ value. For a two-sided test, the z is evaluated against both of these criteria.

This is the percentage of values that are farther away from X than x, regardless of whether they are on the same side of X as x or the opposite side of X. The difference between Equation (1.7) and Equation (1.8) is a factor of two in the calculated probabilities.

Deciding to calculate the one-sided or two-sided probability is sometimes hard. Both tests assess whether a number is different from a population mean. The two-sided test assesses whether the number is different in either direction, up or down. The one-sided test is more restrictive because you are only considering the probability of the number being in only one of the tails. Three important words to consider are "different," "greater," and "lesser"—if the assessment is whether number is *different* from the mean, then a *two-sided* test should be used.

The probabilities from Equation (1.7) and Equation (1.8) are the portion of the Gaussian in the tail beyond that z value. For a one-sided test, this is the probability in only one tail, either below $-z$ or above $+z$. For a two-sided test, this is the probably in both tails, adding the contribution from both below $-z$ and above $+z$.

Remember that there is not a simple analytical functional form for $P(z)$. It is readily calculated though, because it is a built-in function with most spreadsheets or statistical packages, or can be found with one of the several online calculators of these probabilities (one is listed in Further Reading, Section 1.10). The relationships of the probabilities given by Equations (1.7) and (1.8) to z are shown in Figure 1.8. When $z = 0$, the one-sided probability is 0.5 because the left half of the distribution is not being tested, but for the two-sided probability, the value is one, as both sides of the distribution are being included in the test and the x value is exactly on the centroid, X. The values decrease quickly and both probability curves are below 0.1, near $z_{ref} = 1.4$ and 1.8 for the one-sided and two-sided tests, respectively.

Table 1.3 The z_{ref} scores required to reach certain percentages within the cumulative probability distribution of a Gaussian, for both one-sided and two-sided tests.

P(one-sided)	0.10	0.05	0.025	0.01	0.005	0.001
P(two-sided)	0.20	0.10	0.05	0.02	0.01	0.002
z_{ref} score	1.28	1.65	1.96	2.33	2.58	3.09

To further quantify these probabilities, Table 1.3 lists the z_{ref} scores needed to reach a certain probability. Both the one-sided and two-sided probabilities are given above the z_{ref} values. As the z_{ref} value gets larger, the associated probabilities are very difficult to see in Figure 1.8.

What do to with this resulting probability? There is a long-standing tradition within statistics of the significance test, in which a probability of a certain value or lower is declared **significantly different** than the centroid of the population. This probability has historically been set to 0.05, the "5% significance threshold" criterion. The alternate value that

---

Significantly different: The traditional term applied to a z test (or other statistical inference test) with a probability of 5% or less.

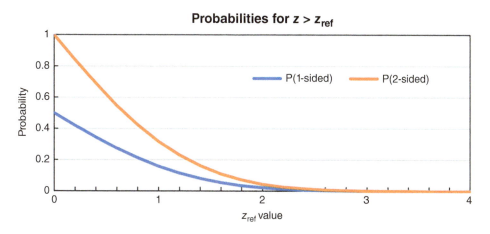

Figure 1.8 The $P(z)$ values as a function of z for one-sided and two-sided z tests.

has received regular usage across many disciplines is 0.01, the "1% significance threshold," a quantity that has traditionally been denoted as a point at which x can be called **very significantly different** from the centroid X of the Gaussian distribution to which it is being compared. These two terms, significant and very significant, carry with them these specific meanings, which have been ingrained into scholarly communities throughout the natural sciences, statistics, and, really, every major research discipline.

These are arbitrarily defined thresholds, though. They do not signify anything particularly conclusive about the inclusion of the value in the full number set except that it is past some round number of 5% or 1% of the full distribution. They were set, long ago, as a nice threshold limit. In fact, there is a debate in the statistics scholarly community on whether to continue to advocate for their use with comparative analyses like the z test (and many others to be defined later in this book). Therefore, consider these as guidelines. In the end, it is up to you as the investigator conducting the analysis to make the call on whether the value should be considered a member of the population to which it is being compared.

Quick and Easy for Section 1.6.3

The z score is a normalized value-to-centroid distance that, assuming a Gaussian distribution, can be converted to a probability.

1.6.4 Comparing Two Values Revisited

So, what about that example of your recording of the temperature? To continue developing the scenario, let us say that everyone is using

---

Very significantly different: The traditional term applied to a z test (or other statistical inference test) with a probability of 1% or less.

---

the same thermometer, an analog dial for its display, and took measurements about the same time. Given these factors, then, let us say that your value is 59 °F and the neighborhood average is 55 °F. It is the estimate that the neighborhood average value has an uncertainty of 3 °F.

To compare these two values with a z test, assume that neighborhood average represents the true population centroid and spread of a Gaussian distribution. In this case, a z score can be calculated for your temperature measurement relative to the population:

$$z_{neighbor} = \frac{|59-55|}{3} = 1.3 \qquad (1.9)$$

The probability that 59 °F belongs to the population centered at 55 °F with a spread of 3 °F can then be calculated. For a one-sided test, the cumulative probability distribution of a Gaussian distribution for $z = 1.3$ is 0.91, meaning that 91% of values within the population are below 59 °F. To put it another way, this is a one-sided probability from Equation (1.7) of $P_{1\text{-sided}} = 0.09$, which means that there are still 9% of the distribution above this value. For a two-sided test, checking both above 59 °F and an equal distance below the centroid (i.e., below 51 °F), the resulting probability from Equation (1.8) is $P_{2\text{-sided}} = 0.18$. That is, there is still quite a bit of the distribution beyond the measured value. Both 9% and 18% are above both the 5% and 1% significance thresholds. Unless there is some extraordinary reason to declare yours a different value, the conclusion should be that your value is from the same population.

We can justify this conclusion because the measurements were made at about the same time of day within very close proximity. With the z test probabilities that high and these circumstances known about the measurements, it is reasonable to conclude that the temperatures were measuring values from the same underlying population of values.

What if the situation were changed so that the thermometers were digital, with an uncertainty on the neighborhood average of 1°F. In this case,

$$z_{neighbor} = \frac{|59-55|}{1} = 4.0 \qquad (1.10)$$

This is a much higher z score, and the probabilities are similarly different. The one-sided probability from Equation (1.7) is $P_{\text{1-sided}}$ = 0.00003, indicating the fraction of the Gaussian distribution that is below 55 °F, and the two-sided probability from Equation (1.8) is 0.00006, revealing the fraction of the Gaussian either less than 55 °F or greater than 63 °F. These are very small, much less than the highly significant designation of 1%. It would be reasonable, then, to declare that your temperature represents a measurement from different underlying population.

The locations of the measurements matter for this comparison. It could be that some of the thermometers are mounted in a permanently shady region, while other thermometers are in direct sunlight. These factors do not necessarily influence the uncertainty, but could influence the interpretation of the probability that they came from the same population; that is, they are "the same."

With a small change in the method, we reached an opposite conclusion about whether your temperature measurement fit within the neighborhood average. Even though we are dealing with specific numbers and well-known equations, statistical inference is a subjective topic.

> **Quick and Easy for Section 1.6.4**
>
> An atmospheric temperature example usage of z-score calculation and the z test is given to show that the application can sometimes be subjective.

1.7 Use and Misuse of Statistics

An excellent resource describing common applications of uncertainties in everyday situations is *Naked Statistics* by Charles Wheelan. There is an insightful quote from it that particularly resonates with the theme of this book: "smart and honest people will often disagree about what the data are telling us." Placing an uncertainty on a measured value is a subjective task; there is no single universally accepted or applicable method for assigning this uncertainty. Furthermore, a wide range of possibilities exist regarding the interpretation and further usage of that uncertainty value. Like the z test, there could be specific probabilities associated with a reported uncertainty value, but the application of this probability toward a decision-making process is subject to a judgment call. It could be that a particular application allows for higher risk than another situation, in which case a different probability might be acceptable. It might even be a single situation, but multiple people being involved in the decision process with one person's risk tolerance is higher than another person's risk tolerance. This could lead to a dispute about the way forward, with "smart and honest people" disagreeing about their understanding of how to use the values. The phrase "what the data are telling us" depends on not only the determination of the uncertainty but also the scenario in which the values are used and the individual people involved in the situation. To summarize, both assigning uncertainty to a value and incorporating that uncertainty into further analysis are uncertain processes.

As will be seen in the following chapters, many equations exist for the calculation of uncertainty and other quantities distilled from the full number set. These equations each have a particular derivation and originally intended meaning. One calculation might serve a particular user of that data set very well, while it might be of little importance to another user, who cares far more about some other aspect of the values. It should be restated that the calculation of uncertainty and analysis of data are subjective processes. The equations can be applied in myriad ways to achieve physical understanding or operational decision-making from the original numbers. This is a main point to grasp about this topic—when examining a number set, not all of the quantities calculated from it will be relevant for every application.

The situation gets even more subjective when considering two number sets. This is perhaps especially true when only one is a measured data set, while the other number set has numerical model output values attempting to reproduce the observations. How "best" to compare these two number sets will receive much attention in the later chapters of this book, but for now, know that there are many data-model comparison metrics available, each with their particular strengths at considering a certain aspect of the relationship between the two sets of numbers.

Also, the various formulas were developed with specific assumptions and intended uses, meaning that the applicability of any one statistical calculation is, by design, limited. This makes it possible to use a statistical procedure in a manner that is beyond its intended use, or outside of the validity of its underlying assumptions. This can be intentional—you might know the distribution is not Gaussian, but calculate and report a standard deviation anyway. The standard deviation will still have its mathematical definition, described in detail in Chapter 3, but no longer will have its specific interpretation relative to percentages of points within each standard deviation range, as it does for a Gaussian distribution. Others might then argue that you should not even bother calculating this quantity if the distribution is not close to Gaussian in form. Why? Because there could be yet another group of people out there who will apply the percentages to the standard deviation ranges, even with the caveat that these percentages might not be true for this number set.

> **Quick and Easy for Section 1.7**
>
> Any statistical equation or method has limitations. It's good to keep these limitations in mind when interpreting the resulting numbers.

1.8 Example: Solar Wind Density and Space Weather

Of growing interest across the world are forecasts of **space weather**. Yes, there is weather in outer space. It is not the same kind of weather we have here near the planet's surface; instead, space weather is defined as the conditions in the local space environment that might adversely influence robotic assets in space, astronauts in space, or ground systems that sensitively respond to near-Earth space conditions. There is a famous example of statistical results yielding opposite findings, that of the Smith et al. (1999) study.

There are many ways in which the space environment can cause harm or damage. A particular example to discuss here is the influx of hot electrically charged particles that cause satellite surface charging in near-Earth space. These charged particles are typically in the kilo-electron-volt (keV) energy range, and they dominate the charged particle pressure in near-Earth space. This pressure, and its associated electric current, distorts the magnetic field, and this change of the field can be detected on the Earth's surface. People have compiled measurements from several magnetic field sensors around the globe in composite indices of geomagnetic activity. One of these is known as the **Dst index**, the "disturbance storm time" index, which is a globally derived measure of the deviation of the Earth's magnetic field due to the near-Earth space currents carried by these keV-energy-charged particles.

Space weather: Conditions in the local space environment that might negatively affect satellites, astronauts, or susceptible ground systems.

Dst index: One measure of space weather calculated from low-latitude magnetic field measurements.

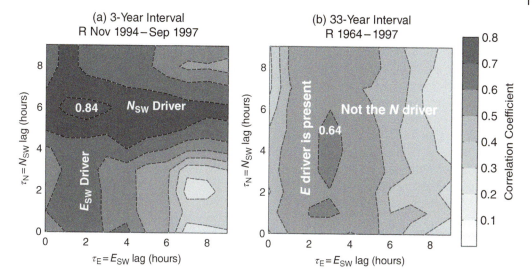

Figure 1.9 The relationship between Dst and two solar wind controlling parameters, motional electric field E_{sw} and density N_{sw}, along the x and y axes, respectively. The left panel (panel (a)) is the interval from the 3-year study of Smith et al. (1999), while the right panel (panel (b)) is from the 33-year study of O'Brien and McPherron (2000). The value plotted is time-lagged correlation coefficient. Adapted from O'Brien and McPherron (2000).

A key parameter that controls the flux of these charged particles, and therefore the Dst index time series, is the state of the solar wind. The solar wind is an electrified, magnetized, supersonic gas flowing away from the Sun. It is energized in the Sun's upper atmosphere and the random thermal motion of the particles is eventually organized into a fast stream of particles radially outward from the Sun. They carry with them the Sun's magnetic field, which we then call the interplanetary magnetic field. Three key values from this are the solar wind density, velocity, and the north–south component of the magnetic field.

A study was published in 1999 by a group of researchers that examined how the solar wind and interplanetary magnetic field relate to the Dst index. They selected 3 years of data, finding 55 space storm events in this interval, from which they recorded the minimum Dst values (i.e., the peak disturbance, as measured by this index). While this might seem like a good number of events for the analysis, it is actually too small, as will be demonstrated in the following text. They calculated the time-lagged correlation between the peak of the Dst disturbance and the value of certain controlling factors, going back for a full day from the storm peak time. The calculation of correlation will be discussed in detail in later chapters, but for now, it only needs to be known that a higher number is better. They found a clear relationship with both the solar wind motional electric field, E_{sw}, and the solar wind density, N_{sw}, as seen in Figure 1.9 as the horizontal and vertical dark stripes. The horizontal dark stripe indicates that there is a correlation with density at a time lag of roughly 5–7 h, while the vertical dark stripe reveals a dependence of peak Dst value on the solar wind electric field with a lag of 0–2 h. This was one of the first studies showing such a clear relationship between the solar wind density at the Dst response.

Because it was such a groundbreaking finding, the study was replicated by other groups. Unfortunately, the result did not withstand scrutiny. Another research group used 33 years of data instead of 3 years, an interval that included 439 geomagnetic storm events. When considering only the 55 events used in the original study, they reproduced the same plots shown by Smith and colleagues. When using their expanded set of 439 storms, they obtained the correlation coefficients shown in the right panel (panel (b)) of Figure 1.9. The horizontal stripe disappears, leaving only the vertical stripe of high correlation between peak Dst and solar wind electric field. The connection between peak Dst and solar wind density is tenuous, at best in this plot; there is a small peak within the vertical stripe, but the full horizontal band is lost with the inclusion of eight times more events.

The original study is not wrong; it was just limited in its applicability. The researchers just happened to choose three particular years that had quite a few geomagnetic storm events with solar wind density variations that aligned well with the eventual Dst peak value. Because the event list was relatively small, these events with similar changes in Dst and solar wind density dominated the correlation and the horizontal stripe of high values appeared in the plot. When many more events were considered, the influence of these outliers on the correlation was diminished, eventually to the point of the horizontal stripe of high correlation blending in with the background noise level of correlation values in the plot.

More about this will be discussed in Chapter 6, when the correlation coefficient calculation will be presented, including how this issue might have been avoided. Its inclusion here is purely illustrative, highlighting a few critical aspects of conducting statistical analysis and making inferences from the results. First, the number of values in the data set, N, matters. For small N, care must be taken in the analysis and interpretation. A few outliers can significantly influence the resulting statistical values from the data set. When N is small, additional analysis should be performed to double-check the veracity of the identified relationships and inferences. Second, extrapolating a study's findings beyond their intended validity is dangerous. The finding that solar wind density controls peak Dst could have become adopted as a prevailing view in the research community, with new models and theories developed based upon this relationship being true for any storm interval. The study only included a 3-year window during a particular phase of the 11-year solar cycle, and, therefore, should not have been applied as fully generalizable to all storm events. Third, it is very useful to repeat other people's work. A foundational element of the scientific process is repeatability. That other research groups thought to conduct their own versions of this correlation analysis is commendable. Remember, the first study was not wrong, it was just limited in its scope and implications for the broader research field.

> **Quick and Easy for Section 1.8**
>
> Sample size can make a big difference in the results you get. A limited sample size might reveal a trend that isn't significant in a larger sample.

1.9 Uncertainty and the Scientific Method

Speaking of the scientific process, let us digress from statistics and discuss this critical aspect of how to conduct research: whether the method includes statistical calculations and inferences or not.

This is a well-trodden path, but here is my version of the **scientific method**. First, you see something strange when perusing through some data. This can be an unexpected wiggle in a line plot, an unexplainable bump in a graph, or a number in a table that does not quite fit. You know the field well enough that you have an idea of what it should look like, but this new thing you are seeing does not quite fit what you think should be there.

Scientific method: The process used to discover previously unknown information about the natural world.

Second, you look it up. Yes, before you march to the laboratory, you investigate whether someone else has already come up with an explanation of why this feature should be here in these numbers. This research could be scanning through journals, reading through books, talking with colleagues down the hall, or sending emails to other experts in the field. Step two can take some time and should be done carefully and thoroughly; it is an essential part of the scientific process. If you find a good explanation, you are done. You learned something, the strange thing is not strange anymore, and you can go back to perusing the data.

If nothing satisfactory arises from this search, then you get to move on to step three, which is guess at an explanation. Scientists have a fancy word for this guess, calling it a **hypothesis**, "less than a full thesis." It is an educated guess, based on what evidence you have collected about this strange thing so far.

Step four is to design an experiment that tests the hypothesis that you devised. This experiment could be an activity in a laboratory, it could be field work in the natural laboratory of the outdoors, it could involve analyzing data that has already been collected, it might require running a numerical model, or it could be a theoretical pencil-and-paper exercise, deriving new equations. Any of these methods counts as a legitimate experiment to test the hypothesis.

Step five is, finally, doing the experiment. It is necessary to spend adequate time on step 4, designing the experiment, before actually doing it. This will help to ensure that the experiment is actually testing your hypothesis, which is not a guarantee.

This brings you to step six, where you compare the results of the experiment you conducted against your hypothesis. Did it explain the strange thing? If so, great, then you get to move on to presenting and publishing this work, sharing it with your research community

---

Hypothesis: An educated guess about the new information you hope to discover.

and claiming credit for advancing humanity's knowledge of this topic. If not, though, you have to go back to step three. Your guess was not correct, so you have to try again with a new guess. Yes, that is right, you toss out your idea. It did not work, so do not hold on to it. It could be that you only have to go back to step four. Perhaps the experiment "failed" to truly test the hypothesis. In this case, you can keep the hypothesis, but you need to rethink your experimental design. Perhaps a small correction will make it properly test your idea, but you might need to start from scratch. Similarly, if you suspect that your hypothesis was way off the mark, then you might want to go all the way back to step two and do some more investigation of what others might have already said about similar phenomena, in order to get a better grasp on what a possible new hypothesis might be.

This methodology of how to conduct science is shown schematically in Figure 1.10. In this version of it (note that there are many others), there are two big decision points in the flow, at which point you could be done or could continue to explore the issue. If the feature has been previously explained, then finding this explanation is the end of your analysis. If the results match the hypothesis, then this finding should be disseminated to others. There are, of course, decisions at each stage along the process. For example, in the second step, the investigator must decide how to conduct your literature search to determine if the strange feature has already been explained by others. This other decisions along the path to discovery, however, are often not ones that stop the progression to the next stage.

Each of these steps can take a short or long time. Each step could take an hour or many years. The literature search in step two might be time-consuming, or conducting the experiment might be a significant investment, but steps three and four can also be sticking points in this process. You could get stuck thinking up a good hypothesis or stuck thinking of a reasonable experiment that will robustly test your hypothesis. Many projects have languished on scientists' desks because of steps three and four.

The Scientific Method

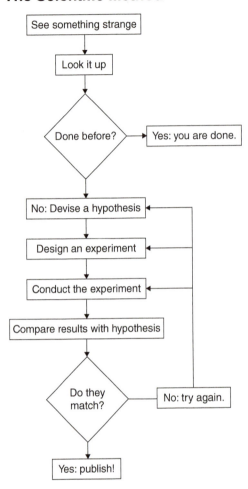

Figure 1.10 A flowchart of the scientific method.

Also, following the scientific method means that you will be wrong. The evaluation in step six might show you that your explanation was not right. Being wrong is not an easy thing to accept, but it is a possibility that scientists face every time they embark down the steps of the scientific method.

> **Quick and Easy for Section 1.9**
>
> The scientific method is a process for discovering new information about the natural world, and it includes being wrong, circling back, and trying again.

Figure 1.11 Chicken Little following up their class session with additional reading of the supplementary references. Artwork by Asher/Anya Hurst.

1.10 Further Reading

Each chapter will end with a section like this one, giving citations and links to other documents with good information and other descriptions and discussions of the material presented in the sections in the earlier text. Like Chicken Little is doing in Figure 1.11, it is highly recommended to further investigate these topics. Mentioned in the following text are the source materials that significantly influenced the development of Chapter 1.

For the traditional story of Chicken Little:
https://www.worldstory.net/en/stories/chicken_little.html

Here is a good description of uncertainty in measurements, written by a chemistry teacher at my old high school:
https://www.auburn.wednet.edu/cms/lib03/WA01001938/Centricity/Domain/1360/1_Uncertainty.pdf

This is a nice refresher on significant figures:
https://www2.southeastern.edu/Academics/Faculty/wparkinson/help/significant_figures/

NASA has a great website about Saturn's moons:
https://solarsystem.nasa.gov/moons/saturn-moons/overview/

Here is another explanation about the Gaussian/normal distribution:
https://encyclopediaofmath.org/index.php?title=Normal_distribution

A good calculator for z test probabilities with the normal distribution:
https://www.hackmath.net/en/calculator/normal-distribution

There is actually quite a debate raging among statisticians about the use of standard hypothesis testing thresholds, to the point of some advocating for removal of the term "statistically significant" from technical usage:
Wasserstein, R. L., & Nicole, A. L. (2016). The ASA statement on p-values: Context, process, and purpose. *The American Statistician*, *70*(2), 129–133. doi: 10.1080/00031305.2016.1154108

Hurlbert et al. (2019) go so far as to recommend that journal no longer allow this term:
Hurlbert, S. H., Levine, R. A., & Utts, J. (2019). Coup de Grâce for a tough old bull: "Statistically significant" expires. *The American Statistician*, *73*(suppl1), 352–357. doi: 10.1080/00031305.2018.1543616

And two papers by Amrhein et al. in 2019 continue this discussion in the context of replication of statistical studies and the difficulty in exactly reproducing another study's p-values:
Amrhein, V., Greenland, S., & McShane, B. (2019a). Scientists rise up against statistical significance. *Nature*, *567*, 305–307. https://doi.org/10.1038/d41586-019-00857-9

Amrhein, V., Trafimow, D., & Greenland, S. (2019b). Inferential statistics as descriptive statistics: There is no replication crisis if we do not expect replication. *The American Statistician*, *73*(suppl1), 262–270. doi: 10.1080/00031305.2018.1543137

The title of Section 1.7 is an homage to a paper by Patricia Reiff that had a significant formative influence on me:
Reiff, P. H. (1990). The use and misuse of statistics in space physics. *Journal of Geomagnetism and Geoelectricity*, *42*(9), 1145–1174. https://doi.org/10.5636/jgg.42.1145

Here is the citation and link to Charles Wheelan's *Naked Statistics* book:
Wheelan, C. (2013). Naked Statistics: Stripping the Dread from the Data. WW Norton & Company, New York.
https://bookshop.org/books/naked-statistics-stripping-the-dread-from-the-data-9781480590182/9780393347777

NOAA's Space Weather Prediction Center has a nice website to learn more about geomagnetic storms:
https://www.swpc.noaa.gov/phenomena/geomagnetic-storms
Or the solar wind:
https://www.swpc.noaa.gov/phenomena/solar-wind
Or the whole space weather glossary of terms:
https://www.swpc.noaa.gov/content/space-weather-glossary

Here is a research article that describes the similarities and differences between the various Dst-type low-latitude geomagnetic indices:
Katus, R. M., & Liemohn, M. W. (2013). Similarities and differences in low- to middle-latitude geomagnetic indices. *Journal of Geophysical Research: Space Physics*, *118*, 5149–5156. https://doi.org/10.1002/jgra.50501

Here is the citation of the Smith et al. (1999) paper:
Smith, J. P., Thomsen, M. F., Borovsky, J. E., & Collier, M. (1999). Solar wind density as a driver for the ring current in mild storms. *Geophysical Research Letters*, *26*, 1797–1800. https://doi.org/10.1029/1999GL900341

And the O'Brien and McPherron (2000) paper:
O'Brien, T. P. O., & McPherron, R. L. (2000). Evidence against an independent solar wind density driver of the terrestrial ring current. *Geophysical Research Letters*, *27*, 3797–3799. https://doi.org/10.1029/2000GL012125

Another view of the scientific method, with a very similar flowchart:
https://www.sciencebuddies.org/science-fair-projects/science-fair/steps-of-the-scientific-method
And another from Britannica:
https://www.britannica.com/science/scientific-method

1.11 Exercises in the Geosciences

1. As will be the case with most of the end-of-chapter exercises throughout this book, there is more than one way to do these problems. For each part of every problem, write out all assumptions made and steps taken.

2. A person takes an outside temperature with a digital thermometer that gives a reading of 33.96 °F. The person determines that the absolute uncertainty is really ±1 °F.
   A How should this temperature be reported, with its uncertainty?
   B What is the fractional uncertainty in Fahrenheit? How should the measured number be reported with the fractional uncertainty (with appropriate significant figures for each)?
   C The conversion from the Fahrenheit to Rankine temperature scale is given by this formula:

   $$T(\text{in °R}) = T(\text{in °F}) + 459.67.$$

3. What is the fractional uncertainty for this temperature on the Rankine scale? How should it be reported with the measured number (with appropriate significant figures for each)?

4. Homing pigeons have magnetic field receptors in their beaks. By feeling the slight torque generated by the magnetoreceptors, they can find magnetic north, helping them navigate very long distances. Let us say that this sense has an angular uncertainty of 2°. One such pigeon is from Ann Arbor, Michigan, a city that is, roughly, 5 miles in diameter.
   A For this pigeon flying the 40 miles from Detroit to Ann Arbor, assuming guidance only from its magnetoreceptors, will the full width of its uncertainty be wider than the diameter of the city? Explain your answer.
   B What if the flight were from New York City to Ann Arbor, a distance of 515 miles? Again, explain your answer.

5. A bird-watcher reports the sighting of a black-bellied whistling duck, a rare bird to be seen in Michigan. They say they spotted it floating in Wamplers Lake, a relatively large body of water in Jackson County that is nearly a mile across in some parts.
   A One person, viewing from the western shore at a section of the lake known to be 4670 ft across, says the bird was 1000 ft out into the water. A group of other people, viewing from the eastern shore directly across from the first, say the bird was 3000 ft out into the water. Did they observe the same bird? Do this by estimating uncertainties and making it a completely qualitative comparison. Explain your answer without calculations, but basing it on a discussion of appropriate significant figures, estimated uncertainties, and a qualitative comparison of the values.
   B Conduct a z score calculation using the measured and uncertainty values from part (a).
   C How would your answer be different if the reported distances to the bird were 1200 ft for the single observer and 3400 ft for the group on the opposite shore? Again, make it qualitative and explain your answer.
   D Conduct a z score calculation from the values in part (c). Compare against the abovementioned qualitative assessment.

2

Plotting Data: Visualizing Sets of Numbers

It is easy to get quantitative with data sets. After all, they are numbers, and numbers can be combined with formulas, from the very simple to the highly complicated, to get different numbers that tell us something about the original set of values. Most of this book is about that quantitative assessment of number sets, not only the processing of data sets but also the comparison of two number sets against each other.

Before getting too quantitative with data sets, it is good to take some time to qualitatively understand the numbers. Your answer could be a flippant, "well, just plot the data," and you would be right that this is often a simple task. At least, simple to get a first look. But what makes for a good plot? What makes for a figure that is worthy of going into a scientific journal, or published by a news outlet, or shared on social media? The final figure could—I would argue *should*—be very different for these three venues. Therefore, before we get into numerical processing of one or two data sets, let us first discuss the ways in which that data could be visually presented.

2.1 Plotting One-Dimensional Data

Consider the values in Figure 2.1. It really does not matter what these values are, so they have been given no defining label. The x axis is equally vague, plotted here against its index within the number set. Here, index refers to the integer value of the measurement's placement within the set. This data set was created by sampling a Gaussian distribution with a random number, around some mean with an arbitrary spread.

A plot like Figure 2.1 often does not tell us very much about the data. It neither has any units along either axis nor **error bars** on the values to indicate their uncertainty. As a first step to analyzing a data set, it is better to make a plot that has more quantitative content. This type of plot is shown in Figure 2.2. Actually, two versions are shown. These are the same data values in Figure 2.1, except that now they have values along both axes and error bars on all of the points. The error bar lengths show two different versions of some assumed uncertainty in the measurement of the values. An initial interpretation of these data would change depending on the uncertainty. In the first case, you might conclude that the quantity is changing rather substantially between observations, while in the second you might say that most of them are within the same uncertainty envelope and are, therefore, about the same.

It is important to start off with a plot like these, even before doing anything quantitative with the analysis of the data set. A display of

---

Error bar: A vertical or horizontal indication on graph of the uncertainty for that data value.

Data Analysis for the Geosciences: Essentials of Uncertainty, Comparison, and Visualization, Advanced Textbook 5, First Edition. Michael W. Liemohn.
© 2024 American Geophysical Union. Published 2024 by John Wiley & Sons, Inc.
Companion website: www.wiley.com/go/liemohn/uncertaintyingeosciences

28 | Data Analysis for the Geosciences

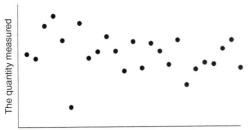

Figure 2.1 Illustrative example of a data set.

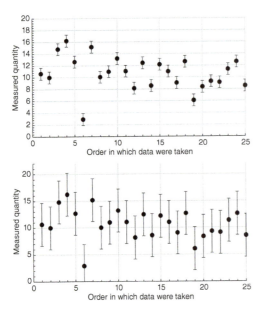

Figure 2.2 The same illustrative example of a data set, this time with error bars. Two different lengths of error bars are shown, one with a small measurement uncertainty and another with a large one.

the data provides us with a good qualitative estimate of many of the values that will be later calculated. Specifically, we can see that the centroid is somewhere around 10 and the spread of the values is probably in the range of 2 or 3, as most of the values lie within a band of 8–12 or perhaps 7–13. Without actually counting the values in intervals or calculating a centroid and spread with standard formulas, we can already make these judgments about the data set.

For the geosciences, there is almost always an implied second set of numbers associated with the first one-dimensional data set. Take, for instance, a collection of measurements of soil acidity. The values of pH themselves are enough for the discussion in the following text, but there is a second set of values that should be reported with it. In space physics, this is known as ephemeris data and this term will be adopted for this book. **Ephemeris data**, also known as **metadata**, includes the location and time of each measurement as well as instrumental information such as temperature and voltages of various components. It might also include the orientation of the instrument, especially if the sensor has certain directions for which it is more sensitive. The ultimate directional sensitivity is something like a telescope, where the field of view might be extremely narrow, in which case the ephemeris data must include the telescope orientation for proper interpretation. A measurement that might be insensitive to orientation is a water-temperature sensor; it simply needs to be immersed in the liquid and the orientation of the sensor does not influence the reading. A similar argument can be made that soil pH readings are insensitive to the sensor orientation. This data set, however, might need different ephemeris, such as depth of the sample being tested and classification of the type of soil from which the sample was taken.

This book does not focus on these other quantities, but it should be understood that they are very important for interpreting the output. The relationship of measurements to their ephemeris data will be discussed occasionally in the following text. For the cases just mentioned, here is an example of how ephemeris data could help interpret the observations. For the telescope, the temperature of its reflecting mirror will alter the optical path, and it could be that your measurement of the brightness of Venus was systematically different than it should have been because of a

---

Ephemeris data or **metadata:** Values accompanying a primary data set that provide context for where, when, and how it was collected.

shift in the mirror temperature reading. For water temperature, local water depth is a big factor, and an undetected shallow spot in the middle of the lake could lead to an unexplained temperature increase in that region. Regarding soil acidity, the nearby vegetation is important to note, as certain plants change the pH of the surrounding dirt. That is, comprehensive ephemeris data often lead to a more robust and more correct diagnosis of systematic or random uncertainties in the measurement set.

It is enough for us here to say that another set of numbers is typically needed when plotting a single data set; we need some value along the x axis to go with the data values along the y axis. This could be one of those ephemeris data sets accompanying the primary observations. By assuming that other ephemeris values are roughly constant, this plotting parameter is often taken as either time, if the measurements are all from the same place, or location, if the measurements are all from the same time. Either of these provides a one-dimensional monotonically increasing array of values for the x axis. To completely remove the ephemeris data from the analysis, another option is simply to plot the data against an integer number set, marking the placement of the observations within the lineup of values. This also provides a monotonically increasing array for the x-axis values and is what will be done for this chapter when an x-axis value set is required. For some plots, however, the axes will be without units completely. These are plots meant to qualitatively illustrate a concept. Do not be alarmed, such plots are perfectly acceptable and sometimes highly useful for conveying information at a more general level. This is where knowing your audience is of key importance.

Quick and Easy for Section 2.1 Intro

Error bars on data give an initial qualitative view of whether differences between values might be significant.

2.1.1 What Makes a Good Plot?

It is important to take a moment and discuss some of the aspects of figures in technical writing that make them impactful. The biggest is to understand several foundational elements of the purpose of the figure. The first is to **know the main point being conveyed**. A second point is to **know the audience for whom it is being made**. A third is to **know the aspects of the plot on which you want the audience to focus**. Keeping this information in mind while developing the layout of a plot will take you a long way toward producing an effective graphic.

Let us examine that first concept. What is the main point of the figure? Once this is known, then focus the content of the plot to highlight this point. Similarly, it is good to de-emphasize other aspects of the plot so that the viewer's attention remains fixed on key features. That is, make that one main aspect of the data the eye-catching component of the plot. A common pitfall of plots in technical writings or presentations is to include far too many distracting elements that clutter the presentation of the information and obscure that main point. The viewer is not thinking about these other features and not paying attention to the section of the figure that really matters for the

Know the main point being conveyed: When approaching the creation of a plot, it is important to decide on the main point of the final version.

Know the audience for whom it is being made: When approaching the creation of a plot, it is important to keep in mind the background and expectations of the intended viewers.

Know the aspects of the plot on which you want the audience to focus: Design the figure to highlight the most important element and de-emphasize other aspects.

present discussion. A common corollary to this pitfall is that the creator attempts to make too many key points with the same figure. That is, they intentionally insert extra annotations, add more panels, and crowd in additional content to maximize the number of findings from a single figure.

The desire to overcomplicate scientific graphics crosses over from the first foundational element (know your main point) to the second one about knowing the audience. In technical communication, one of the biggest splits is with respect to whether the viewers of the figures are a live audience of a presentation or whether they are readers of a document. Figures are often reused between these two venues, but this can lead to big issues in understandability. Particularly, in the former setting of a live audience, the viewer of the figure has no control over how long they get to see the figure. In addition, you are talking while they are looking at it, so they are digesting multiple streams of information. A figure in a presentation, therefore, should be as simple as possible, and the words spoken while the figure is on the screen should direct the audience toward the key feature. Even two key points are distracting, because the audience could be looking at the second feature while you are still talking about the first, lowering their comprehension of both features. You can have a complicated figure in a presentation, but it is wise to build up to the full figure, introducing elements one at a time, so the audience understands each component of the full figure before seeing the entire picture. Animations or separate figures that add in new elements help the viewer comprehend each aspect of the plot. Furthermore, keep words to a minimum unless they reinforce what is being spoken about the plot. Human minds really are not that good at multitasking, so it is best to minimize distractions in order to keep the viewer's focus on the one element of interest at that particular moment. Sometimes, there is more than one feature to discuss in a plot. A good plot, then, will ensure that the viewer stays focused on only the particular element of interest. The easiest way is to have different versions of the same plot, one for each key feature to be discussed. It is perfectly acceptable to repeat figures with a different region of the plot highlighted with a circle, arrow, or shading.

In documents, the reader controls the pace at which they consume the information, including the content of figures. Figures in documents, therefore, can be more detailed than those in presentations. The reader can spend as much time as they need to understand the figure, reading and rereading the accompanying text as many times as necessary to fully comprehend your key points. The reader can magnify the plot to examine a feature up close. They can leave for a while and return to the figure later. Whenever you have multiple panels in the same figure, though, a key element of maintaining clarity is to have a clear connection between panels and, whenever possible, a common axis to align features between the subplots. Complexity is allowed because of the switch of control from presenter to viewer. Do not abuse it, though, as there are still stylistic elements that can make the plot hard to understand. Note that figures in documents do not have to be more complicated; it is perfectly acceptable to have a presentation-worthy figure in a document. Going the other way—using a figure from a document in a presentation—is where caution should be used.

Regardless of this difference, there are some common traits of what makes a good plot. Figure 2.3 illustrates the difference that the words can make on that plot. It highlights the clarity of the text in effectively conveying the information—good axis labeling, readable font, appropriate use of annotations, and a short but descriptive title. The left plot of Figure 2.3 has rather small font and rather cryptic labeling. It requires explanation to fully understand what is important in this figure. The right plot shows the other end of the spectrum, with lots of text across the plot, to point of distraction. The middle panel clearly conveys one key point about the data.

2 Plotting Data: Visualizing Sets of Numbers | 31

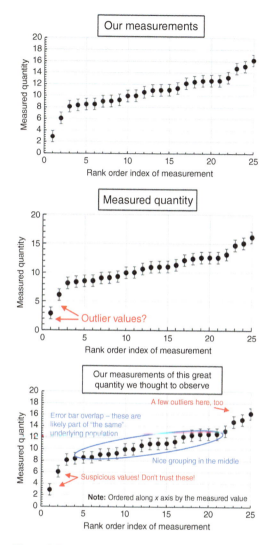

Figure 2.3 The same illustrative example showing the influence of text on the effectiveness of the graphic.

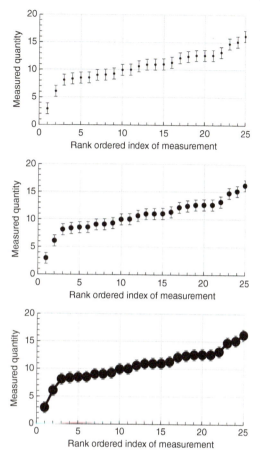

Figure 2.4 The same illustrative example showing the influence of data presentation on the effectiveness of the graphic.

important to correctly scale the size of points or lines on the plot—too small and they are difficult to see, too large and they look cartoonish and perhaps cover the error bars.

A second example about clarity in making the plot is shown in Figure 2.4. This set of plots focuses on the data itself and the need for appropriately displaying the values. The left panel uses a very small point for each data value, while the right panel uses huge dots with a connecting line. For these data, with the sample number as the *x*-axis position, the line is completely superfluous. The middle panel of Figure 2.4 uses a symbol size that is easily seen but not so large as to be distracting. It is also

Quick and Easy for Section 2.1.1

When making a plot, focus its design on the main point being conveyed by it as well as on the audience to whom it is intended.

2.1.2 Exploratory Versus Explanatory Plot Styles

Why do we even bother with figures? As the saying goes, "a picture is worth a thousand words," we include graphics to convey meaning to the audience that otherwise would have needed many paragraphs of explanation. This is true for any kind of information transfer, whether it be at the highly technical level of a peer-reviewed journal article or an illustration accompanying a social media post.

Stylistically, graphics span a vast continuum of options from the highly specific to the very general. The two broadly defined "ends" of this continuum have been given special names—**exploratory graphics** and **explanatory graphics**. These two types of plots are very different, but both can be highly useful in the right circumstances.

Exploratory graphics are best when you have an analytical point to convey with the data in the graphic. This nearly always includes quantitative values and correct units on the axis labels. This seems natural to a scientist or an engineer; the observations have specific values and units, so why not show them? Given a data set, this is the natural place to start. Read the data values into your favorite plotting software and make a figure. Indeed, most figures in technical journal articles are exploratory graphics. It takes effort, though, to make a truly good exploratory plot that effectively draws the viewer's eye to the key feature of interest. That is, the process of honing that original quickly made graphic of the data into the final version for presentation or publication could take many iterations.

Explanatory graphics are best when you want to highlight a particular message. These graphics de-emphasize quantitative rigor in favor of visual esthetics. They often do not even have labels on the axes. These types of plots are used to illustrate concepts without the need to explain the details of the phenomenon being displayed.

Exploratory graphics: A figure that focuses on a quantitative and analytical point.

Explanatory graphics: A figure that conveys a qualitative message.

Let us consider the example data set shown in the earlier text. Figure 2.1 is an example of an explanatory graphic, showing a typical data set without the complication of exact values or units. On the other hand, Figure 2.3 shows some exploratory plots with this same number set, plotting it with these labels in various ways to highlight different aspects of the values. To further exemplify this contrast, Figure 2.5 shows them again, this time as an annotated exploratory version and a highly stylized explanatory visualization. The left plot is similar to the middle plot of Figure 2.4, noting the central clustering and the one very low value. The right plot is a mix of Figure 2.1 and Figure 2.4, with no labels on the axes but instead descriptive annotations that guide the viewer to see certain features in the data set. All quantitative information about the data set has been removed in this case, because it is not meant to impart specific information to viewers but rather only a general impression of the upward trend in the values.

Which is the better one to make? There is no correct answer to that. The analysis process

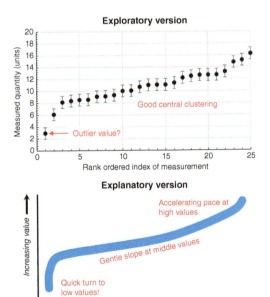

Figure 2.5 The illustrative example of a data set, shown as an exploratory plot and an explanatory plot.

usually begins with a basic exploratory figure, but this is only the starting plot for the analysis and eventual presentation of the data. It depends on the information being conveyed and the expected audience receiving this information. A complex and technical presentation of the data is not appropriate for a general audience, and an artistic rendering of the data might not be rigorous enough for a scholarly audience. Neither end of the visualization spectrum is better than the other. They are both useful and should be used as appropriate. Furthermore, visualization is not a binary choice; this is a continuum with an infinite number of options along the spectrum from one end to the other.

Quick and Easy for Section 2.1.2
There is a continuum of graphical style from the highly quantitative "exploratory" to the highly qualitative "explanatory" figures.

2.2 Example: Earth's Magnetic Field Strength

The Earth's inner core is compressed into solid iron, but the outer core of the Earth is at the correct heat and pressure combination to be molten. In fact, it is not only molten, but electrified—the atoms are so hot that when they strike each other, they sometimes knock an electron free from its orbit around the nucleus. This creates a plasma, a mixture of positively and negatively charged particles which obey not only fluid dynamics but also Maxwell's equations. Moving charges create magnetic fields, or more specifically a *net* movement of charge creates the **geomagnetic field**.

While the outer core is mostly iron, there are enough lighter elements mixed in that they also behave as another constituent within the fluid. In general, iron more easily gives up its outermost electrons compared with the lighter elements, creating a systematic charge separation that loosely follows the elemental breakdown of the molten metal. Positive and negative electrical charges lead to differences in forces and therefore motion, which is further amplified by the differences in mass of the individual particles, exerting different gravitational forces on the particles. The result is that there are net flows of charged particles in the outer core, leading to electric currents and therefore magnetic fields. These currents are mostly azimuthal, flowing either eastward or westward, with a small component flowing latitudinally and another component flowing radially. Most of it flows in a circle around the Earth, creating a dipolar magnetic field. The Earth is slowly giving up some of its other energy reserves, its rotational energy as well as energy from nuclear decay of radioactive materials, to form and then maintain a large-scale magnetic field. The exact process of this energy conversion is still largely unknown. We cannot send a probe to the outer core to measure it locally, and computational resources are still inadequate to fully answer this question with an advanced and accurate numerical simulation. The best that we can do is observe the field on the surface of the planet and apply known physical relationships to understand the processes occurring deep within the Earth's interior. Monitoring these magnetic field can be done with a sensor called a **magnetometer**, essentially a very sensitive three-dimensional compass, which measures the local vector magnetic field.

Figure 2.6 shows plots of magnetic field intensity at two locations, one at a near-equatorial latitude (left column plots) and

---

Geomagnetic field: The magnetic field produced in the outer core of Earth, extending into near-Earth space.

Magnetometer: A device that very precisely measures the local magnetic field magnitude and direction.

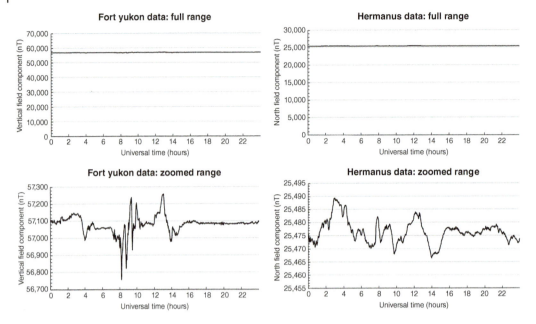

Figure 2.6 Observations of Earth's magnetic field, taken at a low-latitude ground station (left column of plots) and high-latitude ground station (right column). The high-latitude station is more susceptible to space weather influences due to electric currents in near-Earth space.

another a high northern-latitude location (right column plots). Specifically, the two stations are in Hermanus, South Africa, close to the magnetic equator, and Fort Yukon, Alaska, which is usually within the auroral zone. These data are for a single, full 24-h interval, specifically 1 January 2018, which is a randomly selected day of relatively quiet geomagnetic activity. The time along the x axis is **universal time (UT)**, used in much of geoscience and aligned with that of the zero longitude line. The actual values and activity are not the important point here, but rather the focus of this example is the way the values are being plotted.

The upper panels of Figure 2.6 show the values for the day with a y-axis range spanning from zero to the maximum observed value. The lower plots do not use zero as the lower end of the y-axis plot range, but rather zoom in on the range of the values being plotted. The upper plots show that there is a large baseline value to the magnetic field at both of these locations. This is the Earth's geomagnetic field, dominating the signal compared to any magnetic field signatures from space weather. As you can see, plotting it the second way is far more informative than the first regarding the small-scale perturbations caused by electric currents flowing in near-Earth space. In the measurements from the Hermanus observatory, it is seen that there are hours-long undulations throughout the day. The Fort Yukon data show a very strong and quick oscillation in the field between 8 and 10 UT, indicating the presence of strong auroral activity overhead. Neither of these key features in the geomagnetic activity on this day was viewable in the upper plots; the baseline value of the Earth's field is so strong that the features are essentially within the line width. The point is that it is good to make several plots of the data so that your understanding is not clouded by the initial impression from the first plot that you made.

Universal time (UT): A commonly used time for geosciences that aligns with the time at the zero longitude meridian.

> **Quick and Easy for Section 2.2**
>
> An initial plot with "standard" axis ranges might not be particularly useful; it is often good to make more than one version of a figure to assess effectiveness and clarity.

2.3 Probability Distributions—The Histogram

One of the important steps that you should do when analyzing a data set is make a histogram of the values. This is made by defining a set of bins across the range of the data and then counting the number of values within each bin. The plot made from these counts for each bin is usually a vertical bar graph, with counts on the y axis and the bin values across the x axis. It could be done as a line plot or scatterplot, but a bar chart histogram is the normal format for this information. You can transpose the plot and reverse the axes, which is sometimes the convention in the field, convenient to do, or stylistically appealing. Some geophysical quantities, such as altitude or depth above or below the Earth's surface, are often plotted on the y axis, even when it is the independent variable. This practice was started long ago and gives the plots a more natural appearance that helps with the researcher's intuitive sense toward extracting meaning from the plot.

Figure 2.7 shows a histogram of the illustrative example from Section 2.1. As you can see, it is simply the definition of bins along the value range from minimum to maximum and then counting the number of points in the range for each bin. This is a typical histogram, with a peak somewhere in the middle and counts that drop off to zero with distance away from this centroid value. You can draw it in several different ways, as shown in Figure 2.7. The first panel shows a version with a reasonable number of bins and drawn with column widths that span the full range of each bin. By this last point, it is meant that the histogram columns touch each other. By contrast, the

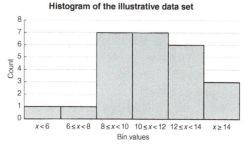
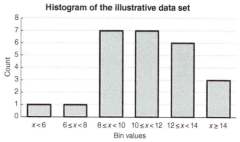
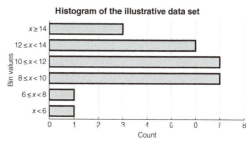

Figure 2.7 Histograms of the illustrative example data set, plotted in several different ways.

second panel shows the same binning but with narrow columns that leave a gap between them. The third panel is the same data but plotted with horizontal columns to show this histogram style.

The fancy name for this plot is the **probability distribution**. Actually, there is one more step in order to make it a probability distribution—divide by N, the total number of data values in the set. This is sometimes convenient because then all of the column heights sum to one. If you then wanted to overplot a known functional form with the data histogram, then scaling the two curves is relatively easy if the data columns

---

Probability distribution: A distribution for which the integral of all x is exactly one.

are already normalized in this way. This type of comparison is very important; many statistical quantities assume that the underlying distribution of the data values fits a Gaussian shape—the normal distribution or bell curve. We'll come back to this in Chapter 5.

One of the challenges with making a good histogram is choosing the right number of bins and, its counterpart parameter, the bin size. If there are too many bins and the bin size is too small, then many bins will have either zero or one count. This makes it impossible to see the distribution of the values and renders the histogram meaningless. If the number of bins is too small and the bin size is therefore too large, then critical features in the data distribution could be lost. For instance, a bump on the tail of a main peak could go undetected.

How many bins? There are several basic rules. The most straightforward is the **square-root rule**. It is what it sounds like, set the number of bins, j, equal to the square root of the number of samples, N, and then round up to the nearest integer,

$$j = \left\lceil \sqrt{N} \right\rceil \tag{2.1}$$

For 10 data values, this means 4 bins, which is just about as low a number of bins as you ever want to go. For 50 samples, it is 8 bins, and for 5000 samples, Equation (2.1) indicates that j should be 71 bins. This bin count, however, sometimes obscures secondary features in the distribution, so I like to think of this rule as providing the minimum number of bins.

The bin width is then the total range of the data values divided by j. It is important that the bin size be even so that the count in each bin is directly comparable to the others.

What is a good upper limit for the number of bins? It depends on the number of samples. For $N < 100$, doubling the square-root value is a reasonable maximum, as many more than this will break down the main peak into too many bins. For $N > 100$, going up to a number of bins that is triple the square-root value might be useful to see secondary features. That said, an upper limit on the number of bins to use in any case is approximately 200. Any more than this and you risk introducing noise near the main peak in the histogram. This is an issue of counting statistics that will be discussed later in Chapter 4, but it introduces difficulty in understanding the features in the histogram. Plus, physically plotting the histogram becomes an issue, as it is difficult to make a plot with 200 bins look good. A good rule about this is to never go over 100 bins unless the features in the histogram demand it.

There are many other options for choosing a number of bins. **Sturges' formula**, developed nearly 100 years ago, set the number of bins as the log base 2 of the number of samples, rounded up, plus one more,

$$j = \left\lceil \log_2 N \right\rceil + 1 \tag{2.2}$$

For 10 samples, this yields 5 bins, while for 50 samples, Equation (2.2) says that j should be 7 bins, and for 5000 samples, this method says that you should use 14 bins. For the smaller sample sizes, this formula yields bin counts roughly equal to the first method. However, for larger data sets, this method gives much smaller bin counts. With only 14 bins for 5000 samples, each bin will have hundreds of data values. While this number of bins for that many data samples should show you the main peak, it will most likely smear together many of the secondary features of the distribution. That is, it might give you a false sense that the distribution is Gaussian when it really is not that close to that shape.

A final bin count method to be discussed here is **Rice's rule**. This is similar to the

---

Sturges' formula: A method of determining the number of bins to use in a histogram, it involves taking the log base 2 of N.

Square-root rule: A method of determining the number of bins to use in a histogram, take the square root of N and then round up.

Rice's rule: A method of determining the number of bins to use in a histogram, it is double the cube root of N.

square-root rule but instead uses a cube root, and then includes a multiplier of two before rounding up to the nearest integer,

$$j = \left\lceil 2\sqrt[3]{N} \right\rceil \quad (2.3)$$

For our three test values of 10, 50, and 5000, Rice's rule yields bin counts of 5, 8, and 35, respectively. Again, for the smaller numbers of samples, Equation (2.3) yields a value about the same as those given by the two methods mentioned in the earlier text. For larger sample sets, Rice's rule gives a number of bins that is in between the other two formulas.

An example of using these three bin count methods is shown in Figure 2.8, using the Hermanus magnetometer data from Figure 2.6 as the illustrative data set. There are 1440 measurements in this data set, which yields bin counts of 38, 12, and 23 from Equations (2.1), (2.2), and (2.3), respectively. With a data set of over a thousand numbers, these three methods yield very different bin counts. The range of the data values is from 25 466.6 nT to 25 489.5 nT; this is a span of 22.9 nT. For this example, the first bin of all three histograms is set to 25 466.5 nT, using bin widths of 0.605 nT, 1.92 nT, and 1.0 nT, respectively, for the three histograms. In Figure 2.8, the value listed below the columns is the upper edge of the bin. While all three histograms show the same basic feature of the distribution, they reveal different levels of detail. In particular, the bin with the most counts shifts to lower values in the three successive plots. The middle panel has a single peak, while the other two panels show double peaks within the main peak. Furthermore, the top panel—the one with the most bins—includes more information about the tails of the distribution that is not seen as well in the lower two plots.

Another feature of Figure 2.8 is the use of a simplified axis label. All of the bin values have a baseline value subtracted from them, making it easier to read the values along the axes. Changes like this are common and should be adopted whenever possible to reduce the complexity of the text on the graphic. In this case, the values have a large offset, but a small variation range. A subtraction of some value, therefore, is a good transformation to make the numbers more readable. The value of 25 400 nT was selected as a convenient amount that is easily added back in, but it is not the only choice. The lowest value, the arithmetic mean, or the median could also have been used. While these other choices are good because they are based on the data set, they are often specific real numbers, so converting the bin values back into the measurement values is slightly more complicated addition.

What is the right number of bins? There is no correct answer. The formulas given in the earlier text are not rules so much as guidelines. It could be that your data set has outliers that make the range incredibly large for the number of samples in the set. In this case, the rules mentioned in the earlier text will yield large bin sizes that do not properly resolve the features of the distribution. If you really want to see the outliers and resolve both the peak and the secondary outlier features around it, then you would need an unusually large number of bins. Alternatively, it might be useful to omit the outliers when making the histogram, focusing only on the main peak of the values. Another way to get around this is to have the lowest and highest bins an infinite width, containing everything below the second lowest bin and everything above the second highest bin, respectively. These bin counts might be disturbingly high compared to their neighbors, but you can take that into account when interpreting the plot.

A special case of making histograms is when the data are discrete. For instance, if the data are counts, and therefore integers, then care must be taken with the bin sizes and some of the rules mentioned in the earlier text need to be modified to account for this. Specifically, you want to include the same number of integers in each bin so that the bin counts are directly comparable with each other. A bin size that is a real number and slightly smaller than one will yield some bins that do not include an integer value, and therefore these bins will have a count of zero. Most likely, the histogram would have gaps, most

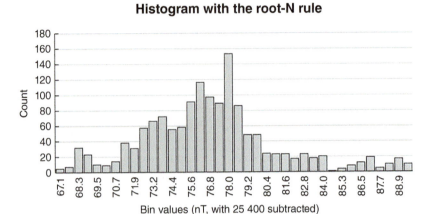

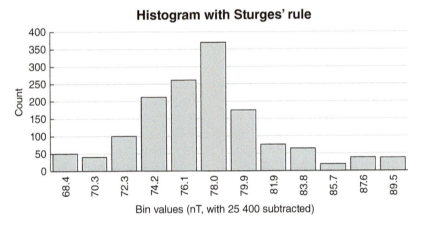

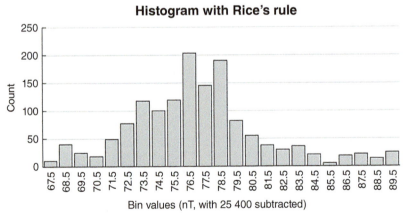

Figure 2.8 Three histograms of the Hermanus magnetometer data shown in Figure 2.6 using the three bin count rules mentioned in this section.

likely irregularly spaced (unless you chose 0.5 for the bin size, in which case exactly every other bin would be zero). If the bin size were slightly larger than one, then some bins would have one integer and others would have two, and the resulting histogram would look like a long row of jack-o'-lantern teeth. Neither of these outcomes is useful for interpreting the true distribution of the values. It is fine to have more than one integer value per bin, but the number of integers per bin should be identical for all bins. In short, if you see patterns in your data like this, then match your bin size to that pattern, regardless of what the formulas tell you to do.

It is highly recommended that you make the histogram of your data set at least twice, if not more times, to explore the nature of the distribution. A good suggestion is to start small, say the value from one of the three methods discussed in the earlier text, and then incrementally increase the number of bins until you reach 100 or 200. Looking at the data in more than one way will reveal different aspects of the distribution. If you do it just once, then you are never quite sure if you missed something important in the distribution.

A question to ask yourself is this: why am I making this plot? The first component of answering this question is to consider the information you are trying to extract from it. Do you only want to test the shape of the main peak? Do you want to see if there are potentially any secondary small-scale features in the data distribution? A yes to one or the other of these questions could steer your bin count to a smaller or larger number. The second component of answering the question is to consider to which audience you might be presenting this plot. If these plots are only for you to better understand your data, then make a systematic set of histograms and compare the features within them. If you are past that stage and now making this plot for a presentation or publication, then you will most likely only show one histogram, so it should convey the message you wish the audience to take away from it. When making a plot to show to others, always have a purpose in mind when formulating the style and content of the graph.

Quick and Easy for Section 2.3

Histograms reveal the probability distribution of the data set. There are a few good rules for how many bins to use; it's good to make several versions.

2.4 Plotting Two Data Sets Against Each Other

The first step is to make a plot of the two data sets against each other. While the analysis at this step is purely subjective, it is a reference to refer to as more quantitative approaches to comparing the two number sets are applied. Several types of plots are discussed in the following text. It could be necessary to remake the plot several times, or to make a variety of different types of plots, in order to get an initial understanding of the relationship between the two data sets and to highlight several of the key features between them.

2.4.1 Overlaid Histograms

Comparing two numbers requires that the two sets be plotted together. An example of this is shown in Figure 2.9 with a histogram of the data set from Figure 2.1 and a corresponding Gaussian distribution from which the 25 values were sampled. Four different versions of the same two histograms are included in Figure 2.9, with a range of overlap between the columns with different shading options.

Declaring one of these histogram plots to be the best is entirely a matter of preference. The upper-left plot in Figure 2.9, with no overlap between the columns, is a standard format, as is the lower-right plot, with full overlap and a semi-transparent shading for the one in the front. The other two offer less common alternatives, one with partial overlap and another with full overlap but no shading at all, only the border outline, for the front histogram. This lower-right option is my personal favorite and the one that will be used later in the book.

This is the first major usage of color in data visualization that we have encountered. While color has been used in the earlier plots, they all could have been made in black-and-white, or at least grayscale. For overlaid histograms, color is often the best choice for clearly separating the two column sets. Color is an important tool in data visualization and, with multiple dimensions or different parameters to display,

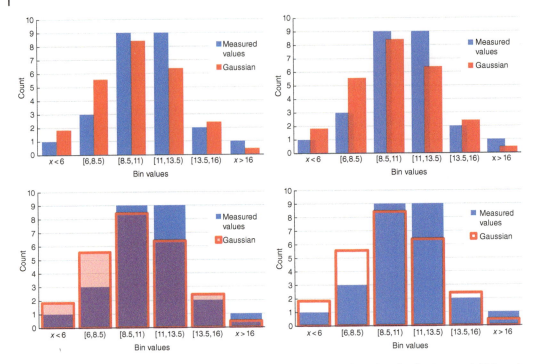

Figure 2.9 Histogram of a number set, in blue, with a corresponding Gaussian distribution overlaid, in red. The comparison is shown in several ways, from no overlap to complete overlap.

color might be the only way to effectively and meaningfully display the values.

> **Quick and Easy for Section 2.4.1**
>
> Overlaid histograms are convenient to directly compare the distributions of two number sets.

2.4.2 The Scatterplot

An initial plot to make with the data sets is often the basic **scatterplot**. This is simply using one value along the x axis, the other for the y-axis values, and putting a point or symbol for each (x,y) pair in the combined number set. Figure 2.10 shows an example of two data sets plotted against each other. As will be shown in Chapter 3, the specific quantities do not matter here, so this plot has been stylized in an

---

Scatterplot: Using two data sets as the x and y values of an ordered pair, plotting the resulting points in the 2D x–y space.

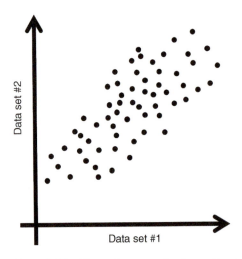

Figure 2.10 Illustrative example of a scatterplot for two data sets.

explanatory format. From this initial look, a qualitative judgment can already be made: is this a "good" relationship?

Figure 2.11 shows similar example comparisons of two data sets, but with several changes

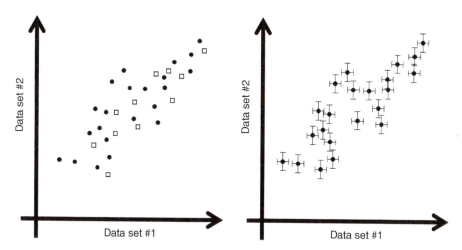

Figure 2.11 Another illustrative example of a scatterplot between two data sets, this time shown differently. On the left, a second set of values are shown with symbols instead of points, while on the right, error bars have been added to the points in both the *x* and *y* directions.

made to the scatterplot. The left plot shows two data sets of the values, one with symbols instead of points. This can be a good or a bad modification to the display, making the points more visible but perhaps making the plot too crowded. Here, with only a limited number of points, the addition of another symbol does not overcrowd the panel. The right panel goes even further, adding error bars to all of the symbols. Does this make it look better? That depends. Sometimes it can be useful to make the plot more stylistic, depending on the expected audience. The inclusion of the error bars is very good for technical audiences, but could be highly distracting for more general audiences.

Sometimes, there are too many points to nicely make a standard scatterplot. Often, the points in some region of (x,y) space significantly overlap and create a fully colored area, from which it is impossible to see trends between the data sets. Sometimes, reducing the size of the dot can separate the values and make the black region into distinct data points again. However, sometimes this makes the points around the fringes of the dense region too small to clearly see, losing information about these outlier points. Other times, reducing the dot size is physically impossible; the plotting software is already using the minimum symbol size. There are several other plot styles that offer ways around this, discussed in the next subsections.

The first option is to add histograms of each data set to the scatterplot. Figure 2.12 shows an example of this type of plot, with not only a scatterplot but also the two histograms of the two data sets included as additional panels. The left panels of Figure 2.12 show the same data sets as in Figure 2.10, but now with histograms showing the density of the data along each of the axes. The right plot of Figure 2.12 includes a large black region in which many points overlap. While it is known that many (x,y) pairs are in this region, the exact location of the peak cannot be deduced, nor can any trends be seen within it. With the inclusion of the histograms, though, a more complete understanding can be developed for how the two data sets relate to each other.

> **Quick and Easy for Section 2.4.2**
>
> The scatter plot is two-dimensional rendering of one data set against another, and provides an excellent first look at their relationship.

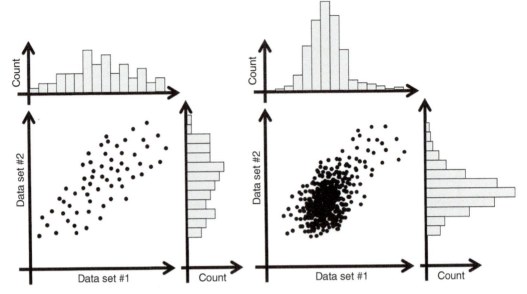

Figure 2.12 Scatterplots with histograms on the top and right to show the density of the values. The left plot is the scatterplot from Figure 2.10, while the right scatterplot shows two data sets for which there is a region with many pairs, resulting in a black region.

2.4.3 The Box Plot

One way to clearly signal that the distribution is not a Gaussian is to use the **box plot** to visualize the number set. It is often called a box-and-whisker plot because of its standard format, with a narrow rectangle indicating the location of the bulk of the distribution and then long thin lines demarking some outlier extent of the values. It is nearly always defined from **quantiles**, which are found by sorting the number set in ascending order and finding the value at a certain percentage of N. That is, quantile is found as the percentage of the rank order of a value divided by N, the total number count in the set. A box plot is usually made with the 50% quantile—the **median**—representing the centroid and other quantiles used for the box extent and even more extreme quantiles as the whisker length.

Figure 2.13 shows a typical box plot. The box is usually, nearly always, defined by something called the **interquartile range**, or IQR, with the 25% and 75% percentile x values defining the

Rank order: The position of a value in a number set that has been sorted from smallest to largest.

Box plot: A method of displaying a number set, often a subset of a larger number set, with the use of quantiles to show the spread and full range.

Median: A measure of a number set centroid, it is the 50% value within the ascending-order sorted number set.

Interquartile range: A measure of a number set spread, it is the difference between the 75% and 25% quartile values within the rank-order listing.

Quantile: Choosing a value from a sorted number set at a particular percentage location within the rank-order listing.

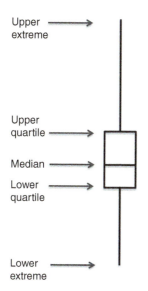

Figure 2.13 Typical box-and-whisker plot of a data set.

upper and lower edges of the box. The box usually also includes another horizontal line at the location of the median. The ambiguous aspect of the box plot is the whisker definition. Many box plots will use the maximum and minimum values of the number set for endpoints of the whiskers. While this is good because it shows the full range of the data, it is susceptible to any extreme outliers in the set. If N is relatively small, then there probably will not be values too far from the center and the whiskers will be a reasonable length. As N increases, though, the probability of including an outlier very far from the median increases. That is, the whiskers will only grow with more numbers in the set. Other options are to use symmetric pairs of quantile values in the tails of the distribution, but not quite at the 0% and 100% levels. That is, other choices for the whisker endpoints are the 10%–90% quantiles, the 5%–95% quantiles, and the 2.5%–97.5% quantiles. These have 20%, 10%, and 5% of the values still outside the whisker extents, respectively. Extreme outliers will be beyond these percentiles and therefore will not influence the length of the whiskers. A final somewhat common method of defining the whisker length is to set the length of each

whisker equal to the IQR. While this shows extra range beyond the actual IQR represented by the box, this provides no extra information and is equivalent to omitting the whiskers. In fact, if one or both of the tails of the distribution are light, then this could actually show whiskers beyond the maximum or minimum of the number set. Whichever of these methods is used to define the whiskers, that method should be explained so that others can correctly interpret the features of the box plot.

A key feature of the box plot is that it clearly depicts any asymmetries in the distribution. The box portion quickly reveals an asymmetry near the centroid and the whiskers expose asymmetries in the wings of the distribution. This is information not provided by a simple, single uncertainty value. In addition, the inclusion of whiskers shows the spread estimate, the IQR, relative to the full range of the data (or nearly the full range). Thus, it provides a qualitative understanding of the distribution.

> **Quick and Easy for Section 2.4.3**
>
> The box plot is an alternative graphic to the histogram that displays several key quantities of the data set's distribution.

2.4.4 The Box-and-Whisker Scatterplot

Now that the box plot has been introduced, it is almost never used as a single box but rather with subsets of a larger number set. That is, it is a common method of simplifying a scatterplot—several box plots summarizing all points within small intervals of the full x-axis domain. Figure 2.14 shows an example of this, showing the data from Figure 2.12 as a set of box-and-whisker bars. Two different versions of this transformation are shown, with the left plot showing a box-and-whisker for each histogram bin in Figure 2.12, while the right plot uses half as many box windows.

A good question to pose as this plot is being made is this: how many bins should be used? Should it be as done in the histograms of

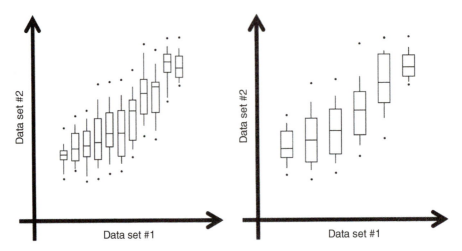

Figure 2.14 Two versions of the right-hand data set shown in Figure 2.12, but this time as box-and-whisker bars, grouping the data into subsets along the x axis. The difference between the two is the grouping. Both plots use the 10%–90% extents for the whiskers, with an extra dot at the most extreme maximum and minimum value in each bin.

Figure 2.12? Or perhaps more bins, in order to see small-scale features of the distribution? Or, alternately, should a smaller number of bins be used so that the statistics of the quantile values reported in each box-and-whisker bar are more robust? In general, you do not want to use more bins that you would have used for the histogram of the x-axis data set. Doing so usually makes the plot too crowded with box-and-whisker bars and leads to confusion as outliers within each subset can lead to strange changes from bin to bin. For clarity, fewer bins are better. This choice, however, can group together large portions of the data and obscure key features of the comparison. The best advice is to iterate this, making the plot several times with different numbers of bins to see what highlights the features of interest in the data set. One suggestion here is to make three versions, with one-third, two-thirds, and the same as the number of bins that you might use in a standard histogram. From there, additional plots can be made as needed.

A related follow-on question is this: how to define the bin widths? As with so much of visualization and statistics, there is no good answer, but there are several good methods. One technique is to keep the bins evenly spaced along the x axis, which is easy to justify but could lead to some bins with very few points, especially near the end of the distribution. A more accepted method is, therefore, to include the same number of points in each subset. This means that the statistical interpretation will be the same for all box-and-whisker bars in the graph. That is, even when using the same number of bins as you would in a standard histogram, the bin widths and edges might be rather different, crowding the box-and-whisker subsets close together in those x-axis regions with many points and spreading out the subsets in other regions with few points. Depending on the number of (x,y) pairs in the full set, a common choice with this method is to pick the intervals to be 5%, 10%, or 20% quantiles in width.

> **Quick and Easy for Section 2.4.4**
>
> Using a series of box plots is a common method of simplifying the scatterplot of a large data set.

2.4.5 The Running Average Plot

Another way to declutter the plot when there are too many points to reasonably include in a single panel is to add an overplot of the **running average** centroids and spreads. This is typically only made when the number of points is very large, well into the thousands or even millions of points. You only want to make this when there can be many intervals along the x axis and having a large number of values allows for good statistics within each window throughout the running average.

This style is similar to the box-and-whisker plots of the previous section, but with a small but important change. In this case, the centroid and spread values are calculated within an incrementally shifting window. This window shifts by a small amount compared to its width, thereby allowing the intervals of neighboring subset regions to overlap. The centroid to be shown is often either the arithmetic mean or the median, but it could be the mode, geometric mean, or some other more informative value for that particular comparison. An appropriate spread value is also calculated, which is usually chosen as standard deviation, σ, for the mean, and IQR for the median (although sometimes larger intervals are chosen). The centroid and upper and lower spread values are then plotted as lines, y values as a function of x, often with a bold line style for the centroid and a thinner line (or even dashed or dotted) for the spread lines. Sometimes these lines are drawn on top of the scatterplot itself and other times the individual values are omitted entirely from the running average plot.

Figure 2.15 shows such plots. The upper-left panel shows 100 paired (x,y) values of points along with an overlaid running average line. This line was created with a 13-point sliding window, so it starts at the 7th value in the list and ends at the 94th. The upper-right panel omits the data points and instead shows the arithmetic mean running average and the ± 1 standard deviation as the spread. The lower-left panel shows the 13-point running median along with the 10% and 90% quantiles as lower and upper envelope values, respectively. The lower-right plot shows three different centroid estimates—the 13-point mean, the 13-point median, and the 7-point mean—without upper/lower envelope lines, so the y-axis range is zoomed in a little bit.

There are several decisions to be made when making this type of plot. The foremost is the bin width, but you also need to decide the shift step and how to handle the endpoints. For a running average plot, the bin widths are defined as a number of points, not by a set x-axis value range. The bin width here can be finer than that used in the box plot visualization. This is because you are not drawing the full box-and-whisker bar for each interval but rather only the centroid and spread values, as continuous lines connecting the values between the bins rather than discrete bars for each bin. You typically want it even finer than you would have made it for the 1D histogram of the x-axis number set. Within each window, 100 points is typically the smallest you want to choose for each bin, but a larger value is fine, depending on the total number of points and the level of robustness desired for the centroid and spread values. This contrasting requirement of lots of small bins but many points per bin is because this plot style is typically only done when there are a very large number of values in the full set.

The shift between the intervals can be one, meaning that only one data point is added to the right side of the interval as a single data point is removed from the left side. This creates a very-fine-scale running average curve, which might be overkill for the point of the plot. It could be that there are so many points that shifting the window by one data point at a time leads to a curve with features that will be obscured by the width of the line itself. You might not need this level of granularity because you might not be able to see it in the plot. You can often take larger step sizes in shifting the

---

Running average: Taking an average value of a subset and then shifting the window, usually by one, from one end of the number set to the other.

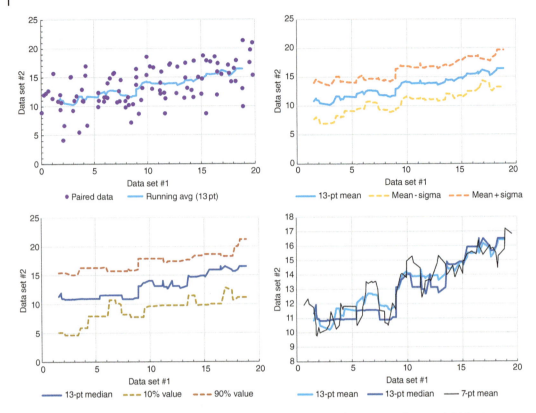

Figure 2.15 The upper-left plot shows a paired data set with a running average line overlaid. In the upper-right panel, an arithmetic mean running average is shown, along with upper and lower spread lines at one standard deviation from the mean. In the lower-left plot, a running median is drawn along with a 10%–90% quantile interval around them. In the lower-right plot, three running centroids are drawn (a 7-point mean, a 13-point mean, and a 13-point median).

window than just a single value at a time. On the other hand, you want the windows to substantially overlap so that the line is smooth. A step size up to 5% of the number points within the window is often a good compromise value, but this should be found with an iterative process of examining the resulting plot until a good balance is reached.

Finally, the beginning and ending of a running average calculation require special care. Three options are presented in Figure 2.16. The easiest way (top panel) to do this is to simply start and end the running average calculation slightly in from the very edge of the data set, specifically half of the window width. This allows all of the windows to be similarly defined as the same number of values per bin.

The downside is that there will be gaps at the beginning and end of the running average lines compared to the actual minimum and maximum of the data set being plotted along the x axis. If the number of points is large and therefore these gaps are small, then this method is probably the best choice. If, however, you want the running average lines to extend to the ends of the data set range, then another method should be used. One technique is to remove points from the interval by the window shift number being used once the interval window reaches an endpoint of the data set (middle panel of Figure 2.16). The final value in the running average at each end will have half the number of points as the bins in the middle of the running average lines. This means that the

Beginning/ending options for a j-point running average

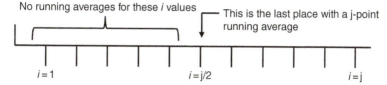

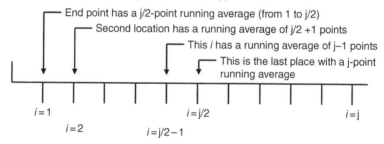

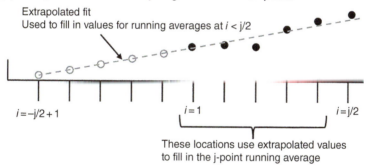

Figure 2.16 Three possible options for how to begin and end a running average.

calculational method is the same as the other bins, but the number of points in the calculation decreases as the line approaches the ends, which could lead to variability in the running average lines in these final few values at each end. Another method is to extrapolate the lines into these end regions by calculating a curve fit for the last full running average window (lower panel of Figure 2.16). Curve fitting will be covered in detail later in Chapter 7. It could be a linear fit, polynomial fit, or some other appropriate curve fit that best matches the functional form leading into the final tip of the running average line. This line fit is then plotted from the last full-window running average value to the end of the x-axis data set range. Whichever method is used, remember the key rule of always explaining your procedure so that others can understand your approach and include this knowledge in their interpretation of your results.

Quick and Easy for Section 2.4.5

Another way to simplify the scatterplot of a large number set is to calculate and plot the running average.

2.5 Example: Temperature and Carbon Dioxide

Modern-era climate change is dominated by the rise of atmospheric concentrations of certain gases, the so-called **greenhouse gases**. These molecules, such as carbon dioxide, methane, the refrigerant HFC (hydrofluorocarbons, which not the ozone-destroying chlorofluorocarbons, CFC, but their replacement), and water, are efficient absorbers and emitters of photons in the infrared wavelength band. This band, with wavelengths of 1–10 µm, which is slightly longer than the visible light range, is where the Earth emits photons, according to its blackbody temperature. Greenhouse gases trap the outgoing longwave radiation and reemit some of it back toward the Earth's surface. The temperature rises as their presence in the atmosphere increases. Over the last 150 years or so, the proportion of carbon dioxide in the atmosphere has been steadily rising. With other influencing factors accounted for, there is a very clear connection between this rise of greenhouse concentrations and a rise in global atmospheric temperature.

But what about the Ice Ages? The temperature of Earth over the last several million years has undergone dramatic swings, much bigger than what has been happening in the last 150 years. So big, in fact, that Earth has oscillated between epochs of vast glacial growth across its land masses, resulting in what are known as **Ice Ages**, interrupted by geologically brief intervals of rapid warming, known as the **interglacial periods**. Is there a greenhouse gas connection to these global temperature variations?

The short answer is yes. The timelines of inferred global temperature and atmospheric carbon dioxide content are shown in Figure 2.17, going back nearly 1 million years. These are both obtained from ice core samples

---

Greenhouse gas: A molecule that resonates with the longwave radiation emitted by Earth, absorbing and reemitting the photons, which sends half of them back down to Earth.

Ice Age: A period in the last several million years where the temperature of Earth dropped and allowed large glaciers to form across much of the high-latitude land mass.

Interglacial period: A short interval, usually less than 10 000 years, of warming of Earth between Ice Ages over the last few million years.

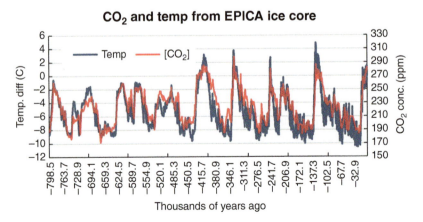

Figure 2.17 Temperatures inferred from δO_{18} levels in an Antarctic ice core (dark blue line, left vertical axis), along with atmospheric CO_2 concentrations obtained from air bubbles in the same ice core sample (red line, right vertical axis).

extracted from Antarctic ice sheets that are almost 2 miles deep. Specifically, these data are from the ice core taken out of the European Project for Ice Coring in Antarctica (EPICA) Dome in Antarctica, the deepest sample taken. The qualitative comparison of these two curves shows that the variations in one seem to closely match the variations in the other.

The full answer is more complicated. The leading theory is that the interglacial warmings are initiated by preferential sunlight conditions on one of the poles. This is brought on by orbital variations in precession, obliquity, and eccentricity, three processes known collectively as **Milankovitch cycles**. These cycles range from roughly 10 000 years on the short end to around 100 000 years for some of the oscillations. When the cycles align in such a way that one of the poles gets extensive and intensive sunlight, there can be significant melting of one of the polar ice caps. This starts a positive **feedback loop** in Earth's climate, with the now-exposed ocean or land mass being much darker—that is, more absorbing—of the incoming shortwave solar radiation. This warms Earth's surface, melts more ice, and exposes more ocean and land, resulting in additional solar-radiation absorption. Much of these polar land masses are what are known as permafrost, land that is frozen year-round. It was not always frozen, though; this land once had forests, marshes, or grasslands across it, rich in biological matter. When they initially froze and became permafrost, the existing biomass was solidified in the ground. As the positive feedback warming continues and the permafrost starts to melt, this biomass starts to release its trapped gases, most notably methane and carbon dioxide. The addition of these greenhouses into the atmosphere then causes the positive feedback loop of the Earth's temperature to accelerate. Eventually, ocean temperatures rise to the level that cloud formation becomes ubiquitous. While water is a greenhouse gas, clouds are very effective reflectors of the incoming solar photons in the visible part of the spectrum. So, usually, water vapor increases the greenhouse effect, but clouds act in the opposite direction by removing some of the solar input to the system. This negative feedback process eventually stops the upward spiral of the global temperature. As the Milankovitch cycles continue to change the orbital characteristics of Earth around the Sun, the polar ice cap is eventually reformed and the climate system eventually goes back, very slowly, toward an Ice Age epoch. The interglacial periods are often much shorter than the Ice Age intervals, with the temperature peak typically only lasting 5000 or 10 000 years, followed by a long progression of dropping global temperatures that could last anywhere from 40 000 to 150 000 years before the next interglacial warming period. The Earth is currently in the midst of one of those temperature peaks, already past the 10 000-year mark and due for a very slow cooling trend. Human activity has kept the temperature high, at first predominantly through biomass burning and deforestation, but now also through the burning of fossil fuels. In fact, the temperature is now rapidly rising—relative to the natural geological cycles—due to human influences, potentially triggering positive feedback loops in the climate system.

To further explore the relationship between carbon dioxide (CO_2) and temperature, Figure 2.18 shows a scatterplot visualization of these two data sets against each other. The first plot is a regular scatterplot of points, while the other two instead show running averages

---

Milankovitch cycles: Oscillations in Earth's orbit, in particular the ellipticity (oval shape) around the Sun, the obliquity (tilt) of the spin axis relative to Earth's orbital plane, and precession (seasonal shift) of the tilt angle relative to Earth's closest approach to the Sun.

Feedback loop: A cycle of processes that eventually influences the original process, either in a positive or a negative way, which either amplifies or dampens the cycle, respectively.

50 | Data Analysis for the Geosciences

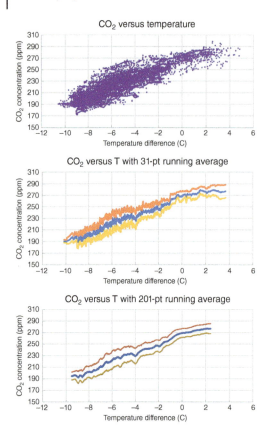

Figure 2.18 Ice core atmospheric CO_2 concentrations versus corresponding inferred temperature values, plotted against each other. The second and third panels show the 31-point and 201-point running averages, respectively, along a spread estimate around each.

(with standard deviation as its spread), using a 31-point and 201-point window for the calculations. There is a definite upward trend in the scatterplot, indicating a probable linear correlation between these data sets. The lower two panels show that the spread is rather tight around the running averages, and that these spreads are roughly the same across temperature values and regardless of the running average interval used. While the lower temperature ranges have slightly larger spread in the corresponding CO_2 concentrations than in the upper temperature range, the difference is quite small.

For this example, four different visualizations were shown that compare CO_2 concentrations and temperature. The first, in Figure 2.17, shows overplotted time series. The second, the first panel of Figure 2.18, is a standard scatterplot of one against the other. The third and fourth are running averages with a small and a large window, respectively, shown in the lower panels of Figure 2.18. A question could be asked: did we learn anything from the additional plots? Here is one brief assessment. From the first, one might conclude that the agreement is excellent, at least qualitatively. The second shows that the agreement is not as spectacular as initially thought, revealing substantial spread in the points around an upward trend. The running average visualizations, though, illustrate that the trend is rather linear with a tight spread; this is a very strong correlation over most of the domain of the temperature difference parameter—all but the highest temperatures, where the curve flattens. The main difference between the two running averages is the small-scale oscillations in the 31-point window, a feature absent in the 201-point window plot. This indicates that there is probably some localized time-lagged correlation between the two parameters existing on the tens-of-point level but not on the hundred-point level. We'll come back to these data sets for a homework problem in Chapter 12 to further explore this time lag.

> **Quick and Easy for Section 2.5**
>
> Paleoclimate atmospheric temperatures and carbon dioxide levels are closely related, therefore providing a good example for some of the plot styles discussed in the earlier text.

2.6 Scientific Visualization: A Sampling from the Literature

There are many great examples of scientific visualization. This section could be its own chapter, or even its own book. A single section cannot provide complete coverage of the many fantastic techniques and important visualization concepts. Instead, this section gives a *very* brief history of scientific visualization, especially some key "firsts" from a century or two ago, as well as a glimpse at a few modern-day graphics from scientific publications that exemplify key features of good visualization.

2.6.1 A Very Brief History of Visualization

Since humans first learned how to draw, we have been making scientifically useful diagrams. We have used these pictures to explain ideas to others, which is the essential element of scientific visualization. The key types of these drawings were maps. Because travel was difficult and dangerous, sometimes the maps were only accurate in the immediate region. Still, maps were vital to the interconnection of people across the globe. With Galileo's invention of the telescope, we even started to make maps of the solar surface, tracking and counting sunspots.

Modern scientific visualization really started in the 1700s. The printing press had reached a state of sophistication to allow for regular publication of newspapers and books, and enhanced graphics rose along with this availability of text. Joseph Priestley is credited with the first publication of a timeline chart in 1765. The real breakthrough came with the 1786 publication of economist William Playfair's influential book, *The Commercial and Political Atlas*, in which he was the first to print a line chart and bar graph. Examples of his plots are shown in Figure 2.19. Using many such graphics to distill the financial standing and trade relations of England, he argued against the funding of colonial wars through government debt. Playfair was also the first to publish a pie chart, a few years later.

Another highly influential person on scientific visualization was Alexander von Humboldt. One of the bold geographic explorers and naturalists of his day, he is credited as the instigator of modern environmentalism. Arguably his most famous graphic is the Naturgemälde, or Chimborazo Map, a full table-top-sized graphic published in 1807, showing the detailed altitude mapping of the vegetation he encountered climbing volcanoes in the South American Andes. This is still examined and studied by historians today as one of the most specific relationships of vegetation to altitude from its time.

Another to celebrate is Florence Nightingale. Yes, that Nightingale whose work inspired the creation of the Red Cross. In her case for better funding of nursing care, she published *Notes on Matters Affecting the Health, Efficiency, and Hospital Administration of the British Army* in 1858, in which she included spiral graphics of deaths throughout the year, like these in Figure 2.20.

These are just a few of the highly influential visualizations that have guided the development of scientific graphics ever since. Many others made significant contributions to scientific visualization over the last 200 years, and this is not meant to be an exhaustive review. Rather, these examples show that it is useful to spend time to carefully craft the display of your number sets (or other information) to reach for the intended audience. This gets us back to the discussion of exploratory versus explanatory plots. For Playfair and Nightingale, for instance, they were communicating with politicians and therefore favored the explanatory graphical style, which emphasizes the general sense of the message, over the exploratory style, which focuses on exact quantities. While explanatory figures are based on specific values, the numbers themselves are often not the main focus of such graphics. For instance, the pink and brown shaded regions in the line plot of Figure 2.19 conveyed the change in the trade balance of imports and exports, a value that could have been shown as its own line plot, but instead Playfair chose this more qualitative means of portraying this shift. Similarly in Figure 2.20, Nightingale used the exact mortality figures to create the sizes of the wedges, but did not include a scale on the actual plot; it was not needed to get her message across to the target audience.

Data Analysis for the Geosciences

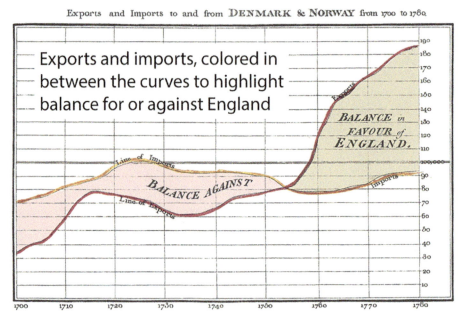

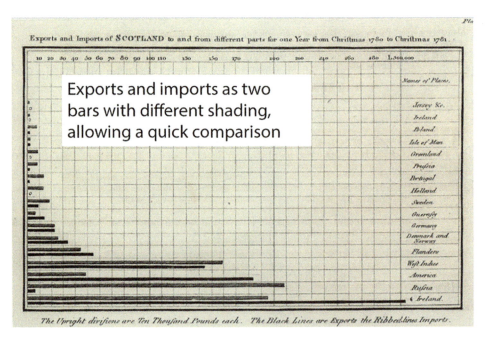

Figure 2.19 Annotated versions of two of the many line charts and bar graphs in William Playfair's *The Commercial and Political Atlas*, the first book to publish such figure types, in 1786.

2 Plotting Data: Visualizing Sets of Numbers | 53

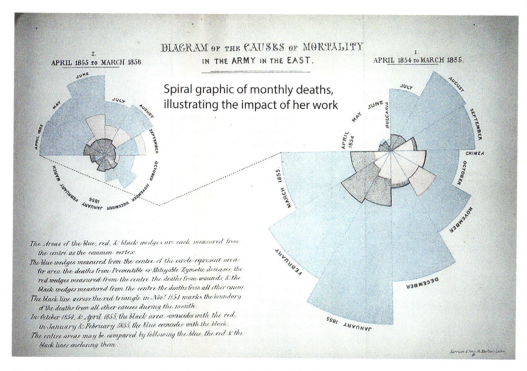

Figure 2.20 An annotated version of one of the spiral graphics from Florence Nightingale's book on the mortality rates in the British army during the Crimean War, published in 1858. Deaths due to preventable diseases are shown as the blue wedges, deaths from wounds in pink, and deaths from other causes in black.

Quick and Easy for Section 2.6.1

Scientific visualization progressed rapidly in the late 1700s and early 1800s as printed material became more readily accessible.

2.6.2 Good Modern-Day Example Visualizations

We do not have to rely on hand-drawn illustrations anymore, so it is useful to show a few examples of good visualizations from recent scientific publications. The following examples are all from the scientific journal *Geophysical Research Letters* (*GRL*), specifically taken from several papers published in early 2017. They show a wide range of options of how to convey a key point in a more explanatory style. Note that all of these papers also included exploratory graphics; line plots or other quantitative presentations of the number sets being analyzed

in the study. Here, the focus is on the explanatory graphics from these papers, because these are often the harder ones to conceive and create.

Some papers begin with an opening schematic. This is often done to introduce the reader to the process being investigated. Figure 2.21 shows an example of this type of graphic, from Piet et al. (2017), who were assessing the consequences of a large impact on a planet. This graphic was used to introduce the problem and justify the choice of modeling approach adopted for the study. This is a highly explanatory graphic, with no axis labels and values, simply annotated regions within the figure itself.

There are several reasons why this is considered an informative illustration. First, it is a set of three, showing the initial and final states along with the agent of change in the center panel. The graphics have some annotation but this is used sparingly, just enough to identify the key features of each plot. The consistency

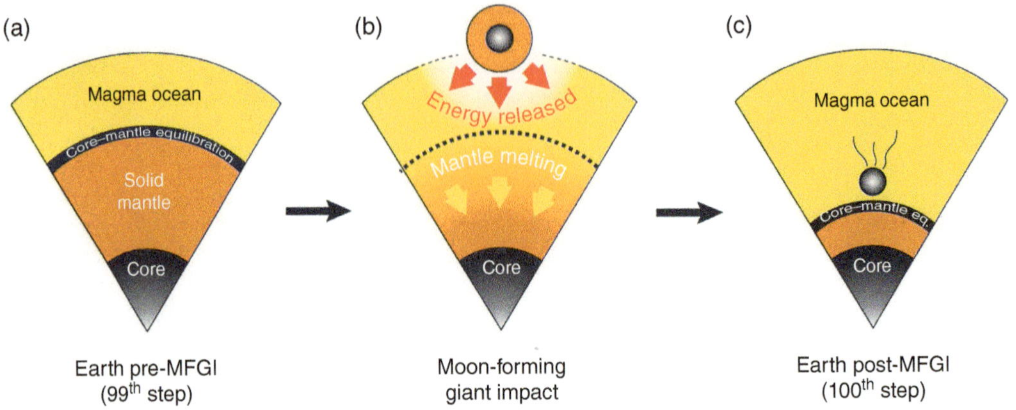

54 | *Data Analysis for the Geosciences*

(a) Earth pre-MFGI (99th step) (b) Moon-forming giant impact (c) Earth post-MFGI (100th step)

Figure 2.21 A schematic used early in the paper by Piet et al. (2017) to introduce the main concept of their study, showing the progression of how a large impact could cause melting deep within a planetary body and change the location of the core–mantle boundary.

between them is also appreciated, so that readers can easily see how the system evolves throughout the event. The simplicity focuses awareness on the clear difference between the start and end states, bringing the reader's attention on the key aspect of the analysis to be conducted in the study.

In some cases, many line plots can be condensed into a single figure using a technique called **superposed epoch analysis**. This is shown in Figure 2.22, from Haugland et al. (2017), who analyzed seismic activity to probe the physical characteristics of the Earth's crust. The top panel shows the seismic wave forms from an array of stations as individual line plots. A similar feature in all of these is a sharp, solitary oscillation, which they have aligned at the zero time reference. By themselves, each of these line plots is difficult to interpret. The next panel, however, clarifies this feature by plotting all of the line plots together, using the reference time to align the signals and then scaling them, so this feature has the same amplitude in each of the individual-station number sets. The

---

Superposed epoch analysis: Aligning time series number sets according to the timing of a particular feature in each individual set, and then plotting all of the time series on the same axes.

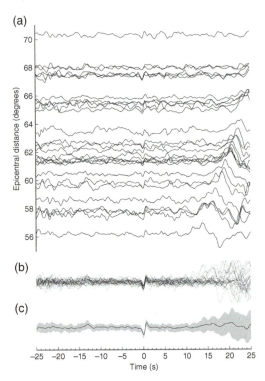

Figure 2.22 Line plots of seismic activity recorded at many stations, arranged according to distance from the epicenter of an earthquake (expressed as degrees across the Earth's surface, these stations are all thousands of kilometers away). Below these line plots is a single plot of their superposition, aligned according to a key feature. The final plot is the same superposed time series but now with a shaded interval showing one standard deviation in each direction from the running average. From Haugland et al. (2017).

feature is not very clear. The final panel shows the same superposed epoch analysis time series, but this time with a running average and a shaded confidence interval of one standard deviation in each direction from that central line. Note the lack of a *y*-axis scale for the seismic recordings; the magnitude is not what matters for this analysis, but rather the alignment of the feature, its identification in all of the individual data sets, and the relative change before and after the feature.

The line plots and grayscale graphics of Figure 2.22 are not flashy, but it was chosen to illustrate that color is not always needed to make the strong scientific point from a complicated data set. The individual seismic data sets are not easy to interpret; there are many small-scale features corresponding to activity across the globe. The use of superposed epoch analysis combines these many wiggly lines into a sharp focus around the particular signature of interest.

Both of the examples provided in the earlier text use multiple panels to progressively advance the scientific and analytical storyline. Another use of multipanel figures is to include seemingly disparate number sets into a synergistic whole. Take, for example, Figure 2.23, a single graphic from Snow et al. (2017) that

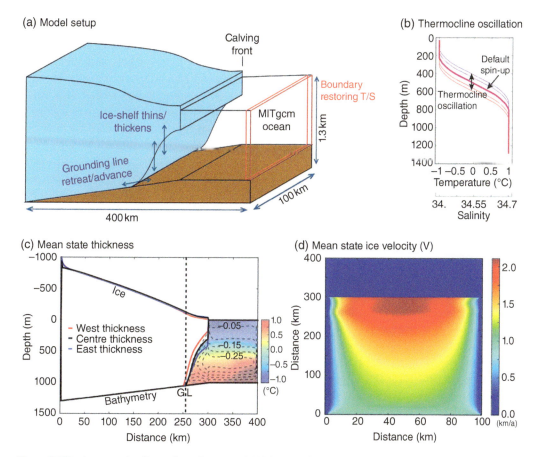

Figure 2.23 A composite figure from Snow et al. (2017) that includes an illustrative schematic, a line plot, and two color graphics of ice-sheet dynamics. The drawing shows the overall physical scenario being modeled, the line plot shows the variability of a critical boundary condition (the ocean thermocline) oscillating up and down during the simulation, the lower-left color plot shows the water-temperature profile at the center of the simulation domain (and lines of ice-sheet thickness), and the lower-right plot shows the velocity of the ice, looking down from the top (the blue region is open ocean).

contains four rather different panels. The upper-left panel is a schematic of the full configuration of the system being modeled, in this case an ice sheet protruding over the ocean. The upper-right panel is a line plot of the boundary condition driving the change in the ice-sheet motion, specifically a warm–cold boundary deep within the ocean that is changing due to climate variability. Two lower plots show colormaps of key parameters from the simulation, with ocean temperature on the left and ice-sheet flow speed on the right. Near water's edge, the ice sheet is moving at over 5 m a day.

Why put all of these together instead of having them as separate figures in the paper? Well, for one, some journals, including *GRL* from which this figure was taken, limit the length of published papers, so this is a way to meet that restriction. Scientifically, though, it makes sense in the context of the study they conducted. This is the first plot in the paper and is used to explore the physical processes governing the issue being studied. The oscillation of the deep-ocean temperature boundary results in ice-sheet motion and iceberg calving, and this series of plots illustrates the connection of this oscillation to increased velocities near the front of the sheet.

Sometimes, authors can be creatively complicated in how they display their data. Take, for example, Figure 2.24, which shows two rows of plots of meteorological measurements. The plots within each row are at different times during an Arctic atmospheric warming event, and the investigation by Binder et al. (2017) seeks to unravel the competing weather phenomena responsible for it. Each row contains four overlaid data sets. That is a lot for a single figure. Actually, there is a fifth overlay, the coastal outlines, also included in each plot. In order to include so much information on a single graphic, they had to carefully choose how to display each one. Specifically, the presentation is simplified for any one data set to convey just enough to be relevant for the analysis without being overly burdensome. In the top row, only eight color levels are used in the background map, only the outer edge of the blockings are included, only a few pressure levels are presented, and only every tenth trajectory intersection is marked. In the lower row, the color scale has 27 unique shades, but really they are all gradations of only two colors, blue at the low end and red at the high end. It is still a complicated figure, but they have reduced the content for any one data set to keep the plot manageable.

This plot allows a multifaceted view of the phenomenon being studied. The overlaid content is, at first, intimidating, and requires time to fully digest the patterns of each quantity included. Because this is a figure in a journal article, the reader can take as long as they like to fully comprehend this intricacy. The overlap of the lines, symbols, and colors allows the viewer to identify relationships between the values. Putting these together on a single plot made it possible for the investigators to piece together a plausible story for how the weather event evolved. While it would not be a good graphic for a presentation—many minutes would have to be spent building each layer—but in a publication, readers can spend as much time as they need to digest the layering and understand the information.

Binder et al. (2017) provide an example of a type of explanatory plot that is sometimes included in scientific articles: a summary schematic. This is shown in Figure 2.25. This is much easier to understand than the complicated and somewhat more exploratory plot of Figure 2.24, but this figure could not have been created without that first set of overlaid data sets. From the series of data-centric figures, they concluded that a lineup of several low-pressure regions combined with a nearby, mostly stationary high-pressure region to the east to create a fast northward flow channel between them. This drove warm air from midlatitudes to the polar region. Unlike the introductory schematic shown in Figure 2.21 by Piet and colleagues in their study, this schematic could not have been drawn by Binder

2 Plotting Data: Visualizing Sets of Numbers | 57

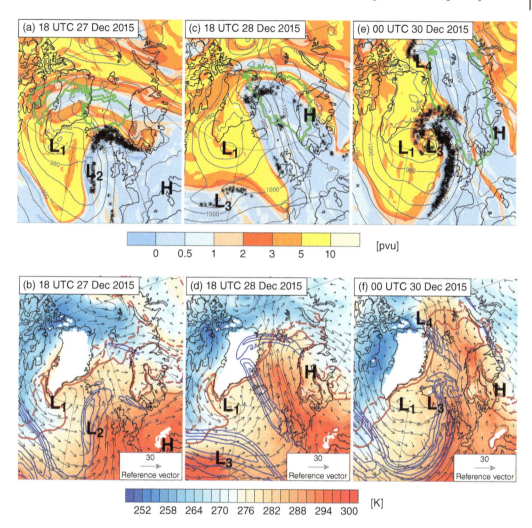

Figure 2.24 Eight different overlaid data sets for three times during a particular weather event. The top panel shows potential vorticity in the color background, outlines of identified blockings in green, pressure contours as thin gray lines, and trajectory intersections from derived flow patterns. The bottom panel shows the potential temperature as the color shading, wind vectors in the lower troposphere as arrows, wind speeds in the upper troposphere as blue contours, and surface temperatures as red contours. All of this information is overlaid on a coastline contour, shown as thin black lines. From Binder et al. (2017).

and colleagues prior to the analysis; they did not know what the qualitative diagram would look like until they made their overlaid data set figures.

This set of example plots is, like the previous section, not meant to be exhaustive, but is here to showcase a few types of visualizations. The split between quantitative exploratory graphics and qualitative explanatory graphics is not a hard boundary but a blended continuum, and these figures discussed in the earlier text reveal some aspects of that spectrum.

> **Quick and Easy for Section 2.6.2**
>
> A skim through a few issues of a high-impact, multidisciplinary journal reveals many examples of effective scientific visualization.

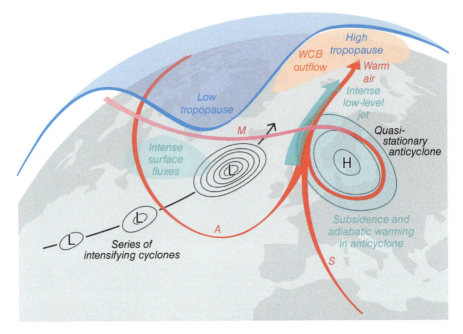

Figure 2.25 A schematic illustration used late in the paper by Binder et al. (2017) to distill the main concept of their study, showing the motion of cyclones in the North Atlantic region and the formation of the Arctic warming event.

2.7 Visualization Best Practices

Visualization influences the way that we "do science." The many choices made in creating the final version of a graphic can alter the way that the information is perceived and therefore interpreted. This includes the (*x*,*y*) domain size, linear versus logarithmic axes, dots versus lines, color versus contours, and the choice of color scheme and color scale, just to name a few. The esthetics of plotting can steer the direction of the scientific outcome.

There are three questions to initially ask yourself:

1) What is my purpose with this figure?
2) Who is my audience for this figure?
3) What do you want to be the first thing that the audience sees?

The graphic is being created for a reason and it is essential to keep that reason in mind while honing it toward its final state. The graphic is being made for a particular group of people that come to the figure with a certain level of prior understanding and a certain expectation about what they will see. Additionally, the graphic can be created in ways that draw attention toward a particular aspect of the plot. Keeping the answers to these questions in mind, there are guidelines for how to approach all of the other choices you need to make to finalize a scientific figure.

> **Quick and Easy for Section 2.7 Intro**
>
> Whenever you finalize a figure for presentation or publication, answer for yourself the three big questions: purpose; audience; key feature.

2.7.1 Levels of Abstraction

A few decades ago, people from statistics and psychology got together to better assess how humans interpret graphics and why certain graphic elements matter more than others. The

2 Plotting Data: Visualizing Sets of Numbers | 59

classic study that dominates the field is that of Cleveland and McGill (1985), which lays out ten elements of data visualization and how they influence our reaction to the data. Specifically, they organized these elements according to how accurately we process the information and form judgments from it. They called them **levels of abstraction** and arranged them in a hierarchy along the exploratory to explanatory continuum.

A summary of their findings is shown in Figure 2.26. The elements at the top, in particular position and length, provide the most quantitative evidence for viewers to make specific conclusions from the values. At the bottom are those elements, in particular the elements related to color, that provide the least specific analysis of the values and therefore allow for a more general conclusion to be drawn. In between these two are other elements of plot styles, such as direction and curvature of

Levels of abstraction: Ten elements of scientific figures that span the range from conveying specific information to more general impressions about the data set.

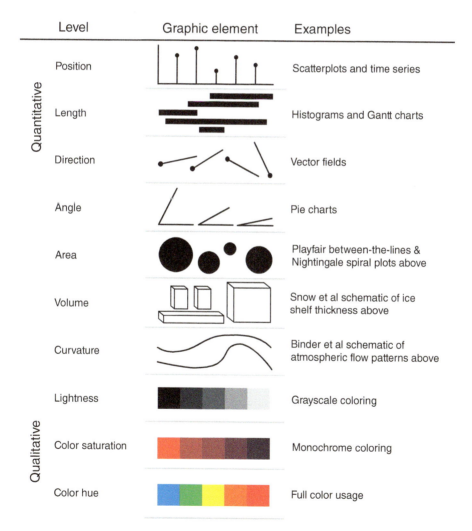

Figure 2.26 The levels of abstraction to consider when making a graphic, adapted from a rendering by Gabriela Plucinska from the Cleveland and McGill hierarchy of graphical elements.

lines on the graph, as well as area and volume of shaded regions of the graph.

The levels of abstraction confirm an initial intuitive guess that line plots and scatterplots are the most quantitative, followed by histograms and vector fields. It also confirms that colormap images are inherently subjective and are best used for making a qualitative impression rather than a detailed finding. These levels of abstraction are useful to remember when contemplating the answers to the questions mentioned in the earlier text about the purpose, audience, and main feature of interest.

Remember the third question to ask when producing a figure: what do you want to be the first thing that the audience sees? To really highlight this feature, adding explanatory elements (those lower on the chart in Figure 2.26) can make viewers focus on this part of the figure. For example, adding color below a line, so that the peaks are filled with a bright and eye-catching color, is a way to highlight the relative maxima.

Before continuing, let us spend some time on those three elements of color: lightness, saturation, and hue. **Hue** is the most obvious change, and in fact is often synonymous with the word "color." In its most basic form, think of the primary and secondary colors—red, orange, yellow, green, blue, and purple—as a very basic list of color hues. The other two color elements work together to vary the color toward black, gray, or white without changing the overall hue, as shown in Figure 2.27. **Saturation** is often referred to as the color intensity. Full saturation is the typical "bright" color, and diminished saturation softens and darkens the color until none of the hue remains

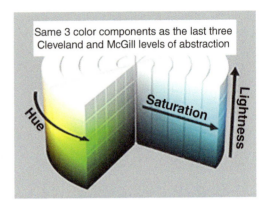

Figure 2.27 The interplay of how lightness and saturation change the perceived color for a single hue. Adapted from Wikipedia.

at 0% saturation. The final category is **lightness**, which is equivalent to adding a grayscale component to the color, changing it from white to black, with the full brightness of the color achieved at medium lightness. At zero lightness, all hues and saturations are black, and at full lightness, all hues and saturations are white. At any other lightness value, saturation varies the color from purely grayscale (zero saturation) to the brightest it can be for that lightness setting (at full saturation).

Together, these define the three-dimensional nature of color, and are often abbreviated as HSL (hue saturation, and lightness). The lightness scale is sometimes mapped to a slightly different scale—with full brightness of the color at top value—and the name is changed to **value**, creating the HSV color scheme (hue saturation, and value).

Lightness or value: The white–dark balance of an object's color, achieving full brightness at medium lightness.

Hue: The color of an object, the dominant photon wavelength emitted or reflected from it.

Saturation: The brightness of an object's color, changing from grayscale to full illumination.

Quick and Easy for Section 2.7.1

Cleveland and McGill's levels of abstraction are a convenient organizational tool for guiding the creation of a scientific figure to be more exploratory or explanatory in nature.

2.7.2 A Process for a Good Graphic

Now that we have an understanding of the levels of abstraction, here is a process for developing a nice scientific figure.

The first step is to answer the questions we have posed: (1) what is the purpose? (2) who is the audience? and (3) on what do you want viewers to focus? These answers identify the key feature of the figure that needs to be highlighted. As an example, if you have a scatterplot but want people to focus on the slight upward trend, then add a running average and make that line thick and colorful. To further emphasize the trend, you could make the individual data points in grayscale rather than black and use a smaller symbol than normal. Yet another addition to the figure that would draw attention to the trend is annotation, either text stating the feature of interest or an arrow parallel to the running average line. All of these changes will make the viewer focus on the feature you want them to spend time contemplating, rather than noticing other features in the scatterplot and being distracted from the main point that you want to make.

The next step is to decide how exploratory or explanatory you want to make the figure. Using the specificity-versus-generality guideline in Figure 2.26, you should decide what elements best convey the information to the intended audience. It could be that, even though you have shown the data as a scatterplot to a previous audience, the next audience will not want that level of detail, and so you change it to a box-and-whiskers format, de-emphasizing the details of the individual data points and focusing on how it is clustered. In fact, let us say that you do not want them to pay attention to the outliers. In this case, you could make the box part of the box-and-whisker elements to be drawn with thick lines and filled with shading, while the whiskers are drawn with very thin gray lines. Or, perhaps you want to even forego the mean or median trend line and simply show a colored region containing 95% of the values.

This brings up the issue of what colors, if any, to use in the plot. Would black-and-white, or grayscale, be sufficient for the intended purpose and audience? Would color be good to include? If color, how best to highlight the features? What are the norms for color regarding the quantity being shown? This last question might seem shallow, but it is important to consider; if there is already an established norm in the field that a certain feature is usually shown with shades of blue, then switching to orange tones for it might startle the intended audience. They would be distracted by the color choice, so that rather than emphasizing a key feature of the plot it detracts from it.

The advice from experts is to choose a dominant color and then accent around it. If the number set spans positive and negative values in roughly equal proportions and ranges away from zero, then two dominant colors are acceptable, one for each sign of the values. Is two colors enough? The word "color" here is equal to hue, the basic color group. Variations of saturation and lightness within that hue are the recommended means of achieving distinct colors within the graphic. For the example of a set of values that are evenly distributed around zero, the choice of a purple-and-orange-color scale might be a good fit, with white at the zero value in the middle (100% lightness of the two colors) and decreasing lightness toward the extreme values. If you do not want to highlight the end values but rather a section at some other range, perhaps where the bulk of the values are, then also vary the saturation, making it brightest (most saturated) at the value range of interest and more muted elsewhere.

If you are blending several color hues, then the advice is to use no more than five hues in the same image. Hues span the "rainbow" of the color wheel. There are different ideas for what constitutes a distinct hue from another nearby color. The standard list has 24 hues, defined every 15° around a color wheel, as shown in Figure 2.28. Some abbreviate this to 12 or even just 8 hues. The guideline of only choosing 5 hues is based on the full assortment of 24.

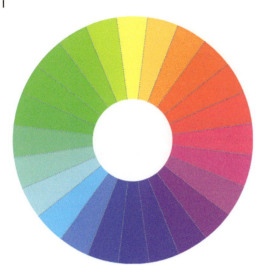

Figure 2.28 The standard 24-hue color wheel. Adapted from Wikipedia.

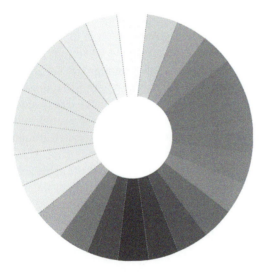

Figure 2.29 The standard 24-hue color wheel, converted to grayscale.

When using color, another consideration is whether to accommodate those that might view it in grayscale. Some colors look identical to others when printed in black-and-white. Figure 2.29 shows the color wheel in Figure 2.28 converted to grayscale. It is seen that the entire section of greens is the same shade of light gray, while most of the red section is blended together as the same shade of dark gray. This blending of greens into a single shade of gray when making a black-and-white version of a color figure is a good reason to avoid green in color graphics.

A final choice to make is the amount of annotation and highlighting features to add on top of the basic plot. This is a subjective call based on the purpose, audience, and main focus of the plot. It could be that nothing extra is needed or, conversely, that many guiding comments and symbols are desired. Layering of this kind can be done for both an exploratory and an explanatory figure. For the former, it could be that the extra annotation calls out the specific discrepancy between two values, or reports other calculated values from the number set (like its centroid and spread). For an explanatory graphic, the additions to the figure could draw attention to the main point that the plot is conveying, rather than noting specific numbers. Among the example visualizations shown in Section 2.6, many contain annotations—both words and symbols—but not all. This is an optional feature set that helps guide the viewer, but is not necessary for every graphic.

The last step in making a good graphic is to iterate and edit to make it as impactful as possible. It is often good to take a break, come back to it, and assess it again with respect to the three big questions we asked earlier. The figure should be edited several times to fully convey the key point, just like the words used to describe it in the presentation or publication in which it will be used. This editing could mean adding annotations, but it could also mean simplification and removal of content to clarify its main point.

> **Quick and Easy for Section 2.7.2**
>
> When creating a scientific graphic, there are several choices to make along the way. And, take a break and come back to it for additional editing.

2.7.3 Types of Colorblindness

Before we get too far advocating the use of color, it is helpful to raise awareness of **colorblindness**. Nearly one in 12 men are colorblind, which means that, mostly likely, someone you know has this condition. Women can also be colorblind, and while the occurrence is less frequent (roughly one in 200), this is still a large number of people across the STEM workforce. Colorblindness is when a set of cones within the eye are insensitive to the color they are supposed to perceive. The issue is that this insensitivity influences many other colors, not simply the one to which they are insensitive, but rather any hue that has a component of that primary color. That is, the whole spectrum is affected.

Figure 2.30 shows several sets of colored boxes. The "normal vision" set shows a typical spread of color hues across the rainbow, from dark blue on the left through purple, and then with a few other common colors added at the far right of the set, including a black box, for reference. The other three sets show the perception of the normal vision colors for people with three common colorblind conditions. Deuteranopia is an insensitivity to green light, which is the most common. Protanopia is an insensitivity to red light; you can see how close this set looks to deuteranopia set; both of them result in reds and greens taking on an olive hue. The final set is for tritanopia, which is an insensitivity to blue, a condition that is very rare. This condition makes blues and greens look similar, and orange hues become pink.

The main point to remember from this is that a significant number of minorities do not see colors the way that most others do. It is good to be aware of this and take it into account when designing a graphic. How best to accommodate this? The easiest is to avoid red and green in the same figure. Yes, that means usually avoiding the "scientific default" rainbow color scale that is commonly used across many disciplines. Many plotting software packages include filters to simulate colorblindness and show you what it would look like to people with these conditions.

> **Quick and Easy for Section 2.7.3**
>
> Colorblindness affects nearly 5% of the population, so the chances are good that someone in your audience has this condition, so please be kind and avoid confusing color choices.

2.7.4 Color Scales

Many plotting software applications have built-in **color scales** (sometimes called color maps or color schemes, depending on the package). Many of these exist because others have found them to be useful for conveying their message. Some, however, are not particularly good for general use. The main piece of advice about color scales is to choose or create one that highlights the aspect of the plot that you want prominently featured.

Take, for example, the two images of ocean current eddies in Figure 2.31. The first uses

Colorblindness: A human condition when a set of cones become insensitive to the color they are supposed to perceive.

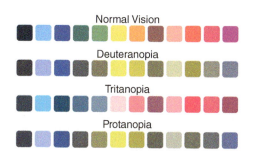

Figure 2.30 Sets of colored boxes as a way of conveying how people with different colorblindness view colors differently than those with normal vision.

Color scale: A mapping of color properties to values for plotting a number set.

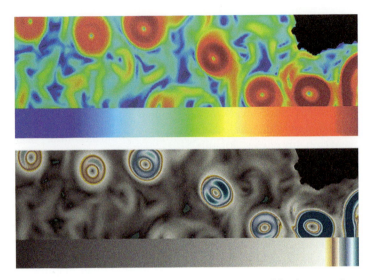

Figure 2.31 Comparison of two visualizations of the same data set—eddies in a prominent current within the Indian Ocean—but with two drastically different color scales. From Zeller and Rogers (2020).

the common rainbow color scale, while the second is a combination of a grayscale for most of the value range and then a series of rapidly varying hues and saturations in the extreme upper end of the scale. What is your eye drawn toward in the two versions? In both, you see the eddy formations. The top panel has lots of color elsewhere in the plot, and so your attention is drawn to the many small-scale features throughout the figure. The lower panel reveals sharp contrasts within the eddy structures themselves, focusing attention on this aspect of the plot. It is clearly seen in the lower plot that there are asymmetries within the eddies and differences between the eddies, a feature that is nearly unrecognizable in the upper panel. The lower panel, however, makes it difficult to discern the amplitude of the value anywhere outside of the eddies, something that the upper panel clearly reveals.

Which of these two plots is better? That depends on the purpose, audience, and key feature of interest. If the main point is the internal structure of the eddies, then the lower plot is clearly the better one. If the main point is the difference between eddies and the smaller-scale features between them, then the lower plot is not the most effective at illustrating that difference. While the upper plot shows that difference better, it is not ideal. We just discussed how the use of too many hues is confusing for the human eye and the multiple highlighted features detract from the intended main focus. So, in general, the upper plot—that uses this color scale—should be avoided. It might be a good first iteration, though, because it highlights so many aspects of the number set being displayed. Further iterations, however, should choose a color scale that uses fewer hues and concentrates the attention of the viewer on the part of the plot that you want them to see first.

> **Quick and Easy for Section 2.7.4**
>
> If a figure uses color, then choose a color scale that highlights the feature of keen interest and de-emphasizes other aspects of the data set so that the viewer's attention stays focused.

2.8 Further Reading

Most of the plotting styles discussed in the earlier text, plus several others, are described and discussed in Chapter 3 of this fantastically comprehensive book by Daniel Wilks:

Wilks, D. S. (2019). *Statistical Methods in the Atmospheric Sciences* (4th edition). Academic Press, Oxford. More about it here: https://www.elsevier.com/books/statistical-methods-in-the-atmospheric-sciences/wilks/978-0-12-815823-4

The square-root rule of histogram binning is something that I have heard for many years. I do not know of an actual reference for it, but it is the default bin setting embedded within Excel.

The original paper by Herbert Sturges (1926) is here:

Sturges, H. A. (1926). The choice of a class interval. *Journal of the American Statistical Association*, 21(153), 65–66. http://www.jstor.org/stable/2965501

And the original paper by Stephen O. Rice (1944) is here:

Rice, S. O. (1944). Mathematical analysis of random noise. *The Bell System Technical Journal*, 23(3), 282–332. 10.1002/j.1538-7305.1944.tb00874.x

The magnetometer data for Fort Yukon and Hermanus were retrieved from the SuperMAG repository:

https://supermag.jhuapl.edu/

Here is an article in Eos about the SuperMAG website:

Gjerloev, J. W. (2009), A global ground-based magnetometer initiative. *Eos, Transactions American Geophysical Union*, 90(27), 230–231. https://doi.org/10.1029/2009EO270002

For more on modern climate change, an excellent website that offers scientific analysis in non-technical language is RealClimate: https://www.realclimate.org/

For more about ice cores, I highly recommend Two Mile Time Machine by Richard Alley:

Alley, R. B. (2014). *The Two-Mile Time Machine*. Princeton University Press. https://press.princeton.edu/books/paperback/9780691160832/the-two-mile-time-machine

A helpful description of ice core data:
http://www.climatedata.info/proxies/ice-cores/

And the site for the ice core data used in this chapter:
http://www.climatedata.info/proxies/data-downloads/

For some funny maps of what people long ago thought the world looked like:
http://www.ancientportsantiques.com/ancientmaps/

A good write-up of Galileo's sunspot drawings from the 1600s:
http://galileo.rice.edu/sci/observations/sunspot_drawings.html

The timeline chart of Joseph Priestley:
https://commons.wikimedia.org/w/index.php?curid=2285294

More about William Playfair:
https://books.google.com/books/about/The_Commercial_and_Political_Atlas.html?id=dgRdAAAAcAAJ

A nice article on the impact of Alexander von Humboldt on ecology and climate change:

Moret, P., Muriel, P., Jaramillo, R., & Dangles, O. (2019). Humboldt's tableau physique revisited. *Proceedings of the National Academy of Sciences*, 116(26), 12889–12894. https://www.pnas.org/content/116/26/12889

And another on his influence on natural science:

Dwyer, P. D. (2021). The invention of nature. In *Redefining Nature* (pp. 157–186). Routledge, New York.

There are also several books about him, like this one:

Wulf, Andrea (2015). The invention of nature: Alexander von Humboldt's new world. New York. https://www.goodreads.com/book/show/23995249-the-invention-of-nature

More about Florence Nightingale:
https://www.history.com/topics/womens-history/florence-nightingale-1

And specifically a link to the original graphic from her book:
https://www.davidrumsey.com/luna/servlet/detail/RUMSEY~8~1~327826~90096398:Diagram-of-the-Causes-of-Mortality-;JSESSIONID=

724e1769-6989-4cf6-809c-b17755146cdd?qvq=q%3Aauthor%3D%22Nightingale%2C%20Florence%22%3Blc%3ARUMSEY%7E8%7E1&mi=1&trs=10

Here is the citation and link to the Piet et al. (2017) paper on large impacts on a planet:
Piet, H., Badro, J., & Gillet, P. (2017). Geochemical constraints on the size of the Moon-forming giant impact. *Geophysical Research Letters*, *44*, 11,770–11,777. https://doi.org/10.1002/2017GL075225

The Haugland et al. (2017) paper on crustal characteristics from seismic data:
Haugland, S. M., Ritsema, J., Kaneshima, S., & Thorne, M. S. (2017). Estimate of the rigidity of eclogite in the lower mantle from waveform modeling of broadband *S*-to-*P* wave conversions. *Geophysical Research Letters*, *44*, 11,778–11,784. https://doi.org/10.1002/2017GL075463

The Snow et al. (2017) paper on the response of ice sheets to climate variability:
Snow, K., Goldberg, D. N., Holland, P. R., Jordan, J. R., Arthern, R. J., & Jenkins, A. (2017). The response of ice sheets to climate variability. *Geophysical Research Letters*, *44*, 11,878–11,885. https://doi.org/10.1002/2017GL075745

The Binder et al. (2017) study of air mass transport during an Arctic warm event:
Binder, H., Boettcher, M., Grams, C. M., Joos, H., Pfahl, S., & Wernli, H. (2017). Exceptional air mass transport and dynamical drivers of an extreme wintertime Arctic warm event. *Geophysical Research Letters*, *44*, 12,028–12,036. https://doi.org/10.1002/2017GL075841

The *Science* paper by Cleveland and McGill on graphical perception and key elements of visualization:
Cleveland, W. S., & McGill, R. (1985). Graphical perception and graphical methods for analyzing scientific data. *Science*, *229*, 828–833. https://doi.org/10.1126/science.229.4716.828

A good follow-on by Stewart and Best (2010) that further examines the categories of Cleveland and McGill:

Stewart, B. M., & Best, L. A. (2010). An examination of Cleveland and McGill's hierarchy of graphical elements. In: Goel, A. K., Jamnik, M., Narayanan, N. H. (Eds.), Diagrammatic Representation and Inference. Diagrams 2010. *Lecture Notes in Computer Science*, vol. 6170. Springer, Berlin, Heidelberg. https://doi.org/10.1007/978-3-642-14600-8_46

A website with a good description of plot-making techniques:
https://www.peteraldhous.com/ucb/2014/dataviz/week2.html
And another with more useful tidbits about visualization:
http://www.thefunctionalart.com/2012/06/few-random-thoughts-on-infographics.html

Gabriela Plucinska's Data in the Spotlight blog post with the original levels of abstraction graphic:
https://www.gabrielaplucinska.com/blog/2017/8/7/pie-charts

The HSL diagram was originally from Wikipedia:
https://en.wikipedia.org/wiki/HSL_and_HSV
As was the color wheel diagram:
https://en.wikipedia.org/wiki/Hue

More about the Colour Blind Awareness organization, with the original pencil set images inspiring the colored block sets:
https://www.colourblindawareness.org/colour-blindness/
The website Color Oracle provides filters that convert your finished graphic into the visual that a colorblind person will see, and provides tips for adjusting your graphic:
http://colororacle.org

The Zeller and Rogers (2020) article, an excellent report on data visualization and the use of color:
Zeller, S., & Rogers, D. (2020). Visualizing science: How color determines what we see. *Eos, Transactions, American Geophysical Union*, *101*. https://eos.org/features/visualizing-science-how-color-determines-what-we-see

2.9 Exercises in the Geosciences

Part 1: Global Temperatures

This set of questions explores the temperature of Earth over time. The two data sets are annual temperature anomalies, one compiled from measurements made over land and the other for readings over the oceans. A temperature anomaly is a difference value relative to a baseline value, which is often taken to be the average over some time interval. This set of questions does not ask about why the temperature is changing with time, but rather explores how to plot it in a few different ways.

1. Import the two data sets labeled:
   `global_land_1880_2018_temp_variants.csv` `global_ocean_1880_2018_temp_variants.csv`
   A The original data set source and format can be found here, and then type in the beginning and ending year of your choosing (could be different, as you prefer, but pick a range over 100 years long): https://www.ncdc.noaa.gov/cag/global/time-series/globe/land_ocean/ytd/
   B Make line plots of the two data sets.
   C Based on these line plots, write a paragraph describing the similarities and differences that you see between these two data sets. Write out your thoughts on why they might be different.

2. For the land temperature data set, make four histograms of the number set.
   A Calculate the number of bins from the three formulas given in Section 2.3.
   B Make histograms of the land temperature data set setting the number of bins to be equal to the values from these formulas.
   C Make a fourth histogram with a number of bins of your choosing (but make it different from the three settings above).

3. Highlight parts of the land temperature data set to different colors by creating a new column for color values. Do this in two different ways:
   A Assign different colors based on year values (different colors before and after a chosen year).
   B Assign different colors based on anomaly value (different colors above and below a chosen anomaly value).

4. Using the two groupings defined in Part 3a, plot the land temperature data in several ways, using a different color for the two groupings:
   A First as a scatterplot (unconnected dots for each yearly value).
   B Next, as a bar chart (vertical bars for each year along the *x* axis).
   C Make another set of 2 plots of the land temperature data using the groupings defined in Part 3b.

5. Write a paragraph comparing and contrasting your data sets plots, including:
   A Why did you choose to highlight what data you did with the colors in Part 3? Justify this choice.
   B Explain your choice of bin number used in Part 2c.

Part 2: Bird Ranges

This set of questions examines the change of ranges that birds typically inhabit. They have been drifting over the last few decades. This plotting exercise set does not get into how the territory for each bird species is shifting, but rather explores ways of plotting the shift for all species.

6. Import two data sets labeled:
   `bird_ranges_coastal_40yavg.csv`
   `bird_species.csv`

7. Comment on the original data source and format in your own words. The original

data are found at the Environmental Protection Agency and the National Audubon Society websites.

8. Create a multipanel plot (two plots) to present:
   A Top plot: the coastal bird range with either error bars or an error envelope.
   B Bottom plot: A scatterplot of your choice of data from the bird species data set with different colors and a legend.

- You must have at least two colors included.
- You should use Boolean indexing to color-code your data.

9. Write two paragraphs analyzing your plot. Review your top plot, describe what this is showing, and what it means. Describe your bottom plot, what it is showing, what you think it means. Why did you choose to highlight the species data that you did?

3

Uncertainty Analysis: Techniques for Propagating Uncertainty

Usually, values with uncertainties are later used in some mathematical operation. This is the case, for instance, when you measure a temperature in Fahrenheit, but then want to report it in Kelvin. This involves an algebraic calculation with multiplication and addition operations to convert from the measured value into the reported value. It could be a much more complicated path to a final value, involving many measured parameters being combined in a nonlinear formulation. Regardless of the complexity of the equation, the process of calculating an uncertainty for a new parameter, one that is a formulaic combination of several other values with uncertainties, is called **uncertainty propagation**.

Calculating the uncertainty of a value is an essential element of scientific analysis. For some numbers, this is an easy formula, while for others it can be an involved, multistep process. This chapter covers the basic aspects of uncertainty calculation and propagation.

Uncertainty propagation: The process of calculating uncertainty for a new parameter based on other variables that have a specified uncertainty.

3.1 Propagating Uncertainty

While the formulas are essentially the same, it is useful to take it step by step through the process of determining how to calculate an uncertainty for a given value up to the point of full generality. The discussion in Section 3.1.1 starts with a single **independent variable** and then later sections extend this to two and eventually an arbitrary number of input variables, which could be independent or **dependent variables**, meaning that the variation of one is correlated with the variation of another.

3.1.1 Calculating Uncertainty with One Independent Variable

Propagating uncertainty is about using a set of values, x, with a known central value x_{best} and a known uncertainty δx, in some equation to

---

Independent variables: Input parameters to a calculation each having uncertainties, but the variation of each of these parameters is independent and random with respect to all of the other input parameters.

Dependent variables: Input parameters to a calculation for which there is correlation in the variation of two or more of these parameters.

Data Analysis for the Geosciences: Essentials of Uncertainty, Comparison, and Visualization, Advanced Textbook 5, First Edition. Michael W. Liemohn.
© 2024 American Geophysical Union. Published 2024 by John Wiley & Sons, Inc.
Companion website: www.wiley.com/go/liemohn/uncertaintyingeosciences

come up with a new parameter, q. At this point, it does not matter how either x_{best} or δx was found. The uncertainty of x, δx, is a positive definite value and represents a spread of values around any x value, including x_{best}, where most of the possibilities for the x values exist. With propagating uncertainty, we are using these values in an equation to get q and its "best guess," q_{best}, and range of possibilities around that best guess, δq. Take the generic formula,

$$q = f(x) \tag{3.1}$$

In Equation (3.1), f is an arbitrary function of x and q is the new variable calculated from that function. The formula for q_{best} is the application of x_{best} in the function,

$$q_{best} = f(x_{best}) \tag{3.2}$$

The specific measurements, however, are not at the x_{best} value, but are spread across a range. That is, any one x measurement is, hopefully, close to x_{best}, but has some variation, or perturbation, around this value. The perturbation around this value could then be found simply by applying the uncertainty of x in both the positive and negative directions around x_{best},

$$\delta q = f(x_{best} \pm \delta x) - f(x_{best}) \tag{3.3}$$

If we do not need to remember that q is a function and we are only doing this once, then we can apply the numbers from our x values directly into Equation (3.3) and calculate δq. If, however, we will be redoing this calculation several times with different x value sets, then this can become tedious. We would like a function for δq. Luckily, there is a path forward for doing just this. Assuming that f does not have any sharp discontinuities, then this deviation away from q_{best} can be directly compared with a fundamental approximation of calculus,

$$f(x+u) - f(x) \cong \frac{df}{dx} u \tag{3.4}$$

In Equation (3.4), u is a perturbation distance, positive or negative, relative to a known location x along the function f, and df/dx is the first derivative of f with respect to x. If we replace x with x_{best} and u with δx, then the subtraction on the left-hand side of Equation (3.4) becomes an equation for δq. Like δx, we need δq to be a positive definite value that reflects the uncertainty in either direction around q_{best}. On the right-hand side of Equation (3.4), we know that u, which is now δx, is positive definite, but we do not know if the derivative is positive or negative. So, we have to take an absolute value of it to ensure that δq is a positive value,

$$\delta q = \left| \frac{df}{dx} \right| \delta x \tag{3.5}$$

Equation (3.5) is the general formula for propagating uncertainty with one independent variable. The special cases to be discussed later come from this equation. Note that some will forego the function designation, f, and simply use |dq/dx| in the right-hand side of Equation (3.5).

> **Quick and Easy for Section 3.1.1**
>
> With one independent variable, propagating uncertainty uses the derivative chain rule, but with an absolute value operation to keep δq positive.

3.1.2 Calculating Uncertainty with Two Independent Variables

What about for two independent variables in the function? Or more? For two variables,

$$q = f(x, y) \tag{3.6}$$

We can still get the best guess for q by simply using the best guess values for both x and y, which we assume we know,

$$q_{best} = f(x_{best}, y_{best}) \tag{3.7}$$

The deviation away from this can be written similarly as in Section 3.1.1,

$$\delta q = f(x_{best} \pm \delta x, y_{best} \pm \delta y) - f(x_{best}, y_{best}) \quad (3.8)$$

If we are only doing this once, then we can apply the numbers to Equation (3.8) and stop here. The fundamental approximation of calculus has a two-variable form, though, that can be useful if we intend to do several calculations of δq. It looks like this,

$$f(x+u, y+v) - f(x,y) \cong \frac{\partial f}{\partial x} u + \frac{\partial f}{\partial y} v \quad (3.9)$$

Again, in Equation (3.9), u and v are small perturbations around x and y, and can be either positive or negative. The script derivative symbol, ∂, indicates a partial derivative, where you hold all other variables constant and just take the derivative of f with respect to the one variable listed in the denominator. Again, the left-hand side looks like δq in Equation (3.8) and, to keep it positive definite, absolute value operators should be applied to the two partial derivatives,

$$\delta q = \left|\frac{\partial f}{\partial x}\right| \delta x + \left|\frac{\partial f}{\partial y}\right| \delta y \quad (3.10)$$

These formulas, either Equation (3.8) or Equation (3.10), will most likely overestimate the uncertainty in q, because it applies the maximum uncertainty for both x and y simultaneously in the calculation. If we are processing x and y values in an equation and they are completely independent, then we should not be applying both of the known uncertainties together like this.

Why? Suppose that two independent measurements, x and y, with uncertainties δx and δy are being added. If one of the values is, in truth, larger than the best guess value, the independence of the two means that this positive offset does not influence the uncertainty in the other value. That is, the true value has a 50% chance of being larger than the best guess value, but also a 50% chance of being smaller. So, the probability of the true summed value being off from our best guess of the summed value, $x_{best} + y_{best}$, by the "extreme" uncertainty of both $\delta x + \delta y$ is much smaller than the probability of the uncertainty being off by either δx or δy alone.

A new rule for variables in an equation for which the uncertainties are independent uses addition in **quadrature**. This is the process in which you square the values, add them, and then take a square root. It is essentially the Pythagorean theorem, $c^2 = a^2 + b^2$, where the sides of the triangle are the functional uncertainties for each variable separately. For one variable, the square and square root cancel and you get back Equation (3.5). For two variables, however, nothing cancels and the formula changes to this,

$$\delta q = \sqrt{\left(\frac{\partial f}{\partial x} \delta x\right)^2 + \left(\frac{\partial f}{\partial y} \delta y\right)^2} \quad (3.11)$$

The absolute value signs can be dropped in Equation (3.11) because of the square, and the final square root should yield a positive value to keep δq positive definite. The reason for this specific formulation will be examined in Chapter 6.

Quick and Easy for Section 3.1.2

There are two formulas for propagating uncertainty with two input variables. The second way, if the uncertainties of the two variables are independent, is called quadrature.

Quadrature: A method of combining terms that squares all the terms, sums them, and then takes the square root.

3.1.3 Calculating Uncertainty with Many Independent Variables

What about for three or more variables? Everything extends to the arbitrary case, with the definition of the new variable q becoming this,

$$q = f(x, y, ..., z) \qquad (3.12)$$

Two formulas for the uncertainty δq can be written, as was done in Section 3.1.2, with additional terms to account for each variable included in the function f. The maximum possible uncertainty of q is given by the sum of the individual terms,

$$\delta q = \left|\frac{\partial f}{\partial x}\right|\delta x + \left|\frac{\partial f}{\partial y}\right|\delta y + \cdots + \left|\frac{\partial f}{\partial z}\right|\delta z \qquad (3.13)$$

If the $x, y, \ldots, z$ variables all have independent uncertainties from each other, then the uncertainty of q can be determined by the quadrature formula,

$$\delta q = \sqrt{\left(\frac{\partial f}{\partial x}\delta x\right)^2 + \left(\frac{\partial f}{\partial y}\delta y\right)^2 + \cdots + \left(\frac{\partial f}{\partial z}\delta z\right)^2} \qquad (3.14)$$

Note that it does not matter how complicated the function f is, Equations (3.13) and (3.14) will work to get the uncertainty δq. In some sense, we are done with uncertainty propagation. These two formulas cover it all. What comes next are special cases of uncertainty and normalizations for it to add to the interpretative value of the uncertainty calculation.

> **Quick and Easy for Section 3.1.3**
>
> With more than two variables, the formulas are simply an extension of the two-variable versions, either adding absolute values or addition in quadrature.

3.2 Example: Atmospheric Density

In the Earth's atmosphere, the density of air is not constant. As one moves higher in the atmosphere, the amount of air per unit volume decreases. How much does it decrease with altitude? This requires a bit of fluid dynamics, but the relationship is readily derived. Start with the full **Navier–Stokes momentum equation**, which is often written like this:

$$\rho\left[\frac{\partial \mathbf{u}}{\partial t} + (\mathbf{u} \cdot \nabla)\mathbf{u}\right] = -\nabla p - \rho g \hat{\mathbf{z}} + \nabla \varphi + \mu \nabla^2 \mathbf{u} \qquad (3.15)$$

In Equation (3.15), ρ is the mass density of the air, $\mathbf{u}$ is its velocity vector, p is atmospheric pressure, g is gravitational acceleration, $\nabla \varphi$ represents any other externally applied forces, and μ is viscosity.

3.2.1 The Hydrostatic Equilibrium Approximation

While this looks intimidating, most of the terms in Equation (3.15) can be neglected in our simple case here. Specifically, we will assume a stationary fluid, so the two terms inside the brackets on the left-hand side are both zero, as is the last term on the right-hand side. Similarly, we can assume that the only force acting on the air is gravity, pointing locally "downward" in the minus z direction. This leaves us with a rather simple expression for the air pressure called **hydrostatic equilibrium**, now only in one dimension:

$$\frac{dp}{dz} = -\rho g \qquad (3.16)$$

Navier–Stokes momentum equation: A formula for a fluid's velocity, with terms for gravity, viscosity, and both internal and external forces.

Equation (3.16) is also known as Pascal's principle. We now need to substitute mass density for pressure using the **ideal gas law**, which can be written most conveniently for this problem like this:

$$p = \frac{R}{M}\rho T \qquad (3.17)$$

In Equation (3.17), R is the ideal gas constant, 8.31 J/(mole·K), and M is the molar mass of the atmosphere, usually close to 29 kg/mole, just a bit above the 28 value of molecular nitrogen, the dominant molecule in the atmosphere, because of the contribution of molecular oxygen at 32. Furthermore, T in Equation (3.17) is the temperature of the air, as measured in Kelvin.

Taking the derivative of Equation (3.17) with respect to z and inserting into Equation (3.16) leave us with this equation for the rate of change of the mass density:

$$\frac{RT}{M}\frac{d\rho}{dz} = -\rho g \qquad (3.18)$$

Equation (3.18) is one in which the change in the parameter—in this case atmospheric density—is directly proportional to the local value of the parameter. This famous class of differential equation is solved by separation of variables, with the solution being a natural exponential function. That is, Equation (3.18) is readily integrated to yield a formula for the variation of mass density as a function of altitude:

$$\rho = \rho_0 \exp\left[-\frac{Mgz}{RT}\right] = \rho_0 \exp\left[-\frac{z}{H}\right] \qquad (3.19)$$

Hydrostatic equilibrium: A formula specifying atmospheric pressure when there is no wind and the only force is gravity.

Ideal gas law: A formula relating the pressure of a gas to its density, temperature, and composition.

To write Equation (3.19), the boundary condition of mass density being ρ_0 at $z = 0$ was applied. The new variable H is called the **scale height** of the atmosphere, defined as RT/Mg, and has units of distance. It is the altitude increment over which the atmospheric density decreases by e^{-1}. Because e is 2.72, this is a drop to roughly 37% of the value just one scale height lower.

Equation (3.19) is the final form needed for our uncertainty calculation. We can now propagate uncertainty in several different ways, assuming that various parameters in this formula are either variable or constant.

Quick and Easy for Section 3.2.1

As an example, the formula for hydrostatic equilibrium will be used, an exponential equation between atmospheric density and temperature.

3.2.2 One Independent Variable

To start with an easy example, let us say that we have measurements of air density at the ground, ρ_0, from which we have compiled a best estimate of this parameter, $\rho_{0,best}$, and a measured uncertainty around this value, $\delta\rho_0$. If every parameter inside the exponential is taken to be a constant, then the uncertainty on the mass density of air at some altitude above the surface, $\delta\rho(z)$, would be an application of Equation (3.5), the formula for one independent variable:

$$\delta\rho = \left|\frac{d\rho}{d\rho_0}\right|\delta\rho_0 \qquad (3.20)$$

Working out the derivative, the final form of the uncertainty for atmospheric density is then this:

$$\delta\rho = \exp\left[-\frac{Mgz}{RT}\right]\cdot\delta\rho_0 \qquad (3.21)$$

Scale height: The vertical distance over which the atmospheric density drops to approximately 37% (e^{-1}) of its original value.

Note that the absolute value signs have been removed because they are not needed; the exponential term is always positive.

> **Quick and Easy for Section 3.2.2**
>
> For a single independent variable, in this example it is the ground-level atmospheric density, only one derivative is conducted.

3.2.3 Two Independent Variables

To make this slightly more complicated, let us now assume that we also made measurements of the temperature at the ground and compiled a best estimate and uncertainty for this value, $T_{best} + \delta T$. If we can assume that the air density and air temperature are independent values with independent and Gaussian spreads about the best estimates, then we can use Equation (3.11) for the uncertainty of the mass density at some altitude:

$$\delta\rho = \sqrt{\left(\frac{\partial\rho}{\partial\rho_0}\delta\rho_0\right)^2 + \left(\frac{\partial\rho}{\partial T}\delta T\right)^2} \quad (3.22)$$

Applying Equation (3.19) and doing the derivatives yield this:

$$\delta\rho = \sqrt{\left(\exp\left[-\frac{Mgz}{RT}\right]\cdot\delta\rho_0\right)^2 + \left(-\frac{\rho_0 Mgz}{RT^2}\exp\left[-\frac{Mgz}{RT}\right]\delta T\right)^2} \quad (3.23)$$

Using the scale height H, Equation (3.22) can be simplified somewhat to this:

$$\delta\rho = \exp\left[-\frac{z}{H}\right]\cdot\sqrt{(\delta\rho_0)^2 + \left(\frac{\rho_{0,best} z}{H T_{best}}\delta T\right)^2} \quad (3.24)$$

> **Quick and Easy for Section 3.2.3**
>
> For two independent variables, in this example both the ground atmospheric density and the local temperature, the quadrature formula is used.

3.2.4 Many Independent Variables

A final version would be if we also add in variability to the parameters of M and z. The molar mass of the atmosphere changes slightly through the lower atmosphere, especially depending on the water composition in the local air. Therefore, it could also have a local measurement on the ground with an associated uncertainty, δM. The altitude at which you want to know the atmosphere could be a variable with uncertainty, as well, especially if this is a place of another measurement, say being made by an aircraft or high-altitude weather balloon. In this case, z has an uncertainty of δz associated with it. The resulting uncertainty for air mass density at some altitude should use the full general formula, Equation (3.14), and have this form:

$$\delta\rho = \sqrt{\left(\frac{\partial\rho}{\partial\rho_0}\delta\rho_0\right)^2 + \left(\frac{\partial\rho}{\partial T}\delta T\right)^2 + \left(\frac{\partial\rho}{\partial M}\delta M\right)^2 + \left(\frac{\partial\rho}{\partial z}\delta z\right)^2} \quad (3.25)$$

Upon evaluating all of the derivatives and simplifying the solution, Equation (3.25) can be rewritten as this:

$$\delta\rho = \exp\left[-\frac{z}{H}\right]\cdot\sqrt{(\delta\rho_0)^2 + \left(\frac{\rho_{0,best} z}{H T_{best}}\delta T\right)^2 + \left(\frac{\rho_{0,best} z}{M_{best} H}\delta M\right)^2 + \left(\frac{\rho_{0,best}}{H}\delta z\right)^2} \quad (3.26)$$

There is one more simplification that could occur in this formula for the air mass density uncertainty—rewriting the values as fractional uncertainties by pulling $\rho_{0,best}$ out of the square-root operator. This would make the equation look like this:

$$\delta\rho = \rho_{0,best}\exp\left[-\frac{z}{H}\right]\cdot\sqrt{\left(\frac{\delta\rho_0}{\rho_{0,best}}\right)^2 + \left(\frac{z}{H}\frac{\delta T}{T_{best}}\right)^2 + \left(\frac{z}{H}\frac{\delta M}{M_{best}}\right)^2 + \left(\frac{\delta z}{H}\right)^2} \quad (3.27)$$

> **Quick and Easy for Section 3.2.4**
>
> When all parameters in the hydrostatic equilibrium equation have uncertainty, the uncertainty for atmospheric density can be found with quadrature, with each term inside the square root looking somewhat similar.

In this case, all of the terms inside the square-root operator are unitless. They each have the same dimensions in their numerator and denominator. More about this form of uncertainty is discussed in the next section.

3.3 Fractional and Percentage Uncertainties

The calculations of uncertainty using the general formulas, Equations (3.13) and (3.14), yield a value that has the same units as the original quantity. If, however, values with different units are to be compared, then it is highly desirable to normalize these uncertainties into a unitless formulation. That is, the dimensional "absolute uncertainty" can be converted into a dimensionless "relative uncertainty." This second value is also known as the fractional uncertainty and, as seen later, is often converted into a percentage uncertainty.

The basic calculation of a fractional uncertainty, which we will call δx_{frac}, is to simply divide by the original variable:

$$\delta x_{frac} = \frac{\delta x}{|x_{best}|} \tag{3.28}$$

The absolute value operation around x_{best} is needed so that δx_{frac} is always a positive value. The extension of this to the functional form, $q = f(x)$, is therefore this:

$$\delta q_{frac} = \frac{\delta q}{|q_{best}|} = \frac{1}{|q_{best}|}\left|\frac{df}{dx}\right|\delta x \tag{3.29}$$

In Equation (3.29), δq in the middle part of the equation has been listed for a function of a single variable in the last part of the equation because it was listed this way, $q = f(x)$, leading into the equation. That transformation, however, can be given by any of the appropriate formulas from Section 3.1, with the most general being Equation (3.14) if the variables within the function are independent of each other with uncorrelated uncertainties, which yields this:

$$\delta q_{frac} = \frac{1}{|q_{best}|}\sqrt{\left(\frac{\partial f}{\partial x}\delta x\right)^2 + \left(\frac{\partial f}{\partial y}\delta y\right)^2 + \cdots + \left(\frac{\partial f}{\partial z}\delta z\right)^2} \tag{3.30}$$

All of the terms inside the square-root operator are squared, so they will be positive, and only the leading q_{best} term requires an absolute value operator to make δq_{frac} always positive. It might be required to use Equation (3.13) if the independent and uncorrelated conditions are not met, however, leaving this equation as a general form for the fractional uncertainty:

$$\delta q_{frac} = \frac{1}{|q_{best}|}\left\{\left|\frac{\partial f}{\partial x}\right|\delta x + \left|\frac{\partial f}{\partial y}\right|\delta x + \cdots + \left|\frac{\partial f}{\partial z}\right|\delta z\right\} \tag{3.31}$$

Absolute value symbols appear around not only the q_{best} term in front but also around all of the derivative operators within the brackets in order to ensure that δq_{frac} is a positive value.

As was just shown for Equation (3.27), a polynomial or exponential functional form will return a new function that can be easily converted to a fractional uncertainty. The absolute uncertainties in the example of atmospheric density in Section 3.2 can all be readily converted to fractional uncertainties. The basic formula is this:

$$\delta \rho_{frac} = \frac{\delta \rho}{\rho_{best}} \tag{3.32}$$

Let us work through two of the cases. First, let us convert the easy one with only a single independent variable. Dividing Equation (3.21) by Equation (3.19) gives this:

$$\delta \rho_{\text{frac}} = \frac{\delta \rho_0}{\rho_{0,\text{best}}} \quad (3.33)$$

The exponential term cancels out and only the density term remains in the denominator. This creates a unitless fractional uncertainty.

The more complicated case of many independent variables is also straightforward. Converting Equation (3.27) into a fractional uncertainty is again simply dividing it by Equation (3.19), which yields this:

$$\delta \rho = \sqrt{\left(\frac{\delta \rho_0}{\rho_{0,\text{best}}}\right)^2 + \left(\frac{z}{H} \frac{\delta T}{T_{\text{best}}}\right)^2 + \left(\frac{z}{H} \frac{\delta M}{M_{\text{best}}}\right)^2 + \left(\frac{\delta z}{H}\right)^2} \quad (3.34)$$

Note that the general form of the density equation in front of the square-root operator cancels in the fractional uncertainty calculation, making this formula easier than the absolute uncertainty formula in Equation (3.27). A final step could be writing the terms inside the square root as fractional uncertainties themselves, like this:

$$\delta \rho_{\text{frac}} = \sqrt{\left(\delta \rho_{0,\text{frac}}\right)^2 + \left(\frac{z}{H} \delta T_{\text{frac}}\right)^2 + \left(\frac{z}{H} M_{\text{frac}}\right)^2 + \left(\frac{\delta z}{H}\right)^2} \quad (3.35)$$

This simplified form is not the case for operations such as trigonometric or logarithmic functions, for instance. In those functional forms of q, the extra step of dividing by $|q_{\text{best}}|$ usually complicates the formula rather than simplifying it. Still, it might be very useful to convert the dimensional uncertainty into a fractional uncertainty for direct comparison, or even combination, with other uncertainties in the processing of the data set.

A final point to note about fractional uncertainty is that it is often never reported directly with the original number. That is, you will not see $q_{\text{best}} \pm \delta q_{\text{frac}}$ being reported; it is too easily confused with $\pm \delta q$. The remedy for this is that fractional uncertainty is usually reported as a percentage uncertainty rather than a fractional uncertainty. This is done simply by multiplying δq_{frac} by 100%, like this:

$$\delta q_{\text{perc}} = \delta q_{\text{frac}} \cdot 100\% \quad (3.36)$$

Writing this formula term of the absolute uncertainty looks like this:

$$\delta q_{\text{perc}} = \frac{\delta q}{|q_{\text{best}}|} \cdot 100\% \quad (3.37)$$

To ensure that readers of a value with a percentage uncertainty understand that it is a percentage sign after the uncertainty value, reporting the number with its units and then the uncertainty with its "units" of percent, transforming the reporting of the value:

$$\begin{aligned} q_{\text{best}} &\pm \delta q \, [\text{units of } q] \\ &= q_{\text{best}} \, [\text{units of } q] \pm \delta q_{\text{perc}} \% \end{aligned} \quad (3.38)$$

The left-hand side of Equation (3.38) is how the absolute uncertainty should be reported, while the right-hand side is how percentage uncertainty should be reported. It is a small difference but one to note with every value being reported with an uncertainty, as this is the standard practice for distinguishing between these two versions of uncertainty for the same base value. Sometimes the values are so different that it is obvious that a conversion to percentage uncertainty has been performed. Sometimes it is not apparent, especially if the base value is of order 100; in this case, the two versions could have somewhat similar uncertainties which might lead to problems later in the analysis. Always report the units of your variables.

Quick and Easy for Section 3.3

While fractional uncertainty is useful within uncertainty propagation calculations, it should never be reported with the base value. For clarity, always convert to percentage uncertainty.

3.4 Special Cases of Uncertainty Propagation

It is useful to note several simplifications to the general formula. This is not meant to be an exhaustive list of all of the special cases that you might encounter, but rather a few of the most common ones. These are all applications of the two general rules, Equations (3.13) and (3.14).

3.4.1 Addition and Subtraction

For the operation of addition between two parameters with associated uncertainties, $q = f(x,y) = x + y$, the formulas are straightforward. The uncertainty assuming correlated uncertainties is the addition of the two uncertainties, because the partial derivative of f with respect to either x or y is simply one. So, we get this formula:

$$\delta q = \left|\frac{\partial f}{\partial x}\right|\delta x + \left|\frac{\partial f}{\partial y}\right|\delta y \quad (3.39)$$

Because the assumed function is the addition of these two variables, the uncertainty is:

$$\delta q = \delta x + \delta y \quad (3.40)$$

In the case where x and y have independent uncertainties from each other, then δq can be determined by the quadrature formula, with, again, the partial derivatives becoming one:

$$\delta q = \sqrt{\left(\frac{\partial f}{\partial x}\delta x\right)^2 + \left(\frac{\partial f}{\partial y}\delta y\right)^2} \quad (3.41)$$

Using our assumed formula for f yields this:

$$\delta q = \sqrt{\delta x^2 + \delta y^2} \quad (3.42)$$

For subtraction of two parameters, we have $q = f(x,y) = x - y$, and the partial derivative with respect to y has a negative sign in front rather than plus sign. This does not matter, though, because in the two general formulas the partial derivative is either squared or an absolute value operator is imposed on it. For subtraction, then, the two δq formulas are exactly the same as those for addition, Equations (3.39) and (3.41).

Why does subtracting two numbers result in an addition of their uncertainties? This seems counterintuitive, but it is correct. The uncertainty of a summation or difference is defined by the two base values being perturbed in either the plus or minus direction, and it is that "either" word that matters here. While the perturbations of the two values could be in the same direction, resulting in a smaller spread, they just as often could be perturbed in the opposite directions, thereby creating a larger final spread for the computed q value.

With this understanding of addition and subtraction of two independent variables, the addition and subtraction formulas can be extended to any number of variables being summed or differenced. That is, for $q = f(x_1, x_2, \ldots, x_N, x_{N+1}, x_{N+2}, \ldots, x_{N+M})$ with the specific formula being entirely addition and subtraction, $f = x_1 + x_2 + \cdots + x_N - x_{N+1} - x_{N+2} - \cdots - x_{N+M}$, then the two versions of the formulas for δq are with correlated uncertainties between the x_i values:

$$\delta q = \delta x_1 + \delta x_2 + \cdots + \delta x_N + \delta x_{N+1} \\ + \delta x_{N+2} + \cdots + \delta x_{N+M} \quad (3.43)$$

This long list of uncertainties can be written in a more compact form:

$$\delta q = \sum_{i=1}^{N+M} \delta x_i \quad (3.44)$$

In the case where x and y have independent uncertainties from each other, then δq can be determined by the quadrature formula, with, again, the partial derivatives becoming one:

$$\delta q = \sqrt{\delta x_1^2 + \delta x_2^2 + \cdots + \delta x_N^2 + \delta x_{N+1}^2 + \delta x_{N+2}^2 + \cdots + \delta x_{N+M}^2} \quad (3.45)$$

Data Analysis for the Geosciences

A summation can be used to write it more succinctly:

$$\delta q = \sqrt{\sum_{i=1}^{N+M} \delta x_i^2} \quad (3.46)$$

> **Quick and Easy for Section 3.4.1**
>
> The special cases of addition and subtraction have the same uncertainty calculation procedure.

3.4.2 Multiplication and Division

Like addition and subtraction, multiplication and division are also closely related in the final form of the uncertainty of a function including these operations. There are several aspects of this that deserve some attention, so it is useful to methodically step through some of the derivation and application of these final forms.

3.4.2.1 Multiplication of Two Parameters

Now let us apply the general formula for the simplified case of multiplication of two parameters that each has their own uncertainties. Given the formula $q = f(x,y) = x \cdot y$, we can put this function into Equation (3.13) to get the general formula for which there is correlation between the uncertainties of x and y:

$$\delta q = \left|\frac{\partial f}{\partial x}\right| \delta x + \left|\frac{\partial f}{\partial y}\right| \delta y \quad (3.47)$$

Using our assumed formula gives:

$$\delta q = |y| \cdot \delta x + |x| \cdot \delta y \quad (3.48)$$

The absolute values must remain around the two variables in Equation (3.48); it is unknown whether they are positive or negative, but the uncertainty should be a positive number.

There is a nice limiting case that should be considered at this point—when one of the two parameters is an exact constant. Assuming that x is a constant a with no uncertainty (so, $\delta x = 0$), then we have $q = ay$ and Equation (3.47) reduces to this:

$$\delta q = |a| \cdot \delta y \quad (3.49)$$

The absolute value sign around the constant remains because we want δq to be a positive value. That is, if the measured quantity (in this case, y) is later multiplied by a constant during processing, then so is the uncertainty of the new quantity.

Equation (3.47) can be written in a slightly more simplified form by dividing both sides by $|q|$. This will leave the left-hand side as a fractional uncertainty, with fractions of absolute values around parameters and functions in each term on the right-hand side:

$$\frac{\delta q}{|q|} = \frac{|y|}{|q|} \cdot \delta x + \frac{|x|}{|q|} \cdot \delta y \quad (3.50)$$

Plugging in the formula for q in each term on the right-hand-side terms in Equation (3.50) allows us to cancel the y and x variables in the numerators, thus converting these terms into similar fractional uncertainties for each parameter:

$$\delta q_{\text{frac}} = \frac{\delta x}{|x|} + \frac{\delta y}{|y|} \quad (3.51)$$

Noticing the functional form of the terms on the right-hand side, Equation (3.51) can be rewritten as:

$$\delta q_{\text{frac}} = \delta x_{\text{frac}} + \delta y_{\text{frac}} \quad (3.52)$$

Equation (3.51) shows that the fractional uncertainty of the processed function is simply the addition of the fractional uncertainties for the two parameters multiplied in the function. Remember that this was for correlated uncertainties of x and y. The case for independent uncertainties will be considered after a quick example.

When reporting this uncertainty value, remember to convert to percentage uncertainty. The unitless nature of fractional uncertainty makes it an ambiguous value to understand.

> **Quick and Easy for Section 3.4.2.1**
>
> Propagation of uncertainties in a formula with multiplication is conveniently done with addition of the fractional uncertainties.

3.4.2.2 Uncertainty of Air Pressure

Equation (3.51) might not be the answer you would have intuitively expected—why does the multiplication of uncertain quantities result in the addition of their uncertainties? To explore this a little more, let us consider an example. In fact, let us continue to build on the example discussed earlier in this chapter (Section 3.2). Using the ideal gas law in Equation (3.17), let us assume that the molar mass M is a constant, but that the mass density ρ and temperature T are variables with known uncertainties. In this case, R and M are always positive numbers and do not require the extra absolute value around them. A straight insertion of values and uncertainties into Equation (3.17) gives this:

$$p = \frac{R}{M}(\rho_{best} \pm \delta\rho)(T_{best} + \delta T) \quad (3.53)$$

Extracting the ρ_{best} and T_{best} values to the front of the equation, and noting that both of these values are always positive numbers and therefore absolute value operators can be added as needed, yields a version with fractional uncertainties for these two parameters:

$$p = \frac{R}{M}\rho_{best}T_{best}\left(1 \pm \frac{\delta\rho}{|\rho_{best}|}\right)\left(1 \pm \frac{\delta T}{|T_{best}|}\right) \quad (3.54)$$

One thing that can be done with Equation (3.54) is to recognize that the values out front constitute the best estimate of the air pressure, p_{best}. The other is that the second terms in each of the parentheses are fractional uncertainties and can be written as $\delta\rho_{frac}$ and δT_{frac}. These substitutions will be used in the equations provided in the following text. Another piece of information is to remember that all of the specific quantities are positive, so it is straightforward to determine the highest and lowest possible deviation from p_{best}, by simply taking the two plus signs and then the two minus signs, respectively.

For the highest possible pressure using the two plus signs, the product can be expanded in Equation (3.54) to yield this:

$$p_{max} = p_{best}(1 + \delta\rho_{frac} + \delta T_{frac} + \delta\rho_{frac} \cdot \delta T_{frac}) \quad (3.55)$$

The last term is the multiplication of the two fractional uncertainties. Assuming that these two values are small, such that $\delta\rho_{frac} \ll 1$ and $\delta T_{frac} \ll 1$, then this final term is much less than either of the other two. A linearization of Equation (3.55) is, therefore, to drop this last term, leaving us with this:

$$p_{max} = p_{best}(1 + \delta\rho_{frac} + \delta T_{frac}) \quad (3.56)$$

For the minimum pressure, the minus signs in Equation (3.54) should be used, resulting in this equation:

$$p_{min} = p_{best}(1 - \delta\rho_{frac} - \delta T_{frac} + \delta\rho_{frac} \cdot \delta T_{frac}) \quad (3.57)$$

While the last term in Equation (3.57) is positive and thus contributes in the opposite direction as the other two fractional uncertainty terms, it is a multiplication of two fractional uncertainties. Again, this term can be considered to be small and a linearization of Equation (3.57) leaves us with this:

$$p_{min} = p_{best}(1 - \delta\rho_{frac} - \delta T_{frac}) \quad (3.58)$$

These two extreme formulas, Equations (3.56) and (3.58), can now be combined again and written in a way that it is easy to recognize the uncertainty from the base value:

$$p = p_{best} \pm p_{best}(\delta\rho_{frac} + \delta T_{frac}) \quad (3.59)$$

The uncertainty of the pressure, δp, is everything after the $\pm$ symbol. Dividing by p_{best} to make it a fractional uncertainty yields:

$$\delta p_{frac} = \delta\rho_{frac} + \delta T_{frac} \quad (3.60)$$

Equation (3.60) is exactly the form we derived for a quantity with two values multiplied together, in Equation (3.51).

This example shows that Equation (3.51) comes directly from plugging in the uncertainties and linearizing the resulting formula. Equation (3.51) is not simply adding the uncertainties, as we did in Equation (3.43), but rather is an addition of the fractional uncertainties. There is no guarantee that the two parameters have the same units, and indeed Equation (3.47) has extra multiplication within each term on the right-hand side to get all terms into the units of the full function of q. Equations (3.51) and (3.60) show that these multiplying terms convert to fractional uncertainties, all unitless values, that are added together to yield the fractional uncertainty of the new functional parameter.

As a side note, this is the type of uncertainty propagation that Chicken Little should be doing (as seen in Figure 3.1), if they were making measurements of density and temperature instead of a direct measurement of pressure. Chicken Little would then have to determine p_{best} from Equation (3.54) using the best values for the measured parameters and then find an uncertainty on this derived quantity from Equation (3.60).

> **Quick and Easy for Section 3.4.2.2**
>
> Atmospheric pressure from the ideal gas law serves as a good example of propagating uncertainty in a formula with multiplication.

3.4.2.3 Division with Correlated Variables

Instead of multiplication, let us assume that our new functional quantity q is the division of two parameters with known but perhaps correlated uncertainties. In this case, $q = f(x,y) = x/y$, and the uncertainty of q can be found with the application of Equation (3.13),

$$\delta q = \left|\frac{\partial f}{\partial x}\right|\delta x + \left|\frac{\partial f}{\partial y}\right|\delta y \qquad (3.61)$$

Using the function, we get:

$$\delta q = \left|\frac{1}{y}\right|\cdot \delta x + \left|-\frac{x}{y^2}\right|\cdot \delta y \qquad (3.62)$$

This looks rather different from our initial functional form of δq for multiplication in Equation (3.47). The same transformation can be applied, however, with the conversion to a fractional uncertainty for q by the division of both sides by $|q|$:

$$\delta q_{\text{frac}} = \frac{\delta q}{|q|} \qquad (3.63)$$

The general form is written like this:

$$\delta q_{\text{frac}} = \left|\frac{1}{y\cdot q}\right|\cdot \delta x + \left|-\frac{x}{y^2\cdot q}\right|\cdot \delta y \qquad (3.64)$$

Substituting in the formula for q, and noticing that the negative sign inside the absolute value operator goes away, yields this:

$$\delta q_{\text{frac}} = \left|\frac{1}{x}\right|\cdot \delta x + \left|\frac{1}{y}\right|\cdot \delta y \qquad (3.65)$$

Noticing that the right-hand-side terms are also fractional uncertainties, we could write it as this:

$$\delta q_{\text{frac}} = \delta x_{\text{frac}} + \delta y_{\text{frac}} \qquad (3.66)$$

Figure 3.1 Chicken Little, hard at work propagating uncertainty for atmospheric pressure. Artwork by Asher/Anya Hurst.

Equation (3.65) is identical to Equation (3.51). The fractional uncertainty for division of two quantities is simply the addition of the fractional uncertainties.

Generalizing both Equation (3.51) and Equation (3.65) to the case with an arbitrary number of parameters in the numerator and denominator means a long list of N variables being multiplied and M variables being divided:

$$q = f(x_1, x_2, \ldots, x_N, x_{N+1}, x_{N+2}, \ldots, x_{N+M}) \quad (3.67)$$

Assuming all variables are combined with multiplication or division, we have the functional form of q as this:

$$q = \frac{x_1 \cdot x_2 \cdot \ldots \cdot x_N}{x_{N+1} \cdot x_{N+2} \cdot \ldots \cdot x_{N+M}} \quad (3.68)$$

After applying Equation (3.13) and similar processing steps as was done in Section 3.1.3, the resulting uncertainty formula δq is then this:

$$\delta q_{\text{frac}} = \frac{\delta x_1}{|x_1|} + \frac{\delta x_2}{|x_2|} + \ldots \frac{\delta x_N}{|x_N|} + \frac{\delta x_{N+1}}{|x_{N+1}|} + \frac{\delta x_{N+2}}{|x_{N+2}|} + \ldots + \frac{\delta x_{N+M}}{|x_{N+M}|} \quad (3.69)$$

This can be written in compact form with a summation:

$$\delta q_{\text{frac}} = \sum_{i=1}^{N+M} \delta x_{i,\text{frac}} \quad (3.70)$$

> **Quick and Easy for Section 3.4.2.3**
>
> A formula with division of correlated variables is handled the same way as with multiplication—addition of the fractional uncertainties.

3.4.2.4 Multiplication and Division with Independent Variables

In the case of independent uncertainties for all of the variables in the function for q, the quadrature general formula, Equation (3.14), can be used. Similar to Section 3.4.2.1, let us start with the easiest case of two parameters multiplying each other, $q = f(x,y) = x \cdot y$. The uncertainty for q is then from Equation (3.11):

$$\delta q = \sqrt{\left(\frac{\partial f}{\partial x}\delta x\right)^2 + \left(\frac{\partial f}{\partial y}\delta y\right)^2} \quad (3.71)$$

Applying our assumed functional form of f, we get:

$$\delta q = \sqrt{(y \cdot \delta x)^2 + (x \cdot \delta y)^2} \quad (3.72)$$

This is similar to the case with correlated uncertainties in Equation (3.47), and so it is natural to try the same conversion to fractional uncertainties. Dividing Equation (3.71) by $|q|$ has this form:

$$\delta q_{\text{frac}} = \frac{\delta q}{|q|} \quad (3.73)$$

Using our specific function for δq gives this fractional uncertainty:

$$\delta q_{\text{frac}} = \sqrt{\left(\frac{y \cdot \delta x}{q}\right)^2 + \left(\frac{x \cdot \delta y}{q}\right)^2} \quad (3.74)$$

Insertion of our formula for q reduces the equation a bit:

$$\delta q_{\text{frac}} = \sqrt{\left(\frac{\delta x}{x}\right)^2 + \left(\frac{\delta y}{y}\right)^2} \quad (3.75)$$

Application of a final simplification gives us this:

$$\delta q_{\text{frac}} = \sqrt{\delta x_{\text{frac}}^2 + \delta y_{\text{frac}}^2} \quad (3.76)$$

This is analogous to the final form of δq with multiplication of parameters with correlated uncertainties, in Equation (3.51), with the summation of fractional uncertainties contributing to the fractional uncertainty of the new functional parameter.

The same process can be conducted for division of two parameters, and the result will follow the same math and yield a formula for δq

identical to Equation (3.75). This formula for two variables can then be extended to an arbitrary number of variables in the numerator and denominator, as given in Equation (3.67), yielding this general formula for multiplication and division of parameters with independent uncertainties:

$$\delta q_{\text{frac}} = \sqrt{\left(\frac{\delta x_1}{x_1}\right)^2 + \left(\frac{\delta x_2}{x_2}\right)^2 + \cdots + \left(\frac{\delta x_N}{x_N}\right)^2 + \left(\frac{\delta x_{N+1}}{x_{N+1}}\right)^2 + \cdots + \left(\frac{\delta x_{N+M}}{x_{N+M}}\right)^2} \quad (3.77)$$

Equation (3.77) can be written as a summation of fractional uncertainties:

$$\delta q_{\text{frac}} = \sqrt{\sum_{i=1}^{N+M} \delta x_{i,\text{frac}}^2} \quad (3.78)$$

> **Quick and Easy for Section 3.4.2.4**
>
> When all of the input variables are uncorrelated, then uncertainty propagation with multiplication and division can be done with quadrature addition of the fractional uncertainties.

3.4.3 Power Laws

A specific application of the multiplication and division general formulas for δq is the case of a power law relationship. There are many applications in the geosciences when a new parameter is needed to be calculated that has this basic functional form: $q = f(x) = x^n$, where n is a known constant. Note that n does not have to be only an integer but can be any real number, for example Kolmogorov's famous $x^{-5/3}$ power law spectrum for turbulence in a fluid. This can be expanded to be a multiplication of the same variable by itself, as many times as needed, to produce the desired exponent n in the power law relationship. In this case, the uncertainties of the multiplied parameters are clearly not independent but rather perfectly correlated; they are identical. Therefore, Equation (3.13) should be applied to our functional form:

$$\delta q = \left|\frac{\partial f}{\partial x}\right| \delta x \quad (3.79)$$

Using the power law formula for f, we get:

$$\delta q = \left|nx^{n-1}\right| \cdot \delta x \quad (3.80)$$

This can be further processed as was done in Section 3.4.2.1 with a division by $|q|$ to yield this relationship:

$$\delta q_{\text{frac}} = \frac{\left|nx^{n-1}\right|}{|q|} \delta x \quad (3.81)$$

Much of the numerator of Equation (3.81) will cancel with the terms in q in the denominator, leaving this:

$$\delta q_{\text{frac}} = |n| \frac{\delta x}{|x|} \quad (3.82)$$

A final simplification allows us to write it this way:

$$\delta q_{\text{frac}} = |n| \delta x_{\text{frac}} \quad (3.83)$$

The absolute value operator is needed around the n term in front because the original power could be negative, as is the case for a Kolmogorov spectrum. Note that Equation (3.81) can also be obtained from Equation (3.69), allowing for $N+M$ to be a real number instead of an integer.

> **Quick and Easy for Section 3.4.3**
>
> Power law formulas are essentially repeated multiplication of many (very highly) correlated variables.

3.4.4 Exponentials and Logarithms

Exponential and logarithmic functions are common in the natural sciences, and so these are also special cases that deserve attention in our discussion of specific uncertainty formulas.

3.4.4.1 Exponential Functions

What if the situation were reversed from the last special case? That is, instead of a power law with the measured variable being raised to a power, what if the new function has the variable in the exponent? This could be raised to any base value, but the most common are base 10 and the natural exponential, e. In the case of $q = f(x) = e^x$, application of the general formula, Equation (3.5) in this case of only one parameter in the function, is:

$$\delta q = \left|\frac{df}{dx}\right|\delta x \qquad (3.84)$$

Using the formula we are assuming gives:

$$\delta q = e^x \cdot \delta x \qquad (3.85)$$

Equation (3.85) can be written in a slightly simpler form:

$$\delta q = q \cdot \delta x \qquad (3.86)$$

Equation (3.86) reveals that, for this special functional form, the fractional uncertainty of the new parameter is simply the uncertainty of the initial parameter, $\delta q_{frac} = \delta x$. The definition of q is that this new parameter is dimensionless. Because it is used in an exponent, the initial variable x is also dimensionless.

The more general version of this special case is when the exponential argument is a function of x, rather than simply x itself. This was exactly the case in Section 3.2.1, in Equation (3.19), when we derived formulas for the uncertainty of the mass density of air and some of the variables with uncertainties were inside the exponent. That is, we have already examined this special case. The general functional form is $q = e^{f(x)}$, in which case its uncertainty is given by:

$$\delta q = e^{f(x)} \cdot \left|\frac{df}{dx}\right|\delta x \qquad (3.87)$$

Dividing by q, Equation (3.87) can be written as a fractional uncertainty:

$$\delta q_{frac} = \left|\frac{df}{dx}\right|\delta x \qquad (3.88)$$

If there are several initial variables in the exponential, as was the case for atmospheric mass density, then $q = e^{f(x_1, x_2, \ldots, x_N)}$. If the uncertainties of the input variables could be correlated, then the δq should follow Equation (3.13) and look like this:

$$\delta q = e^{f(x_1,x_2,\ldots,x_N)} \cdot \left|\frac{df}{dx_1}\right|\delta x_1 + e^{f(x_1,x_2,\ldots,x_N)} \cdot \left|\frac{df}{dx_2}\right|\delta x_2$$
$$+ \cdots + e^{f(x_1,x_2,\ldots,x_N)} \cdot \left|\frac{df}{dx_N}\right|\delta x_N \qquad (3.89)$$

Equation (3.89) simplifies in the same way to yield a formula for the fraction uncertainty of the new parameter:

$$\delta q_{frac} = \sum_{i=1}^{N}\left|\frac{df}{dx_i}\right|\delta x_i \qquad (3.90)$$

If the initial variables have independent uncertainties, then Equation (3.14) can be used instead, yielding a slightly more complicated uncertainty:

$$\delta q = \sqrt{\left(e^f \frac{\partial f}{\partial x_1}\delta x_1\right)^2 + \left(e^f \frac{\partial f}{\partial x_2}\delta x_2\right)^2 + \cdots + \left(e^f \frac{\partial f}{\partial x_N}\delta x_N\right)^2} \qquad (3.91)$$

Again, remembering that e^f is q and dividing by this provides the exponential rule for independent uncertainties:

$$\delta q_{frac} = \sqrt{\sum_{i=1}^{N}\left(\frac{\partial f}{\partial x_i}\delta x_i\right)^2} \qquad (3.92)$$

As with all of the independent variable uncertainty formulas, Equation (3.92) yields a δq value that is smaller than that from Equation (3.90).

Quick and Easy for Section 3.4.4.1

Because the derivative of a natural exponential is itself, the *fractional* uncertainty from an exponential function looks like the *general absolute* uncertainty propagation formulas in Section 3.1.3.

3.4.4.2 Logarithmic Functions

The opposite case of exponentials is that of logarithms. Again, we will consider the case of natural logarithms, starting with the single variable functional form of $q = f(x) = \ln(x)$. First, apply Equation (3.5):

$$\delta q = \left|\frac{df}{dx}\right|\delta x \tag{3.93}$$

Then, use the assumed formula:

$$\delta q = \left|\frac{1}{x}\right|\delta x \tag{3.94}$$

Finally, simplify with the fractional uncertainty formula:

$$\delta q = \delta x_{\text{frac}} \tag{3.95}$$

This looks backward from the case of the exponential; Equation (3.86) had $\delta q_{frac} = \delta x$, while Equation (3.95) has $\delta q = \delta x_{frac}$. We might have even guessed that the form of this relationship would reverse for the natural logarithm from what it was for the natural exponential.

To generalize this, let us first expand to the case of a function within the natural logarithm, so that q has this form: $q = \ln(f(x))$. In this case, the formula changes to be:

$$\delta q = \left|\frac{1}{f(x)}\right|\left|\frac{df}{dx}\right|\delta x \tag{3.96}$$

To keep it general, the functional form of $f(x)$ is unknown, and therefore the right-hand side of Equation (3.96) cannot be reduced to a fractional uncertainty of x. The form given is as simplified as it gets.

The next step is a further generalization to a functional form that depends on several variables, such that $q = \ln(f(x_1, x_2, \ldots, x_N))$. As with all of the other special cases, we have two options for proceeding, either with correlated or independent variables. For the case of correlated uncertainties, Equation (3.96) then expands into many terms as follows:

$$\delta q = \left|\frac{1}{f}\right|\left|\frac{df}{dx_1}\right|\delta x_1 + \left|\frac{1}{f}\right|\left|\frac{df}{dx_2}\right|\delta x_2 + \cdots + \left|\frac{1}{f}\right|\left|\frac{df}{dx_N}\right|\delta x_N \tag{3.97}$$

This can be written in summation form:

$$\delta q = \left|\frac{1}{f}\right|\sum_{i=1}^{N}\left|\frac{df}{dx_i}\right|\delta x_i \tag{3.98}$$

For the case of independent uncertainties of the initial variables, the quadrature formula can be applied with the following result:

$$\delta q = \sqrt{\left(\frac{1}{f}\frac{\partial f}{\partial x_1}\delta x_1\right)^2 + \left(\frac{1}{f}\frac{\partial f}{\partial x_2}\delta x_2\right)^2 + \cdots + \left(\frac{1}{f}\frac{\partial f}{\partial x_N}\delta x_N\right)^2} \tag{3.99}$$

Like the other formulas for an arbitrary number of variables, this can be written in summation form:

$$\delta q = \left|\frac{1}{f}\right|\sqrt{\sum_{i=1}^{N}\left(\frac{\partial f}{\partial x_i}\delta x_i\right)^2} \tag{3.100}$$

> **Quick and Easy for Section 3.4.4.2**
>
> The uncertainty from a natural logarithm function also looks like the general absolute uncertainty propagation formulas in Section 3.1.3, but with an inverse of the function multiplier out front.

3.4.5 Trigonometric Functions

A final special case to consider is that of trigonometric functions. It is a regular occurrence to have periodic signals in natural systems and it could be the case that a sine, cosine, tangent, or other similar operator is applied to a measured angle and its uncertainty. This can occur, for instance, when measuring the change in path of light as it passes through a refractive medium, or at the boundary between two materials (or fluids) of different refractive indices.

The most essential rule to follow in this case is the conversion of any angle values provided in degrees into radians before any calculations are conducted. The basic number to three significant digits is 57.3 degrees per radian, but a more accurate value can be used as needed. That is, apply Equation (3.49) for multiplication by a constant to get a new variable and uncertainty in this new unit. When converted to radians, all angles become dimensionless parameters and can multiply other parameters.

Doing the simplest trigonometric formula of $q = f(x) = \sin(x)$ as an example, with $x \pm \delta x$ already converted to radians, we should use Equation (3.5):

$$\delta q = \left|\frac{df}{dx}\right| \delta x \qquad (3.101)$$

Application of our specific formula gives this:

$$\delta q = |\cos(x)| \cdot \delta x \qquad (3.102)$$

The x value applied inside the cosine operator is the best estimate of x, x_{best}, in radians.

A problem with this formula could arise if x is exactly on a zero of the cosine function, $x = \pi/2 \pm i\pi$. When this is the case, the original sine function is at a relative maximum or minimum of ± 1, and its slope goes to zero. The q function is essentially insensitive to δx at this angle and the uncertainty δq does, indeed, approach zero. This is not only a problem with a sine function relationship; most trigonometric functions include zero-slope regions, and if the measured quantity being processed by this operator is at that special value, then δq vanishes. Some trigonometric functions have the opposite problem of an infinite slope, such as at $\tan(\pi/2)$, where its derivative explodes to large positive or negative values. In these cases, δq also becomes large, perhaps to the point of rendering any further processing and interpretation impossible.

The way around this problem is that the uncertainty calculation is often only valid for certain angle ranges. In particular, this angle range is usually far from the angle values for which the function's slope goes to zero or infinity. To calculate an uncertainty for angles very close to these problem values, a more sophisticated approach—beyond the linearization in Equation (3.4) used throughout this chapter—is needed. A higher-order approximation to the local derivative should be applied in such cases. These instances are quite rare and will not be covered here.

> **Quick and Easy for Section 3.4.5**
>
> Trigonometric functions pose a special problem for uncertainty propagation because the function's slope periodically is zero or infinite, so there is usually a limited angle range of validity to the uncertainty propagation calculation.

3.5 Stepwise Uncertainty Propagation

While it is always preferable to use the general formulas in Section 3.1.3, there are some functional forms that are not easily differentiated. Other times, it is useful within the context of the specific application to have formulas for intermediate δq values, perhaps for evaluating sensitivity to specific processing of the measurements. In these very special cases, the calculation of uncertainty can be conducted with a step-by-step procedure.

There is not a one-size-fits-all process for conducting a stepwise uncertainty calculation. It is done by breaking down the big function into smaller chunks and applying our special case rules. The uncertainty is then calculated for each part of the main function, sequentially, until you get through the entire formula. You might use both absolute and fractional uncertainties in the process of getting to the final form, which could be in either absolute or fractional form. Different paths can be taken through the various steps toward the final form; there is no general rule about the order in which the steps should be conducted. In fact, these steps might be dictated by the desired analysis, should these intermediate δq values be of interest for the assessment.

Let us go through an example to show the process of a stepwise uncertainty calculation. The Earth's space environment is driven by the electromagnetic forces of the supersonic solar wind that continuously flows past it. One of the key quantities of this interaction is the electric potential difference across the high-latitude upper atmosphere. This quantity is thought to vary linearly with solar wind driving up to some threshold and then vary less than linearly beyond that driving level. One construction for this nonlinear functional form of the total potential drop, Φ_T, is as follows:

$$\Phi_T = \frac{\Phi_L \Phi_S}{\Phi_L + \Phi_S} \quad (3.103)$$

Here, Φ_L is the "linear" component of the electric potential that is directly proportional to the solar wind driving conditions, while Φ_S is the "saturated" component of the electric potential that does not depend on the solar wind. Usually, when the activity is low, $\Phi_S \gg \Phi_L$, but when driving is large, then the reverse could be true and Φ_T is no longer dependent on the solar wind. In general, the input values have independent uncertainties.

Let us assume that our functional form of $q = f(x,y)$ is the same as that in Equation (3.103), but we will write it with x and y instead of Φ_L and Φ_S, like this:

$$q = \frac{x \cdot y}{x + y} \quad (3.104)$$

Let us also assume that the uncertainties for x and y are independent, so that we can use the quadrature formulas for the uncertainty of q. To conduct the calculation of δq with a stepwise approach, let us define two new functions, $q_1 = x \cdot y$ and $q_2 = x + y$. This gives us the new form for the full function q, which can now be written as $q = q_1/q_2$.

We can then find the uncertainties for each of these based on the special case δq formulas mentioned in Section 3.4. For q_1, we can use Equation (3.75), getting this:

$$\delta q_{1,\text{frac}} = \sqrt{\delta x_{\text{frac}}^2 + \delta y_{\text{frac}}^2} \quad (3.105)$$

For q_2, we can use Equation (3.41), getting this:

$$\delta q_2 = \sqrt{\delta x^2 + \delta y^2} \quad (3.106)$$

The uncertainty for q is then the application of Equation (3.77):

$$\delta q_{\text{frac}} = \sqrt{\delta q_{1,\text{frac}}^2 + \delta q_{2,\text{frac}}^2} \quad (3.107)$$

For the right-hand-side terms, the first is directly from Equation (3.105) and the second can be found from the absolute uncertainty in Equation (3.106):

$$\delta q_{\text{frac}} = \sqrt{\delta q_{1,\text{frac}}^2 + \left(\frac{\delta q_2}{q_2}\right)^2} \quad (3.108)$$

Equation (3.107) can now be expanded with our two stepwise uncertainties, giving:

$$\delta q_{\text{frac}} = \sqrt{\delta x_{\text{frac}}^2 + \delta y_{\text{frac}}^2 + \frac{\delta x^2 + \delta y^2}{(x+y)^2}} \quad (3.109)$$

Remembering that $\delta x_{\text{frac}} = \delta x/x$ and $\delta y_{\text{frac}} = \delta y/y$, Equation (3.109) can be written out in terms of only the two variables, x and y:

$$\delta q_{\text{frac}} = \sqrt{\frac{\delta x^2}{x^2} + \frac{\delta y^2}{y^2} + \frac{\delta x^2 + \delta y^2}{(x+y)^2}} \quad (3.110)$$

Our functional form of q, Equation (3.104), is simple enough that we can go directly to δq with the application of Equation (3.14) or, even more specifically, Equation (3.11) for two variables with independent uncertainties. In this method, we get the following for δq:

$$\delta q = \sqrt{\left(\frac{\partial f}{\partial x}\delta x\right)^2 + \left(\frac{\partial f}{\partial y}\delta y\right)^2} \quad (3.111)$$

The two partial derivatives have very similar forms:

$$\frac{\partial f}{\partial x} = \frac{y^2}{(x+y)^2} \text{ and } \frac{\partial f}{\partial y} = \frac{x^2}{(x+y)^2} \quad (3.112)$$

Inserting these partial derivatives into Equation (3.111) gives this:

$$\delta q = \frac{1}{|x+y|}\sqrt{(y^2\delta x)^2 + (x^2\delta y)^2} \quad (3.113)$$

Converting Equation (3.113) into a fractional uncertainty for direct comparison with Equation (3.109), we get:

$$\delta q_{\text{frac}} = \sqrt{\left(\frac{y}{x \cdot y}\delta x\right)^2 + \left(\frac{x}{x \cdot y}\delta y\right)^2} \quad (3.114)$$

Converting to fractional uncertainties inside the square root, we get:

$$\delta q_{\text{frac}} = \sqrt{\delta x_{\text{frac}}^2 + \delta y_{\text{frac}}^2} \quad (3.115)$$

Comparing Equation (3.114), δq_{frac} from the general formula, with Equation (3.109), δq_{frac} from our chosen path through a stepwise process, we see that the stepwise procedure introduced a new term to the fractional uncertainty for q. It is often the case that a stepwise approach to calculating the uncertainty creates additional terms in the final functional form of the uncertainty. For this simple example, there was a relatively clear path to take in terms of breaking up the full function, but we could have done it a different way, such as letting $q = q_3 \cdot q_4$ and defining $q_3 = x/\sqrt{x+y}$ and $q_4 = y/\sqrt{x+y}$. This type of split could be useful to examine the separate uncertainties of x and y relative to this combined square-root denominator term. This yields a rather complicated function for δq_3:

$$\delta q_3 = \sqrt{\left[\frac{(2y+x)\delta x}{2(x+y)^{3/2}}\right]^2 + \left[\frac{x \cdot \delta y}{2(x+y)^{3/2}}\right]^2} \quad (3.116)$$

There is similarly complicated function for δq_4:

$$\delta q_4 = \sqrt{\left[\frac{y \cdot \delta x}{2(x+y)^{3/2}}\right]^2 + \left[\frac{(2x+y)\delta y}{2(x+y)^{3/2}}\right]^2} \quad (3.117)$$

Combining this with a multiplication rule uncertainty formula gives this for the full uncertainty of δq_{frac}:

$$\delta q_{\text{frac}} = \sqrt{\left(\frac{\delta q_3}{q_3}\right)^2 + \left(\frac{\delta q_4}{q_4}\right)^2} \quad (3.118)$$

Equation (3.118) can be expanded out as this:

$$\delta q_{\text{frac}} = \sqrt{\frac{(2y+x)^2(\delta x)^2 + (x \cdot \delta y)^2}{2x(x+y)} + \frac{(y \cdot \delta x)^2 + (2x+y)^2(\delta y)^2}{2y(x+y)}} \quad (3.119)$$

Equation (3.119) simplifies to this final form:

$$\delta q_{\text{frac}} = \sqrt{\frac{(4y^3 + 5xy^2 + x^2y)(\delta x)^2 + (4x^3 + 5x^2y + xy^2)(\delta y)^2}{2xy(x+y)}} \quad (3.120)$$

Equation (3.120) is atrociously complicated compared to Equation (3.114) and more complex than the other stepwise version in Equation (3.109). This example was intentionally chosen to show that there is not a unique answer to a stepwise uncertainty calculation because the path from beginning to ending is not unique. The function can be split in any way desired.

> **Quick and Easy for Section 3.5**
>
> Stepwise uncertainty propagation breaks down the full-function derivative into smaller calculations. The partitioning and order of calculation are subjective and often yield larger uncertainties than the single-step method. You will *very rarely*, if ever, have to use this technique.

3.6 Example: Planetary Equilibrium Temperature

Let us do an example of calculating a **planetary equilibrium temperature**. For this calculation, there are two main terms that balance each other—the incoming stellar radiation

---

Planetary equilibrium temperature: The average temperature across a whole planet's surface found by a balance of energy inflow and outflow.

---

and the outgoing planetary radiation. The first of these is usually a constant and the second is usually dependent on the temperature of the planet. Setting these two functions equal to each other provides us with an equation that includes the planetary temperature as the only unknown, a value that will rise until the outgoing energy balances the incoming energy. Solving for this variable, we will see that it is dependent on several input parameters, each with its own uncertainty. We will then explore methods of calculating the uncertainty of the derived planetary equilibrium temperature.

Energy input to a planet is most easily understood as power, energy per unit time, so the starting equation for this exercise is this:

$$P_{in} = P_{out} \qquad (3.121)$$

We will assume that both are emanating photons according to blackbody radiation, for which the power density, S, on the surface of the object is written as:

$$S_{blackbody} = \sigma_{SB} T^4 \qquad (3.122)$$

In Equation (3.122), σ_{SB} is the Stefan–Boltzmann constant, 5.67×10^{-8} W/(m^{2}K^4), and T is the temperature of the surface of the object. The stellar power is not needed at the surface of the star but rather at the location of the planet, and therefore an area conversion is needed:

$$S_{star-at-planet} = \frac{R_{star}^2}{D_{s-p}^2} \sigma_{SB} T_{star}^4 \qquad (3.123)$$

In Equation (3.123), R_{star} is the radius of the star and D_{s-p} is the distance between the star and planet. For most stars, the temperature is fairly stable and everything on the right-hand side of Equation (3.123) can be combined as a stellar constant, S_0.

This power density S_0 is only absorbed on half of the surface, and, furthermore, it is at an angle to the surface everywhere on the dayside except for the one spot where the star is directly overhead. That is, the full value of S_0 should not be used. A simple correction is to use the disk that the planet blocks out of the full surface area at the star–planet distance. Because the planetary radius is usually much smaller than D_{s-p}, this area can be assumed to be a flat circle.

One more correction to the incoming radiation is that not all of it is absorbed; some of the incoming radiation is reflected back out to deep space. This process can be distilled into a single number, a reflection coefficient known as planetary albedo, A_p, and so the fraction of radiation absorbed by the planet is $1 - A_p$. Given all of this, the incoming power can be written as this:

$$P_{in} = (1 - A_p) \pi R_p^2 \frac{R_{star}^2}{D_{s-p}^2} \sigma_{SB} T_{star}^4 \qquad (3.124)$$

The outgoing power is simpler because we want the total energy flux leaving the planet, so we can integrate Equation (3.122) over the surface of the planet. Not all of the radiation released this way from the planet escapes to deep space; some of it is reflected back to the surface. This is the case at Earth, as its outgoing radiation is peaked in the microwave band of the electromagnetic radiation spectrum and certain atmospheric molecules are efficient at absorbing and then reemitting the photons in this band. This process redirects the outgoing radiation in all directions, some of it back down to the surface. This is the same process as putting on a coat or wrapping yourself in a blanket to stay warm, with some of your body's outgoing heat captured and redirected back toward you. In the atmosphere, this is universally known as the greenhouse effect, and the molecules doing this process are called greenhouse gases. This is a bit of a misnomer because much of the heat buildup inside of a greenhouse is because it blocks the wind, but the similarity is that, because the air is stagnant inside the greenhouse, it can interact with the plants, people, and ground inside the structure and reach a higher equilibrium temperature. This reflection of some of the outgoing radiation is analogous to the reflection of some of the incoming radiation, and therefore can be included in our equations as a greenhouse albedo, A_{GH}, through an attenuation factor of $1 - A_{GH}$. Putting these terms together, our outgoing power can be written like this:

$$P_{out} = (1 - A_{GH}) 4\pi R_p^2 \sigma_{SB} T_p^4 \qquad (3.125)$$

Using Equation (3.121), we can set Equation (3.124) equal to Equation (3.125) and, after some algebra, arrive at an equation for the equilibrium temperature of the planet:

$$T_p = \left[\frac{S_0(1 - A_p)}{4\sigma_{SB}(1 - A_{GH})} \right]^{1/4} \qquad (3.126)$$

There are three values in Equation (3.126) that typically have uncertainties attached to them: S_0, A_p, and A_{GH}. These are often independently varying parameters and therefore we can use the quadrature formulas for determining an uncertainty on T_p, δT_p.

We will use several methods to find δT_p, the general formula, and two variations of the step-by-step approach. For the first, applying Equation (3.14) to Equation (3.126), a few steps of algebra yield this:

$$\delta T_p = \frac{T_p}{4} \sqrt{\frac{(\delta S_0)^2}{S_0^2} + \frac{(\delta A_p)^2}{(1 - A_p)^2} + \frac{(\delta A_{GH})^2}{(1 - A_{GH})^2}} \qquad (3.127)$$

Note that the T_p variable in front of the square-root operator is Equation (3.126) and includes the base values of the three parameters but not their uncertainties.

For the stepwise approach, we have several options of how to break down the full T_p formula, but let us do it as a numerator and a denominator:

$$T_p = \frac{\left[S_0(1 - A_p) \right]^{1/4}}{\left[4\sigma_{SB}(1 - A_{GH}) \right]^{1/4}} = \frac{T_{p1}}{T_{p2}} \qquad (3.128)$$

In this case, the uncertainty for T_{p1} becomes this:

$$\delta T_{p1} = \sqrt{\frac{(1 - A_p)^{1/2}}{16 S_0^{3/2}} (\delta S_0)^2 + \frac{S_0^{1/2}}{16(1 - A_p)^{3/2}} (\delta A_p)^2} \qquad (3.129)$$

The equation for T_{p2} only includes one variable, so it has this form:

$$\delta T_{p2} = \frac{(\sigma_{SB})^{1/4}}{(1 - A_{GH})^{3/4}} \delta A_{GH} \qquad (3.130)$$

These two uncertainties then should be combined with the division rule in Equation (3.77) from Section 3.4.2.4, but with an extra multiplication by T_p to convert it to an absolute uncertainty for direct comparison with Equation (3.127). This gives us a general formula like this:

$$\delta T_p = \frac{T_{p1}}{T_{p2}} \sqrt{\left(\frac{\delta T_{p1}}{T_{p1}} \right)^2 + \left(\frac{\delta T_{p2}}{T_{p2}} \right)^2} \qquad (3.131)$$

Applying Equations (3.129) and (3.130) for the two uncertainties, we get this:

$$\delta T_p = \frac{T_{p1}}{T_{p2}} \sqrt{\frac{1}{16 S_0^2}(\delta S_0)^2 + \frac{1}{16(1 - A_p)^2}(\delta A_p)^2 + \frac{1}{16(1 - A_{GH})^2}(\delta A_{GH})^2} \qquad (3.132)$$

Converting the leading fraction into T_p and extracting the constant value of 16 from the square-root operator reveals that Equation (3.132) is identical to Equation (3.127). That is, in this case, the stepwise approach yielded the same uncertainty formula. This is because the separation of variables was clean; the three parameters with uncertainties appeared in either T_{p1} or T_{p2}, but not both.

Because of the complete separation of input variables between the two intermediate parameters, the stepwise uncertainty matched the general formula solution. That is, it was a fortuitously easy example. In the problems provided in Section 3.10, a different separation into intermediate values is conducted, with an unfortunately messy final uncertainty formula.

> **Quick and Easy for Section 3.6**
>
> Planetary equilibrium temperature serves as a good example for a stepwise uncertainty propagation.

3.7 Multistep Processing

Perhaps you have noticed that the daily high temperature may be forecasted to be, say, 75 °F, and, oddly, the forecast for the hourly temperatures for the day only went up to 73 °F. Why don't they match? This is a common complaint about the weather forecast.

The difference arises as a matter of methodology and uncertainty. First, the forecast is made from a set of model calculations. It is not made from one run of the weather forecasting model but several. The National Weather Service conducts several runs with its models and provides all of this information to local or national weather providers. The daily high value is often calculated as the average of the high values from these forecasts, while the hourly values are usually reported as the average of the model results for each hour throughout the day. Each model run, however, used slightly different initial or boundary conditions, or perhaps different grid configurations or numerical techniques to solve the equations. They will most likely predict a peak in the daily temperature at different times during the day; some will find a peak in the early afternoon, while other runs will forecast the highest temperature to occur late in the afternoon. For any particular hour, therefore, the values might include a peak temperature from one of the model runs but most likely not from all of the model runs. Unless all of the model runs peak at the same time, the hourly average values will always be less than the peak values for any one model, and most likely less than the average of those peak values from each model.

Figure 3.2 shows this with temperature data from 4 model runs on a rather boring weather day. The temperature reaches a minimum just before dawn, rises to a peak in the afternoon, and then drops, slowly in the late afternoon and then quickly after sunset. The peak temperatures from the 4 model runs are 75, 75, 77, and 73 for the Models 1 through 4, respectively. The average of this is easily calculated to be 75. These occur throughout the afternoon, though, from 12:30 in Model 3 to 15:30 for Model 2. Averaging the temperature values at each hour of the day yields an hourly temperature forecast for the day with a peak of 74 reached

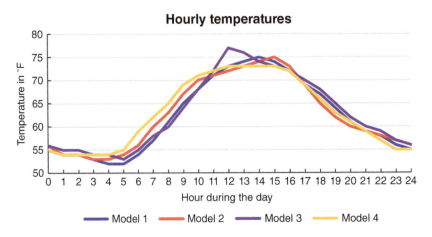

Figure 3.2 Four model results for the hourly temperatures across a 24-h interval.

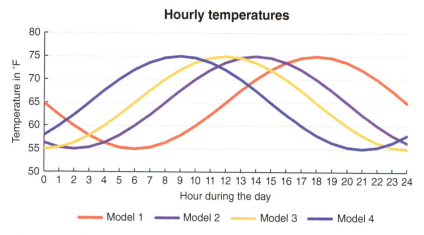

Figure 3.3 Four idealized hourly temperature time series across a 24-h interval. They are exactly the same sine curves, but offset from each other so that the peak occurs at a different time during the day.

both at 13:30 and 14:30. This is a full degree lower than the average of the maxima, yet came from the same set of numbers. The methodology is different.

To make this point even more poignantly, let us assume that the temperature profiles throughout the day are sine waves, as in Figure 3.3. All of the 4 models have exactly the same curve, just shifted in phase so that the peak temperature occurs at a different time. They all have exactly the same maximum of 75 °F, but the maximum of the hourly averages is only 71.7 °F. This is now several degrees lower than the average of the peak values.

A key point to make here is that both methods of obtaining a forecasted high temperature for the day are correct. Both used standard mathematical procedures to process the data and determine a peak value. In fact, they both used the same functions, taking a maximum of a number set and taking an arithmetic average of a number set. The only difference was that the order of these two functions was reversed. In one method, the "maximum value" function was applied first, to each model output set, and then the "average value" was applied to those maxima. In the other method, the "average value" function was applied first to the values from all models at each hour, and then the "maximum value" was found of these hourly averages.

The point is that order matters. Both processing paths are legitimate, but lead to different answers with different information. You should always be very clear about your methodology, so that others are able to not only repeat it but also properly interpret it.

> **Quick and Easy for Section 3.7**
>
> The order in which a processing calculation is conducted could change the final result, as illustrated by the reported daily high temperature.

3.8 Final Advice on Uncertainty Propagation

The most fundamental uncertainty propagation formula is Equation (3.13), the general formula for multiple input variables which are not entirely independent. Its partner is Equation (3.14), the general formula when the input variables are known to be independent. It might overestimate the uncertainty, but it should give you a value in the right neighborhood. All of the other equations in this chapter are either simplifications or extensions of these two.

Table 3.1 Easy guide to error propagation formulas.

Case	Input variables are independent	Input variables are dependent
General formula	$\delta q = \sqrt{\left(\dfrac{\partial f}{\partial x}\delta x\right)^2 + \left(\dfrac{\partial f}{\partial y}\delta y\right)^2 + \cdots + \left(\dfrac{\partial f}{\partial z}\delta z\right)^2}$	$\delta q = \left|\dfrac{\partial f}{\partial x}\right|\delta x + \left|\dfrac{\partial f}{\partial y}\right|\delta y + \cdots + \left|\dfrac{\partial f}{\partial z}\right|\delta z$
One input variable	$\delta q = \left\|\dfrac{df}{dx}\right\|\delta x$	$\delta q = \left\|\dfrac{df}{dx}\right\|\delta x$
Fractional uncertainty	$\delta q_{\text{frac}} = \dfrac{1}{\|q_{\text{best}}\|}\sqrt{\left(\dfrac{\partial f}{\partial x}\delta x\right)^2 + \cdots + \left(\dfrac{\partial f}{\partial z}\delta z\right)^2}$	$\delta q_{\text{frac}} = \dfrac{1}{\|q_{\text{best}}\|}\sqrt{\left(\dfrac{\partial f}{\partial x}\delta x\right)^2 + \cdots + \left(\dfrac{\partial f}{\partial z}\delta z\right)^2}$
Percentage uncertainty	$\delta q_{\text{perc}} = \delta q_{\text{frac}} \cdot 100\% = \dfrac{\delta q}{\|q_{\text{best}}\|}\cdot 100\%$	$\delta q_{\text{perc}} = \delta q_{\text{frac}} \cdot 100\% = \dfrac{\delta q}{\|q_{\text{best}}\|}\cdot 100\%$
Only add and subtract operations	$\delta q = \sqrt{\sum_{i=1}^{N+M}\delta x_i^2}$	$\delta q = \sum_{i=1}^{N+M}\delta x_i$
Only multiply and divide operators	$\delta q_{\text{frac}} = \sqrt{\sum_{i=1}^{N+M}\delta x_{i,\text{frac}}^2}$	$\delta q_{\text{frac}} = \sum_{i=1}^{N+M}\delta x_{i,\text{frac}}$
Only a power law	$\delta q_{\text{frac}} = \|n\|\delta x_{\text{frac}}$	$\delta q_{\text{frac}} = \|n\|\delta x_{\text{frac}}$
Only an exponential function	$\delta q_{\text{frac}} = \sqrt{\sum_{i=1}^{N}\left(\dfrac{\partial f}{\partial x_i}\delta x_i\right)^2}$	$\delta q_{\text{frac}} = \sum_{i=1}^{N}\left\|\dfrac{df}{dx_i}\right\|\delta x_i$
Only a logarithm function	$\delta q = \left\|\dfrac{1}{f}\right\|\sqrt{\sum_{i=1}^{N}\left(\dfrac{\partial f}{\partial x_i}\delta x_i\right)^2}$	$\delta q = \left\|\dfrac{1}{f}\right\|\sum_{i=1}^{N}\left\|\dfrac{df}{dx_i}\right\|\delta x_i$

This relationship is seen when all of the major uncertainty propagations introduced in this chapter are listed together, as they are done in Table 3.1. The two columns give the formulas for the case of independent input variables or not. Each row provides the equation for that case, as listed in the left column and collected from the sections of this chapter. If the function form $q = f(x, y, \ldots, z)$ does not allow for a straightforward use of the general formula, then the simplified cases lower in Table 3.1 can be combined in a stepwise approach to uncertainty propagation.

When reporting uncertainty, two key aspects should be taken into account: units and significant figures. The uncertainty δq has the same units as q, the new variable, and the units of both should be listed once after the uncertainty: $q_{\text{best}} \pm \delta q$ [units of q]. For percentage uncertainty, the convention is different: q_{best} [units of q] $\pm \delta q_{\text{perc}}\%$. Remember that reporting the unitless fractional uncertainty value can be confusing; reporting q_{best} [units of q] $\pm \delta q_{\text{frac}}$ is technically correct, but could be misinterpreted by others. It is better to convert fractional uncertainty to percentage uncertainty and report it with distinct units of percentage.

The other important reminder about reporting uncertainty is to be mindful of significant figures. Rarely is it needed to report more than one significant figure for an uncertainty value. It is recommended to apply the guidelines given in Chapter 1, with the usual advice being to round the uncertainty to one digit (unless that digit is a one, then keep the second) and round the base value to same decimal place. If another convention is used, then it is good to explain the reasoning for keeping the additional digits on the uncertainty value.

> **Quick and Easy for Section 3.8**
>
> When in doubt about uncertainty propagation, use one of the two general formulas; they will yield a reasonable number for further analysis.

3.9 Further Reading

An excellent explanation of uncertainty propagation is given in John R. Taylor's book:
Taylor, J. R. (1997). *An Introduction to Error Analysis*. University Science Books, Mill Valley, CA, USA.
https://uscibooks.aip.org/books/introduction-to-error-analysis-2nd-ed/
And here is another good book on the topic of error analysis:
Drosg, M. (2007). *Dealing with Uncertainties: A Guide to Error Analysis*. Springer-Verlag, Berlin, Germany.
https://www.springer.com/gp/book/9783540296089

A good text on the physics of atmospheric dynamics is Holton's famous book:
Holton, J. R., & Hakeem, G. J. (2013). *An Introduction to Dynamic Meteorology* (5th edition). Elsevier, Waltham, MA, USA.
https://www.elsevier.com/books/an-introduction-to-dynamic-meteorology/holton/978-0-12-384866-6

If you do not want to sift through a whole book, then here is a simplified version of hydrostatic equilibrium:
https://www.e-education.psu.edu/meteo300/node/7

Here is a good educational website about planetary equilibrium temperature:
https://scied.ucar.edu/planetary-energy-balance-temperature-calculate

3.10 Exercises in the Geosciences

1. Two students have magnetometers and measure the field strength at a particular location as 53 μT and 55 μT, both with a $\pm 1\ \mu T$ uncertainty.
   A What is the uncertainty of the discrepancy between these two measurements? Remember that discrepancy is the difference of two numbers.
   B A method to calculate a z score is with this formula:

   $$z = \frac{|\Delta B - \Delta B_{perfect}|}{\sigma_{\Delta B}}$$

   C In this formula, ΔB is the discrepancy, $\Delta B_{perfect}$ is zero, and the denominator is the uncertainty found in Part a. Find this z score. Is this descrepancy statistically significant?
   D An alternative method to calculate a z score is by assuming that one measurement is the "true value" and its uncertainty is the standard deviation in the z score calculation. Calculate this z score and determine its significance. Comment on the similarity or difference with your answer to Part b.

2. The vertical extent of sea ice can be approximated with the following formula:

   $$h = \frac{k(T_s - T_w)}{Q^*}$$

   Here, k is the thermal conductivity, T_S is the surface temperature, T_W is the water temperature, and Q^* is the net heat flux.
   A Assuming the thermal conductivity is a constant but the other 3 parameters are independent variables, determine

the formula for the uncertainty in vertical extent of the sea ice. Explain any assumptions applied in working through this calculation.

B Assuming k is the conductivity for pure ice (2.2 W m^{-1} K^{-1}), the surface temperature is well below freezing (varying between -10 °C in the day and -16 °C at night, so 260 ± 3 K), the water temperature is very close to freezing (274 ± 1 K), and the heat flux is about average for the arctic region (-0.8 ± 0.3 W m^{-2}). Calculate the sea ice vertical extent h and its uncertainty.

3. In Section 3.6, the formula for planetary equilibrium temperature was used as an example for a stepwise uncertainty calculation. The stepwise separation of T_p into components could have been chosen differently. For this problem, let us choose it this way:

$$T_p = \left(\frac{S_0}{4\sigma_{SB}(1-A_{GH})} - \frac{S_0 A_p}{4\sigma_{SB}(1-A_{GH})} \right)^{1/4}$$
$$= (T_{p1} - T_{p2})^{1/4}$$

A Find the uncertainty formulas for T_{p1} and T_{p2} using general quadrature formula.

B Find the uncertainty formula for T_p using a stepwise propagation approach by combining these two intermediate values.

C Assuming $S_0 = 1361$ W m^{-2}, $A_p = 0.3$, $A_{GH} = 0.4$, and $\sigma_{SB} = 5.67 \times 10^{-8}$ W m^{-2} K^{-4}, calculate the base value of T_p using Equation (3.126) and then uncertainty using both the general formula method from Equation (3.127) and the stepwise formula just derived. Discuss the similarity or difference between these results.

4

Centroids and Spreads: Analyzing a Set of Numbers

The term "data processing" often refers to the act of distilling a set of numbers into a much smaller and often more useful set of numbers. In this chapter, the basics of processing a single data set are discussed. The content focuses on quantitative processing—calculated values that describe the data set—rather than the visualizations of these values, some of which have been covered in Chapter 2.

A single set of numbers without other context is considered a one-dimensional array of data. For the methods discussed in this chapter, no other information is needed about the data. Putting one data set in the context of another set is the subject of later chapters. As discussed in Section 2.1 (Chapter 2), ephemeris data—other numbers that provide context for the primary measurement—will not be considered here, but please remember that it is important for a robust understanding of the number set, especially with regard to the possible uncertainties of the values. For the most part, though, the analysis techniques provided in this chapter do not require the ephemeris data accompanying a set of observations. When discussing measurements of the same phenomenon, it is assumed that the ephemeris data for these measurements are similar enough to allow all of the observations to be grouped together.

There are lots of ways to describe a set of numbers. You can use descriptive language, such as "the data are very spread out" or "the data are tightly confined to similar values." You can use relative language, comparing a data set to another by stating that it appears to be systematically "higher" or "lower" than the other value set. While such descriptions are helpful as an initial evaluation and assessment, this is highly qualitative and subjective. It is useful to conduct quantitative calculations with the number set to distill it into values that have particular meaning to others, so that they can quickly and readily understand the key features of the number set.

Here, we focus on two quantitative descriptor values—centroid and spread. There are multiple definitions of each of these two concepts. While some derivations will be shown, most will not be shown, instead focusing on how to use and interpret the statistic rather than the detailed mathematics of how it came to be the way it is.

4.1 Quantitatively Describing a Data Set: The Centroid

You probably already know the three most common measures of the centroid of a number set: mean, median, and mode. Let's quickly go through how to determine each of these centroid values.

Data Analysis for the Geosciences: Essentials of Uncertainty, Comparison, and Visualization, Advanced Textbook 5, First Edition. Michael W. Liemohn.
© 2024 American Geophysical Union. Published 2024 by John Wiley & Sons, Inc.
Companion website: www.wiley.com/go/liemohn/uncertaintyingeosciences

4.1.1 Three Versions of Mean

The usual option for the centroid of a number set is the mean, or average, of the values. There are, however, several different formulas for calculating a mean value. By far the most common is the **arithmetic mean**, often written as $\bar{x}$, which is the sum of the values divided by the number of values in the set:

$$\bar{x} = \frac{x_1 + x_2 + \cdots + x_N}{N} \quad (4.1)$$

Equation (4.1) can be written more compactly with summation notation:

$$\bar{x} = \frac{1}{N} \sum_{i=1}^{N} x_i \quad (4.2)$$

Note that $\bar{x}$ is sometimes called μ, especially when referring to a full population rather than just a sample data set from that population, but it is called μ occasionally even for the sample mean.

While Equation (4.2) is the most ubiquitous formula for mean, others exist. Another frequently used value is the **geometric mean**, $\bar{x}_G$:

$$\bar{x}_G = \sqrt[N]{x_1 \cdot x_2 \cdot x_3 \ldots x_N} \quad (4.3)$$

Equation (4.3) can be written in shorthand form like this:

$$\bar{x}_G = \left(\prod_{i=1}^{N} x_i \right)^{\frac{1}{N}} \quad (4.4)$$

The Π symbol in Equation (4.4) denotes a product of all of the elements. This has a particular usefulness in finance, where the x_i values are interval growth rates, resulting in $\bar{x}_G$ as the average compound growth rate over all of the intervals. For the natural sciences, this is simply an alternative to the arithmetic mean. It is useful when the values of the number set are highly variable, spanning several orders of magnitude. In this case, the arithmetic mean will be dominated by the values at the upper end of the range. The geometric mean will be in the center of the values as plotted on a logarithmic scale.

This last statement about the geometric mean is more readily understood by converting it into a formula with logarithms and exponentials. Specifically, this is done by applying logarithm and exponential operators to both sides of Equation (4.3). The left-hand side will still be $\bar{x}_G$, but the right-hand side will undergo a transformation. Any log and exponential pair can be used, but the most convenient is the natural logarithm and natural exponential operators. Applying the natural logarithm operator converts the radical into a $1/N$ multiplier in front of the natural log operator. Furthermore, the natural log of a series of numbers multiplied together is equal to the summation of the natural log of each number individually. This yields the following formula:

$$\bar{x}_G = \exp\left(\frac{1}{N} \sum_{i=1}^{N} \ln(x_i) \right) \quad (4.5)$$

Note the similarity between Equation (4.5) and Equation (4.2). It is seen that the geometric mean is the exponential of the arithmetic average of the natural logs of the values in the number set. When the values span multiple orders of magnitude, then the geometric mean will be in the center of the orders of magnitude, rather than systematically favoring the values in the upper-most order.

A third and final average to mention here is the **harmonic mean**. This mean, $\bar{x}_h$, is the

---

Arithmetic mean ($\bar{x}$): A measure of a number set centroid, it is the sum of all values in the set divided by the total count of numbers.

Geometric mean ($\bar{x}_G$): A measure of a number set centroid, it is the multiplication of all values raised to the 1/N power, often used for highly variable number sets.

Harmonic mean ($\bar{x}_h$): A measure of a number set centroid, it is the inverse of the arithmetic mean of the inverses of the values, often used for combining rates.

---

inverse of the arithmetic average of the inverse of each value, written like this:

$$\bar{x}_h = \frac{N}{\frac{1}{x_1} + \frac{1}{x_2} + \cdots + \frac{1}{x_N}} \quad (4.6)$$

This can be written in summation form:

$$\bar{x}_h = \frac{N}{\sum_{i=1}^{N} \frac{1}{x_i}} \quad (4.7)$$

At first glance, this is a rather strange formula. It is the correct one, however, for certain types of averaging, such as combining the efforts of unequal rates. It is often used in finance to combine the rates of return from different investments. Using the harmonic mean instead of the arithmetic mean provides a better estimate of the combined growth of the original investment. It is especially used for price-to-earnings ratios. In the natural sciences, the harmonic mean is the right choice for combining different speeds to get an average rate of travel. It is faster than using the weighted arithmetic average, which we'll discuss in the next chapter.

Similar to the geometric mean, the harmonic mean is also useful when the values span more than one or two orders of magnitude. In particular, it is good when there are a few extreme outliers in the number set. These outliers will heavily influence the arithmetic mean, but will only have a marginal influence on the harmonic mean.

For a clear and easy example of the differences between the arithmetic, geometric, and harmonic means, let's consider the case of wanting to find the centroid of a set of three numbers: 1, 10, and 100. Applying Equation (4.2), the arithmetic mean is 111/3 = 37 (to two significant digits). This is higher than two of the three values because the third is a factor of ten larger than its closest neighbor. Using Equation (4.3) (or Equation (4.5) if you prefer), the geometric mean is the cube root of 1000, so it is 10. This is significantly lower than the arithmetic mean. Finally, Equation (4.6) gives us this:

$$\bar{x}_h = \frac{3}{\frac{1}{1} + \frac{1}{10} + \frac{1}{100}} \quad (4.8)$$

This is 3/1.11, which is 2.7. The three means give values that are an order of magnitude different from each other. Of course, this is an extreme case and a very simple example with only a few numbers in the set. It shows, however, the variation possible for the definition of centroid for a number set.

There is a natural ordering of these three mean values. It goes like this: the arithmetic mean will be the largest, the harmonic mean will be the smallest, and the geometric mean will be between these other two. The only time that they will be exactly equal is when all values in the number set are equal, in which case the centroid is a trivial and rather meaningless concept.

There are even more means in existence, formulas that have been developed using various combinations of the values to create a distinct measure of the centroid. All of these were developed with a specific purpose and application. As an "essential set" of centroid options, though, those presented here provide a good list of options.

Quick and Easy for Section 4.1.1

The arithmetic mean emphasizes the values in the highest order of magnitude within a number set, the harmonic mean emphasizes the values in the lowest order of magnitude within the number set, and the geometric mean falls in between these other two.

4.1.2 More Centroids: Median and Mode

The **median** is "the 50% mark" in the set. It is found by sorting the values from lowest to highest and then counting upward to the middle value. Find the middle value and you are done; that is the median.

There is one small snag to this simple definition, and that is when there is no value that is exactly at the 50% level in the data set. Given an odd number of values in the set, one of the numbers is at the center of the list. For example, consider a list of nine values, as given in the first column of Figure 4.1. With an odd number of values, there is a clear "middle value" for which there are four others above and below it. You must sort it first, though—the fifth entry in the first column of numbers is not the median because the sorting has not yet occurred. The second column of Figure 4.1 shows the same nine values, but this time sorted, and the number in the fifth place is, now, the median value of the set. If the values are in even number in the set, then the 50% level falls between two of the entries in the list.

This is the case in the right two columns of numbers in Figure 4.1, for which the fourth and fifth entries are equally close to the 50% mark in the list length. If the two values are identical, as is the case in the first of these two examples, then the median is clear. If not, like the last column, then the median is unclear. There are at least three options for defining the median—choose the lower value, choose the higher value, or choose the average of the two values. Mathematically, the third option is the most logical, as this gives a value that is closest to "the 50% mark" of the number set. Sometimes it is useful, though, for the median to be one of the numbers in the set, in which case choosing either of the adjacent values is appropriate.

Next, let's consider **mode**. The basic definition of this centroid is that it is the most-repeated value in the set. For a number set of integers with a tight spread so that there are multiple entries of exactly the same value, this definition is straightforward. If two or more integers have the same maximum count, then the number set is said to be multimodal and all of them are equally defined as the mode of the data set.

Median: A measure of a number set centroid, it is the 50% value within the ascending-order-sorted number set.

Mode: A histogram-dependent measure of a number set centroid, it is the bin with the most counts.

---

9 random values	9 values sorted	8 values sorted	8 other values
37.6	30.0	40.1	40.1
33.3	30.5	42.7	42.7
30.6	30.6	43.0	43.0
32.5	31.5	44.7 ⎫ The median	43.8 ⎫ The median
35.0 ← Not the median	32.5 ← The median	44.7 ⎭	44.7 ⎭
31.5	33.3	48.3	48.3
30.0	35.0	48.3	48.3
30.5	36.1	49.6	49.6
36.1	37.6		

Figure 4.1 Number sets with easy and challenging median definition. The first two columns have the same nine values, one unsorted and the other sorted, with the median noted. The third and fourth columns have eight values, already sorted, with entries in the fourth and fifth locations noted as the median.

When the number set consists of real values, or the integers are spread apart so that repetitions of exactly the same value are rare or nonexistent, then another step must occur to identify the mode—a histogram should be created. This is often the case with geoscience data sets. Relying on a histogram for the centroid value, however, makes it a subjective determination. That is, when a histogram is employed to determine the mode, then the mode is susceptible to the bin size and also to the exact break points between the bins. Take, for example, the histograms in Figure 4.2. The histograms in each panel are of exactly the same data. The mode in the top panel is clearly the 10-to-11.5 bin; it is the bin with the highest count value. The difference between the first and the second panel is that the bin width is slightly larger in the latter graph. The change of bin width shifted the values so that now two bins are the mode—both the 8.5-to-11 bin and the 11-to-13.5 bin—as they have exactly the same highest count value. In the third panel, the bins were reduced in width relative to the first plot. In this case, the mode shifted quite dramatically up to the 12-to-13 bin. Remember, all of these histograms in Figure 4.2 are created from the same number set, and all of these values are valid mode determinations that could be reported as the centroid. That is, unlike the median, the mode of a real-valued number set is entirely determined by the format of the histogram.

Reliance on a histogram for the determination of the centroid of a distribution should only be done if the peak of the distribution is a critical factor to the analysis. This is sometimes the case; it is the basis of the principle of maximum likelihood to be used in Chapters 5 and 7. When the distribution has a clear central peak, then changing the histogram will not shift the mode by more than a bin width. When the distribution does not have a clear central peak, like the example in Figure 4.2, then the mode might not be the best choice for the centroid.

> **Quick and Easy for Section 4.1.2**
>
> The centroid values of median and mode require no calculations, they are found by sorting or binning the number set.

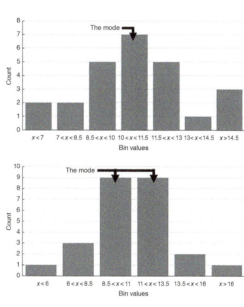

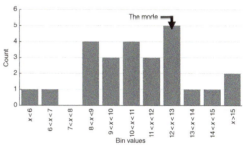

Figure 4.2 Histograms of a number set, created with slightly different bin widths and bin edge values. The mode in each histogram is marked.

4.1.3 Histograms and the Arithmetic Mean

Before moving on to measures of spread, it is useful to connect the arithmetic mean with the histogram of a data set. Specifically, an estimate of the mean can be calculated from the information in the histogram. All of the data within a particular bin, k, can be assigned the value at the bin center, x_k, and, then using the count n_k within each bin, an arithmetic mean can be created from this slightly revised set of values:

$$\bar{x} = \frac{1}{N}\sum_{k=1}^{M} x_k n_k \quad (4.9)$$

The M at the top of the summation is the number of bins in the histogram. Depending on the width of the bins and the placement of the exact values within each bin, this estimate given by Equation (4.9) could be quite accurate or rather rough. Equation (4.9) can be slightly rewritten by moving the number of values, N, inside the summation, making a new variable $F_k = n_k/N$, the fraction of the total number set within bin k:

$$\bar{x} = \sum_{k=1}^{M} x_k F_k \quad (4.10)$$

The summation over only F_k would be one, as the n_k bin counts add up to the total number set count, N. The inclusion of x_k within the summation weights each of the counts by its value along the x axis, therefore making this equivalent to an average of the number set.

The limit as $M \to \infty$ means that the bin width becomes infinitesimally small. This changes the summation into an integral, and the fraction F_k becomes a distribution function, $f(x)$. This function is known as the limiting distribution of the population, and it has the same feature as F_k in that its integral over all x is one:

$$\int_{-\infty}^{+\infty} f(x)dx = 1 \quad (4.11)$$

The inclusion of x within the integrand converts this equation into the mean of the population:

$$\bar{\bar{x}} = \int_{-\infty}^{+\infty} x \cdot f(x) dx \quad (4.12)$$

Equation (4.12) is not applicable to a number set with a finite N, but rather represents the idealized case of being able to accurately and completely sample the full population. It is useful, though, when comparing a number set against an analytical functional form, as will be done later in this chapter, because it gives the average value of the analytical function.

> **Quick and Easy for Section 4.1.3**
>
> An estimate of the arithmetic mean can be obtained from the histogram.

4.2 Quantitatively Describing a Data Set: Spread

In considering the figures in Section 4.1, it is seen that understanding the centroid of the distribution is only part of the quantitative description that could be determined. The other big piece of information that must accompany the estimate of the centroid is an estimate of the spread of the values around this centroid. There are several common ways to calculate spread, and these are often paired with particular centroid values.

4.2.1 Measures of Spread: Standard Deviation and Mean Absolute Difference

The first, the one that matches well with the mean, is the **standard deviation**, σ. In this chapter, where there is only one data set being considered, it is fine to simply use σ, but this symbol is often given a subscript to denote which data set is being analyzed and even which formula is being used. The general formula for standard deviation is this:

$$\sigma_x = \sqrt{\frac{1}{N}\sum_{i=1}^{N}(x_i - \bar{x})^2} \quad (4.13)$$

Standard deviation (σ_x): A measure of a number set spread, it is the square root of the arithmetic average of the squares of the difference of the individual data values with the number set mean.

Equation (4.13) is similar to Equation (4.2) but with additional operations included. The procedure conducted in Equation (4.13) even gets its own name—**root mean square**, or RMS—because of its functional form.

Why does standard deviation have these extra operations? Why not simply use the **discrepancy**, defined as the difference of two values. Let's consider what would happen without the square and square root operations, instead trying to define a measure of spread with the difference of the points to the average? In this case, a first thought might be to define a "mean discrepancy" like this:

$$\text{Mean discrepancy} = \frac{1}{N}\sum_{i=1}^{N}(x_i - \bar{x}) \quad (4.14)$$

The term inside the summation of Equation 4.14) can be separated into two summations, like this:

$$\text{Mean discrepancy} = \frac{1}{N}\sum_{i=1}^{N}x_i - \frac{1}{N}\sum_{i=1}^{N}\bar{x} \quad (4.15)$$

The first summation in Equation (4.15) is identical to Equation (4.2), so it can be replaced by $\bar{x}$. The term inside the second summation is independent of i, so it is simply $N\bar{x}$, and then this N cancels the N in the denominator. This means both the first and second terms are $\bar{x}$ and our definition of mean discrepancy is, *always*, zero. This is because, by definition, the mean value is located so that there is a summed discrepancy of all numbers in the set below it equal to the summed discrepancy of all numbers above it.

In order to avoid this pitfall, the discrepancy term must be transformed to be always positive. The two most basic operations to achieve this are to either square the discrepancy or take its absolute value. The first option, squaring the discrepancy, is what is done for standard deviation. The second option provides us with another measure of spread, the **mean absolute difference**, or MAD:

$$\text{MAD} = \frac{1}{N}\sum_{i=1}^{N}|x_i - \bar{x}| \quad (4.16)$$

This is a less commonly used measure of spread than standard deviation, but still a perfectly valid measure.

In general, MAD will be smaller than σ_x because the square operator exaggerates the influence of the outlier values that are far from the mean. To illustrate this, consider the set of numbers 9, 10, and 20. The mean of this set is 13. Inserting this value into our formulas for the standard deviation and mean absolute difference, Equations (4.13) and (4.16), these two measures of spread are 4.97 and 4.67, respectively. To one significant digit, these two are the same, but to more significant figures, you can see that σ_x is larger than MAD because of the amplified influence of the outlier point. Let's add another point to the example number set: 11. With this new value, the mean is now 12.5 and the two measures of spread become 4.47 and 3.75. With the inclusion of another data value near the mean, the gap between σ_x and MAD increased. Still, to one significant digit, the two values of

---

Root mean square: A simple distillation of the calculation procedure for the standard deviation.

Discrepancy: The difference between two values, usually between an individual data value and its mean, or between the means of two number sets.

Mean absolute difference (MAD): A measure of a number set spread, it is the arithmetic average of the absolute value of the difference of the individual data values with the number set mean.

spread are the same. We would have to add a fifth number, 12, which changes the mean value to 12.4, in order to get σ_x and MAD to be 3.93 and 3.04, which would round to 4 and 3, respectively, and therefore be reported as different values.

A highly useful trait of both MAD and σ_x is that they have the same units as the base value. As a measure of spread with the same dimensions, they are prime candidates for use as the uncertainty on the values in the number set. Which of these would be better to use as an uncertainty? There are reasons in favor of using either one, but most people will pick the standard deviation.

This is because assuming that the distribution of points is roughly Gaussian, then the standard deviation has a particular meaning in terms of probabilities of the closeness of other values to it. Specifically, remembering back to the definition of the normal distribution in Chapter 1, the standard deviation spread around the arithmetic mean centroid estimate will contain approximately two-thirds (more specifically, 68%) of the numbers in the set. Stated the other way around, an interval of ± 1 σ_x around any value of the number set has a two-thirds probability of including the mean. This relationship to specific probabilities has led to the adoption of standard deviation as the default measure of spread in most natural science fields and is often reported as the uncertainty for all of the individual numbers in a data set.

MAD could also be adopted as the uncertainty for the individual numbers in the set. There is nothing particularly magical about 68% as the ultimate definition of uncertainty. The main driver is the prolific existence of the Gaussian distribution in nature and the easy multiplier of ± 1 in front of the standard deviation. Why not an interval expected to contain 50% of the values? In this case, the multiplier would be ± 0.67. For a large number set, MAD will typically be about 80% of σ_x, so we could adopt ± 0.8 as our multiplier. Any of these choices are fine. The crucial step is to report which of these choices is being adopted so that others know how to interpret this choice and, if needed, say, for consistency with other uncertainties, convert to a different choice.

These two measures of spread are not the only ones in use in the natural sciences. Dropping the outermost square-root operator from Equation (4.13) converts standard deviation into another famous measure of spread known as the **variance**:

$$\sigma_x^2 = \frac{1}{N}\sum_{i=1}^{N}(x_i - \bar{x})^2 \quad (4.17)$$

Note the resemblance of Equation (4.17) to that for the arithmetic mean, Equation (4.2). This is not by coincidence, this is by design—the variance is defined as the arithmetic average of the squares of the discrepancies:

$$\sigma_x^2 = \overline{(x_i - \bar{x})^2} \quad (4.18)$$

The variance is less useful than standard deviation as a direct measure of spread because it is not in the same units as the base value. It is, however, used extensively in statistical calculations of data sets and this functional form of spread will appear several times in later sections and chapters.

Quick and Easy for Section 4.2.1

Standard deviation and mean absolute difference are measures of spread that pair well with mean as the centroid.

4.2.2 Another Measure of Spread: Quantiles

Yet another approach to determining the spread of a number set follows the procedure

---

Variance (σ_x^2): A measure of a number set spread, it is the square of the standard deviation.

---

used to determine median in Section 4.1.2. Specifically, after sorting all of the numbers in the set in ascending order, it is a straightforward task to delineate values within the full range for which certain percentages of the numbers are found below it. That is, the values represent specific quantiles of the distribution. Another word for this that is used quite often is **percentile**. The quantile/percentile value is equivalent to the cumulative probability distribution value; it is the fraction of the total number set below that particular value.

As an example of this, quantile values for each standard deviation of the Gaussian distribution are shown in Figure 4.3. The percentage of the total number set within each specific σ interval is listed along the bottom, while the quantiles are given across the upper portion of the figure. For a Gaussian distribution, with a known functional form, this is readily calculated.

Quantile is the generic word, but if you are choosing 25%, 50%, or 75%, then the word changes to **quartile**, and if you are finding the value for 10%, 20%, etc., then the word switches to **decile**. These are the two most often used special designations, but the term quantile applies to any selected percentage within the sorted value set.

All points can be assigned an exact quantile at which they are located. There are several ways to do this, but the two most common ways are shown in Figure 4.4. The values ($N = 25$) are randomly distributed in a Gaussian around a mean of 40. They have been sorted in ascending order and assigned an index number, i. The two typical methods are **exclusive percentiles** and

---

Quartile: The quantile values at the specific locations of 25%, 50%, and 75% within the rank-order listing.

Decile: The quantile values at the specific locations of every 10% within the rank-order listing.

Percentile: Another name for quantile, choosing a value from a sorted number set at a particular percentage location within the rank-order listing.

Exclusive percentiles: A method of converting a sorted number set into percent values that assigns a percent at the center of the number-specific interval.

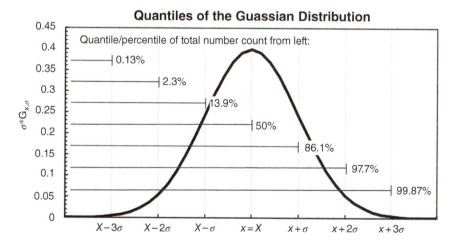

Figure 4.3 A listing of the quantile values at several standard deviation increments from the mean of the Gaussian distribution.

Value	4.6	20.5	30.7	31.8	32.7	32.9	35.3	35.6	36.6	40.0	40.2	43.9	44.9	45.1	45.3	46.8	50.9	52.3	53.2	53.3	53.4	56.0	64.0	65.8	71.2
Index	1	2	3	4	5	6	7	8	9	10	11	12	13	14	15	16	17	18	19	20	21	22	23	24	25
$Q_{i,exc}$	2	6	10	14	18	22	26	30	34	38	42	46	50	54	58	62	66	70	74	78	82	86	90	94	98
$Q_{i,inc}$	0	4.2	8.3	12.5	16.7	20.8	25	29.2	33.3	37.5	41.7	45.8	50	54.2	58.3	62.5	66.7	70.8	75	79.2	83.3	87.5	91.7	95.8	100

Figure 4.4 Alignment of percentages with the sorted values of a number set, for both the exclusive and inclusive conversion methods.

inclusive percentiles. In the exclusive case, the formula for determining the quantiles at each of the numbers is given by this:

$$Q_{i,exc} = \frac{i-0.5}{N} \cdot 100\% \qquad (4.19)$$

In the exclusive case, each number occupies a "space" along 0%–100% (in this case, with $N = 25$, a space of 4%), and Equation (4.19) places the value at the center of that space. This means that the first and last values are located at 2% and 98%, respectively, as seen in the third row of Figure 4.4. The inclusive conversion to quantiles places the first and last values at 0% and 100%, in which case we have the formula:

$$Q_{i,inc} = \frac{i-1}{N-1} \cdot 100\% \qquad (4.20)$$

Because the 0% and 100% ends of the percentile range are used, both the numerator and denominator need to be reduced by one to give the proper percentile of the values in between these endpoints.

The example in Figure 4.4 shows that, for $N = 25$, the values are only different by 2% at most. For very small number sets of 10 or less, the difference between inclusive and exclusive could be substantial. For large number sets, though, this difference becomes tiny and is often negligible for most uses of percentiles.

It will often be the case that a chosen quantile value does not specifically align with a number in the set. In this case, there are three commonly used methods for determining the value for this quantile. The first is to use the value nearest to the desired quantile. The quantile value is then identical to one of the numbers in the set, which is perhaps a useful feature for the analysis. In the example shown in Figure 4.4, a quantile choice of 15% would fall within the bin of the fourth sorted number in the set, in this case 31.8. Similarly, the second option is to use the value closest to the median value of the set. This systematically "rounds down" with "down" being toward the middle of the distribution. This option is useful when the distribution is sparse, preventing a rounding that exaggerates the distance of the quantile from the median. In our example in Figure 4.4, the 15% quantile would then be assigned to the fifth sorted entry in the number set, which is 32.7. The third option is perhaps the most frequently used—calculating a linear interpolation between the center quantiles of the nearest two numbers in the set. The general formula for **linear interpolation** looks like this:

$$x_{Qchosen} = (1-\beta)x_i + \beta x_{i+1} \qquad (4.21)$$

In Equation (4.21), $x_{Qchosen}$ is the value for the desired chosen quantile of interest and x_i and x_{i+1} are the two closest numbers in the ordered list to this quantile. The coefficient β in Equation (4.21) is the fractional distance of the

---

Inclusive percentiles: A method of converting a sorted number set into percent values that assigns the first and last values to 0% and 100%, with the others evenly distributed between.

Linear interpolation: A simple method for determining a location between two given values, based on the fractional distance from each value.

chosen quantile to the quantile value of the x_i number in the list, given by this expression:

$$\beta = \frac{Q_{\text{chosen}} - Q_i}{Q_{i+1} - Q_i} \quad (4.22)$$

The Q values in Equation (4.22) are quantiles, either the chosen quantile or the quantiles at the center of the bin for the two numbers neighboring the chosen quantile. If the chosen quantile is either below the first bin center or above the last bin center, then Equation (4.22) becomes a linear extrapolation using the first two or last two numbers in the list, respectively. Note that if the chosen quantile falls between two repeated numbers in the list, then the two x values in Equation (4.21) are equal, so β will cancel from the equation and $x_{Q\text{chosen}}$ is the repeated value. For our example in Figure 4.4, the 15% quantile is between the fourth and fifth entries, so Equation (4.22) yields a β value of 0.25 and an $x_{15\%}$ of 32.0, reported here to the same number of significant figures as the original data set.

As you can see, when there are only a few numbers in the set, then these three options could yield rather different values for the chosen quantile. When N is large, though, the numbers are usually very close to each other and the shifts in quantile value between the three methods shrink to small differences, eventually becoming negligible on the level of significant figures of the number set.

This is useful as a measure of spread because any quantile can be chosen, or, more commonly, a difference between two quantiles. A common choice is the difference between the 75% and 25% quantiles, known as the **interquartile range**, or IQR. The IQR contains 50% of the numbers in the set. If the numbers in the set form a Gaussian distribution, then the IQR is equivalent to choosing values of $\bar{x} \pm 0.67\sigma_x$. Actually, the IQR is the difference of these $\pm$ values around the mean, so IQR is equal to $1.34\sigma_x$. So, using half of the IQR as a measure of uncertainty typically yields a number that is smaller than the standard deviation, while using the full IQR as the uncertainty is typically larger than if you had used the standard deviation. The qualifier "typically" is included in the previous sentence because this comparison assumes that the number set is close to a Gaussian distribution. Any of these options is fine as a definition of uncertainty, but please state which option is being used.

Remember back to Chapter 2 when the box plot was introduced. Nearly all box plots are defined using quantiles, not the mean and standard deviation. A box plot can be made for a Gaussian distribution, of course. Because the IQR has, by definition, 50% of the numbers within it, the box is slightly smaller than a $\pm 1\sigma$ interval. For a Gaussian distribution, the box would end at $\pm 0.67\sigma$, typical whiskers of 2.5%–97.5% extent would end at $\pm 1.96\sigma$. That is, based on this whisker definition, the whiskers should extend 1.29σ beyond the end of the box, while the box itself is a nearly equal length of 1.34σ. With this definition, each of these elements—the lower whisker, the box, the upper whisker—is essentially the same length for a Gaussian.

If, instead, the whiskers were defined to have 5%–95% extent, a Gaussian distribution would have them end at $\pm 1.65\sigma$. This is 0.98σ beyond the extent of the box. That is, in this case, the whiskers represent a standard deviation width if the distribution is Gaussian.

Interquartile range: A measure of a number set spread, it is the difference between the 75% and 25% quartile values within the rank-order listing.

Quick and Easy for Section 4.2.2

The interquartile range is a measure of spread that pairs well with median as the centroid.

4.2.3 Spread Via Full Width at Half Maximum

Yet another method of estimating spread is based on the mode. This is the "half maximum" rule, finding the width of the distribution at the location where the amplitude of the histogram drops to 50% of the peak value. It is very common in some disciplines and essentially never used in others. Regardless of whether your chosen field uses these values, it is good to be aware of them and how they are found.

There are two common measures of spread that use this technique. By far the more common is the **full width at half maximum**, FWHM, defined as shown in Figure 4.5. For FWHM, the amplitude level that is 50% of the peak should be determined in both directions away from the mode. For the example in the diagram, the FWHM spans from 91 to 107, so the value is 16. The other is the **half width at half maximum**, HWHM, usually taken in the positive direction away from the mode. The definition of HWHM is ambiguous, though, because it could be defined in two other ways, either as the width to the 50% amplitude x value below the mode or as half of the FWHM value. In the example shown in Figure 4.5, these three options yield different numbers: the HWHM found in the positive direction from the mode is 5, the HWHM in the negative direction is 11, and the final option of using half of the FWHM yields 8. All of these are possibilities and so it is always good to list the calculation method used to determine HWHM whenever that value is used in the analysis.

Note that this method of finding the spread only works for smooth distributions that

---

Full width at half maximum: A measure of a number set spread, it is the difference between the histogram locations, moving outward from the mode, where the bin counts are half of the mode value.

Half width at half maximum: A measure of a number set spread, it might have one of the three definitions: the difference of the mode value with the bin location with half as many counts moving left from the mode; the same thing but found moving right from the mode, or half of the full width half maximum value.

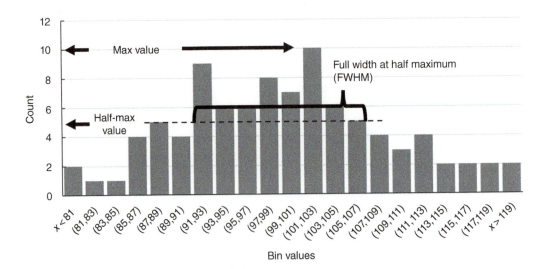

Figure 4.5 FWHM determined from a number set histogram.

monotonically decrease from a single peak, because multiple peaks can lead to ambiguity about which width to choose as the spread estimate. The distribution does not need to be symmetric, but it should be decreasing away from the mode in both directions. The exception to the monotonicity requirement is when the amplitude of the secondary peak is small. Far from the main peak, when the histogram bin counts are much less than 50% of the peak value, then local maxima in the distribution do not contribute to the selection of the FWHM spread value.

There is a further caveat to using either FWHM or HWHM as the spread estimate for a number set—they are values determined from the histogram, and therefore subject to change given a different histogram creation process. Choosing an optimal number of bins for the histogram is important. Too few bins will obscure non-Gaussian features of the number set, while too many bins will reduce the counts per bin to the level of noise. If you are using mode for the centroid and would like to pair it with one of these spread estimates, then it is strongly advised to make several histograms and iterate to a reasonable value for these quantities.

There is no set proportion of the number set contained within the FWHM spread value. If the distribution is close to a Gaussian function, then we can use this distribution to determine estimates for FWHM and HWHM. The height of the mode of the Gaussian distribution, located at $x = X$, is given by this formula:

$$y_{\text{Mode,G}} = \frac{N}{\sigma\sqrt{2\pi}} \quad (4.23)$$

The inclusion of N in the numerator of Equation (4.23) converts the normalized Gaussian height into the count value at the peak, which could be compared against the amplitude of the mode from the number set, if desired. The x value at which the Gaussian distribution reaches half of this peak value, x_{HM}, can be found with a bit of algebra:

$$x_{\text{HM}} = X \pm \sigma\sqrt{2\ln(2)} = X \pm 1.18\sigma \quad (4.24)$$

Equation (4.24) shows that the HWHM of a Gaussian is quite close to one standard deviation away from the Gaussian mean. The FWHM is, therefore, 2.36σ. If the data set is similar to a Gaussian, then percentages can be found for the proportion of values within certain ranges based on these measures of spread. Specifically, each HWHM contains 38% of the distribution and the FWHM contains 76% of the values, leaving 12% in each of the two tails of the distribution beyond this interval. In summary, to repeat the guideline: either HWHM or FWHM are legitimate values to report as the spread of the data set, just be sure to report how it was found.

> **Quick and Easy for Section 4.2.3**
>
> The full or half width at half maximum are measures of spread that pair well with mode as the centroid.

4.2.4 Spread as an L-p Norm

In Section 4.2.1, we found σ_x and MAD as two measures of spread based on the difference of the values against the mean of the numbers in the set. The first has a "power" integer of two, square then square root, while the second has a power of one. There is nothing special about these two formulas and, really, any integer power p can be applied as the polynomial power and the radical root power. For example, setting $p = 3$, we get:

$$\sigma_x^{p=3} = \sqrt[3]{\frac{1}{N}\sum_{i=1}^{N}\left(|x_i - \bar{x}|\right)^3} \quad (4.25)$$

Note that the absolute value operator is needed for all odd values of p, otherwise the summation will partially or fully cancel out. With the cube operator applied to the difference quantities, Equation (4.25) puts even more emphasis on the outliers within the set than does σ_x. For the above-mentioned example of the three-valued set of 9, 10, and 20, Equation (4.25) yields 5.24, larger than the spread estimates from Equations (4.13) and (4.16).

Data Analysis for the Geosciences

For each increment of p, the calculation increases the weighting in favor of the measurements farthest from the mean. Taking this notion to its extreme, the limit as $p \to \infty$ can be written like this:

$$\sigma_x^{p \to \infty} = \sqrt[\infty]{\frac{1}{N}\sum_{i=1}^{N}\left(|x_i - \bar{x}|\right)^{\infty}} \qquad (4.26)$$

The infinity power and root gives infinite weighting to the one value farthest from the mean, so Equation (4.26) can be rewritten as the maximum difference with the mean:

$$\sigma_x^{p \to \infty} = \max\left(|x_i - \bar{x}| \forall i\right) \qquad (4.27)$$

For our example with the mean at 13 and the farthest value from this being 20, σ_x^{∞} is 7. With a larger N and a Gaussian distribution to the values around the mean, this would be half of the full range of the number set.

All three of these formulas are part of a family of equations in which p is arbitrary. The **L-p norm** functional form refers to the power-sum-root combination of a function z, written in general form like this:

$$\overline{|z_i|^p} = \sqrt[p]{\frac{1}{N}\sum_{i=1}^{N}\left(|z_i|\right)^{p}} \qquad (4.28)$$

In Equation (4.28), the absolute value operation is needed for the odd values of p and does not change anything for the even p values, so it is included for completeness and generality. For calculating spread around a mean value, $z_i = x_i - \bar{x}$. With this definition, using $p = 1$ yields MAD and $p = 2$ yields σ. That is, we could refer to MAD as the "L1 norm spread" and σ_x as the "L2 norm spread."

L-p norm: A family of formula in which the arithmetic mean of a particular quantity—the absolute value of the difference between the individual data value and the number set mean raised to the power p—is then raised to the power $1/p$.

> **Quick and Easy for Section 4.2.4**
>
> Standard deviation and mean absolute difference are members of a family of spread calculations known as L-p norms.

4.2.5 Sample Versus Population

There is a caveat to all of the L-p norm calculations of spread, including standard deviation and mean absolute difference. There is an alternative form for them. Consider the standard deviation formula, Equation (4.13), from Section 4.2.1. Another version of this formula is written as follows:

$$\sigma_x = \sqrt{\frac{1}{N-1}\sum_{i=1}^{N}(x_i - \bar{x})^2} \qquad (4.29)$$

The only difference between Equation (4.13) and Equation (4.29) is the extra minus one (-1) in the denominator. Specifically, this extra term is the **degree of freedom**, d, of the number set. There is one degree of freedom, the count of numbers in the set, N, for a single set of numbers like this.

These two forms of the standard deviation have specific names. The first, Equation (4.13), is known as the population standard deviation and is often written as σ_p. The second, Equation (4.29), is known as the sample standard deviation, designated as σ_s or even as s instead of σ. In practice, this difference rarely matters, especially for large N. This is seen by taking the ratio of the two standard deviation formulas:

$$\frac{\sigma_s}{\sigma_p} = \sqrt{\frac{N}{N-1}} \qquad (4.30)$$

If $N = 10$, then Equation (4.30) yields a ratio value of 1.054. That is, the two standard

---

Degrees of freedom: The number of independent parameters that could influence the result of a calculation.

deviation values will be 5% off from each other. For $N = 11$, this drops to 1.048. Because these spread calculations are often used as uncertainties, they are usually rounded to a single significant digit. With a relative difference of less than 5%, the extra minus one (-1) in Equation (4.29) will have no influence on the first significant digit of the resulting standard deviation. That is, if you have more than 10 numbers in the data set, then the formulas are essentially identical for the purposes of reporting uncertainty. If you have 10 or fewer values in the number set, then Equation (4.29) is the proper form of standard deviation because the influence of the degrees of freedom could be important.

There are instances when several significant figures are needed on standard deviation, for a specific calculation in which such accuracy is warranted. One of these is calculating an exact probability of getting a particular value in the number set, a calculation to be conducted later in this chapter. Even for this, if the distribution is not exactly a Gaussian, then the specific probabilities are estimates and the calculation of such probabilities with means and standard deviations known to several significant figures probably is not justified. That is, usually, such precision is not needed for a measure of uncertainty, and therefore the extra minus one (-1) in the denominator can be ignored.

> **Quick and Easy for Section 4.2.5**
>
> Technically, there is a difference between population and sample spread formulas, but the correction is usually negligible.

4.3 Random and Systematic Error of a Data Set

The above-mentioned discussion of using spread as the uncertainty on the values is incomplete. In fact, this is only one of the possible uncertainties that should be adopted for the values. Our discussion in Chapter 1 of different types of uncertainties discussed systematic and random uncertainties in data values.

The spread, as calculated in Section 4.2, is one kind of random uncertainty. It might encompass all of the random uncertainty, but there could be additional random uncertainty due to the measurement technique. That is, each value in the number set already has an uncertainty from the way in which it was obtained, and then this spread around the mean is an additional uncertainty due to the randomness of the quantity being measured.

The other uncertainty that could be added in is a systematic uncertainty. This is related to the uncertainty of the measurement technique but results in an offset of the values away from the true value. This could be due to some bias in the instrumentation, if the instrumentation is not calibrated accurately. It could be due to processing in the recording of the measurement, for instance, due to a person misreading a dial or digital reading if it is a manual entry of an observed event, or a software error. Unless you are able to make the measurement with another instrument, you probably will not know that the systematic offset exists in the values.

There are three other usual ways to identify systematic uncertainty. The first is to use several measurement techniques, devices, or experimenters taking the data. Calculating the means and standard deviations of these separate and hopefully independent sets of measurements should allow you to identify systematic differences between them. The second is to compare against a well-known theoretical value, or against a computed value from a relevant analytical formula. The third is to use the measurement device with a "known source," something that has been observed many times by others and therefore provides a community-accepted standard against which your sensor can be calibrated. Regarding conducting this comparison, we will come back to this in the next chapter, which has a lengthy discussion on testing two number sets against each other.

A famous geoscience example of systematic error differences is with our measurements of total solar irradiance. The research community has settled on a centroid value of 1361 W m^{-2}. As seen in Figure 4.6, the values from various spacecraft that have made this measurement span the range of 1360–1375 W m^{-2}. This is a difficult measurement to make; the telescope on the satellite is attempting to observe all photons from the Sun and then integrate over wavelength to obtain a total energy per second per unit area value. Telescopes do not capture every photon; some invariably scatter or absorb within the optical path in the telescope. There is loss of signal at the detector itself; the process of converting a photon into an electrical signal that can be digitally recorded, as a count in a particular wavelength bin is not perfect. Because the energy of the photon is a function of its wavelength, the detector must have very high spectral resolution in order to do an accurate integration. The measurement is also susceptible to spurious counts, mostly from energetic particles that penetrate through the walls of the instrument and flip a bit inside the detector. Before flight, each instrument undergoes extensive calibration testing, shining known sources into the telescope to determine its sensitivity. Even with this effort, differences arise in the final numbers. Committees of scientists then convene to deliberate which of these values is the correct one, to which the others will be scaled.

The concatenation of these three uncertainties—in the measurement itself δx_{meas}, in the random variation of the physical quantity being measured δx_{rand}, and in some systematic deviation δx_{syst}—should be done with the techniques shown in Chapter 3. In fact, it is one of the more straightforward uncertainty propagation calculations, because these should be added together. Usually, these three uncertainties are independent of each other, and therefore the quadrature summation formula should be used to arrive at the total uncertainty:

$$\delta x_{\text{total}} = \sqrt{\delta x_{\text{meas}}^2 + \delta x_{\text{rand}}^2 + \delta x_{\text{syst}}^2} \quad (4.31)$$

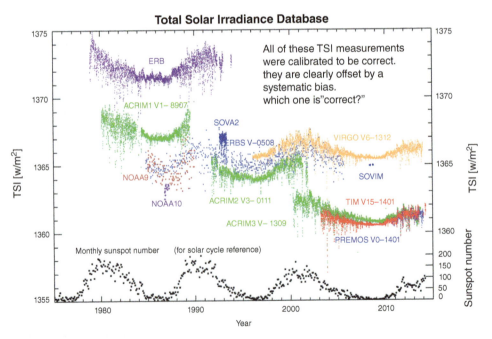

Figure 4.6 Measurements of total solar irradiance from various satellites over the last 4 decades. The differences are systematic errors due to instrument calibration. Adapted from Kopp (2014).

If, for some reason, some or all of them are correlated with each other, then straight addition should be used:

$$\delta x_{total} = \delta x_{meas} + \delta x_{rand} + \delta x_{syst} \quad (4.32)$$

Also, note that this summation of uncertainties can still be applied if the set is numerical data. Model output is also susceptible to systematic errors, uncertainties in the coefficients within the physical laws being modeled, and uncertainties due to numerical diffusion and uncertainty due to the initial and boundary condition values being applied. Any, or all, of these can arise in a single numerical model calculation, and so the output from the code, although often reported as very precise, is also accompanied by an uncertainty value, just like observed values. This summation, either linear or in quadrature, of the various uncertainties, must be applied.

How do you determine the uncertainty on a model output? Chapter 12 gives a brief overview of a process called uncertainty quantification, which is the process of determining spread on model values. One simple way is to adjust the uncertain inputs and coefficients with randomly determined perturbations. This perturbation value can be found with a random number generator, as weighted by the Gaussian distribution discussed in Chapter 1. These perturbations, which could be positive or negative, can then be added to the values used in the numerical simulation, resulting in a new model output value. This sensitivity test should be conducted enough times (dozens if not hundreds) to create a distribution of output values, from which a mean and a standard deviation can be calculated.

Quick and Easy for Section 4.3

The various types of uncertainties should be combined using the uncertainty propagation techniques discussed in Chapter 3.

4.4 Which Centroid and Spread to Use and Other Tidbits of Advice

With all of these choices for the centroid and the spread, a good question to ask yourself is which one to use for a particular situation? Like so many topics in statistics, there is no universally applicable rule. There are, however, some common guidelines to follow.

By far the most common choices for the centroid and spread of a number set are the arithmetic mean and the standard deviation. In fact, these are so common that the specific names are not even used. When you see a value for the average of a number set, it is most likely the arithmetic average that was calculated. Similarly, when an uncertainty or "margin of error" is listed for a number set, it is most likely the standard deviation.

Choosing the median and a definition of spread from the quantile method is useful when the distribution of the number set is not Gaussian. When a standard deviation is reported (or a multiplicative factor of σ_x), others might interpret this as implying that the distribution of the number set is Gaussian. Using a quantile-based definition of spread does not carry this implication of a Gaussian; in fact, it implies the opposite. We will go over this question of whether a number set follows a Gaussian distribution in great detail in the next chapter.

The choice of the mode for centroid is good when the peak is important. The mode is the histogram's "most probable value" bin, and sometimes this is an important feature to highlight about the number set. If the number set has a long tail in one direction, then the mean will be substantially influenced by the outliers in the extreme reaches of that part of the distribution. Even the median could be well away from the peak. Of the three centroids discussed, only the mode invariably reports the peak of the distribution.

Calculating an uncertainty value from the spread of a subset of measurements is useful when the observation will be repeated many times. Let's say that you are taking a humidity measurement reading at a field site. The humidity changes systematically throughout the day, plus it varies due to the local weather conditions and regional weather patterns. While it is perhaps easy to combine all of the measurements for all time, calculate the standard deviation, and use this as the uncertainty on any one observation value, this would be a big overestimation of the true uncertainty of each measurement because there are physical drivers causing a humidity change included in this uncertainty value. That is, because of the environmental changes, you should not expect all of the humidity values to be the same. This method of calculating uncertainty for any one value in the set, then, is flawed.

Instead, you want to quantify the uncertainty for any particular time against a mean of several values for that same time and location. One way to do this is to use several humidity sensors and make many measurements in a localized zone within a short time interval. You expect the humidity to be the same in this small spatial and temporal window. The sensors should be moved around so that you have multiple measurements from each sensor at each location within the local zone. You now have a sample set that should be from the same population. You can also test your equipment, comparing the means and standard deviations from each humidity sensor. Differences in sample means or standard deviations from the sensors can then be taken into account when determining the full same mean and standard deviation.

The resulting standard deviation is then a good measure of the uncertainty for any one of the measurements in the set. If you think that the sensor works essentially the same across the range of expected humidity values for the field site, then you can use this standard deviation not only at the particular time of the extra samples but also at all times throughout the day and probably over the entire time interval of the field site experiment.

> **Quick and Easy for Section 4.4**
>
> If a number set's distribution is Gaussian, then arithmetic mean and standard deviation are generally used for the centroid and spread. If the distribution is not Gaussian, the median and interquartile range are common alternatives. Mode is a good choice when the peak of the distribution is important.

4.5 Standard Deviation of the Mean

So far, the spread values introduced in Section 4.2 are applicable to a single value in the number set. That is, they quantify the randomness of the specific numbers. Sometimes, though, we want to know the uncertainty on the mean value of the number set. This is particularly useful when considering whether the mean of the samples is really the mean of the full population. In this case, it is useful to assign an uncertainty on the mean itself, not on the individual values of the set.

This is usually defined as the **standard deviation of the mean**, $\sigma_{\bar{x}}$:

$$\sigma_{\bar{x}} = \frac{\sigma_x}{\sqrt{N}} \qquad (4.33)$$

This formula is also known as **standard error** or sometimes as the standard error of the mean. The σ_x term in the numerator of Equation (4.33) is the standard deviation, found with either Equation (4.13) or Equation (4.29). When reporting x_{best} or $\bar{x}$, Equation (4.33) should be used, not standard deviation.

There is an interesting feature that distinguishes standard deviation from standard

---

Standard deviation of the mean or standard error: A measure of spread for the value of the mean of a number set, it is the standard deviation divided by the square root of the count of numbers in the set.

error. It is this: once N is large enough to reduce the statistical noise of the randomness from a small sample size, the standard deviation does not shrink with additional numbers being included in the set. For large N, adding more samples will, in general, reinforce the previously calculated standard deviation. This is expected; if you are sampling from the same population, then the spread of an individual value from that population should be the same.

With σ_x asymptotically achieving a value that is independent of N, Equation (4.33) reveals that the standard error, $\sigma_{\bar{x}}$, will continue to decrease with increasing N. That is, more samples reduce the uncertainty on the mean. This indicates that a target uncertainty can eventually be reached with the inclusion of additional numbers to the set. Equation (4.33) can, therefore, be rearranged to get a formula for the desired N:

$$N_{required} = \left(\frac{\sigma_x}{\sigma_{mean,desired}}\right)^2 \quad (4.34)$$

Given a quasi-constant standard deviation, Equation (4.34) is a quick calculation to determine the number of values needed to reach a particular standard error.

Quick and Easy for Section 4.5

The standard deviation is the variability of an individual value about the mean, while the standard error is the spread of possible mean values.

4.6 Counting Statistics

There are many times where the data being collected are discrete counts. To pick an example of this, let's say we are estimating the number of bald eagles living on a particular nature preserve. Let's say that you go out each day and tally the eagles that you see on your rounds through the grounds. Assuming that you do a thorough search through the region, you might expect the number of eagles to be the same each day. It most likely is not, though. You might miss some of the eagles, or count some of them twice as they fly around during your survey.

This type of counting with uncertainty leads to what is known as a Poisson distribution, and the related variation in counted values is called **Poisson uncertainty**. The **Poisson distribution**, $P\mu(\nu)$, quantitatively describes the probability of a particular count value, ν, given a true count value μ,

$$P_\mu(\nu) = \frac{\mu^\nu}{\nu!}e^{-\mu} \quad (4.35)$$

By definition of counting up from zero, μ is a positive integer. The count value of interest, ν, is either a positive integer as well or could be zero. The term in the denominator is the factorial function, written as the multiplication of all positive integers equal to and below the argument:

$$\nu! = \nu(\nu-1)(\nu-2)...(2)(1) \quad (4.36)$$

For small values of ν, this is readily calculated. For large ν, the factorial can become extraordinarily big. Also, note that the factorial function is equal to one when ν is zero.

Equation (4.35) has competing terms in its numerator and denominator. In the numerator, there is a number raised to a power, which can be very large. In the denominator, we have a factorial, which also can be very large. Finally, there is the exponential operator in the numerator, which is always less than one and decreases

---

Poisson uncertainty: The uncertainty associated with counting, for which the standard deviation is the square root of the counts.

Poisson distribution: The special function that yields the probability that a certain count will arise, given a true count value.

as the true count value increases. The general rule of the Poisson distribution is that the numerator dominates until ν is the same as μ, and then the denominator dominates at values above this. That is, the curve increases to a peak at $\nu = \mu$ and then decreases as $\nu \to \infty$.

Example curves from the Poisson distribution formula in Equation (4.35) are shown in Figure 4.7. Given the true count values of $\mu = 3$, 10, and 25, the probabilities for getting particular observed counts are shown to peak at these values for each curve and decrease symmetrically away from this maximum. The Poisson formula is a probability distribution curve, so the full area under each of these curves is one.

Continuing our example of counting bald eagles, let's assume that we know the true number to be 40. One day, you counted 46 eagles. You could then ask this question—was there an influx, or can I ascribe this increase to "noise" in my counting?

The Poisson distribution can be calculated for these values of $\mu = 40$ and $\nu = 46$:

$$P_{40}(46) = \frac{(40)^{46}}{46!} e^{-40} \qquad (4.37)$$

This becomes

$$P_{40}(46) = \left(4.25 \cdot 10^{-18}\right) \frac{4.95 \cdot 10^{73}}{5.50 \cdot 10^{57}} \qquad (4.38)$$

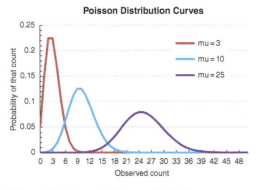

Figure 4.7 Example Poisson distribution curves, showing the probability as a function of observed count for μ = 3, 10, and 25.

The final answer is $P_{40}(46) = 0.0383$. The Poisson distribution is a powerful tool for obtaining the probability of getting a particular count value. However, if the question were asked differently requiring several such probabilities, then the calculation becomes tedious. In fact, it can approach values that are computationally impractical, as the three terms become very small or large, beyond the capabilities of a simple calculator.

A useful fact about the Poisson distribution is that, when μ is large, this function asymptotically approaches the Gaussian distribution:

$$\lim_{\mu \to \infty} P_\mu(\nu) = G_{X,\sigma}(\nu) \qquad (4.39)$$

This limit is seen in Figure 4.7 in the purple $\mu = 50$ curve, which looks very much like the Gaussian distribution.

The Poisson distribution is truncated at zero because the count can never be negative. That is, the full tail of the left side of the distribution is missing. However, when the peak of the Poisson distribution is far from zero, then the left-side tail is almost entirely at zero or above, and the terms within the Poisson distribution can be approximated as exponentials and the Poisson distribution resembles a Gaussian. If this truncation is very small, then it can be ignored, and we can use the Gaussian distribution instead of the Poisson distribution. This exchange of functional forms is highly beneficial because the Gaussian only includes exponential operators instead of powers and factorials.

In order to use this replacement of the Poisson distribution by the Gaussian distribution, we need to know the centroid and spread of the Poisson distribution. That is, we need to know X and σ for use in the Gaussian formula. It was stated that the true count value is μ. A quick derivation readily shows that this is indeed the case. The key to this proof is to realize that $P\mu(\nu)$ is a fractional value that integrates to one. In this case, Equation (4.10) can be used to find the average value, $\bar{v}$, of all possible count values, and can be written like this:

$$\bar{v} = \sum_{v=0}^{\infty} v P_\mu(v) \tag{4.40}$$

Inserting the Poisson distribution formula then gives

$$\bar{v} = \sum_{v=0}^{\infty} v \frac{\mu^v}{v!} e^{-\mu} \tag{4.41}$$

Because of the v multiplier, the first term of the summation (when $v = 0$) is zero. This leaves us with this expression:

$$\bar{v} = e^{-\mu} \sum_{v=1}^{\infty} v \frac{\mu^v}{v!} \tag{4.42}$$

This can be further rewritten so that the argument is a function of $v - 1$:

$$\bar{v} = \mu e^{-\mu} \sum_{v=1}^{\infty} \frac{\mu^{v-1}}{(v-1)!} \tag{4.43}$$

The summation term in Equation (4.43) should then be written out,

$$\sum_{v=1}^{\infty} \frac{\mu^{v-1}}{(v-1)!} = 1 + \mu + \frac{\mu^2}{2!} + \frac{\mu^3}{3!} + \cdots \tag{4.44}$$

The infinite series in Equation (4.44) is equal to e_μ, which can be substituted back into Equation (4.43) to yield $\bar{v} = \mu$. That is, calling μ the true value is correct because it is the expected value from all possible v options.

The other critical value that needs to be derived from the Poisson distribution is the uncertainty around any particular count value. It is useful to start from the formula for variance, Equation (4.17), with v in place of x:

$$\sigma_v^2 = \frac{1}{N} \sum_{i=1}^{N} (v_i - \bar{v})^2 \tag{4.45}$$

Remembering Equation (4.18), there is a convenient mathematical identity to apply here:

$$\overline{(v_i - \bar{v})^2} = \overline{v^2} - (\bar{v})^2 \tag{4.46}$$

The second term on the right-hand side of Equation (4.46) is μ^2. The first term on the right-hand side requires another mathematical identity:

$$\overline{v^2} = \mu^2 + \mu \tag{4.47}$$

Substituting in the μ versions of the two terms in Equation (4.46) yields $\sigma_v^2 = \mu$. Taking the square root of this leaves us with the standard deviation of a counted value, $\sigma_v = \sqrt{\mu}$.

This is an important point to emphasize: Poisson statistics of a counted values gives the standard deviation as the square root of the true count value. The absolute uncertainty of the count value increases nonlinearly with μ. The fractional uncertainty, found by dividing the absolute uncertainty by the base value, is therefore $1/\sqrt{\mu}$. That is, the fractional uncertainty will decrease as μ increases. This is a foundational aspect of Poisson statistics—increasing the collection time to increase the counts will improve the **signal-to-noise ratio**, a quantity that is the inverse of the fractional uncertainty.

Now that we have values for both the centroid and spread of the Poisson distribution, we can apply these for the Gaussian distribution for our example of counting bald eagles. Let's say that the question you ask yourself is this—what is the chance that I would see 46 or more bald eagles, given a true population of 40?

To answer this, we need to calculate a z score and convert the resulting score into a probability. The probability associated with a z score is based on an underlying Gaussian distribution, and it requires the mean and standard deviations of that population. For the eagle count, we have $X = 40$ and $\sigma = \sqrt{40} = 6.3 \approx 6$. Plugging this into the formula for a z score yields this:

$$z = \frac{|40 - 46|}{6} \tag{4.48}$$

Signal-to-noise ratio: The base value divided by its uncertainty (the inverse of fractional uncertainty), this quantity improves with more counts due to Poisson statistics.

This leaves $z = 1$. The probability associated with this is a one-sided integral of the Gaussian probability distribution function, yielding $P = 0.16$. This is well above the 0.05 threshold for statistical significance, so it can be declared that this is most likely not a real increase and instead a result of the process of counting the eagles.

A final point to make about Poisson counting statistics is that, sometimes, there is a background signal to our counts. This is the case, for example, in measuring the aurora of Jupiter with the Hubble Space Telescope (HST). The process of counting photons received in the telescope is subject to other sources of a signal in the detector. Specifically for the HST, there are galactic cosmic rays that can penetrate the outer walls of the telescope and pass right through the detector plate, often a highly sensitive charge-coupled device (CCD) chip. We'll come back to galactic cosmic rays in the next section, but for now, know that they create a rather uniform background to the photon count from the CCD. For the HST, they quantify this by closing the shutter door to block out all photons and record the counts still occurring in the CCD. This noise level can now be removed from the photon counts when the shutter door is open and the HST is recording measurements of scientific interest, such as images of the auroral oval at Jupiter.

To remove this background noise from the physical source of counts, the trick is to convert everything into count rates. The background value might have been found over a very long interval (several orbits), while the measurement of Jupiter might be conducted for only minutes during one orbit. The background count total over those several days could be large, but the count rate—counts per unit time—is of unknown size relative to the desired source signal. To directly compare them, both should be converted to a standard time interval. This interval could be anything (seconds, minutes, hours, days, years), whatever is most convenient for the problem. The isolation of the physical source is then a matter of subtracting rates:

$$R_{source} = R_{total} - R_{background} \qquad (4.49)$$

It is not the counts but the rates that should be subtracted, with all rates given as counts within the same unit time interval. To find the uncertainty on R_{source}, the uncertainties of the other two rates must be found by propagating the original Poisson counting uncertainties of the original count values. Note that the final step of the combination of the two rate uncertainties should be additive and not by quadrature. This is because the uncertainty of $R_{background}$ is included within R_{total}, so the two uncertainties are definitely not independent.

Quick and Easy for Section 4.6

Counts usually follow Poisson statistics, which adds a square root of N uncertainty to other sources of uncertainty.

4.7 Example: Galactic Cosmic Rays

Galactic cosmic rays (GCRs) pose an excellent example for illustrating the usefulness of counting statistics. GCRs are nuclei of atoms, from which all of the electrons have been stripped, accelerated far away in deep space by colliding magnetic structures or supernova shock fronts. They reach velocities up to nearly the speed of light, so fast that relativity needs to be taken into account. At these relativistic speeds, they enter our solar system and spiral along the interplanetary magnetic field. Actually, this is a double spiral motion. Electrically charged particles travel in a helix around the

---

Galactic cosmic ray: An extremely high energy nucleus of an atom, often from beyond our solar system, accelerated across astrophysical scales to close to the speed of light.

local magnetic field. The interplanetary magnetic field is connected to the Sun at one end, which is rotating once every 27 days or so. The interplanetary magnetic field is carried by the solar wind out from the Sun through the solar system at a radial velocity of hundreds of kilometers per second, with a typical speed of around 400 km/s. While this appears to be a very fast number, it is not relativistic (the speed of light is 300 000 km/s) and the solar wind takes roughly 4 days to travel one astronomical unit (AU), the 150-million-kilometer distance from the Sun to the Earth. The edge of the solar system is 100–200 AU, which means that the solar wind takes a year or two to reach this outer boundary. All that time, the Sun is rotating, and therefore the interplanetary magnetic field is being wrapped up one or two dozen times in a spiral configuration around the Sun. GCRs will follow the magnetic field, and this is the second spiral motion that GCRs go through our solar system. They are gyrating around the magnetic field and spiraling in circles around the Sun, eventually getting very close to the Sun before being deflected and sent back out along the magnetic field spiral to deep space again.

Some of these incoming GCRs are reflected off perturbations in the solar wind and interplanetary magnetic field. When the Sun is at "solar minimum," its magnetic field is well structured, and the interplanetary magnetic field is rather smooth. This allows for easy transit of GCRs into the inner solar system. When the Sun is very active, however, its magnetic field is rather complicated, and the interplanetary magnetic field is able to reflect a higher proportion of the incoming GCRs.

There are always some GCRs that will make it to one of the planets. Our local magnetic field generated from inside the Earth protects us somewhat from these incoming GCRs, creating a higher flux of GCRs near the magnetic poles than at the magnetic equator. The real protection is from our atmosphere. Most GCRs have enough energy to zip down to the lower stratosphere or even the upper troposphere before reaching an atmospheric density thick enough to stop it. That is, they are stopped at an altitude of 5–20 km, well above the surface. Essentially no GCR makes it to the ground. Because they are going so fast, though, stopping them requires collisions with many molecules, some resulting in catastrophic disintegration of the atmospheric particle. For every GCR that reaches Earth, a shower of cascading collisions creates a debris field within the atmosphere. Millions of atmospheric gas molecules are affected by a single GCR. An illustration of this air shower process is shown in Figure 4.8.

One of the bits of debris created in this process is a fast neutron. These secondary particles are able to reach the ground. In fact, these neutrons are still going so fast, and they are rather ineffective at interacting with other matter, that the neutron monitors built to detect them, large tanks of heavy water, are underground and very big just to register a tiny signal. One such neutron monitor is in the Newark, Delaware. It is sensitive only to very fast neutrons from the high-energy tail of the GCR distribution and records anywhere from 0 to about 500 counts per hour, depending on the timing within the solar cycle.

As an example of counting statistic usage, let's say that in one particular hour during solar maximum the measurement was 9 counts. An hour later it was 12 counts. The question could be asked—does this represent a real increase in counts or is this due to random fluctuations from counting statistical noise?

To do this, one of the two measurements should be taken as "correct." Let's assume the first value, 9, is the correct one. In this case, we have a Poisson probability given by Equation (4.35) as this:

$$P_9(12) = \frac{(9)^{12}}{12!} e^{-9} \qquad (4.50)$$

This works out to be:

$$P_9(12) = \frac{(9)^{12}}{12!} e^{-9} = (1.23 \cdot 10^{-4}) \frac{2.82 \cdot 10^{11}}{4.79 \cdot 10^{8}} \qquad (4.51)$$

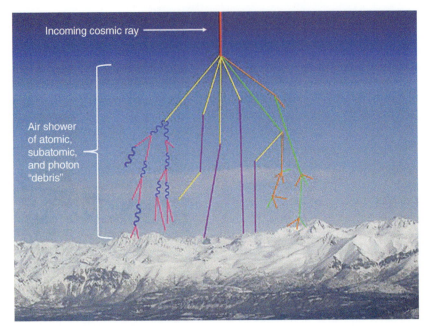

Figure 4.8 A simplistic illustration of a galactic cosmic ray creating an air shower of debris particles. Some of these will be fast neutrons that can be detected by special sensors on the ground. The colors represent different types of particles (or particle fragments) influenced or released during the stopping of one GCR. Adapted from artwork by Randy Russell for Windows to the Universe, using a photo courtesy of University Corporation for Atmospheric Research (UCAR, Nicole Gordon).

The final answer is $P_9(12) = 0.072$. This probability is more than the 5% significance level, so according to that guideline, the difference is not significant. A single number's probability should not be compared against the 5% or 1% thresholds, however. These thresholds are based on a summation of all probabilities in the tail of the distribution. So, to be more complete in this analysis, we should assume the Gaussian approximation and calculate a z score, with $X = 9$ and $\sigma = \sqrt{9} = 3$:

$$z = \frac{|12-9|}{3} \tag{4.52}$$

This is $z = 1$, which has a one-sided probability of $P = 0.16$, well above the 5% threshold. The difference between these two numbers is not significant and the change can be reasonably ascribed to counting uncertainty.

A few years later, at solar minimum, the count in a particular hour is 90, then the next hour it is 120. The numbers look essentially the same, but shifted by an order of magnitude. That is, the relative difference between the two values is the same; the second value is still a third higher than the first value. So, what about now—can this difference be considered to be statistical noise?

Using the Poisson distribution formula is not practical, because 120! is such a huge number that most calculators will not return a value. It is best to move directly to the Gaussian approximation and calculate a z score, this time with $X = 90$ and $\sigma = \sqrt{90} \approx 9$:

$$z = \frac{|120-90|}{9} \tag{4.53}$$

This z score of 3.3 has a probability of $P \approx 5 \cdot 10^{-4}$, which is very small and much lower than not only the 5% threshold but also more than an order of magnitude below the 1% threshold. This change should, therefore, not be ascribed to counting statistics. There was probably a physical reason for this change, either in the natural environment that altered the GCR flux or in the monitoring system itself.

Even though the uncertainty is larger in the second example—it grew from 3 to 9—the uncertainty became smaller as a proportion of the base value. In the first example, the uncertainty was a third of the base value, while in the second, it was only a tenth. One key feature of Poisson counting statistics is that as the counts increase, the relative uncertainty on the value decreases.

These two comparisons highlight the value of additional counts to reduce uncertainty due to counting statistics. We did not have to wait 5 years for the new set of numbers with higher counts, either. Simply waiting more time and adding together the counts from several hours achieve the same result at the low-count-rate interval. The trade-off is, therefore, between value uncertainty and time resolution. You can achieve higher signal-to-noise ratios by observing over a longer period, adding or integrating all of the measurements in that window.

> **Quick and Easy for Section 4.7**
>
> As counts increase, the absolute uncertainty from Poisson statistics increases, but the percentage uncertainty decreases.

4.8 Further Reading

There are many good books out there about processing a single data set, most notably the plethora of introductory statistics books. Here are a few that I have found particularly useful.

William Navidi's statistics book for engineers and scientists:
Navidi, W. (2019). *Statistics for Engineers and Scientists* (5th edition). McGraw-Hill, New York.
https://www.goodreads.com/book/show/1518982.Statistics_for_Engineers_and_Scientists

Alan Chave's book on statistics for Earth sciences:
Chave, A. D. (2017). *Computational Statistics in the Earth Sciences*. Cambridge University Press, Cambridge.
https://www.cambridge.org/core/books/computational-statistics-in-the-earth-sciences/6352412A284512AF4F38766E3717FE38

Sheldon Ross's introduction to probability and statistics:
Ross, S. M. (2014). *Introduction to Probability and Statistics for Engineers and Scientists* (5th edition). Academic Press, Burlington, MA.
https://www.sciencedirect.com/book/9780123948113/introduction-to-probability-and-statistics-for-engineers-and-scientists

Ian Jolliffe and David Stephenson's forecast verification monograph:
Jolliffe, I., and Stephenson, D. (2011). *Forecast Verification: A Practitioner's Guide in Atmospheric Science* (2nd edition). John Wiley & Sons, Hoboken, NJ.
https://www.wiley.com/en-us/Forecast+Verification%3A+A+Practitioner%27s+Guide+in+Atmospheric+Science%2C+2nd+Edition-p-9780470660713

Daniel Wilks' book on statistical methods for atmospheric science:
Wilks, D. (2019). *Statistical Methods in the Atmospheric Sciences* (4th edition). Academic Press, Oxford.
https://www.elsevier.com/books/statistical-methods-in-the-atmospheric-sciences/wilks/978-0-12-815823-4

And, of course, John Taylor's textbook on error analysis:
Taylor, J. R. (1997). *An Introduction to Error Analysis*. University Science Books, Mill Valley, CA, USA.
https://uscibooks.aip.org/books/introduction-to-error-analysis-2nd-ed/

This is a fantastic little write-up by Jasmin Komić about the harmonic mean:
Komić, J. (2011). Harmonic Mean. In Lovric, M. (Eds.), *International Encyclopedia of Statistical Science*. Springer, Berlin, Heidelberg.
https://doi.org/10.1007/978-3-642-04898-2_645

The total solar irradiance plot is adapted from Kopp (2014):

Kopp, G. (2014). As assessment of the solar irradiance record for climate studies. *Journal of Space Weather and Space Climate, 4*, A14. https://doi.org/10.1051/swsc/2014012

More on galactic cosmic rays can be found at this NOAA site:
https://www.swpc.noaa.gov/phenomena/galactic-cosmic-rays
Or at Windows to the Universe, the location of the original figure on this topic:
https://www.windows2universe.org/physical_science/physics/atom_particle/cosmic_rays.html

4.9 Exercises in the Geosciences

Part 1: By-Hand Calculations

1. A magnetometer onboard a satellite makes measurements of the magnetic field in the near-Earth space environment. The spacecraft it is on is spinning, with the sensor taking 10 measurements per spin. While the vector components should oscillate around in a sinusoidal manner, the total field should be relatively constant. The 10 measurements during one particular spin, in nanoTesla (nT), are as follows:

   17.1, 17.3, 17.1, 17.0, 16.8, 16.9, 17.1, 17.4, 17.3, 17.0

   A What is the arithmetic mean of the magnetic field during this spin?
   B What is the sample standard deviation of the magnetic field during this spin?
   C How many measurements are within one standard deviation of the mean? Is this what we expect? Explain your answer.

2. Commercial spacecraft in low-Earth orbit (between 500 and 1000 km above the Earth) often use "engineering-grade" magnetometers with resolution of 50 nT, which is fine for finding orientation relative to the geomagnetic field (which is tens of thousands of nanoTesla), but is often not good enough for most science applications. However, with repeated measurements, you can get the standard error down to whatever you want.

   A How many measurements are needed to give a final uncertainty on the mean field of ± 3 nT?
   B How many are needed for a final uncertainty of ± 0.1 nT?

3. The Mackinac Straights Raptor Watch (MSRW) makes a thorough count of birds of prey crossing over to and from the Upper Peninsula every spring and fall. For golden eagles, one year the count was 315, which at first glance seems higher than the multi-year average count of 298. Can MSRW report that the eagle count increase was statistically significant? Assume that the golden eagle counts follow a Poisson distribution.

Part 2: Jupiter Magnetic Field Data

The Juno mission was inserted into orbit around Jupiter on 4 July 2016, making a highly elliptical orbit close to Jupiter and then very far away from the planet. One of the instruments onboard is a magnetic field detector. This set of exercises explores this data set, calculating centroid and spread, including plots with the addition of error bars on the graphs.

4. Download three consecutive days of fluxgate magnetometer data from the Juno mission data server on the Planetary Data System.
   A The data are here: https://pds-ppi.igpp.ucla.edu/search/view/?f=yes&id=pds://PPI/JNO-J-3-FGM-CAL-V1.0
   B Write an introductory comment paragraph about the data, the time interval, and your method of reading the data.

- C Convert the spacecraft location data into units of Jupiter Radii, and convert the magnetic field vector component data into units of nanoTesla.
- D Calculate the magnitude of the magnetic field in nanoTesla (square root of the sum of the squared components).

5. Analyze the distribution of the magnetic field magnitude data. For each day within your data:
   - A Make a histogram of the magnetic field magnitude values.
   - B Determine the mode of the distribution and the full width half maximum.
   - C Determine the median.
   - D Determine the interquartile range.
   - E Calculate the standard deviation.
   - F Calculate the arithmetic, harmonic, and the geometric means.

6. Comment on your calculated values in at least one paragraph.
   - A Did you use *sample* or *population* standard deviation and why? How do the harmonic, geometric, and arithmetic means describe the daily magnitudes? Are there differences day to day?
   - B To help answer these, make some exploratory graphics of the daily magnitudes. Specifically, make two additional histograms with different bin widths or bin centers. It is not necessary to make finalized figures here; these are just for a quick look at the distribution of the data. However, any figures that you make should still be readable and interpretable.

7. Make finalized, explanatory figures of the magnetic field magnitude (in nT units) and of the spacecraft location for the entire 3-day interval. That is, plot |B| and the x, y, and z positions as a function of time. They position values could be plotted against each other, showing orbit segments for this time interval. The figure can be in any format/style that you find appropriate, but should be informative and clear to your audience. Use the *standard deviation* as an illustrative range of error on the magnetic field magnitude.

8. Comment on your full data set over the 3 days in a paragraph. Include statements about where was the spacecraft going and what it measured for magnetic field at those locations.

5

Assessing Normality: Tests for Assessing the Gaussian Nature of a Distribution

As has been mentioned several times already in the previous chapters, the Gaussian or normal distribution introduced in Chapter 1 is a fundamental quality underlying much of statistics. This is because there is a natural progression toward a Gaussian when individual phenomena are combined. The succession toward a Gaussian distribution is a feature of the **central limit theorem**. The central limit theorem, a key tenet of probability theory, states that when independent variables are added, the distribution of the final summed parameter tends toward a Gaussian distribution. This is true no matter how asymmetric or non-Gaussian the original variable distributions were at the beginning of the combination process. It is, therefore, useful to conduct a few tests to determine if the distribution of a given number set can be considered to be Gaussian. If a number set has a distribution that matches the form of a Gaussian, then certain quantities defined in previous chapters have specific meanings with respect to percentages of points, which, as we have seen, can be useful for interpretation. When applied to a non-Gaussian distribution, the interpretation of some statistical values becomes far more difficult, perhaps totally meaningless.

Central limit theorem: Distributions, when combined together, eventually trend toward a Gaussian distribution, no matter the initial distributions involved.

Remember the functional form of the Gaussian distribution—it is the natural exponential operator with a negative argument that is squared with respect to the x axis variable:

$$G_{X,\sigma}(x) = \frac{1}{\sigma\sqrt{2\pi}} e^{-\frac{(x-X)^2}{2\sigma^2}} \quad (5.1)$$

In Equation (5.1), X is the centroid of the Gaussian distribution and σ is its spread. The coefficient in front of the exponential is a normalization factor that makes the integral of the distribution, from x values of $-\infty$ to $+\infty$, equal to one.

If the distribution of the number set is declared to be different than a Gaussian, what should be done differently in the processing of this data set? The assumption of a Gaussian distribution is embedded throughout most of the standard statistical measures. When applied to a number set that has a non-Gaussian distribution, these formulas will still yield results, but their interpretation—especially probabilities used for hypothesis testing—are no longer valid. If a number set is far from a Gaussian, then it is usually better to not even report such statistics, so readers are not tempted to apply the Gaussian-based interpretation. There are several special calculations and visualizations that might be necessary to best process and understand the values of a non-Gaussian distribution. One way to

---

Data Analysis for the Geosciences: Essentials of Uncertainty, Comparison, and Visualization, Advanced Textbook 5, First Edition. Michael W. Liemohn.
© 2024 American Geophysical Union. Published 2024 by John Wiley & Sons, Inc.
Companion website: www.wiley.com/go/liemohn/uncertaintyingeosciences

clearly signal that the distribution is not a Gaussian is to use the box plot to visualize the number set. It is often called a box-and-whisker plot because of its standard format, with a narrow rectangle indicating the location of the bulk of the distribution and then long thin lines demarking some outlier extent of the values. It is nearly always defined from quantiles, with the median representing the centroid and other quantiles used for the box extent and whisker length.

This chapter takes another step along the path for the much larger topic of comparing two number sets. The first quantitative step was way back in Chapter 1—the z score and related z test. The comparison strategy presented in this chapter provides a robust method of comparing the shape of two histogram sets against each other. Later chapters will introduce yet more techniques, presenting additional quantitative methods for comparing additional aspects of two different number sets against each other.

5.1 Histogram Check

The first and most qualitative is to consider the shape of the histogram. We have already considered overlaid histograms, so a first step in checking to see if a distribution is normal or not is overplot the expected Gaussian distribution. Two values need to be chosen—X and σ. If the distribution is normal, then it should qualitatively match a Gaussian distribution histogram based on the mean and standard deviation of the data set itself. A good initial choice for a Gaussian fit, therefore, would be the arithmetic mean and standard deviation of the data set, respectively. To adjust the height of the Gaussian to correctly compare with the number set histogram, it should be multiplied by N. This is shown in Figure 5.1 with a histogram of an example data set and a corresponding Gaussian distribution as defined by the number set's arithmetic mean and standard deviation ($\bar{x} = 10.7$, $\sigma = 2.9$, $N = 25$). Do they match? Some columns with it are quite close and others are not close at all. Note that the example number set is positive definite, but that is not a limitation of histograms; the x axis should span the range of the values regardless of whether they are positive or negative.

This question can be assessed further with additional examples. Four more distributions ($N = 40$) are shown in Figure 5.2: one that is close to a Gaussian, another that is skewed to the right, and two that have heavy and light tails away from the central peak of the distribution.

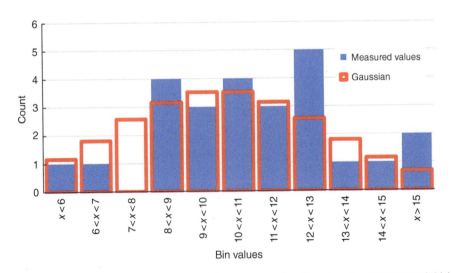

Figure 5.1 Histogram of a number set, in blue, with a corresponding Gaussian distribution overlaid, in red.

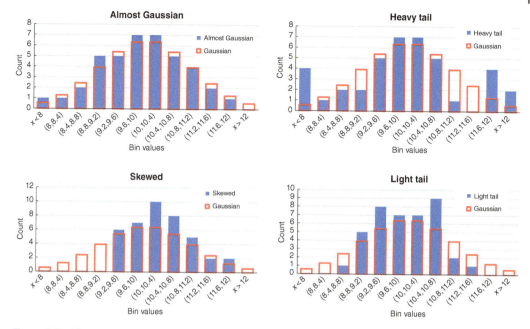

Figure 5.2 Histograms of number sets with a particular qualitative look, with a corresponding ideal Gaussian distribution overlaid on each. The data set counts are in blue and the Gaussian histogram is in red.

To do an initial normality check, a Gaussian distribution histogram is overlaid in all of the panels. In examining the panels of Figure 5.2, the first distribution appears to be quite close to the Gaussian histogram. All of the others, however, have some of the bins with large differences to the Gaussian histogram counts for that bin, either lower or higher. With only a qualitative reaction to these overlaid histograms, based on these differences, the first example histogram is a good match and the other three are not. In the first histogram, only one bin—the highest value bin—is off by a large percentage, and even this is only off by a count of less than one. The other three have at least five bins with significant percentage differences compared to the Gaussian histogram.

It is good to be a bit more quantitative with our analysis. The panels in Figure 5.2 offer some guidance in creating this rubric. There are two criteria to consider in this assessment: the size of the differences relative to the Gaussian histogram and the number of these large differences. It is necessary to define, at least roughly, what is a "large" difference, and also what is a "large number of large differences." In examining Figure 5.2, the first criterion of a large difference could be defined as the data histogram count being 50% or more off from the Gaussian value for that same bin. For the second, a substantial fraction of the bins could be defined as being third or more of the total number of bins. With these definitions, a large number of large differences would be strong evidence that the distribution might not be Gaussian. Note, however, that these criteria definitions are only guidelines. At this point, there is no calculation, these criteria were determined by eye and the assessment is done by eye; everything about this first-look judgment is a rough estimate.

> **Quick and Easy for Section 5.1**
>
> Overplotting a Gaussian distribution on the data set histogram provides a quick qualitative check on whether the set's distribution is normal.

5.2 Comparing Centroid and Spread Measures

Another check that is approaching a quantitative comparison is to assess the similarity or difference among the centroids of the distribution. For a Gaussian distribution, the arithmetic mean should be identical to the median and mode. Determining all of the centroids discussed in Chapter 4 is useful because the variation of these estimates for centroid provides information about the number set. If they are all close to each other, then this indicates that the distribution could be, and probably is, close to that of a Gaussian. If they are far from each other, then, depending on which of the centroid estimates is the outlier of the group, this also informs you about the nature of the distribution.

From our example number sets shown as histograms in Figure 5.2, the median, mode, and the three different means (arithmetic, geometric, and harmonic) are listed in Table 5.1. In this table, the numbers across a given row should be compared against each other (that is, the various centroid values for a given distribution). Are they similar? Some rows have values that are extremely close to each other—both the almost Gaussian and skewed distributions have a centroid discrepancy of at most 0.1—while others have a larger discrepancy among the centroids—the light tail distribution goes up to 0.7 difference between the centroids. A large discrepancy between the centroids is an indication that the distribution is not Gaussian. However, the reverse is not true; discrepancies between these centroids must necessarily be small if the data are close to a Gaussian, but this is not a sufficient measure by itself.

We do not know if these discrepancies are large, though, without a second piece of information: the spread of the distribution. As we learned earlier, uncertainty values help us interpret base values.

Because these are centroid values, the proper spread to consider is the standard error (see Section 4.5 in Chapter 4). These are listed in the last column of Table 5.1 for each of the distributions. At this point, only a qualitative comparison is being made of whether the various centroids are within one standard deviation of each other. This threshold is met for both the almost Gaussian and skewed distributions, but is not met for the heavy tail and light tail distributions.

Spread estimates are the variation of any one point in the set away from the centroid of the distribution. Because there are several estimates of spread, a comparison of these provides another assessment of the distribution. The important aspect of this check is to calculate comparable spread values of the data set. That is, for a Gaussian distribution, a range of plus or minus one standard deviation from the mean should include 68% of the numbers in the set. This range can be readily calculated, as can a comparable value from the quantile values of spread, specifically those with an identical range of 68% of the distribution centered about the median. For a Gaussian distribution, these two should be equal, not

Table 5.1 Centroid estimates for the number sets shown as histograms in Figure 5.2.

Distribution	Median	Mode	Arithmetic mean	Geometric mean	Harmonic mean	Standard error
Almost Gaussian	10.0	10.0	10.0	9.9	9.9	0.14
Skewed	10.3	10.2	10.3	10.3	10.3	0.10
Heavy tail	10.0	10.2	9.9	9.9	9.8	0.21
Light tail	10.0	10.6	10.0	10.0	9.9	0.10

only the 68% range but also the two endpoints of this range. If the distribution is far from Gaussian in shape, then differences will arise in these values. Similarly, the full width at half maximum (FWHM) for a Gaussian distribution is 2.36σ, so 85% of the FWHM should be equal to 2σ.

For the example distributions shown in Figure 5.2, these estimates of spread are listed in Table 5.2. To continue the interpretation of the centroids, the column marked $2\sigma_x$ is double the spread that we should be considering. To be a Gaussian distribution, the centroid discrepancy should be much less than the standard deviation. For most of the distributions, this is the case. For the light tail distribution, however, σ_x is 0.65, which is comparable to the 0.7 discrepancy among the centroids. This is another indication that this distribution is not Gaussian.

Next, let's assess the various spread estimates against each other. The last three columns are the measures of spread that should be directly comparable to each other. Are they? For the almost Gaussian distribution, the numbers have a maximum discrepancy of 0.3, which is rather small compared their magnitudes. The same can probably be said about both the skewed and light tail distributions, which have a maximum discrepancy between the spread measures of 0.4. The heavy tail distribution, however, has one spread value at 1.4 and the others at 2.3 and 2.6. This is a large difference that indicates that the heavy tail distribution probably is not a Gaussian.

Like the centroid test mentioned earlier, this finding does not mean that the other three are Gaussian. Close spread measures are necessary but not sufficient to declare a distribution to be Gaussian. Therefore, from this test of the spreads, we can only make an assertion about the light tail distribution and not the others.

It is an interesting result that the two assessments here (comparisons of means and then spreads) identified different distributions as being non-Gaussian. Because each test is focused on a particular aspect of the distribution, the determination of a Gaussian shape based on either one is not conclusive but rather another piece of evidence leading to a robust case for or against the similarity.

Note that the proportion of the distribution chosen for the comparison in Table 5.2 (a width of $\pm 1\sigma$ around the mean) is 68%, but this is merely because it is convenient, not because it has a special assessment quality. Any percentage could have been chosen. If assuming a Gaussian distribution, then an α multiplier can be applied to the standard deviation, $\bar{x} \pm \alpha\sigma_x$, to select any percentage of the full distribution for comparison against a similar quantile range.

> **Quick and Easy for Section 5.2**
>
> Comparing the differences among the various measures of centroid and spread are quantitative methods of identifying that some distributions are not Gaussian.

Table 5.2 Equivalent spread estimates for the number sets shown as histograms in Figure 5.2.

Distribution	$\bar{x}-\sigma_x$	$\bar{x}+\sigma_x$	FWHM	16% quantile	84% quantile	$2\sigma_x$	85% of FWHM	16%–84% quantile range
Almost Gaussian	9.1	10.8	2.4	9.1	10.8	1.8	2.0	1.7
Skewed	9.7	11.0	2.0	9.7	10.9	1.3	1.7	1.3
Heavy tail	8.6	11.3	1.6	8.6	10.9	2.6	1.4	2.3
Light tail	9.3	10.6	2.0	9.4	10.8	1.3	1.7	1.4

5.3 Skew

We referred to a couple of the histograms in Figure 5.2 as "skewed." The qualitative definition of this word is that the distribution is not symmetric about its centroid but rather shifted in one direction or the other, with a short tail on one side and a long tail on the other. **Skew** has a formal mathematical definition, too, designated by the **skewness coefficient** γ and looking something like the L3 norm but with some important modifications:

$$\gamma = \frac{1}{(N-1)\sigma_x^3} \sum_{i=1}^{N} (x_i - \bar{x})^3 \quad (5.2)$$

Note that the argument of the summation in Equation (5.2) is not subjected to an absolute value operator. This is intentional and in fact a desirable aspect of the definition of γ. That is, if the distribution is perfectly centered on the arithmetic mean, then the summation will have equal contributions left and right of the mean and the summation will become zero. This is the "perfect" score for γ, indicating there is no skew in the distribution of the number set around the mean. A Gaussian distribution is symmetric about its centroid and therefore has a skewness coefficient of zero. The $N-1$ factor dividing the summation indicates, as discussed in Chapter 4, that this way of writing γ is for a sample and not the population. The extra minus one is a small correction factor that is only important when N is small and can often be ignored. It is also seen in Equation (5.2) that the denominator has the cube of the standard deviation, σ_x, of the number set. This is a normalizing factor that puts the summation value into context of the expected spread of the values if the distribution were a Gaussian.

The skewness coefficient can be either positive or negative. This sign indicates the direction of the skewness of the distribution. If, for example, γ is negative, the left side of the distribution (x values below the mean) has the elongated tail and the right side (x values above the mean) is truncated. For a positive γ, the reverse situation is the case with a long tail above the mean.

Going back to the example data sets introduced in Section 5.1, the calculated skewness coefficients are listed in Table 5.3. It is seen that three of the values are quite close to zero, within 0.1 of the Gaussian skewness coefficient of zero. The other, the data set labeled "skewed," has a coefficient of 0.62, well away from the others. We knew this one was different; it was designed to be skewed right, so this score is expected. What we have learned from this, though, is that a score of 0.6 is big enough to be considered far from that of a Gaussian.

The interpretation of the skewness coefficient is a subjective call, but, like the meaning of so many other values in statistics, there are guidelines. Because of the normalization, a γ score that is one or more away from zero (i.e., $|\gamma| \gtrsim 1$) is indicative of a large skew. A small skew is a value of the γ score that is close to zero, specifically $|\gamma| \ll 1$. In between these extreme ranges for γ, the skew can be called moderate. As we saw in Table 5.3 and Figure 5.2, a γ of 0.6 is clearly skewed, but is not severely

---

Skew: The asymmetry of a distribution, revealing the "one-sidedness" of the tails of the distribution relative to each other.

Skewness coefficient: A quantitative measure of skew, it is the arithmetic mean of the cube of the discrepancies between the individual data values and the mean, normalized by the cube of the standard deviation.

Table 5.3 Skewness coefficients for the number sets shown as histograms in Figure 5.2.

Distribution	Skewness
Almost Gaussian	−0.05
Skewed	0.62
Heavy tail	−0.09
Light tail	0.01

asymmetric. There is no clear cutoff between these categories.

To illustrate this, Figure 5.3 shows distributions with incrementally increasing values of the skewness coefficient. Overlaid on the histograms is a histogram in red outline showing the expected Gaussian distribution with the same mean and standard deviation (mean of 0.5 and standard deviation of 0.1). There are 100 values in these number sets; 12 bins were chosen for the histograms so that the bin width was exactly 0.05. The three distributions are illustrations of a low, medium, and high skewness coefficient (of 0.24, 0.73, and 1.23, respectively). In order to have the same mean and standard deviation, the mode of the distribution shifts to lower values, moving just under a bin width for each chosen increment in the skew. The blue histogram in the top panel of Figure 5.3 does not look particularly skewed. The blue histogram in the last panel, however, clearly looks skewed. This hopefully provides some context for the skewness coefficient values that you might encounter, as well as some justification for the guidelines given in this section.

Note, however, that a γ score of zero does not imply that the distribution is a Gaussian. It means that the distribution has equal contributions to γ of values above and below the mean, with a cubic weighting to the $x_i - \bar{x}$ discrepancy. Figure 5.4 shows several histograms,

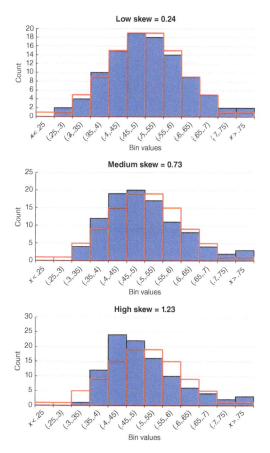

Figure 5.3 Histograms that have progressively larger values of the skewness coefficient (in blue) with the corresponding normal distribution overlaid (in red, same mean and standard deviation).

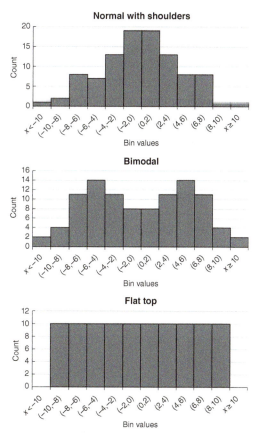

Figure 5.4 Histograms of non-Gaussian distributions that have a skewness coefficient of zero.

all of which have a skewness coefficient of essentially zero. The first one is close to a Gaussian distribution but exhibits enhanced "shoulders" along the downward slopes that increase the number of values at two equidistant locations away from the mean. These two features, however, will cancel within the skewness coefficient summation, and therefore the resulting γ score for this distribution will be zero. The second one also has equal contributions from two "bumps" away from the mean, but this time there is no relative maximum at the center of the distribution; it is actually the superposition of two equal Gaussian distributions. The third is a distribution with equal counts in several bins and zero counts in all others. Because the bin counts are equal, the skewness coefficient calculates to a value of zero; for every value above the mean, there is one of equal distance below the mean that cancels its contribution to the summation in Equation (5.2). All of these show that the skewness coefficient assesses only a specific aspect of the distribution—its "lopsidedness." Other features of the distribution are not quantified by this calculation. We need other assessments for features like shoulders and bimodal peaks, like that covered in the next section.

> **Quick and Easy for Section 5.3**
>
> The skewness coefficient of the data set reveals asymmetries relative to the mean, which should not be present in a Gaussian distribution.

5.4 Kurtosis

Another term used in Section 5.1 to describe the distributions in Figure 5.2 is the "weight of the tail." Specifically, the adjectives of heavy and light were applied to characterize the shape of the histogram relative to a Gaussian distribution. Like skewness, this also has a quantitative definition, called **kurtosis**, again defined similarly to an L4 norm but with important variations. The standard form of the **kurtosis coefficient**, k, is as follows:

$$k = \frac{1}{(N-1)\sigma_x^4}\sum_{i=1}^{N}(x_i - \bar{x})^4 \quad (5.3)$$

The only change between Equations (5.3) and (5.2) is the increase in the power integer from 3 to 4. This has the effect of making k positive definite. This also means that the x values above and below the mean do not cancel each other within the summation. Because a Gaussian always has some spread, σ, to the distribution, the kurtosis of a Gaussian is not zero. Rather, it is 3, which can be proven by a lengthy derivation.

For the example histograms shown in Figure 5.2, the calculated kurtosis values are listed in Table 5.4. As these data sets were designed, the first two have kurtosis values quite close to the ideal Gaussian value of three, but the other two have kurtosis values that seem pretty far from that of a Gaussian. The histograms for these distributions definitely appear to have a

Table 5.4 Kurtosis values for the number sets shown as histograms in Figure 5.2.

Distribution	Kurtosis	Alternate kurtosis
Almost Gaussian	2.9	−0.12
Skewed	3.0	0.008
Heavy tail	3.6	0.63
Light tail	2.1	−0.9

Kurtosis: The heaviness of the tails of a distribution, revealing the content of the data set far from the centroid relative to the content near the centroid.

Kurtosis coefficient: A quantitative measure of kurtosis, it is the L4 norm divided by the fourth power of standard deviation.

heavy and a light tail, so these kurtosis values, more than 0.5 but less than 1.0 away from the Gaussian value, can be considered as "big differences" for this measure of the distribution.

The interpretation of kurtosis is seemingly straightforward. The basic rule is that values of k below 3 indicate that the number set has fewer values at large distances from the mean than would a Gaussian with the same σ. We need to ask, though, what constitutes a good value? How close does k have to be to 3 in order for the number set to be considered to have a Gaussian-like tail distribution? To explore this, Figure 5.5 shows a few distributions with different k values. Overplotted is the expected Gaussian. All of the distributions have a mean of 20, standard deviation of 10, and skewness coefficient of zero. The two plots on the left have heavy tail distributions (blue histograms) with kurtosis values greater than 3, while the two plots on the right have light tail distributions with $k < 2$. The upper-row histograms look fairly close to the expected Gaussian, while the lower-row histograms are quite different from the Gaussian histograms. Remember, though, that Equation (5.3) contains a difference quantity raised to the fourth power, which means that kurtosis is highly sensitive to the outliers of the distribution. For the goodness of a kurtosis value, then, we should really only examine the tails of the distributions, not the central peak. It is seen that the upper-row plots are different only in the last bin on each end. The lower panels of Figure 5.5 show four, or even five, bins with substantial differences to the expected Gaussian. From all of these plots, a rather qualitative guideline for evaluating kurtosis can be made: if k is within one of the Gaussian value of three, then it is a moderate difference, at most. At values farther away from three than this rough threshold of ± 1, the kurtosis of the distribution can be declared to be highly different.

Because of the $p = 4$ power in Equation (5.3), the kurtosis coefficient is extremely sensitive to outliers. In the plots on the left half of Figure 5.5, the blue columns are taller than the red columns in the outermost bins. These two distributions have $k \geq 4$. In the plots on the right, the outermost have higher values in the red columns than the blue columns. For these distributions, k is well below 3. The bins near

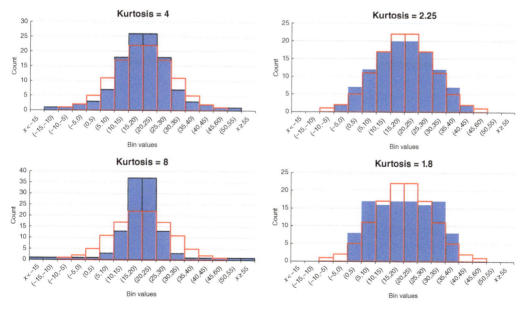

Figure 5.5 Histograms that have the listed values of kurtosis (blue histograms), with an overlaid expected Gaussian (red outline histograms).

the centroid contribute very little to k, and even the bins of moderate distance from the centroid only contribute a minor amount to the summation. The fourth power in the definition of the kurtosis coefficient makes it very efficient at comparing the outliers of the distributions.

It should be pointed out that an alternative form of the kurtosis coefficient exists. Some users of kurtosis dislike that the Gaussian distribution has a nonzero k value, an offset that could cause confusion to the interpretation of k. To make it analogous to γ, another version of k can be written with this shift included to set the Gaussian value equal to zero:

$$k_{alt} = \left[\frac{1}{(N-1)\sigma_x^4}\sum_{i=1}^{N}(x_i-\bar{x})^4\right] - 3 \quad (5.4)$$

The definition of kurtosis given in Equation (5.4) is the dominant form in some disciplines, to the point that sometimes the "alt" subscript is dropped. So, unfortunately, rather than alleviating confusion, this alternate formula for k might cause interpretation issues if the functional form of k is not specified. Always be sure to reveal this designation, either by listing the formula used or by stating the k value for a Gaussian (3 or 0).

Note that, because of the even power of the summation argument, the kurtosis coefficient does not report any information about the lopsidedness of the distribution. That is the role of the skewness coefficient. Similarly, γ can be zero even when the number set has a very heavy tail compared to a Gaussian distribution. These two metrics have distinct and complementary contributions to the determination of whether a number set follows a Gaussian distribution or not. Neither will, by itself, answer that question. Taken together, Figure 5.3, Figure 5.4, and Figure 5.5 reveal not only the information that can be gleaned from skew and kurtosis but also the limitations of these two descriptive parameters. There are, however, a few statistics that specifically target the quantification of how closely one distribution matches another. We address two of these in the next sections.

> **Quick and Easy for Section 5.4**
>
> Comparing the kurtosis of a distribution against the ideal value for a Gaussian is a quantitative method of identifying that some distributions are not Gaussian.

5.5 The Chi-Squared Test

The **chi-squared coefficient**, written with the Greek letter as χ^2, is a measure of comparison between two histograms. It is often conducted with one of the two histograms being an observational number set and the other being an analytical function that the measured values are supposed to follow. Some examples of what that second, hypothetical distribution could be are as follows:

- A Gaussian distribution from the observational mean and standard deviation.
- Some other complicated statistical distribution.
- A second set of observations.
- A set of measurements estimated by a numerical model.
- A number set of a proposed causal driver of the observations.

They do not even have to be of the same units, because they can be normalized by subtracting the mean and dividing by the standard deviation. In all of these cases, a χ^2 coefficient can be calculated between the two histograms and a **chi-squared test** (χ^2 **test**) conducted on the resulting value.

Chi-squared (χ^2) coefficient: A quantitative measure of the similarity of two histograms.

Chi-squared (χ^2) test: The conversion of the χ^2 coefficient into a probability that the two distributions were sampled from the same population.

For our specific case here, it is desired to compare the histogram of observed values with a Gaussian distribution, calculating the expected counts for each bin from the analytical function. For a Gaussian distribution, the expected values for the hypothetical histogram to which the data would be compared can be readily calculated. That is, probabilities within known ranges relative to the centroid and spread—found via the cumulative probability distribution integral of the Gaussian—can be multiplied by the total count of the number set, N, to yield Gaussian values in each histogram bin. These can then be directly compared with the observed counts in those bins. That is, for each bin k, the observational values in that bin, O_k, should compare well with the expected value in that bin, E_k, if the observational samples are truly being taken from an underlying Gaussian distribution.

To what should this difference be compared? Like a z score, the discrepancy between the observed and expected values in any bin, $O_k - E_k$, should be compared against the spread of the population. An underlying assumption of the χ^2 coefficient is that these are counts and so the uncertainty of the expected values should follow a Poisson distribution. Therefore, the z score equivalent for this assessment is as follows:

$$z_k = \frac{O_k - E_k}{\sqrt{E_k}} \tag{5.5}$$

The χ^2 coefficient is then calculated from the sum of the squares of these z score equivalent values:

$$\chi^2 = \sum_{k=1}^{M} \frac{(O_k - E_k)^2}{E_k} \tag{5.6}$$

Note that χ^2 from Equation (5.6) is unitless. While it appears that all of the values in the summation are counts, the denominator matches the "counts squared" units of the numerator because it is a Poisson uncertainty value, which has the same units as the base value.

For the observations to be considered "close" to the expected values, O_k should lie within the uncertainty spread of E_k. This should be true for all of the bins, or at least all of the bins on average. Because χ^2 includes a summation of all of the bins, it is desired that $\chi^2 < M$ for the observed distribution to be considered similar to a Gaussian distribution. If, instead, this calculation yields $\chi^2 \gg M$, then the underlying distribution of the observations is probably not a Gaussian function.

That definition is rather qualitative and more quantitative assessments of χ^2 exist. Specifically, the complexity of the function used to create the expected values should be considered. A complicated function might be chosen specifically to fit a bimodal distribution or some other feature in the data set. These additional coefficients in the function make it easier to fit the features of the observational histogram, which means that it is easier to get a lower χ^2 score with a more sophisticated function designed to fit it. To account for this, χ^2 can be converted to a "reduced chi-squared" coefficient, $\tilde{\chi}^2$, via division by the degrees of freedom of the expected value set, d:

$$\tilde{\chi}^2 = \frac{\chi^2}{d} \tag{5.7}$$

The degrees of freedom coefficient of the comparison is given as the difference between the number of bins in the histogram, M, and the number of free parameters used in the calculation of the expected values, known as the number of constraints, c:

$$d = M - c \tag{5.8}$$

A histogram from a Gaussian distribution uses three parameters. The first is the arithmetic mean of the observations, taken as the centroid value X of the Gaussian. The second is the standard deviation of the observations, assumed to be the spread σ of the Gaussian function. The third is N, the total number of values in the observational number set, used to scale the Gaussian probabilities within each bin into an equivalent count value. The degrees of freedom must be at least one, otherwise $\tilde{\chi}^2$ is undefined. So, for an expected Gaussian distribution with $c = 3$, the number of bins in the histogram used for the comparison should be at least 4. Note that the number of constraints could be zero, if no free parameters, including nothing about the data, are used in the calculation of the expected values. In most cases, though, c will be a small number, and in fact it is difficult to get it lower than one. Even if the observations are being compared against a well-known value with a predetermined spread, this function should be scaled to the observed bin counts with the use of the total count in the number set. So, c is rarely smaller than one.

This analysis can be conducted for the example data sets from Section 5.1. The resulting χ^2 and $\tilde{\chi}^2$ values are listed in Table 5.5. For a very rough interpretation, we should compare the bin count, $M = 12$, against the χ^2 value. While the almost Gaussian value is well below the bin count, two of the distributions—skewed and heavy tail—are over this quick-check assessment, and light tail is close. This is how these distributions were created, so this is expected. We can be even more quantitative with the χ^2 test, though.

The $\tilde{\chi}^2$ value from Equation (5.7) can be converted into a probability that the observational number set came from the expected value set. This probability is given by a complicated function that we will not try to calculate ourselves. Rather, if an approximate value is needed, then the look-up chart in Table 5.6

Table 5.5 Chi-squared and reduced chi-squared values for the number sets shown as histograms in Figure 5.2.

Distribution	χ^2	$\tilde{\chi}^2$	P values
Almost Gaussian	1.7	0.18	0.995
Skewed	13.0	1.45	0.16
Heavy tail	36	3.9	0.00004
Light tail	10.4	1.15	0.32

Table 5.6 $\tilde{\chi}^2$ values as a function of specific probability values and the degrees of freedom in the comparison.

d	$\tilde{\chi}^2$ for $P = 0.1$	$\tilde{\chi}^2$ for $P = 0.05$	$\tilde{\chi}^2$ for $P = 0.01$	$\tilde{\chi}^2$ for $P = 0.001$
1	2.71	3.84	6.63	10.8
2	2.31	3.00	4.61	6.91
3	2.08	2.61	3.78	5.42
4	1.94	2.37	3.32	4.62
5	1.85	2.22	3.02	4.10
10	1.60	1.83	2.32	2.96
20	1.42	1.57	1.88	2.27
50	1.26	1.35	1.52	1.77
100	1.18	1.24	1.36	1.49

is adequate, otherwise an online calculator should be used to determine this probability (two are provided in Section 5.12). Table 5.6 shows $\tilde{\chi}^2$ as a function of given probabilities, with a few often-used probability values along the top axis, to show the exact $\tilde{\chi}^2$ values needed to meet these specific probabilities. Table 5.7 is a "reversed" version of this information, with $\tilde{\chi}^2$ along the top axis, showing the probability associated with that particular $\tilde{\chi}^2$ score. In both tables, the values down the first column indicate the degrees of freedom as calculated from Equation (5.8).

This probability can be assessed to test whether the two distributions are the same. For our case of checking a histogram against a Gaussian distribution, this is a probability that the histogram can be declared to be a reasonable approximation of a Gaussian. If this is the case, then the arithmetic mean and standard deviation have the interpretations discussed in Chapter 4, with intervals of one standard deviation width having particular proportions of the total distribution within them.

For our example data sets, probabilities can be found for the $\tilde{\chi}^2$ values with $M = 12$. The probabilities are listed in the last column of Table 5.5. Three of the four P values are above

Table 5.7 Probabilities as a function of specific $\tilde{\chi}^2$ values and the degrees of freedom for the comparison.

d	$\tilde{\chi}^2$ = 1.0	$\tilde{\chi}^2$ = 1.5	$\tilde{\chi}^2$ = 2.0	$\tilde{\chi}^2$ = 3.0	$\tilde{\chi}^2$ = 4.0	$\tilde{\chi}^2$ = 6.0
1	0.317	0.221	0.157	0.0833	0.0455	0.0143
2	0.368	0.223	0.135	0.0498	0.0183	0.0025
3	0.392	0.212	0.112	0.0293	0.0074	0.0004
4	0.406	0.199	0.0916	0.0174	0.0030	8×10^{-5}
5	0.416	0.186	0.0752	0.0104	0.0013	$<10^{-5}$
10	0.440	0.132	0.0293	0.0009	2×10^{-5}	$<10^{-5}$
20	0.458	0.0699	0.0050	$<10^{-5}$		
50	0.473	0.0126	3×10^{-5}	$<10^{-5}$		
100	0.481	0.0009	$<10^{-5}$			

the traditional 5% significance threshold. According to the chi-squared test, only the heavy tail histogram is significantly different from a Gaussian distribution.

As discussed previously, the two traditional thresholds for the probability calculation are 0.05 and 0.01, representing a 5% and 1% chance that the two distributions (observed and expected) are the same. The 5% value is called a significant difference and the 1% value has the label of highly significant. These are qualitative descriptions and somewhat arbitrary cutoffs, though. The probability is simply another piece of information, along with the others discussed in the earlier text, to inform you about how to interpret the meaning of the statistical values calculated from the number set. If the probability is below 0.05, then the distribution of the number set is pretty far from a Gaussian and the statistics should not be taken with the Gaussian interpretations. A probability above 0.05, however, does not necessarily mean that you can definitively conclude the distribution is Gaussian.

When conducting a $\tilde{\chi}^2$ calculation and test, the number of bins in the histogram does not have to be the same number of bins as done for the qualitative examination of the histogram. Nor do the bin widths have to be equal; they can be whatever width is needed to span the range included in the comparison. For a Gaussian distribution with $c = 3$, the minimum number of bins needed for this calculation is $M = 4$, resulting in $d = 1$. The simplest calculation for a Gaussian is to use four bins with the bin edges located at the arithmetic mean and ± 1 standard deviation from the mean. The first and last bins extend to $\pm\infty$, therefore including all observed values in the comparison. In this case, the proportions for the four expected value bins are simple: 0.16, 0.34, 0.34, and 0.16, respectively. Multiplying these proportions by N yields the expected values for the four bins, which can then be compared against the observed value count within each of these intervals. While this case of the lowest M for a Gaussian has unequal bin widths, it could be the case that the extreme tails of the distribution have so few counts that you do not want to include them in the comparison. It is acceptable to truncate the range used in this test and only use some of the bins in the χ^2 test, which can be of equal or unequal widths, whatever is desired or convenient.

In fact, this feature becomes particularly useful if the distribution has multiple relative maxima, then it might be useful to test each peak separately. First, conduct a χ^2 test on the largest peak, and only in the range near that particular peak. If this biggest peak resembles a Gaussian distribution, then the expected values can be calculated for the rest of the bins and subtracted, essentially removing this peak from the distribution. Then the next largest peak can be tested. If it can be declared to be similar to a Gaussian, then subtract its expected value set from all bins. This can be done for all peaks in the distribution, isolating each one with its own χ^2 test.

There is not agreement on an ideal number of bins for a χ^2 test. While 4 is a minimum for a check of normality, it does not allow for any features in the histogram except the central peak. The guidelines for histogram creation, as discussed in Section 2.3 (Chapter 2), could be followed here, but this number might be overkill for assessing normality.

A feature of the $\tilde{\chi}^2$ test is that, as the number of bins used in the comparison increases, it first becomes easier and then harder to declare two distributions to be the same. This is not intuitive and requires a bit of explanation. It is perhaps best seen in the value from Equation (5.5), the z_k score for the contribution of a single histogram bin to the χ^2 coefficient. For the very simple case of only four bins in the comparison against a Gaussian distribution, the $\tilde{\chi}^2$ value needed to reach the 5% level is 3.84. Because $d = 1$, this is also the χ^2 coefficient value, which means that each z_k score from Equation (5.5) may contribute, on average, 0.96 to the χ^2 coefficient summation in Equation (5.6). Adding one bin (so that M is now 5), the $\tilde{\chi}^2$ value for the 5% level is 3.00. Because $d = 2$ for this case, the χ^2 coefficient is actually double the $\tilde{\chi}^2$ value, and the average contribution from each bin can be 1.20. That is, the distribution can be farther away from a Gaussian and yet still meet the 5% threshold. Table 5.8 shows the relationship of these values as a function of bin count. It is assumed that $c = 3$ for a Gaussian distribution. The third and fourth columns show that $\tilde{\chi}^2$ monotonically decreases and that χ^2 monotonically increases as M and d increase. The final column, however, reveals that the average z_k score at first increases and then decreases with increasing M and d.

In reality, this increase in the average z_k contribution to remain "similar to the expected

Table 5.8 Average z_k scores as a function of number of bins used in the comparison to exactly match the 5% "significantly different" threshold.

Bins (M)	d	$\tilde{\chi}^2$ for $P = 0.05$	χ^2 for $P = 0.05$	Average z_k for $P = 0.05$
4	1	3.84	3.84	0.96
5	2	3.00	5.99	1.20
6	3	2.61	7.82	1.30
7	4	2.37	9.49	1.36
8	5	2.22	11.1	1.39
9	6	2.10	12.6	1.40
10	7	2.01	14.1	1.41
13	10	1.83	18.3	1.41
20	17	1.62	27.6	1.38
30	27	1.48	40.1	1.34
100	97	1.25	121	1.21

distribution" is not much of a help. As more bins are included in the χ^2 test, the statistical noise of counting within each bin becomes relatively larger. So, individual bins most likely have more variation, and the z_k scores could increase because of uncertainty in the decreasing O_k value in each bin, not just the E_k values, for which the uncertainty is included in the denominator of Equation (5.5). Whether this feature of the χ^2 test helps for declaring the distribution to be similar to the expected distribution depends on N and whether the uncertainty of the counts within each bin remains small.

As a final note in this discussion, remember that the assumed distribution did not have to be a Gaussian function. It could have been a Poisson distribution or other analytical formula. While the Poisson distribution approaches a Gaussian function, many measurements are needed to reach that limit in order to have the mean be several standard deviations above zero. A χ^2 test can still be conducted even if this condition is not met; the formula for this special distribution can be used to calculate probabilities within each bin. Really, any distribution can work with the χ^2 test. It could be a half-life formula for radioactive decay, it could be a power law formula for a spectral density decrease with frequency, it could even be a sine wave or a square wave.

> **Quick and Easy for Section 5.5**
>
> The χ^2 test is a quantitative method of comparing two histograms, resulting in a probability that the two distributions were sampled from the same population.

5.6 The Kolmogorov–Smirnov Test

A related assessment of the distribution is the **Kolmogorov–Smirnov test**, which compares a particular value of the cumulative probability distributions, that is, of the running summation of the histograms rather than the histograms themselves (introduced in Section 1.6.2 in Chapter 1). Like the chi-squared test, it can be used for any two distributions, and it is not limited to just this comparison of an observed set against an expected Gaussian distribution.

The test is actually very easy: the **Kolmogorov–Smirnov statistic**, $D_{n,m}$, is the maximum difference between the two cumulative probability distributions, CPD_1 and CPD_2, at any x value along the range of either number set,

$$D_{n,m} = \max\{|CPD_1 - CPD_2| \forall x\} \quad (5.9)$$

Kolmogorov–Smirnov test: An assessment comparing the maximum discrepancy between two cumulative probability distributions.

Kolmogorov–Smirnov statistic ($D_{n,m}$): The maximum discrepancy between two cumulative probability distributions, found by searching through all possible x values within the range of either data set.

This should then be compared against the **critical value**, a number that depends on the selected probability of the hypothesis test you are conducting and the sizes of the two number sets:

$$D_{n,m,P}^{crit} = \sqrt{-\ln\left(\frac{P}{2}\right) \cdot \left(\frac{n+m}{2nm}\right)} \quad (5.10)$$

In Equation (5.10), n and m are the sizes of the two samples and P is the desired probability for the hypothesis test. Because P is always less than 1, the argument of the natural logarithm operator is also always less than one, making the value from this operator always negative. Thus, the negative sign in front of it makes it always positive. If $D_{n,m} < D_{n,m,P}^{crit}$, then the hypothesis that the two distributions are the same is accepted; if the reverse is true, then the hypothesis is rejected.

Let's conduct the Kolmogorov–Smirnov test for the histograms shown in Figure 5.2. First, we should convert the histograms into cumulative probability distributions. This is done in Figure 5.6, with a second curve in each panel showing the Gaussian CPD. For the almost Gaussian distribution, the CPD closely follows the expected Gaussian CPD. The other three, however, all show what could be substantial deviations from the Gaussian curve. The skewed distribution has very few values at the low end of the range, and therefore its CPD starts later than the Gaussian curve. The heavy tail histogram has extra points far from the centroid of the distribution, and this is reflected by its CPD rising earlier than the Gaussian and arriving at the highest value of one later than the Gaussian CPD. Conversely, the light tail distribution has opposite features, rising later away from zero and arriving earlier at one than the Gaussian curve.

Critical value of the K–S test ($D_{n,m,P}^{crit}$): Part of the Kolmogorov–Smirnov test, it is the K–S statistic value above which the two distributions are considered statistically different.

138 *Data Analysis for the Geosciences*

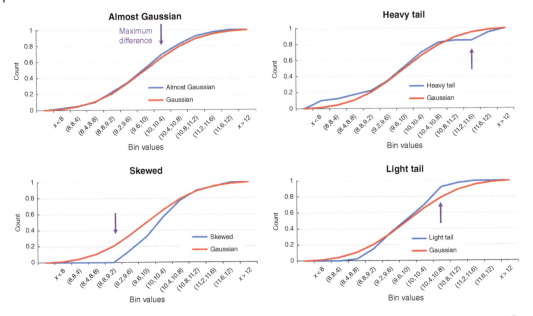

Figure 5.6 Cumulative probability distributions for the number sets shown in the histograms in Figure 5.2, along with the CPD for the expected Gaussian distribution.

From these CPD curves, the Kolmogorov–Smirnov $D_{n,m}$ value can be found. These numbers are listed in Table 5.9. For all of these distributions, $n = m = N = 40$, and if we assume a probability of 5% for the test that the distributions are the same, then $D^{crit}_{n,m,P}$ from Equation (5.10) is 0.20. According to these criteria, only one of the distributions is significantly different from a Gaussian—the one labeled skewed. This test does not reveal that either the heavy tail or light tail distributions are statistically significantly different from a Gaussian.

Table 5.9 Kolmogorov–Smirnov $D_{n,m}$ values for the number sets shown as histograms in Figure 5.2.

Distribution	$D_{n,m}$
Almost Gaussian	0.041
Skewed	0.21
Heavy tail	0.10
Light tail	0.13

There are a few points to note about these two D values. The first is that all x values should be checked. If comparing a histogram to an analytical CPD, then the edges of each bin should be included in the x values checked for the maximum difference. If two histograms are being compared, then having the bin widths and centers at exactly the same locations for the two histograms makes the comparison easier, but this simplification is not necessary. You can still conduct the test even when the bins are different; this just makes the examination to find the maximum difference a bit more tedious. Another point to note is that the $D^{crit}_{n,m,P}$ value resulting from Equation (5.10) is always greater than zero and less than one. Some values for $D^{crit}_{n,m,P}$ are listed in Table 5.10 for typical P choices under the assumption of equal-sized samples ($n = m = N$). This is a good assumption if you are creating a normal distribution from the mean and standard deviation of the sample. As seen in the table, the critical difference drops to below 0.1 for data sets of a few hundred points. With this many values in the number sets, the CPDs are very

Table 5.10 Values for $D^{crit}_{n,m,P}$ for values of $N = n = m$ and a few common P choices.

	$D^{crit}_{n,m,P}$ for $P = 0.10$	$D^{crit}_{n,m,P}$ for $P = 0.05$	$D^{crit}_{n,m,P}$ for $P = 0.01$	$D^{crit}_{n,m,P}$ for $P = 0.001$
$N = 10$	0.361	0.400	0.480	0.575
20	0.255	0.283	0.339	0.406
50	0.161	0.179	0.215	0.257
100	0.114	0.127	0.152	0.182
200	0.0807	0.0895	0.107	0.128
500	0.0510	0.0566	0.0678	0.0813
1000	0.0361	0.0400	0.0480	0.0575
2000	0.0255	0.0283	0.0339	0.0406
5000	0.0161	0.0179	0.0215	0.0257
10 000	0.0114	0.0127	0.0152	0.0182

reliable in revealing their underlying population's distribution and therefore the two CPDs must be very close to be labeled as "from the same population." With smaller data sets, especially of only 10 or 20 points, the difference can be quite large.

This is where we will end the litany of tests for comparing a given data set against a normal distribution. There are many more tests that could be brought into the assessment, but this set is enough to provide a nice collection that examines many aspects of the distribution. Each test hones in on a certain facet of the data set and provides a measure of how it relates to the expected Gaussian curve. Each test only considers certain features and is formulated in a particular way; it is perfectly reasonable that the different tests might yield opposite results, as they did for our histograms in Figure 5.2.

> **Quick and Easy for Section 5.6**
>
> The Kolmogorov–Smirnov test is another quantitative method of assessing the similarity of two histograms, being a check on the maximum difference between the cumulative probability distributions of the two data sets.

5.7 Example: pH in a Lake

As a natural science example, let's take 30 water acidity measurements from a small body of water taken every few days over a summer. A histogram of the values is shown in Figure 5.7. All of the values are given to one decimal place, that is, two significant figures. The full range of the values is shown, with the lowest being 7.4 and the highest being 8.3.

Does the histogram in Figure 5.7 look like a normal distribution? It has a nice central peak, but is slightly lopsided toward low values. It is hard to tell just by examining this distribution whether or not it is a Gaussian distribution. Let's do the calculations described

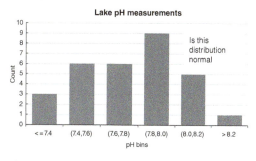

Figure 5.7 Histogram of the water acidity values.

in Sections 5.1 – 5.6 to explore its similarity to a Gaussian.

The first thing is to calculate the mean and standard deviation and use these to overplot a Gaussian distribution with the histogram of observed values. For this data set, the arithmetic mean is 7.81 and the standard deviation is 0.26 (which should then be reported as 7.8 ± 0.3). Because there are only 30 values in the set, the sample standard deviation was used, with the $N-1$ denominator. The histogram in Figure 5.7 can now be redrawn with Gaussian distribution expectation values on the same axes, shown in Figure 5.8. The upper panel is exactly the same binning of the observations as in Figure 5.7. Conducting a "by-eye" assessment, three of the six bins in the top panel are more than 50% different from the Gaussian value, which is beyond our rough estimate threshold for being "close."

The lower panel of Figure 5.8 shows the same data, but this time binned in the simplest 4-bin grouping, with bin interfaces at −1, 0, and +1 standard deviations from the mean. Here, we can see that the middle two bins are slightly undersampled relative to the expected Gaussian and the outer two bins are slightly oversampled. None of the bins, though, is more than 50% away from the Gaussian count.

The two panels of Figure 5.8 highlight the usefulness of making multiple histogram plots from the same data. A qualitative conclusion could be that the two value sets in the upper panel are not particularly close to each other, but the two value sets in the lower panel are closer and represent a reasonable match. Because one of the two comparisons is not high quality, it could be argued that a defensible initial assessment is that the distribution does not closely follow a Gaussian. Let's see what the rest of the analysis shows.

As a side note, it should be pointed out that the mean and standard deviation listed in the previous paragraph have been reported with an extra significant figure. This is one of the very few times when inclusion of extra significant figures *could* be justified. This allows a more specific Gaussian curve to be overdrawn on the histogram and a more nuanced evaluation of the various centroids and spread estimates. In addition, when calculating expected values for each bin for the χ^2 test to be conducted in a moment, the use of an extra significant figure is often very helpful to achieve a more accurate E_k for comparison to O_k in each bin. While it is often done in this context, the extra significant figure is, technically, not justified, and the values within the bins *should* be calculated based on the coarse values where there is only one significant figure in the uncertainty estimates. Furthermore, an additional argument is that this assessment of the similarity between the observed distribution and a Gaussian distribution is open to a subjective interpretation, which means that numbers do not really have to be calculated with high specificity. For this calculation and example assessment, an extra significant figure has been included in all calculations. Whichever your choice, it should be explained

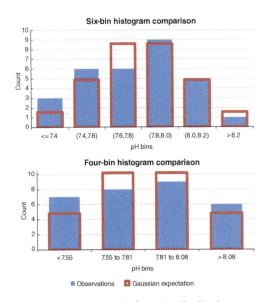

Figure 5.8 Overlay of a Gaussian distribution expectation for each bin on the histogram of the water acidity values. The upper plot uses exactly the same bins as Figure 5.7, while the lower panel shows the same data clustered into four bins, with bin interfaces at the mean, $\bar{x}$, and $\bar{x} \pm \sigma_x$.

so that those reading, viewing, or evaluating your assessment understand the methodology used to obtain the reported numbers.

The next step in the analysis of this data set is to calculate the other centroid estimates and spread estimates. For this set, both the median and mode are 7.9 (reported to the level of the data). These numbers match the central peak of the histogram. The geometric mean is 7.81 and the harmonic mean is 7.80, which are very close to the arithmetic mean. There is a discrepancy of nearly 0.1 between these estimates. This should be compared against the standard error of the data set, which is 0.05. The variation of the centroids, therefore, could be considered relatively large.

Following the plan described in Section 5.2, the spread estimates are given in Table 5.11. The two "double spread values" are 0.53 and 0.6, respectively. By comparing the end of these ranges, also reported in the table, it is seen that most of the difference arises from the shift in the lower-end value. One standard deviation below the mean is 7.55, while the equivalent quantile of a value with 16% of the numbers below it is 7.5. This is not a large difference, but one that quantifies the influence of the lopsidedness toward low values.

The next check is to calculate the skewness and kurtosis coefficients of the distribution. For our measurements, the values are $\gamma = 0.0011$ and $k = 2.0$. The skewness coefficient is nearly zero. This might seem counterintuitive because there is a clear asymmetry to the distribution. There are, however, just enough in the uppermost bin to balance out the influence of the lowest bin. Moreover, the mean is at the low end of the central peak in the histogram, making the uppermost bin farther from the mean and therefore contributing more per number than those in the lowermost bin. The net result is a skewness coefficient very close to zero. The kurtosis coefficient indicates a light tail. This is because the distribution is truncated with no values beyond a small range from the centroid. A Gaussian distribution with the observed mean and standard deviation would have one or more of the 30 points farther away from the mean than is included in this set.

Next, χ^2 tests can be conducted for the two histograms shown in Figure 5.8. For the upper panel with six bins, the χ^2 coefficient from Equation (5.6) is 4.63. Because $M = 6$ for this histogram and $c = 3$ for a Gaussian fit, $d = 3$ and Equation (5.7) yields a $\tilde{\chi}^2$ score of 1.54. This value is below the $P = 0.05$ threshold value of 2.61 for $d = 3$. For the 4-bin histogram in the lower panel of Figure 5.8, M is now 4 so $d = 1$ and the two chi-squared values are equal at $\chi^2 = \tilde{\chi}^2 = 4.83$. This is well above the 3.84 threshold for $P = 0.05$ with $d = 1$. Yes, the two chi-squared tests yielded opposite answers, and furthermore these answers are opposite of the qualitative analysis earlier in this section.

Finally, let's conduct the Kolmogorov–Smirnov test. First, we need to examine the cumulative probability distribution, which is shown in Figure 5.9. This is presented with the

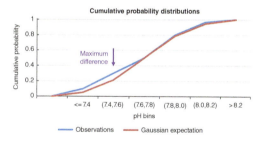

Figure 5.9 Cumulative probability distributions for both the observed lake pH values and an expected normal distribution.

Table 5.11 Equivalent spread estimates for lake pH example.

Spread quantity	$2\sigma_x$	$\bar{x} - \sigma_x$	$\bar{x} + \sigma_x$	16%–84% quantile range	16% quantile	84% quantile
Value	0.53	7.55	8.08	0.6	7.5	8.1

same bin steps as in the earlier figures. The biggest difference between the observed-value CPD and the expected Gaussian CPD occurs in the second nonzero bin, yielding $D_{n,m} = 0.09$. For $N = n = m = 30$, a probability threshold of 0.05 yields a $D_{n,m,P}^{\text{crit}}$ of 0.23. This leads us to accept the null hypothesis that they are from the same underlying distribution.

How to interpret all of this information? Let's summarize the findings. Qualitatively, the 6-bin histograms were deemed to be not particularly close, but the 4-bin histograms were okay. Some of the centroid and spread values are quite close to each other, but were divergent enough to leave doubt about the Gaussian nature of the distribution. The skewness coefficient is very close to zero, which potentially implies a Gaussian shape. The kurtosis coefficient was rather small, indicating a light tail, which is a strike against a Gaussian functional form to the distribution. The χ^2 tests were mixed, with one histogram revealing a statistically significant deviation from a Gaussian and the other not reaching this threshold. The Kolmogorov–Smirnov test indicated that it is close to a Gaussian distribution. There is no clear decision for or against similarity to a Gaussian distribution (Table 5.12).

Remember, the interpretation is a personally subjective judgment, so in this case, with mixed results from the various checks and tests, there is no right answer. A good argument could be made that this is far enough away from an ideal Gaussian distribution that it is better to use the median as the centroid and a quantile range as the spread, rather than using the arithmetic mean and standard deviation as the centroid and spread. You might come to the opposite conclusion, though, and you would be justified in making that conclusion.

The point of this example is that there are many tests that you can conduct to answer the question—"is it normal?"—when examining a data set. These assessments test different aspects of the distribution and therefore no single test should be relied upon to make this decision. When confronted with the issue of testing normality, it is advisable to use many metrics and conduct a robust assessment. If they all agree, then the answer is clear. The outcomes might conflict, though, as they do in this example case, leaving you with the subjective call regarding the yes or no answer to the posed question.

> **Quick and Easy for Section 5.7**
>
> The pH in a lake is used here as an example of how to conduct a robust assessment of whether a data set's distribution is Gaussian.

5.8 Asymmetric Uncertainties

If it is known that the uncertainties in the two directions away from the centroid are different and it is desirable to report this difference, then there are several methods for conducting this calculation. When used, they are often reported as superscript and subscript to the base value. Here is the general form as well as an example:

$$x_{-\delta x_{dn}}^{+\delta x_{up}} \rightarrow 14_{-3}^{+8} \quad (5.11)$$

Note that the use of **asymmetric uncertainties** is not standard and should be implemented only when there is a logical and defensible

Table 5.12 Summary of the assessment of the Gaussian nature of the pH measurement example.

Test	Assessment
Centroid check	Not Gaussian
Spread check	Gaussian
Skewness check	Gaussian
Kurtosis check	Not Gaussian
χ^2 tests	Yes and no
K–S tests	Gaussian

Asymmetric uncertainties: Reporting different spreads above and below a listed quantity, sometimes desired to indicate a large skew.

reason to do so. That said, asymmetric uncertainties are used in some studies to indicate a large skew to the distribution and so it is useful to be aware of them.

There are several reasons for using them, but they all come down to the skew of the number set distribution. One way to convey skew to readers is to report the uncertainty as an asymmetric pair of values. This catches their attention and alerts them to the fact that the distribution is skewed. One reason for a skewed distribution could be that the base value is positive definite, such as snowfall amounts. The uncertainty below the centroid could be small, but above the centroid it could be large due to some very high values in the measurement set. To reflect this asymmetry in the distribution, two uncertainties could be reported, one applicable to the lower half of the distribution below the centroid and the other for the high end of the values above the centroid. Another reason is that the distribution is Gaussian with a logarithmic x axis, which means that it is skewed right in linear space. It could also be that you have reason to suspect an asymmetric uncertainty due to a systematic issue with the methodology of determining the number set.

The easiest and arguably the most straightforward method of determining asymmetric uncertainty is to report the 25% and 75% quartiles. While smaller than the standard deviation, they are a well-known and defensible measure of spread. When using quartiles for the uncertainty, it is best to report the median as the centroid, for calculational consistency.

Another simple method is to report two HWHM values, one determined at the 50% peak amplitude x value below the mode and the other one found at the 50% amplitude x value above the mode. When using this measure of spread, it is best to report the mode as the centroid of the number set.

Another way is to calculate the standard deviation twice, using the same arithmetic mean but subdividing the number set into the values above and below the mean for the two σ calculations.

There is also the issue of propagating uncertainty when asymmetric uncertainties are reported for one or more of the parameters in the calculation. One way is to calculate the uncertainty twice, once with only the lower uncertainty values and then again with only the upper values, with either additive or quadrature formulas. This is only reasonable with simple formulas in the easiest cases. In general, the two sides of the asymmetric uncertainties will mix, with the small-side uncertainty from one parameter combining with the large-side uncertainty from another. Eventually, given a complicated-enough formula or multiple steps, this leads to a symmetric Gaussian distribution for the computed parameter and its range of possible values. Remember the central limit theorem from the very beginning of Chapter 5; the combination of distributions will eventually yield a Gaussian.

A more complicated method of propagating uncertainty given input parameters with asymmetric uncertainties is to construct a propagated value distribution by sampling from the full distributions of the input parameters. If the distributions are not known, and only the asymmetric uncertainties are given, then define two half-Gaussians on either side of the reported centroid. There will be a discontinuity at the middle, one side broad and lower and the other narrow and taller. Using the cumulative probability distribution from each of these input parameter distributions, you can then select a random value from each input parameter number set. These can be used in the formula to obtain a value for the new computed parameter. This process should be repeated many times, conducting many calculations through the formula with different input parameter values. From all of these calculations, you get a distribution of outcomes, from which you can make a histogram, which is perhaps Gaussian but maybe not. From this, a centroid and spread can be determined using any of the methods from Chapter 4, with the uncertainty reported as either symmetric or asymmetric, as desired.

> **Quick and Easy for Section 5.8**
>
> There are several methods for calculating asymmetric uncertainties as well as propagating them through a formula that uses the base value.

5.9 Outliers—Tests for a Single Data Value

One of the problems with the whiskers of a box plot, or a justification for reporting an asymmetric uncertainty, is that the distribution has **outliers** from the centroid. Sometimes, though, the outlier is not a physically meaningful data value, but a spurious measurement that should be discarded. Usually, we do not know why one value is different from the rest, but we can guess at plausible reasons why a measurement might be tainted and therefore omitted from later analysis. Sometimes we do know the reason, though. This might happen when a helicopter flies over your field experiment station, completely dominating the wind measurements from the anemometer. It could happen when someone is smoking next to your volatile organic compound monitor. It could be that your magnetometer signal is dwarfed by the engine running in the logging truck passing by on a nearby road. It could be a flashlight inadvertently swept across the field of view of your all-sky camera, polluting your observations of the overhead auroral emissions. It could be galactic cosmic rays creating false counts in your space telescope's imaging array. Spurious outliers can happen.

When is it acceptable to discard one of your measurements? This is reasonable only when it is very improbable that the measurement came from the same population as the other values. This is a very hard fact to know for sure. The Gaussian distribution extends to infinity in both directions, so it can be argued that *nothing* should be thrown out. This is also an acceptable choice. The same argument can be used to discard outliers, though, because the Gaussian distribution should have very few values that are far from the mean. Remembering the Gaussian CPD, 95.4% of values are within two standard deviations from the mean, and this percentage increases to 99.7% within three standard deviations. Depending on N, the number of samples in the set, these percentages imply that there are probably few if any values this far from the mean.

One method for deciding when to discard a value from the set is **Chauvenet's criterion**. This uses the power of the error function to estimate the probability of naturally obtaining the suspect value, x_{sus}. The process is rather simple, and one with which we are already familiar. The first step is to calculate the mean and standard deviation, $\bar{x}$ and σ_x, of the number set, including x_{sus}. The second step is to calculate a z score for the suspect value:

$$z_{sus} = \frac{|x_{sus} - \bar{x}|}{\sigma_x} \quad (5.12)$$

The two-sided probability for this z score, P_{sus}, should then be found. It should be the two-sided probability because the outlier could just have easily been on the other side of the mean. One of the only reasons to use a one-sided probability would be if the value were positive definite and this z score would be an impossible negative value on the other side. With this probability, you then calculate the expected count of values in the number set with this z score or larger away from the mean:

$$n_{sus} = N \cdot P_{sus} \quad (5.13)$$

Outliers: Data points far from the centroid, they are sometimes rejected as valid members of the set and removed.

Chauvenet's criterion: A test regarding the acceptability of an outlier within a data set, it is defined as the z score at which the expected number of observations drops below 0.5.

Chauvenet's criterion states that if n_{sus} is less than 0.5, then the value should be rejected as a plausible entry in the set, and it should be removed.

This threshold of 0.5 is justified because it rejects data that, according to a Gaussian distribution, should not exist in the number set. That is, there should be less than half of one sample in this range, and therefore any value in this range is probably not from the same population but rather a spurious measurement. As N increases, the P_{sus} needed to have n_{sus} below 0.5 decreases, which means that z_{sus} must increase. That is, with a larger N, a broader spread of points is allowed because the probability of naturally obtaining those outliers increases above the 0.5 threshold.

Chauvenet's criterion is not universally accepted as the right process for all situations. For one, the threshold of 0.5 has been argued to be rather high, allowing for the rejection of data that just barely cross this threshold. Some adjust Chauvenet's criterion with a different threshold for rejection, reducing it to 0.25, 0.1, or even lower. Another filter is the "three sigma rule," in which everything beyond $\pm 3\sigma$ is removed from the data set, but only values at this distance from the centroid. It should, on average, remove three values for every 1000 numbers in the set. This criterion will remove very few data points from a small sample, but, if the number set is in the millions, then it could remove thousands of values, which might be unreasonable. Some argue that collected data should never be discarded, but instead all values should be included in the full analysis unless there is a legitimate external factor that is known to have influenced the measurement. While this is justifiable—the values were in the observation set, after all—this could lead to the inclusion of truly spurious measurements of unknown cause, which could then lead to incorrect physical interpretations.

This process of considering the removal of data points is always a subjective call by the analyst. The particular rule just discussed, Chauvenet's criterion, is only a guideline. The other methods just discussed are also guidelines. If another method is preferred, then apply it. The key point is to convey the process to the eventual recipients of the analysis, so that they can include this information in their interpretation of the results and conclusions. Furthermore, in any case, do not discard more than a few percent, at most, of the sample set. If there is a cluster of anomalous values, then a careful examination needs to be conducted to decide if the values should be rejected. It could be that the cluster is spurious. For instance, on a scientific satellite, a circuit in a satellite subsystem might cause a repeatable signature in a data stream for one of the scientific sensors. It could be, though, that the cluster is real, representing a rare but stable state of the observed parameter. For a cluster, all or none of the values should be discarded. If your rejection method would cut off only some of the cluster, then ignore or revise the method.

After one or more of the data values have been rejected, then the statistics of the data set should be recalculated. These new values are the ones that should be used for further analysis. Always clearly state what values were discarded and what criteria were used to determine what to discard. Furthermore, arguments should be included that might justify the removal of the values, even if these are guesses. Also, only apply Chauvenet's criterion, or any data rejection technique, once. Do not iteratively apply these methods or you could end up eroding the data set and removing physically meaningful values from the sample.

> **Quick and Easy for Section 5.9**
>
> Chauvenet's criterion is a common method for identifying outlier data that could be removed from the number set. Remove the values and recalculate the centroid and spread.

5.10 Combining Centroid and Spread: The Weighted Average

What about the case where the values going into the estimate of the centroid or spread have different spreads? For instance, consider the case of taking several sets of samples. Each set has its own mean and standard deviation. How should these be combined to obtain an aggregate mean and uncertainty value?

A possible path forward is the additive formula of uncertainty propagation discussed in Chapter 3. This would lead to a combined mean value with the functional form of equation of the arithmetic mean:

$$\overline{x}_{comb} = \frac{\overline{x}_1 + \overline{x}_1 + \cdots + \overline{x}_{N_{sets}}}{N_{sets}} \qquad (5.14)$$

The combined uncertainty, assuming that the measurements were independent and could be propagated with a quadrature method, would have the form of quadrature addition:

$$\delta q_{comb} = \frac{1}{N_{sets}} \sqrt{(\delta q_1)^2 + (\delta q_2)^2 + \cdots (\delta q_{N_{sets}})^2} \qquad (5.15)$$

While this is fine for a quick calculation, it is not correct and could lead to a rather inaccurate estimate of both $\overline{x}_{comb}$ and δq_{comb}. This is not a robust calculation because it treats all sample sets the same. If the uncertainties are different between the sample sets, then those with smaller uncertainties should be trusted more in the creation of a combined average value and uncertainty. Even if we believe that the underlying population has a Gaussian distribution, simply combining the values to get an arithmetic mean from Equation (5.14) and a standard deviation from Equation (5.15) is not appropriate because each of the measurements has its own uncertainty.

A **weighted average** should be conducted to obtain these estimates of centroid and spread of the combined set. To work this out, let's assume that two measurements, x_A and x_B, come from the same population but were recorded with different uncertainties. Let's further assume that the underlying population from which these measurements were taken has a Gaussian probability distribution with an unknown centroid of X. In this case, we can write these two equations for the probabilities that the recorded values came from the same population:

$$P_A = \frac{1}{\sigma_A} e^{-\frac{(x_A - X)^2}{2\sigma_A^2}} \qquad (5.16)$$

$$P_B = \frac{1}{\sigma_B} e^{-\frac{(x_B - X)^2}{2\sigma_B^2}} \qquad (5.17)$$

The combined probability of obtaining these two specific measurements is the multiplication of P_A and P_B:

$$P_{AB} = \frac{1}{\sigma_B \sigma_B} e^{-\frac{\xi^2}{2}} \qquad (5.18)$$

In Equation (5.18), the ξ symbol is a convenient placeholder for the full exponential argument:

$$\xi^2 = \frac{(x_A - X)^2}{\sigma_A^2} + \frac{(x_B - X)^2}{\sigma_B^2} \qquad (5.19)$$

We can now apply the principle of maximum likelihood to determine the best estimate of X. This principle assumes that the probability distribution in Equation (5.18) has one peak located which can be determined by finding the x value where the slope of P_{AB} is equal to

---

Weighted average: A method of finding the centroid of a number set that takes into account the different uncertainties for each value in the set.

zero. The derivative of Equation (5.18) with respect to the true mean, X, is this:

$$\frac{dP_{AB}}{dX} = \frac{-1}{2\sigma_B \sigma_B} e^{-\frac{\xi^2}{2}} \frac{d(\xi^2)}{dX} \quad (5.20)$$

This derivative should be set equal to zero. The two spreads in the denominator are nonzero values, as is the exponential term, so only the last term can be zero. Taking the derivative of Equation (5.19) with respect to X yields this:

$$\frac{d(\xi^2)}{dX} = \frac{2(x_A - X)}{\sigma_A^2} + \frac{2(x_B - X)}{\sigma_B^2} \quad (5.21)$$

This should set equal to zero and solved for X, giving us a final formula for the combined centroid:

$$X = \frac{\left(\dfrac{x_A}{\sigma_A^2} + \dfrac{x_B}{\sigma_B^2}\right)}{\left(\dfrac{1}{\sigma_A^2} + \dfrac{1}{\sigma_A^2}\right)} \quad (5.22)$$

Comparing Equation (5.22) with the simple version of Equation (5.14), it is seen that the difference is the inclusion of σ_i^2 in the denominator of every term. If the two σ values are equal, then Equation (5.22) reduces to Equation (5.14). Because they are in the denominator, a larger σ leads to a smaller contribution of that value to the combined average. That is, that value is weighted less in the overall average, and the x_i values with small σ_i are weighted more heavily in this calculation. These extra terms can be considered as the **weights**:

$$w_A = \frac{1}{\sigma_A^2} \text{ and } w_B = \frac{1}{\sigma_B^2} \quad (5.23)$$

Weights: The coefficients applied within a weighted average are defined as the inverse of the variances for each number in the set.

Equation (5.22) can now be written in a slightly more compact form using the weights defined by Equation (5.23):

$$X = \bar{x}_w = \frac{w_A x_A + w_B x_B}{w_A + w_B} \quad (5.24)$$

In Equation (5.24), X is now defined as the weighted average, $\bar{x}_w$, of the values.

The two-value simplistic form of Equation (5.24) can be generalized for the combination of any number of values, each with their own uncertainty:

$$\bar{x}_w = \frac{\sum w_i x_i}{\sum w_i} \quad (5.25)$$

In Equation (5.25), the two summations are over all of the values contributing to the weighted average. The weights are defined in the same way as those in Equation (5.23).

The spread estimate for this weighted average also gets a new formula. Skipping the derivation, the final form of a combined weighted-average standard deviation is this:

$$\sigma_w = \sqrt{\frac{N}{\sum w_i}} \quad (5.26)$$

Without the N in the numerator, Equation (5.26) becomes the weighted-average standard error. Because w_i is defined as the inverse of the square of the individual σ_i values, the weighted value, σ_w, has the same units. It resembles a harmonic mean, but includes a square and square-root operator that makes it a harmonic quadrature calculation.

While the weighted-average formula in Equation (5.25) is commonly used throughout many disciplines, it should be remembered that certain assumptions were made in its derivation. Specifically, it was assumed that the samples (or sets of samples) all came from the same overall population, and that this underlying population from which all values were drawn has a Gaussian distribution. If the full population is not Gaussian, then this form of the weighted average is not correct. For a skewed or multimodal distribution, the weighted-average formula should be derived based on the expected functional form of the underlying population's probability distribution. This could be the case for values that are positive definite, with a truncated left-side tail at x values below zero. It could also be the case for stratum-specific data, for which there are hard endpoints to the range. Alternate versions of the weighted average are hardly ever used, though, even when these limitations are known and clearly invalidating of Equation (5.25). Usually, the analysis simply uses the Gaussian-based weighted average given above and any caveats about the applicability of this formula are noted.

> **Quick and Easy for Section 5.10**
>
> The weighted average calculation takes into account a different uncertainty for each value within the number set.

5.11 Example: pH in a Lake Redux

To make the last few concepts more tangible, let's continue the example from Section 5.7 of the pH readings in a lake. To examine the case of outliers, let's say that two more measurements are included in the data set, one at 7.1 and another at 6.6. The new mean and standard deviation are 7.753 and 0.355, which should be reported as 7.8 ± 0.4. The original data set had a lowest value of 7.4, so both of these two new measurements are well below the previous lower limit of the range. They should be checked against an outlier criterion to determine if they should be retained in the set or rejected as spurious and removed from the set.

Before z scores can be calculated for these two new values, the uncertainty needs to be slightly adjusted. In particular, we need to remember back to Section 4.3 (Chapter 4) and the combination of various sources of uncertainty. Three were discussed: uncertainty among the values within the data set itself, random uncertainty in the measurement technique, and systematic uncertainty in the measurements. Unless there is a reason to suspect its existence, systematic uncertainty is often assumed to be zero. Furthermore, it is often assumed that the uncertainty from the technique is independent of the uncertainty in the values themselves, which is supposedly from external environmental conditions. Therefore, we can apply Equation (4.28),

$$\delta x_{\text{total}} = \sqrt{\delta x_{\text{meas}}^2 + \delta x_{\text{rand}}^2} \qquad (5.27)$$

For δx_{meas}, we have 0.4. For δx_{rand}, we should assign an uncertainty to each of the values because of the measurement itself. Without other information, let's assume that each value has an uncertainty of 0.05, half-way between the increments of the values in the list (all are given to one decimal place, i.e., two significant figures). In this case, Equation (5.27) becomes $\delta x_{\text{total}} = \sqrt{(0.4)^2 + (0.05)^2}$ which yields 0.403, which should be reported as 0.4. This extra calculation did not change the reported uncertainty value of 0.4, but it could have and needs to be done.

We can now calculate z_{sus} values for both of these new measurements using Equation (5.12). This yields scores of 1.75 and 3, respectively. This can be used to determine an expected number of values at this z score or beyond, multiplying the two-sided probability of z_{sus} against 32, the number of points in the data set, as done in Equation (5.13). This yields n_{sus} values of 2.6 and 0.086, respectively.

Chauvenet's criterion states that the cutoff for n_{sus} is 0.5. Below this threshold, the value should be rejected, while n_{sus} above this threshold indicates that the value should remain in the set. For the reading of 7.1, the expected number of 2.6 is well above the criterion of 0.5, so this value should remain in the set. For the reading of 6.6, however, the expected number of 0.86 is well below the 0.5 cutoff, so this value should be rejected.

With the rejection of this value, a new mean and standard deviation should be calculated. These new values are 7.79 and 0.29, respectively. The uncertainty, which should be rounded to 0.3, replaces the δx_{meas} in Equation (5.27), so a new δx_{total} from this equation is 0.304, and the final value of the centroid and spread is reported as 7.8 ± 0.3.

Using this new set, let's do one more calculation with it, a weighted average. Let's assume that the measurement technique was more accurate for higher pH levels. For convenience, let's use the four-bin histogram in Figure 5.8 to define the groupings for different measurement uncertainties. In the top bin (those values above 8.08, so those that are 8.1 or higher), the readings have 0.05 uncertainty, for next lower bin, it is 0.1, then 0.2, and finally 0.3 for those measurements below 7.55 (that is, those with 7.5 or lower, including the new value at 7.1).

The weighted average of the number set can be calculated. The average is found with Equation (5.25). This formula yields a new average value of 8.08. This has to be compared with a new uncertainty value. Of the two sources of uncertainty in Equation (5.27), Equation (5.26) replaces δx_{rand}. This formula yields δx_{rand} of 0.093, which is substantially larger than the 0.05 uniform value used earlier. A recalculation of Equation (5.27) yields 0.32. So, the overall uncertainty did not change, but the mean value shifted upward because of the weighted-average calculation, and the new centroid and spread should be reported as 8.1 ± 0.3.

Quick and Easy for Section 5.11

The pH in a lake is used again, this time as an example of how to conduct a test for outliers and a weighted average calculation.

5.12 Further Reading

Because the Gaussian distribution is the focus of this chapter, it is useful to restate a website where cumulative probability distribution integrals for z-score ranges can be calculated:
https://www.hackmath.net/en/calculator/normal-distribution

There are many good books that provide explanations of these same concepts. For these topics, I like these two:

Alan Chave's statistics book also has essentially all of this chapter's content covered, sprinkled throughout the text:

Chave, A. D. (2017). *Computational Statistics in the Earth Sciences.* Cambridge University Press, Cambridge.
https://www.cambridge.org/core/books/computational-statistics-in-the-earth-sciences/6352412A284512AF4F38766E3717FE38

Some of this content, like the chi-squared test, Chauvenet's criterion, and weighted averages, can be found in John Taylor's error analysis book:

Taylor, J. R. (1997). *An Introduction to Error Analysis.* University Science Books, Mill Valley, CA, USA.
https://uscibooks.aip.org/books/introduction-to-error-analysis-2nd-ed/

Here is a reference that defends the value of a skewness coefficient of 1 indicating high skew:

Goodliff, M. R., Fletcher, S. J., Kliewer, A. J., Jones, A. S., & Forsythe, J. M. (2022). Non-Gaussian detection using machine learning with data assimilation applications. *Earth and Space Science, 9,* e2021EA001908. https://doi.org/10.1029/2021EA001908

Here is a reference for the derivation of the kurtosis of a Gaussian distribution:
https://proofwiki.org/wiki/Kurtosis_of_Gaussian_Distribution

A quick calculator for probabilities from chi-squared values (*not* reduced chi-squared values):
https://homepage.divms.uiowa.edu/~mbognar/applets/chisq.html

And here is another one:
https://www.socscistatistics.com/pvalues/chidistribution.aspx

A good write-up of the Kolmogorov–Smirnov test can be found here:
https://www.itl.nist.gov/div898/handbook/eda/section3/eda35g.htm
And another one, from a rather different and informal perspective, is here:
https://towardsdatascience.com/kolmogorov-smirnov-test-84c92fb4158d

5.13 Exercises in the Geosciences

These questions explore monthly averaged temperatures compiled across the United States. By focusing on these months, there is some hope that the distribution might be Gaussian. Some intervals are close to this functional form, others are not.

1. Evaluate the *winter* temperatures of the last century for the contiguous United States, specifically values only from the months of January and February, not the full year. The data are available here: https://www.ncdc.noaa.gov/cag/national/time-series
   A Make a time series plot of these temperature values from the beginning of the set (1885) through the present.
   B Make a histogram of these temperatures with an appropriate number of bins.
   C Make a cumulative probability distribution of these temperatures.
   D Calculate the mean and standard deviation of this data set.
   E Calculate anomaly values (discrepancy of individual values against the mean) and the z values for each number in the set.
   F Apply a criterion for rejecting any outliers in the data set. It is okay to determine that none should be removed. Recalculate the mean and standard deviation, if necessary.

2. Write a paragraph or two answering the following:
   A Explain the method used for setting the number of bins in the histogram.
   B Explain the method of considering the rejection of individual data values from the set, and justify the exclusion of any values.

3. Create a second histogram with a Gaussian distribution of the same mean and standard deviation.
   A Calculate skew and kurtosis of the full set.
   B Conduct a chi-squared test between the histograms of the observed temperatures and this related Gaussian distribution.
   C Conduct a Kolmogorov–Smirnov test between the cumulative probability distribution of the observed temperatures and the related Gaussian.

4. Write a paragraph or two answering the following:
   A Explain why your skew and kurtosis calculations of the data set are, or are not, consistent with a normal distribution.
   B Assess the outcomes of the chi-squared and Kolmogorov–Smirnov tests.

5. Let's assume that we have different uncertainties for each year of the data set that follows this functional form (Y is the near, the four out front has units of temperature):

$$\delta T_Y = 4 \cdot \exp\left(-\frac{Y-1895}{50}\right)$$

This formula produces an uncertainty that is 4 °C for the year 1895 and only 0.5 °C for the year 1999. Note that this reduction in uncertainty with time is purely illustrative and, while correctly decreasing with year, is not the true uncertainty of these compiled temperature values. Calculate a weighted average and weighted standard deviation using this formula as an uncertainty on each individual yearly value in the set, for the years 1895–1999.

6. Write a paragraph that discusses the difference between the mean and standard deviation of the uniform weighting compared to that using the year-dependent uncertainty and weighting.

7. Now consider only the most recent years, from 2000 to the present. Again, consider the data set that includes only the months of January and February.
   A Calculate the mean, standard deviation, skew, and kurtosis for this subset of the most recent years.
   B Create a plot of your choice comparing the base period (1999 and earlier) to the most recent years of data (from year 2000 to present).

8. Write a paragraph answering the following:
   A Compare the years 2000 onward with the base period of 1999 and earlier. What does this tell you about the recent temperatures in the wintertime period?
   B In this assignment, you are reviewing the contiguous US temperatures. What limitations must you consider if you want to discuss the impact such analysis has on the winter temperature anomalies in the state of Michigan? What changes to the analysis would be necessary to conduct a more appropriate assessment of temperature change in only Michigan?

9. From the *full* data set of all months:
   A Make a box plot of temperature in Fahrenheit for each month. Put these box plots all in the same figure, with labels as appropriate to distinguish each month. There will be 12 box plots in total.
   B Make sure each box plot clearly shows:
      i) Median.
      ii) Interquartile range.
      iii) Any outliers.

10. Write a paragraph answering the following:
    A Based on your box plots, do you suspect any data are from a different population than monthly temperature averages from the US? Why or why not? (Are there outliers? How did you determine what is an outlier?)
    B Discuss the need to use an asymmetric uncertainty for any of the months.

6

Correlating Two Data Sets: Analyzing Two Sets of Numbers Together

Up to now, the focus has been on processing one data set. With mastery of that content, it is now time to add in a second one. We have already done this with the Gaussian distribution as our second number set in Chapter 5. While the previous two chapters provide a foundational collection of graphs, equations, and procedures to understand a single number set, this chapter builds on that foundation, presenting and discussing the critical elements of analyzing two data sets together. The combination of two data sets together provides insight into the connection between them, leading to new physical understanding and new knowledge for humanity.

Processing two data sets together is about examining the relationship between them. Are the two distributions similar? Are they linearly correlated? Is there a nonlinear relationship between them? This chapter and the next provide a battery of methods that are essential for this assessment.

While important, how the two data sets were obtained is not the point of this analysis. The two number sets could be measured at the same place or very far from each other. The main concern is to connect the two data sets with a possible physical process. Without this, you are fishing for answers where none exists and the correlations that you find might have no meaning. This is part of the scientific process—have a hypothesis, an educated guess, about what might be going on in the data set of interest, and then find other data sets that might be the driving parameter that governs the physical, chemical, or biological process causing the unusual feature that you noted. That is a critical step in the scientific process and, if left out, can lead to serious misinterpretations.

6.1 Comparing Two Number Sets

Sometimes, two data sets are taken of the same quantity, but perhaps in different ways, in different locations, or at different times. In this case, it might be very useful to compare the similarity of the two sets of measurements. Various techniques have been developed to do this. In fact, two such methods have already been discussed in detail in Chapter 5—the χ^2 test and the Kolmogorov–Smirnov test. Another major method for comparing means between two number sets is known as the t test. There are a few different t tests, but the two that are described in Section 6.1 are arguably the most famous and most widely used in the natural sciences. These are the Student's t test and Welch's t test.

None of these tests may be needed for an assessment of two data sets. They are only needed if the two sets are thought to be the same and a statistical assessment of this

Data Analysis for the Geosciences: Essentials of Uncertainty, Comparison, and Visualization, Advanced Textbook 5, First Edition. Michael W. Liemohn.
© 2024 American Geophysical Union. Published 2024 by John Wiley & Sons, Inc.
Companion website: www.wiley.com/go/liemohn/uncertaintyingeosciences

similarity is desired. If, instead, the two data sets are related through a cause-and-effect pairing, they are then most likely measurements of completely different quantities, with different ranges and different units. None of these tests should, therefore, apply. These tests are only useful when the two data sets are of the same units, and it is desired to a probability regarding their statistical equivalence.

6.1.1 Chi-Squared and Kolmogorov–Smirnov Tests

These two tests were already covered in detail in the previous chapter, so only a very brief recap is given here. Specifically, these two tests should compare the two measured number set histograms, rather than comparing one data set against its expected Gaussian distribution. The swap of the second data set for the expected Gaussian is straightforward, but there are a couple of small issues to discuss. Also, please remember the caveats associated with these tests discussed in Chapter 5.

The χ^2 test (see Section 5.5) assesses whether the shape of one distribution is similar to another. By considering a histogram of observed values against expected values of the count within each bin, the χ^2 test uses z scores from each bin and combines them into a single metric score, the reduced chi-squared value, $\tilde{\chi}^2$. This value can be converted into a probability that the two histograms—observed and expected—both describe the same population.

In this context, the χ^2 test could be used to compare the distributions of the two data sets—the x and y values of a scatterplot—against each other. The small issue here is that one of the two data sets has to be assumed to be the "expected value," with the histogram counts from that set used to define the denominator spread in the z scores. This is usually the x-axis value set, but it does not have to be and just as easily could have been the y-axis number set instead, because our hypothesis is that the two data sets are the same. So keep in mind this potential swapping of the x and y variables

in these equations. Using M bins, the z score for a particular bin k is therefore this:

$$z_k = \frac{|y_k - x_k|}{\sqrt{x_k}} \qquad (6.1)$$

The $\tilde{\chi}^2$ score for the comparison is therefore this:

$$\tilde{\chi}^2 = \frac{1}{d}\sum_{k=1}^{M}\frac{(y_k - x_k)^2}{x_k} \qquad (6.2)$$

In Equation (6.2), the degrees of freedom, d, is simply the number of bins in the comparison, M, because no information from the y number set was used in the creation of the "expected" value set. This is another change from using the original data set to define the expected normal distribution. This $\tilde{\chi}^2$ score can then be converted into a probability, either using one of the tables in Chapter 5 or an online calculator.

The number of bins used in this comparison can be as high or low as desired, and the resolution of the binning can be as fine or coarse as desired. There is no set rule on this and there is no obligation to use the full range of either data set in the comparison.

The Kolmogorov–Smirnov test (see Section 5.6) is also readily converted to use with two data sets that we think might be measurements of the same thing. The Kolmogorov–Smirnov statistic, $D_{n,m}$, should be calculated from the two cumulative probability distributions, CPD_x and CPD_y, choosing the largest difference at any value along the range of either number set,

$$D_{n,m} = \max\{|CPD_x - CPD_y| \forall x, y\} \qquad (6.3)$$

The other number to calculate is the critical difference value:

$$D_{n,m,P}^{crit} = \sqrt{-\ln\left(\frac{P}{2}\right)\cdot\left(\frac{n+m}{2nm}\right)} \qquad (6.4)$$

In Equation (6.4), remember that n and m are the sizes of the two data sets and P is the desired probability for the hypothesis test.

If $D_{n,m} < D_{n,m,P}^{\text{crit}}$, then it can be stated that the two distributions are the same, otherwise this test shows that they are different.

> **Quick and Easy for Section 6.1.1**
>
> The chi-squared test and Kolmogorov–Smirnov tests are two quantitative methods of assessing whether two distributions might have originated from the same underlying population.

6.1.2 The Student's t Test

The **Student's t test** is an assessment of the mean of a data set against the population mean. Note that this is a professional-grade statistical test, but one originally published under the pseudonym "Student." It is directly related to the z score and z test, but now raised to the level of the sample mean rather than being for an individual value. Versions of the Student's t test are also applicable for two samples, even samples of unequal total counts and sample standard deviations. The similarity of all of these Student's t test versions is that they assume that the underlying populations from which the samples are taken all have the same variance, σ^2. If you are claiming that they come from the same population, then this is a reasonable assumption to make, as the underlying population should have a single variance associated with it.

The first version of the Student's t test is for a single sample, with mean $\bar{x}$ and total number count N, against a known mean and variance, X and σ^2. The formula for Student's t test takes this form:

$$t = \frac{|\bar{x} - X|}{\left(\sigma/\sqrt{N}\right)} \quad (6.5)$$

Student's t test: An assessment of two sample sets to determine whether they might have originated from the same underlying population.

Equation (6.5) has a very similar form to that for a z score, but this time the denominator is the standard error rather than the standard deviation. This is because the test is for the mean of the sample, $\bar{x}$, rather than a single value within the sample, x_i.

The degrees of freedom for this t score in Equation (6.5) is $d = N - 1$. This value is needed in order to convert the t score into a probability that the sample mean is similar to the population mean. Table 6.1 shows the t test values needed to reach certain probabilities that the number sets are the same. There are two probabilities listed across the top of the columns, for both the one-sided and two-sided tests. Most of the time, a two-sided test is needed, but there are some instances when a one-sided test is necessary. Note that the t score required to meet a certain probability is a function of d, listed in the first column. For $d = 1$, rather large t scores are needed, because the small sample size increases the uncertainty of the mean. As d increases, the t score needed to reach a given probability decreases. The value asymptotes to the z score probabilities, given in the final row of Table 6.1.

When you want to compare two samples instead of a sample against a known mean, then the formulas change slightly. In this case, the population mean and standard deviation do not need to be known, but instead are approximated by the sample parameters. In the case of equal sample sizes, $N_1 = N_2 = N$, but unequal sample means, $\bar{x}_1$ and $\bar{x}_2$, and unequal sample standard deviations, σ_1 and σ_2, the Student's t test formula changes to this:

$$t = \frac{|\bar{x}_1 - \bar{x}_2|}{\sigma_p \sqrt{\dfrac{2}{N}}} \quad (6.6)$$

In Equation (6.6), σ_p is an approximation for the population standard deviation. This approximation assumes that the population variance is the arithmetic average of the two sample variances:

$$\sigma_p^2 = \frac{\sigma_1^2 + \sigma_2^2}{2} \quad (6.7)$$

Table 6.1 The *t* score values needed in order to meet certain probabilities that the number sets are the same (that score or lower for similarity), listed with respect to the degrees of freedom, *d*.

One-sided	10%	5%	2.5%	1%	0.5%	0.01%
Two-sided	20%	10%	5%	2%	1%	0.02%
$d = 1$	3.08	6.31	12.7	31.8	63.7	318.
2	1.89	2.92	4.30	6.97	9.93	22.33
3	1.64	2.35	3.18	4.54	5.84	10.21
4	1.53	2.13	2.78	3.75	4.60	7.13
5	1.48	2.02	2.57	3.37	4.03	5.89
6	1.44	1.94	2.45	3.14	3.71	5.21
7	1.41	1.90	2.37	3.00	3.50	4.79
8	1.40	1.86	2.31	2.90	3.36	4.50
9	1.38	1.83	2.26	2.82	3.25	4.30
10	1.37	1.81	2.23	2.76	3.17	4.14
20	1.33	1.73	2.09	2.53	2.85	3.55
100	1.29	1.66	1.98	2.36	2.63	3.17
200	1.29	1.65	1.97	2.35	2.60	3.13
∞	1.28	1.65	1.96	2.33	2.58	3.09

The 2 in the denominator of Equation (6.7) will cancel the 2 in the denominator of Equation (6.6), but it is included in both equations because it is needed for the definition σ_p if this value is ever desired for use separately from Equation (6.6). The degrees of freedom for this calculation is $d = 2N - 2$, which can be used with Table 6.1 to convert the *t* score into a probability that the two samples were or were not derived from the same population.

A final version of Student's *t* test to be discussed here is the case when the two sample sets are of unequal size; that is, when N_1 is not equal to N_2. In this case, the two variances are combined with a weighted average in order to get the population variance and the formula is changed as follows:

$$t = \frac{|\bar{x}_1 - \bar{x}_2|}{\sigma_p \sqrt{\frac{1}{N_1} + \frac{1}{N_2}}} \quad (6.8)$$

The formula for σ_p also changes to account for the unequal sample sizes:

$$\sigma_p^2 = \frac{(N_1 - 1)\sigma_1^2 + (N_2 - 1)\sigma_2^2}{N_1 + N_2 - 2} \quad (6.9)$$

This weighted averaging process also very slightly changes the degrees of freedom calculation to be $d = N_1 + N_2 - 2$. The same conversion can then be done, using Table 6.1, to obtain a probability that the two samples came from the same population.

Quick and Easy for Section 6.1.2

Student's *t* test is a method for assessing the similarity of the means of two number sets.

6.1.3 The Welch's *t* Test

There are times when it cannot be assumed that the underlying population is the same, but nevertheless it is desired to test two sample means against each other to assess the chance that they can be declared to be "the same." By "the same" it is meant that the two samples are statistically equivalent. The Student's *t* test is no longer appropriate for this case, but fortunately another *t* test has been derived to fill this need—**Welch's *t* test**. The result is similar in that a *t* score can be compared against probabilities that such a score would arise from two samples that are actually equivalent. Welch's *t* test still takes a form very close to a *z* score function:

$$t = \frac{|\bar{x}_1 - \bar{x}_2|}{\sigma_{\bar{\Delta}}} \quad (6.10)$$

The combined standard deviation is also relatively straightforward to calculate:

$$\sigma_{\bar{\Delta}} = \sqrt{\frac{\sigma_1^2}{N_1} + \frac{\sigma_2^2}{N_2}} \quad (6.11)$$

The expression for Welch's *t* test degrees of freedom, *d*, though, is quite a bit more complicated:

$$d = \frac{\left(\frac{\sigma_1^2}{N_1} + \frac{\sigma_2^2}{N_2}\right)^2}{\left(\frac{\sigma_1^2}{N_1}\right)^2\left(\frac{1}{N_1-1}\right) + \left(\frac{\sigma_2^2}{N_2}\right)^2\left(\frac{1}{N_2-1}\right)} \quad (6.12)$$

Welch's *t* test uses the same probability conversion table as Student's *t* test, so Table 6.1 can still be used to assess the *t* score from Equation (6.10) against a probability that the two samples can be declared statistically equivalent.

Welch's *t* test: An assessment of two sample sets to determine whether they might have originated from the same underlying population in which a weighted average is used for both the combined spread and the degrees of freedom.

When in doubt, the general recommendation is to use Welch's *t* test instead of Student's *t* test. It is a more robust version of Student's *t* test. Comparing Equation (6.8) with Equation (6.10), it is seen that the numerators are identical; the difference is in the combined spread and the degrees of freedom for the conversion to a probability. For the spread, Welch's method uses a weighted quadrature combination, weighted by the number of values in each set. Student's version uses an arithmetic weighted average to estimate the population variance and then uses this as the spread for both samples. That is, they both use Equation (6.11) for the weighted spread calculation, but Welch's method simply uses the variances of the two data sets, while Student's method uses identical variances for them, its estimate of the population variance. The *d* value is also different. For the Student's *t* test, *d* is independent of the variances of the two number sets. For Welch's test, *d* is found with a harmonic mean weighted by the sample variances. For certain combinations of counts and spreads for the data sets, the two tests will have the same degrees of freedom, but, in general, Welch's test *d* value is lower than that from Student's *t* test. If the variances are far from each other, then the *d* value from Welch's test could be dramatically smaller than Student's *t* test *d* value. This makes it more difficult to declare that the two data sets are from the same underlying population. In short, the common advice is to go with Welch's *t* test whenever such a test is needed.

> **Quick and Easy for Section 6.1.3**
>
> Welch's *t* test is the commonly preferred method for assessing the similarity of the means of two number sets.

6.2 Linear Correlation

A common assessment between two data sets is to examine whether there is a linear relationship between them. This is a process called

correlation analysis, which is really about comparing the linearity between two number sets. **Linearity** here refers to the relationship between two number sets closely following a single straight line. The first step is to examine the scatter plot. Does it look like a linear relationship? This initial assessment is qualitative and highly subjective, though. There are methods for making it more quantitative.

We start with an introduction to **covariance**. As shown in Section 6.2.1, this is the mathematical term for the difference between the quadrature and additive uncertainty formulations. We then move on to two forms of linear correlation, introducing coefficients that quantitatively describe the linearity of the relationship between the two number sets. The last part of correlation analysis is the uncertainty of these correlation coefficients.

6.2.1 Covariance of Two Data Sets

This section starts with a quick derivation of covariance. You do not have to know these details in order to use it, but it is shown because it highlights the assumptions on which it is based. We'll come back to those later.

Suppose that we have a function, q, of two variables, x and y, which are both measured values. We can write this general expression for any one value of q:

$$q_i = q(x_i, y_i) \text{ for } i = 1,\ldots,N \quad (6.13)$$

Correlation analysis: The process of assessing the linearity between two number sets.

Linearity: The closeness of the relationship between two number sets to follow a straight line.

Covariance: A measure of the relationship between two data sets, it is the arithmetic average of the product of the discrepancies of each x and y data point relative to its own mean.

Let's further assume that these N measurements are all from a similar location in (x,y) space, rather than a big range along the x values and a corresponding range in the y values. That is, these two input number sets have distributions with well-defined centroids, $\bar{x}$ and $\bar{y}$, and spreads, σ_x and σ_y.

This formula for q_i can then be rewritten as a linear expansion around the mean values of the two input number sets:

$$q_i \approx q(\bar{x},\bar{y}) + \frac{\partial q}{\partial x} \cdot (x_i - \bar{x}) + \frac{\partial q}{\partial y} \cdot (y_i - \bar{y}) \quad (6.14)$$

This allows us to rewrite the formula for the mean of the q_i values:

$$\bar{q} = \frac{1}{N}\sum_{i=1}^{N} q_i = \frac{1}{N}\sum_{i=1}^{N}\left[\begin{array}{l}q(\bar{x},\bar{y}) + \frac{\partial q}{\partial x}\cdot(x_i - \bar{x}) \\ + \frac{\partial q}{\partial y}\cdot(y_i - \bar{y})\end{array}\right] \quad (6.15)$$

The partial derivatives in Equations (6.14) and (6.15) are evaluated at the mean values, $\bar{x}$ and $\bar{y}$. The first term, q, at the mean values, and these two derivatives are, therefore, the same for all q_i values. The only dependence on i in the bracket within Equation (6.15) are the x_i and y_i values in the differences multiplying the fractions. That is, Equation (6.15) can be rewritten like this:

$$\bar{q} = q(\bar{x},\bar{y}) + \frac{1}{N}\frac{\partial q}{\partial x}\sum_{i=1}^{N}(x_i - \bar{x})$$
$$+ \frac{1}{N}\frac{\partial q}{\partial y}\sum_{i=1}^{N}(y_i - \bar{y}) \quad (6.16)$$

The two summations in Equation (6.16), though, are zero; the sum of the difference of values around their arithmetic is, by definition of the arithmetic mean, zero. This leaves only the first term:

$$\bar{q} = q(\bar{x},\bar{y}) \quad (6.17)$$

To find the mean value of a function where the two independent variables have well-defined centroids and spreads, we only need to find the mean value of the two independent variables and use this in the function.

The variance of this q number set has this functional form:

$$\sigma_q^2 = \frac{1}{N}\sum_{i=1}^{N}(q_i - \bar{q})^2 \qquad (6.18)$$

Applying Equation (6.14) for q_i and Equation (6.17) for $\bar{q}$ leaves us with this new form:

$$\sigma_q^2 = \frac{1}{N}\sum_{i=1}^{N}\left[\begin{array}{c} q(\bar{x},\bar{y}) + \dfrac{\partial q}{\partial x}\cdot(x_i - \bar{x}) \\ + \dfrac{\partial q}{\partial y}\cdot(y_i - \bar{y}) - q(\bar{x},\bar{y}) \end{array}\right]^2 \qquad (6.19)$$

It is seen that the first and last terms within the bracket are identical and therefore cancel. We should also remember that the partial derivatives are constant and can be moved outside of the summation. There is a square operator applied to the bracket, though, which should be expanded. This results in the following:

$$\sigma_q^2 = \left(\frac{\partial q}{\partial x}\right)^2 \frac{1}{N}\sum_{i=1}^{N}(x_i - \bar{x})^2 + \left(\frac{\partial q}{\partial y}\right)^2 \frac{1}{N}\sum_{i=1}^{N}(y_i - \bar{y})^2$$
$$+ \frac{2}{N}\left(\frac{\partial q}{\partial x}\right)\left(\frac{\partial q}{\partial y}\right)\sum_{i=1}^{N}(x_i - \bar{x})(y_i - \bar{y}) \qquad (6.20)$$

The first two terms on the right-hand side of Equation (6.20) include all of the components of variance for the x and y independent variables, respectively. The third term, the cross product of the value differences against their means, is what is known as the covariance, written as cov(x,y):

$$\text{cov}(x,y) = \frac{1}{N}\sum_{i=1}^{N}(x_i - \bar{x})(y_i - \bar{y}) \qquad (6.21)$$

The covariance formula in Equation (6.21) is a quantitative assessment of the correlation between two number sets. The sign of each term in the summation depends on the relationship of x_i and y_i to their respective means. If both of the differences are positive, then the term is a positive contribution to the sum. This is also the case when both differences are negative; the contribution to the sum of such a term is positive because the multiplication cancels the two minus signs. If one is positive and the other is negative, though, then the contribution to the summation is negative. Because there are no absolute value operators in Equation (6.21), such terms will counteract the contributions of the positive terms. It is possible, therefore, for cov(x,y) to be highly positive, highly negative, or close to zero.

Figure 6.1 shows three example scatterplots with different types of correlation between the number sets. The left panel is a uniform cloud

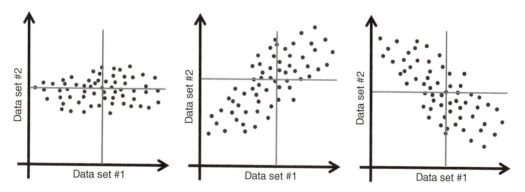

Figure 6.1 Correlation scatterplots showing essentially zero covariance in the left panel, positive covariance in the middle panel, and negative covariance in the right panel. Lines are drawn on each panel for the mean of the x values and the mean of the y values.

of dots with essentially no correlation. With the mean value lines drawn on the plot, the x–y space is divided into four quadrants. For every point in one quadrant contributing positively to the summation in Equation (6.21), there is a point contributing a nearly identical negative value to the summation. For this point spread distribution, therefore, we have cov(x,y) ~ 0. The other way to think of this relationship is that if x_i is above the $\bar{x}$ line, then the y_i value could be anywhere with respect to the $\bar{y}$ line.

The other two panels of Figure 6.1 show the case of nonzero covariance. In the middle panel, x and y are positively correlated, as the two quadrants that contribute positive values to the summation in Equation (6.21) have many points, while the other two quadrants have very few points. The summation is, therefore, lopsided toward a positive value and cov(x,y) ≫ 0. That is, if x_i is above $\bar{x}$, then usually y_i is also above $\bar{y}$. The opposite case is shown in the third panel of Figure 6.1, with the quadrants that contribute negatively to cov(x,y) being heavily populated with points and the other two quadrants being nearly empty.

Using the definitions of variance and covariance for the terms in Equation (6.20) yields this:

$$\sigma_q^2 = \left(\frac{\partial q}{\partial x}\right)^2 \sigma_x^2 + \left(\frac{\partial q}{\partial y}\right)^2 \sigma_y^2 \\ + 2\left(\frac{\partial q}{\partial x}\right)\left(\frac{\partial q}{\partial y}\right)\text{cov}(x,y) \quad (6.22)$$

If the x and y variables are indeed independent, then the covariance will be very small relative to the variances of the individual data sets and the last term of Equation (6.22) can be neglected. This leaves us with the quadrature summation formula of the two uncertainties for the uncertainty of the new q variable:

$$\sigma_q = \sqrt{\left(\frac{\partial q}{\partial x}\right)^2 \sigma_x^2 + \left(\frac{\partial q}{\partial y}\right)^2 \sigma_y^2} \quad (6.23)$$

It can be shown that this equality is always true:

$$\left|\text{cov}(x,y)\right| \leq \sigma_x \sigma_y \quad (6.24)$$

The fact that the absolute value of covariance is also less than or, at most, equal to the multiplication of the two standard deviations for the sample sets is known as the **Schwarz inequality**. This can be used to place an upper bound on the influence of the covariance in Equation (6.22). That is, we can replace cov(x,y) in the final term with this upper limit on its amplitude to write an expression for the maximum variance for the q number set as purely a function of the standard deviations:

$$\sigma_q^2 \leq \left(\frac{\partial q}{\partial x}\right)^2 \sigma_x^2 + \left(\frac{\partial q}{\partial y}\right)^2 \sigma_y^2 + 2\left|\frac{\partial q}{\partial x}\right|\left|\frac{\partial q}{\partial y}\right|\sigma_x \sigma_y \quad (6.25)$$

Equation (6.25) looks like an expanded version of a "squared summation." This leads to a formula for the maximum standard deviation σ_q, in the case where the two input variables are highly correlated:

$$\sigma_q \leq \left|\frac{\partial q}{\partial x}\right|\sigma_x + \left|\frac{\partial q}{\partial y}\right|\sigma_y \quad (6.26)$$

Equation (6.26) is the additive formula assumed as the maximum uncertainty back in Chapter 3. Now, using the concept of covariance, we have shown that this is, indeed, the case.

> **Quick and Easy for Section 6.2.1**
>
> The covariance parameter quantifies the independence or interconnection of two data sets.

Schwarz inequality: A relationship that states that the covariance between two data sets is, at most, equal to the product of their standard deviations.

6.2.2 Pearson Linear Correlation Coefficient

While covariance is a powerful concept for evaluating two data sets against each other, it is a difficult number to interpret. This is because the units are a multiplication of the units of the x and y variables. In the best case, the x and y units are equal and the covariance units are simply that unit squared. This does not have to be the case, though. The units of the two could be completely different, in which case their multiplication could result in a confusing mixture of units without a good physical connection.

This confusion regarding the interpretation of covariance is addressed by employing the terms of the Schwarz inequality to obtain a dimensionless number:

$$R = \frac{\text{cov}(x,y)}{\sigma_x \sigma_y} = \frac{\Sigma(x_i - \bar{x})(y_i - \bar{y})}{\sqrt{\Sigma(x_i - \bar{x})^2 \Sigma(y_i - \bar{y})^2}} \quad (6.27)$$

This unitless value, R, is known as the **Pearson linear correlation coefficient**. It is sometimes written as R_P or r. Because of the Schwarz inequality, R is always a value between -1 and $+1$. These two extremes indicate perfect correlation, with all of the (x_i, y_i) points arranged along a straight line. For $R = +1$, the line is tilted upward with a positive slope, while for $R = -1$, that line would have a negative slope. The only way to get $R = 0$ is to have $\text{cov}(x,y) = 0$, which means that the two variables are uncorrelated and the scatterplot of (x,y) pairs is a formless cloud. It is hard to get an R score that is exactly equal to zero, so independence is defined when R is "close" to zero.

Pearson linear correlation coefficient (R): A measure of how well the values in one data set trend up and down with those in another data set, it is defined as the covariance divided by the product of the standard deviations.

A question to ask yourself is this: how close to ± 1 should R be in order to call it a "good" correlation? Conversely, the question could be worded like this: how close to 0 should R be in order to say that the variables are uncorrelated (i.e., independent)? This can be answered quantitatively, with a statistical significance assessment, as well as qualitatively, with rules of thumb about good R values.

The R value can be converted into a probability, as we have already done several times with various quantities. The probability of interest is whether two sets of random and uncorrelated values would produce the R value obtained from the two number sets of interest (more specifically, the obtained R value or greater). This probability can be written as a combination of several integrals and calculated numerically. Table 6.2 lists these probabilities as a function of not only the absolute value of R but also N, the number of (x,y) pairs in the correlation evaluation, on which this probability also depends. This table is only meant to show the general trend of these probabilities; online calculators exist that can provide a much more specific number than listed here (one is listed in Section 6.6).

There are few key features of Table 6.2 that should be pointed out. First, for low N, it takes a very large $|R|$ value to achieve statistical significance at the 5% or 1% level. For $N = 5$, for instance, it takes $|R| = 0.96$ in order to reach the 1% very significant threshold. With only 5 (x,y) values in the data set, it is rather pointless to assess the statistical significance of the correlation. The $|R|$ threshold for significance, however, drops very rapidly with increasing N. By $N = 20$, the 5% significance threshold only requires $|R| = 0.45$, and by $N = 100$, this drops to $|R| = 0.20$. If you have a few hundred or more, then R values within 0.1 of zero can reach statistical significance.

While the check of the probability that the R value could have arisen from random chance is necessary, it loses its usefulness as an assessment of a good value of R. One rule of thumb about R that has been used is a

Table 6.2 The probability that N pairs of random and uncorrelated x–y pairs of values would produce a correlation coefficient value of $|R|$ or larger.

$\|R\|$	0.1	0.2	0.3	0.4	0.5	0.6	0.7	0.8	0.9
$N = 5$	0.873	0.747	0.624	0.505	0.391	0.285	0.188	0.104	0.0374
10	0.783	0.580	0.400	0.252	0.141	0.0667	0.0242	0.0055	0.0004
20	0.675	0.398	0.199	0.081	0.0248	0.0052	0.0006	2×10^{-5}	$<10^{-5}$
30	0.599	0.289	0.107	0.0285	0.0049	0.0005	2×10^{-5}	$<10^{-5}$	
40	0.539	0.216	0.060	0.0105	0.0010	4×10^{-5}	$<10^{-5}$		
50	0.490	0.164	0.0343	0.0040	0.0002	$<10^{-5}$			
100	0.322	0.0460	0.0024	3×10^{-5}	$<10^{-5}$				
200	0.159	0.0045	2×10^{-5}	$<10^{-5}$					
300	0.0838	0.0005	$<10^{-5}$						
500	0.0253	$<10^{-5}$							
1000	0.0015	$<10^{-5}$							

value of $|R| > 0.5$. This should not be taken alone but in conjunction with the significance test described in this section; *both* conditions should be met in order to declare the R value to be "good." This threshold of 0.5 is only defensible in that it is a nice round number halfway between perfect correlation and no correlation. For a more reasoned approach to choosing a specific value of $|R|$ greater than the significance test value, we need to introduce the cousin to R, the **coefficient of determination**.

The coefficient of determination is, quite simply, the square of the correlation coefficient, and in fact is known as R^2:

$$R^2 = \frac{\left[\sum(x_i - \bar{x})(y_i - \bar{y})\right]^2}{\sum(x_i - \bar{x})^2 \sum(y_i - \bar{y})^2} \quad (6.28)$$

R^2 is always between 0 and 1 and always closer to zero than R.

Coefficient of determination (R^2): A measure of the variance in one data set captured by the variation in another data set, it is the square of the Pearson linear correlation coefficient.

It will be shown in Chapter 7 that this is related to linear fitting, when these same summations in the numerator and denominator appear again as part of the slope and offset formulas. For now, however, we can state the meaning behind R^2: it is the ratio of the variance of the fit of the y value to the variance of the y values themselves. It is a quantitative measure of the closeness of these two variances to each other. For a linear fit to the y values, which is a straight line passing through $(\bar{x}, \bar{y})$ and attempting to match the y values elsewhere, if the values are indeed arranged along a straight line, then the variance of the linear fit and the original values will be equal and this R^2 value achieves its perfect score of one. If the relationship is a uniform cloud of points in (x,y) space, then the fit is a horizontal line, its variance is zero, and R^2 will also be zero. Because the linear fit is based on the variances of the two data sets, it is often interpreted as a measure of how well the variance in the x-axis number set captures the variance in the y-axis number set. Some will even say R^2 indicates the amount of variance in one of the number sets that can be explained by the other number set.

This fact can be related to our original question of the goodness of R. So, given our rule of thumb of |R| = 0.5 for a good value, this means that R² = 0.25 and so a quarter of the variance of one data set is explained by the other. This is often taken as a reasonable lower limit to good correlation. Another threshold that is sometimes applied is an R² value of 0.5, which means that half of the variance of the *y*-axis number set is being captured by the *x*-axis data set. This equates to |R| ≥ 0.707. Again, this might not be enough to achieve "goodness" in the correlation value if *N* is very small; both criteria should be met.

Both of these rule of thumb thresholds for declaring R to be good are guideline values. There is nothing particularly sacred about either setting and it depends on the context of the situation on whether either of these numbers should be chosen. There are times when the correlation is expected to be extremely good and even an |R| value of 0.9 is not particularly high. For other quantities, a value below 0.5 could still be considered a worthwhile and useful correlation.

Which quantity should you report, R or R²? Because they are directly related, this is entirely up to you. Coefficient of determination has the advantage of representing the variation in one data set that is captured and explained by the other data set. Correlation coefficient has the advantage of reporting the sign of the relationship between the data sets. Both are limited in that they are derived from an assumption of a Gaussian distribution of points around the linear fit. There is no preferred usage of one over the other and you will find either or both used in many geoscience studies.

6.2.3 Spearman Rank-Order Correlation

Another quantity used to assess the relationship between two data sets is the **Spearman rank-order correlation coefficient**. It uses the same formula as R, but discards the actual data values and instead uses the rank order of the number within its own data set. To calculate this, the values are each sorted according from lowest to highest, separately within each set. The pairing with the other data set, though, needs to be remembered, but the values themselves are not used in the calculation. This metric is sometimes called R_S or ROCC:

$$R_S = \frac{\sum(\text{rank}(x_i) - \overline{\text{rank}})(\text{rank}(y_i) - \overline{\text{rank}})}{\sqrt{\sum(\text{rank}(x_i) - \overline{\text{rank}})^2 \sum(\text{rank}(y_i) - \overline{\text{rank}})^2}}$$

(6.29)

The "rank" operator indicates that the value is not used, but rather its rank order within the full data set. The rank "variables" in Equation (6.29) are the same for both number sets—integer values from 1 to *N*—and therefore the averages are the same, with rank = (*N* + 1)/2. A perfect R_S correlation coefficient has all of the ranks identical between the two sets, and therefore $R_S = 1$.

The numerator still has a sum of two terms multiplied together: the difference between the observation value from its rank mean and the difference of the modeled value from its rank mean. The denominator has these terms in separate summations, and they are squared, so they only contribute a positive quantity to the coefficient. In fact, these two terms in the

---

Quick and Easy for Section 6.2.2

The Pearson linear correlation coefficient is one of the most famous statistics quantifying the relationship between two data sets and whether the values rise and fall together or not.

Spearman rank-order correlation coefficient (R_S): A measure of how well the values in one data set trend up and down with those in another data set, it is defined the same way as the Pearson correlation coefficient but with the exact numbers replaced by their rank orders.

denominator are equal, assuming no repetitions in either value set.

A problem arises in handling repeated values in either data set. Identical values are usually handled by averaging the ranks of the repeated value and assigning this new, often real number (as opposed to an integer) rank to all of the repeated values. If the number sets are large and the repetitions are few, then assigning sequential rank values to the repeated values is acceptable, as this will have a minimal influence on the final metric score.

To better visualize this transformation to rank order, Figure 6.2 shows an example schematic scatterplot and its conversion to rank values. The values for the first and second data sets are listed in the table at the top and are plotted in the panel in the lower left. The first data set values are random numbers from 0 to 10 and the second data set values are a linear relationship to the first values with the addition of a randomly distributed Gaussian spread. The values of each number set are then converted to rank-order numbering, listed above and below the values in the table. The first data set values are already ordered, so the numbering is sequential. The second data values, however, are not. As you can see, the first value in each set is, coincidentally, also the lowest value in each set, so the rank order of both numbers is one. This is also true of the next value for each set. The third value, however, is not the third-lowest value of the second set but rather the fourth lowest, so instead it is labeled accordingly. These rank orders are then plotted against each other in the lower-right panel of Figure 6.2. The first two pairs were rank ordered as (1,1) and (2,2), and they lie along the unity-slope diagonal line. For all of the others, however, the second data set's values were not in order relative to the rank order of the first data set, and so the points do not fall along this diagonal line. The Spearman rank-order correlation coefficient, given in Equation (6.29), is a Pearson linear correlation coefficient calculation of this rank-order plot. This particular example produces a Pearson coefficient of 0.78 and a Spearman coefficient of 0.71; the two are close, but the Pearson version is slightly better. This is seen qualitatively by comparing the two plots in Figure 6.2; the scatter of the dots is tighter to the line in the left panel of the actual values compared to the scatter of the rank-order points in the right panel.

While this coefficient is still assessing the correlations between the two data sets, it does it in a fundamentally different way. The power of R_S is that all of the values are now "spread" evenly in both number sets. This transformation reduces the influence of outliers and exaggerates the influence of a large mass of values, which often occurs near the centroids of the distributions. This is especially useful if there is a small cluster of outliers that would significantly sway the linear correlation coefficient. It is also a good choice if the value sets span several orders of magnitude, and therefore the linear correlation is dominated by the values in the highest order. The downside is that the values themselves are lost and could result in misleadingly good or bad R_S scores. For example, if the two data set pairs are monotonic but not perfectly aligned along the unity-slope, zero-intercept line, then R_S will have a perfect score even though the correlation is not linear. Conversely, a scatterplot with a nonmonotonic "undulation" in the relationship of the values from the two number sets might produce a rather bad R_S value, even though the trends are reproduced very well for the rest of the range. The point is that R_S assesses a different quality than does R, and both could be used.

To better see this relationship between R and R_S, Figure 6.3 includes scatterplots of two more data set comparisons and the corresponding rank-order scatterplots. The upper panels show the actual values, while the lower panels show the rank-order scatterplots of the same data sets. In the upper-left panel of Figure 6.3, the points are arranged along a curve with $y \propto x^7$, and then the last value even higher than this (raised to $y = 90$ instead of 48, as the curve would have dictated). The Pearson linear correlation coefficient for this scatterplot is R = 0.64.

Data set #1 rank	1	2	3	4	5	6	7	8	9	10	11	12	13	14	15	16	17	18	19	20
Data set #1 values	0.79	2.10	2.14	4.02	4.34	4.37	5.39	6.11	6.46	6.76	7.27	7.30	7.79	8.39	8.56	8.67	8.81	9.43	9.53	9.84
Data set #2 values	0.03	2.33	4.17	6.39	3.74	9.59	5.59	7.38	5.15	5.04	9.93	4.54	9.40	12.35	6.60	8.38	9.20	12.21	10.84	9.16
Data set #2 rank	1	2	4	9	3	16	8	11	7	6	17	5	15	20	10	12	14	19	18	13

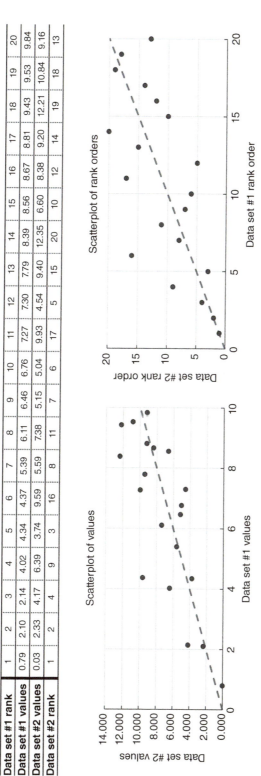

Figure 6.2 Schematic illustration of how to convert from a data-model scatterplot into a rank-order scatterplot. The top table lists model and observation values, each converted to rank order within their respective number sets. The lower-left scatterplot shows the values, while the lower-right scatterplot uses the rank orders.

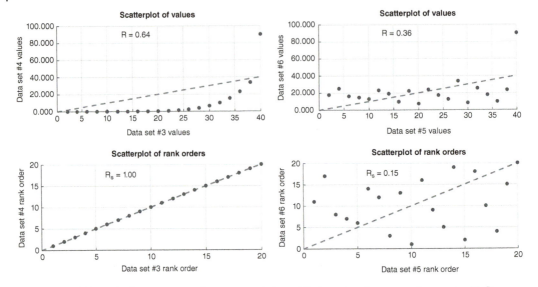

Figure 6.3 Schematic data set comparison scatterplots and their rank-order conversions, one with R_S better than R (left column) and another with R_S worse than R (right column).

The points are monotonic, however, so the rank ordering of these numbers perfectly aligns with the unity-slope reference line, as seen in the lower-left panel of Figure 6.3. This yields Spearman rank-order correlation coefficient of $R_S = 1.00$. In the right-side panels, the data set values are designed as a randomly determined Gaussian spread around a very slowly rising linear trend. The last point, however, is set to $y = 90$. Neither of the right-hand scatterplots yields a particularly good correlation coefficient, but there is a big difference between the values. This last point being so high has a large influence on R—increasing it substantially—while it has no influence on R_S. In summary, the Pearson coefficient assesses the linearity of the values, while the Spearman coefficient assesses the monotonicity.

This issue of linearity versus monotonicity brings up a good question: when would monotonicity matter more than linearity? One answer to this is when you can see from the scatterplot that the relationship between the two data sets is nonlinear and you do not really need to know the specific functional form of the connection, just the quality of the relationship. The pattern cannot be nonmonotonic—that is, one with relative extrema in the middle of the distribution—but rather the scatterplot should trend either upward or downward, like the upper-left panel of Figure 6.3. This occurs in atmospheric science for density and pressure as a function of altitude; these parameters typically follow an exponential decrease with altitude. We expect the correlation to be negative. A Pearson correlation between atmospheric pressure and altitude might not be particularly good, but a Spearman correlation, which replaces their value with rank order, would be very close to -1. If we simply wanted to show that the values decrease monotonically with altitude (and not search for the true functional form with a curve fit procedure), then the Spearman correlation clearly demonstrates this fact.

> **Quick and Easy for Section 6.2.3**
>
> The Spearman rank-order correlation coefficient replaces the values with their placement in a sorted list of the number set in the correlation coefficient formula, reducing the influence of outliers and assessing the monotonicity of the relationship.

6.2.4 Correlation with Logarithms

There are times when the correlation between two data sets is not represented well by a linear relationship of the values. This is especially true if the values span several orders of magnitude. In this case, the Pearson linear correlation coefficient in Equation (6.27) will be dominated by the values in the uppermost order and the other values will contribute very little. If this topmost order has a good fit between the two number sets, then the Pearson coefficient will be high, even though there is a dense cloud of points near the origin. It is often the case, though, that when the values span several orders of magnitude, the linear correlation is not particularly good.

A better method for assessing the correlation between these two data sets is to take the logarithm of all values in the correlation equation. This yields a **correlation coefficient of the logs**:

$$R_{\log} = \frac{\sum(\log y_i - \overline{\log y})(\log x_i - \overline{\log x})}{\sqrt{\sum(\log y_i - \overline{\log y})^2 \sum(\log x_i - \overline{\log x})^2}} \quad (6.30)$$

In Equation (6.30), the logarithm operator could be of any base. This coefficient has the same meaning as R, with a perfect score being plus or minus one. The calculation, however, is no longer dominated by those values in the highest order of the number set; orders of magnitude are converted to integer differences. It has the advantage of retaining the actual values of the two number sets, unlike the replacement done in R_S, but it should only be used if the full span of values across all orders of magnitude is important.

Correlation coefficient of the logs ($R_{\log}$): A measure of how well the values in one data set trend up and down with those in another data set, it is defined the same way as the Pearson correlation coefficient but with the exact numbers replaced by their logarithms.

A related formula is the semi-log correlation coefficient. In this case, only one of the two number sets is converted with the logarithm operator, while the other is left in its original form. There are two versions of this coefficient, the more commonly used one with the y-axis values converted:

$$R_{\log-y} = \frac{\sum(\log y_i - \overline{\log y})(x_i - \overline{x})}{\sqrt{\sum(\log y_i - \overline{\log y})^2 \sum(x_i - \overline{x})^2}} \quad (6.31)$$

and another with only the x-axis values converted:

$$R_{\log-x} = \frac{\sum(y_i - \overline{y})(\log x_i - \overline{\log x})}{\sqrt{\sum(y_i - \overline{y})^2 \sum(\log x_i - \overline{\log x})^2}} \quad (6.32)$$

The first semi-log formula is useful when the response data set (on the y axis) is highly variable, while the driver data set (on the x axis) is not. This is the case, for example, with atmospheric pressure as a function of altitude above a planetary surface. The second semi-log formula is useful when the control parameter is highly variable, while the response does not change that much. This is the case, for instance, with solar wind parameters near a comet; the influence of the comet extends millions of kilometers from the comet itself, which is typically a few kilometers or less in scale. In fact, examining a semi-log correlation coefficient is a way to test whether the relationship of a response quantity to a driving parameter is either exponential (highest R with Equation (6.31)) or logarithmic (best with Equation (6.32)).

Figure 6.4 shows an illustrative example of this conversion to log space in order to examine the trends in the lower orders of magnitude and include these more equitably in the correlation calculation. The top panel shows the values with linear scales for both the x and y axes. The values span from 1000 to 160 000, but on this linear scale it appears as a cluster of points near the origin and a few points sprinkled at

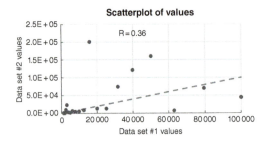

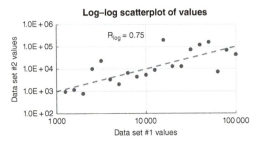

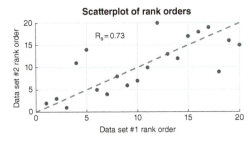

Figure 6.4 Scatterplot conversion to logarithmic space and the influence it has on correlation coefficient.

higher values. The middle panel shows the same values but with logarithmic x and y axes. Now it is possible to see the lower range extent of each number set. The final panel shows the rank-order scatterplot, for comparison. The rank ordering is the same whether ordered by the original values or logarithmic values.

Each panel of Figure 6.4 lists a correlation coefficient for that scatterplot. For the top panel, R is calculated from the original values using Equation (6.27). For the middle panel, R_{log} is calculated using Equation (6.30). The final panel lists R_S from Equation (6.29). The linear correlation is not particularly good, but the R_{log} is quite reasonable. This is because all of the values within the R_{log} calculation are between 3.0 and 5.2—all within the same order of magnitude—rather than from 1000 to 160 000.

The calculation of R from the raw values is dominated by those in the uppermost order of magnitude; the values between 1000 and 10 000 contribute very little to the summations in Equation (6.27).

Quick and Easy for Section 6.2.4

The log version of the correlation coefficient is an alternative that evenly weights values across orders of magnitude, rather than favoring the values in the highest order.

6.3 Example: Atmospheric Ozone and Temperature

One example of two number sets measured close to the same place is the case for a field experiment with multiple sensors. A specific case is measuring tropospheric **ozone** (O_3), a compound created from a chemical reaction occurring in the atmosphere in the presence of both **nitrogen oxides**, also known as NOx (NO + NO_2), and **volatile organic compounds** (VOCs). Two of the leading causes of

---

Ozone: The molecule comprises of three oxygen atoms, it is harmful to living things and generally considered a pollutant in the lower atmosphere, even though its production and loss in the middle atmosphere protect us from ultraviolet light from the Sun.

Nitrogen oxides: Known as NOx, they are a group consisting of the molecules NO and NO_2, these are in general considered atmospheric pollutants.

Volatile organic compounds: Known as VOCs, they are a group of molecules released by processes within living things, formed through combustion, or released by the use of certain products, they are often considered an air pollutant, unless the smell is intentional and desired.

lower-atmospheric ozone are vehicle exhaust emissions and wildfires, which release both of these compounds into the air and subsequent **atmospheric chemistry** leads to the formation of ozone. The ozone could also be produced because of the VOCs released by the plants, but then, once created, the ozone is harmful to the plants. In general, ozone near the surface of the planet is a bad chemical to have in the air. We want the ozone high up in the atmosphere, in a layer called the stratosphere, where it is created by ultraviolet light breaking up diatomic oxygen, after which those single oxygen atoms then find one of the plentifully available oxygen molecules and form ozone. All of this UV light absorption is a good thing; the stratospheric ozone layer is essentially global sunscreen.

Let's say that you have ozone concentration measurements in a rural forest canopy. Let's say that you also have an instrument nearby that measures the air temperature. The hypothesis is that temperature is one of the dominant controlling factors in the production and buildup of ozone. To determine this relationship, measurements are taken many times a day over several weeks.

This is exactly what occurred at the University of Michigan Biological Station (UMBS) in the summer of 2014, for the Program for Research on Oxidants: Photochemistry, Emissions and Transport (PROPHET) field campaign experiment. The PROPHET project was much bigger than just this topic, but as part of the experiment, ozone concentration data were collected every hour from late on 13 July 2014 until midafternoon on 29 July 2014. Hourly temperature data are available from the close by Pellston weather station of the National Weather Service. Many other measurements were taken, and this connection was not the main focus of the PROPHET study, but it is convenient data to examine for this example.

The data from these two sources are shown in Figure 6.5. There is a clear daily signature, especially in the temperature values but also present in the ozone concentrations. A cold front moved through late on the 22nd, dropping the temperatures and, with the wind and rain, clearing out the ozone from the air. There are several times where the ozone concentration dropped due to strong wind and rain, but there is still a qualitative match between these two time series. Note that most of the ozone values are elevated over the background level of approximately 10 ppb usually seen at this location. There are two main sources for the increased ozone concentration, located far away from the monitoring site. The first is flow of air from the Milwaukee and Chicago urban centers on the western shore of Lake Michigan. If the wind is blowing in the right direction, then the NOx from these regions can be transported over the UMBS field site, combined

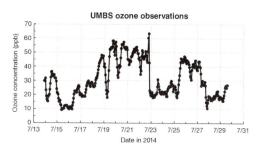

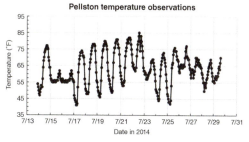

Figure 6.5 Time series of the hourly atmospheric ozone concentration (in parts per billion) in the forest canopy at the UMBS and accompanying temperature data (in degrees Fahrenheit) from the nearby Pellston, MI weather station.

Atmospheric chemistry: Chemical reactions that take place in the air, sometimes when two free-floating molecules bump into each other but often on the surfaces of airborne particulate matter, water droplets, or ice crystals.

with the biological VOCs from the nearby forest, and resulting in increased ozone. The other main source is a large wildfire in northwestern Canada at this time. Yes, from thousands of kilometers away, NOx and ozone were transported in the smoke plume across the North American continent. Again, if the wind carried this plume over Michigan, then an increase in ozone was recorded. In both cases, though, local temperature is thought to have some influence on the specific concentration of ozone recorded on any particular day, and we qualitatively see that relationship, to some degree, in Figure 6.5, so we will proceed with the analysis.

It is useful to take a brief side step and consider the histograms of these two data sets. The correlation coefficient can be calculated for any pairing of two data sets; there is no assumption of normality behind the derivation of R. The interpretation of R, and more specifically the conversion to a probability that a calculated R value could arise from random chance, is, however, based on this assumption. So it is good to at least qualitatively examine the distributions. These are shown in Figure 6.6. While the temperature histogram looks like it could be Gaussian, the ozone concentration is clearly bimodal and far from a Gaussian shape. We'll keep this in mind as we proceed.

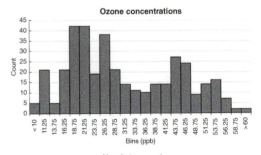

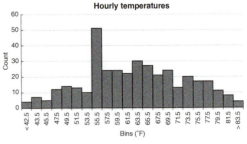

Figure 6.6 Histograms of forest ozone concentrations (ppb) and air temperature (°F) at the UMBS in the summer of 2014.

Additional statistics about these two distributions are listed in Table 6.3. To the hundredths decimal place, the standard deviations are 13.21 and 9.85. Rounding these to one significant digit, and keeping in mind our caveat to that rule when that significant digit is one, the reported base values are both significant only to the ones digit. This leaves two significant digits in the base values, so two significant

Table 6.3 Statistics about the individual data sets in this example analysis.

	Ozone concentrations	Air temperatures
Arithmetic mean	31 ppb	63 °F
Standard deviation	13	10
Mean absolute difference	11	8
Median	27	63
75% quantile	43	70
25% quantile	20	56
Interquartile range	23	14
Skewness coefficient	0.37	0.04
Kurtosis coefficient	1.9	2.3

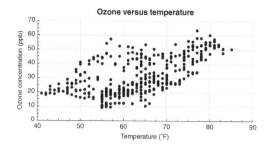
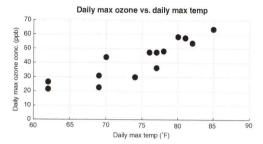

Figure 6.7 Scatterplot of forest ozone concentrations (ppb) against air temperature (°F) at the UMBS in the summer of 2014.

Figure 6.8 Scatterplot of the maximum forest ozone concentration from each day of the field campaign (ppb) against the maximum air temperature from each day (°F).

digits were included on the skewness and kurtosis coefficients, as well. The right-sided skew for the ozone concentration is moderate, and both distributions have a somewhat low kurtosis, indicating light tails.

All of the data are at an hourly cadence, so it is straightforward to plot the two number sets against each other. This is shown in Figure 6.7, plotting the ozone concentrations against the temperature measurements. There are 378 individual points on this scatterplot. At first glance, it looks like a decent relationship between them. We can follow up this scatterplot creation with the calculation of the Pearson correlation coefficient from Equation (6.27). This yields an R value of 0.63, for an R^2 of 0.39. These are above our general rule of thumb guideline for indicating a good correlation, and, comparing this for $N = 378$ with Table 6.2, they are well past the threshold for being labeled highly significant for the statistical probability of being correlated. Remember that the ozone data set was clearly not Gaussian, but since the probability is $<10^{-5}$, there is no need to worry about that issue. If the probability were close to the selected threshold (of 0.05 or 0.01, usually), and conclusions or decisions were being formed based on this correlation, then this would be a point of concern. In that case, a full analysis of how close each of these is to a normal distribution, as described in the previous chapter, should be done, to justify or reject the use of the probability calculation. The calculated R value, however, is well beyond any threshold that is regularly applied, there is no strong need to probe this further.

In examining Figure 6.5, it is seen that the temperature often exhibits a wider swing in values for any particular day than does the ozone concentration. The peak values for any day of these two quantities, though, seem to rise and fall together. That is, it could be that the daily maxima are better correlated than the full time series. This is easy enough to determine, and the scatterplot of these daily max values is shown in Figure 6.8. The connection between them is much clearer than in Figure 6.7, and the quantitative assessment bears this out; the R value for the daily peaks is 0.89, with a corresponding R^2 of 0.79. With $N = 14$, this is also at the approximately 10^{-5} probability of being the same as random chance. Because the ozone concentration is subject to additional source terms and other factors, its concentration is not particularly sensitive to the specific temperature at the same time as the ozone measurement. This buildup and lag time indicate that the more appropriate connection is between the daily maximum values of ozone and temperature.

> **Quick and Easy for Section 6.3**
>
> The relationship between atmospheric ozone and temperature provides a nice illustration of correlation.

6.4 Uncertainty of R

While the goodness of the correlation can be interpreted solely from the value of R, this is only part of the information needed to make a robust assessment about R. The other half is the uncertainty of R. If each of the x and y data has uncertainty values associated with them, then a propagation of error can be conducted according to the rules in Chapter 3. This becomes tremendously tedious to write out because of the nested operators in the definition of R, but it is a straightforward numerical calculation and a result can be obtained. This is the best path, in my view, because it is based on the uncertainty of each data value used in the calculation and follows the standard procedures of error calculation and uncertainty propagation.

There are other methods, however, for determining uncertainty values on statistical quantities for which uncertainties are otherwise difficult or impossible to derive. These are based on resampling of the values to create a new data set from the original two data sets. In the case of correlation between two variables, the resampling selects pairs of (x_i, y_i) values together, so that the correlation is maintained between the original x and y data sets.

6.4.1 The Jackknife Method

The replacement technique that will yield an exact value for the variance is the **jackknife method**. This method forms many new data sets by systematically removing one value from the set. This is done repeatedly for all values in the set, creating N new sets of numbers, each with $N - 1$ values. These new data sets are called resamples, and the process is known as resampling. For the jackknife method, this resampling is as easy as it gets: one pair of values is removed. The resampled new data set is nearly identical to the original data set, with all but the removed pair still there and in the same order as the original set.

With a bit of math, it can be shown that the average of the averages of the resampled number sets obtained from the jackknife method is equivalent to the average of the original values. Similarly, the average of the standard deviations of the jackknife number sets is equivalent to the standard error of the original data set.

These features are directly applicable to the correlation coefficient. For the situation here with two data sets being compared against each other, we do not have a single value but a pair of values that should be removed to create each of the new sets. For each of the N new data set pairs being created, each with a missing (x_i, y_i) pair where i was systematically swept from 1 to N, Equation (6.27) can be used to calculate an R value for that new paired set. These N new R values can be treated as a number set itself, from which a mean and standard deviation can be calculated. The mean of the R values from the jackknife-resampled sets should be equal to the original R. The standard deviation of the jackknife resamplings, though, is actually equivalent to a standard deviation of the mean (the standard error) of the original R value. That is, there is an extra $1/\sqrt{N}$ multiplying the resulting standard deviation of the R values from the jackknife resamplings. If you want an uncertainty on R, then you must multiply this σ from the jackknife method by $\sqrt{N}$ to get σ_R.

Note that the jackknife method works not only with the correlation coefficient but also with any statistic. Furthermore, the statistic does not have to be one that calculates a comparison of two number sets, but could be based solely on a single number set. You could, for example, use it to find the uncertainty of skew and kurtosis for a single data.

Jackknife method: The specific resampling method of omitting one of the values from the sample to create a "new" data set, done with each of the numbers to create N new sets each of length $N - 1$, from which statistics can be calculated.

The problem with the jackknife method is that it is cumbersome for large N. The method dictates that N resamples be created, leaving out each of the (x_i, y_i) pairs from the new sets. As N gets into the thousands or millions, then this process might take a large amount of computing power to calculate. There is no good way around this; the jackknife method in its proper form requires all N new data sets for the full calculation of the standard error of the desired statistic. One method that has been created is the **delete-d jackknife**, in which more than one of the values (i.e., $d \geq 2$) are removed with each resampling. You do not have to randomize the removal. The delete-d jackknife method still removes all of the pairs, so you can simply remove d adjacent pairs at a time from the list to create N/d resampled sets.

Another way around this is to use the fact that the mean of the resamples should be achieving **convergence** toward the original mean, and therefore not all of the N resampled sets need to be created to reach a reliable value for the standard error of R. This is checked by calculating the standard deviation of R as new resample sets are created, and then going until some convergence criteria are met. If the (x_i, y_i) pairs are in a truly random order, then proceeding in order with the systematic removal is fine. If not, then a random number generator should be used to randomize the selection of which pair is removed for any one resample, keeping track so that an already removed pair is not removed again in a later resample data set. The change in the standard error of R from the resamples should be monitored as each new resample set is added. At some point, you can declare the standard error of R to be converged and stop adding new resample sets. The problem with this is that the jackknife method only removes a single-data-set (x_i, y_i) pair from the full set with each resampling, so the R values might not be that different from each other for any two adjacent resampled sets. That is, you might declare convergence too soon, just because the (x_i, y_i) pair removed at resample j was very close to the pair removed in resample $j - 1$. This means that the new standard error of R should really be compared against an average of the standard error values from the last several (or many) resampled sets.

Quick and Easy for Section 6.4.1

The jackknife method is a robust procedure for determining the standard error of a statistic, whether calculated from one number set or from the comparison of two number sets, using a systematic "leave-one-out" process to create "new" sample sets.

6.4.2 The Bootstrap Method

Among the resampling techniques, the most famous and arguably the best choice for large N is the **bootstrap method**. The bootstrap method creates a new data set by resampling the original data, or in this case the original N pairs of (x,y) values, with replacement. This creates a new set of $N(x,y)$ pairs, from which a new correlation coefficient can be calculated. The replacement feature means that some values will be included in the set multiple times, while others, perhaps many others, will

---

Delete-d jackknife method: Similar to the jackknife method except that d values, with $d \geq 2$, are removed in each resampled set.

Convergence: The process of iterating a procedure until the change between any two iterations is below some predefined threshold.

Bootstrap method: A resampling method in which many, typically 50 to 500, "new" N-length data sets are created by randomly selecting, with replacement between selections, values from the original set, and from which statistics can then be calculated.

be omitted from that resampled set. The power of the bootstrap is that it speeds up the convergence process relative to the jackknife method because it changes many values in the resampled set instead of only one (x_i, y_i) pair per resample. This means that we do not have to go all the way to N resampled data set constructions, but rather a smaller number M. When N is very large, it can be the case that M is much smaller than N, but the resulting standard deviation is just as robust as that from the jackknife method.

For each of the bootstrap-resampled data sets, Equation (6.27) can again be used to calculate R. With some number M of these resamples, we have a number set of R values which can be used to calculate a mean and standard deviation of R. The mean should converge to the original R and the standard deviation of this set should converge to an asymptotic value. When the mean of the resamples reaches a specified threshold of closeness to the original R, this is a signal that you have enough resamples and can stop the calculation. Because all of the (x_i, y_i) pairs are possibly subject to omission or multiple usage in any one resampling, the convergence should be fast and, in general for large N, computationally more efficient than the jackknife method.

As with the jackknife method, the bootstrap technique can be applied to determine a spread on any statistic. It is being introduced here for R, but could be used for single-data-set statistics or multi-number-set statistics. It is a robust and widely used method for determining spread.

You can also simply do a set number of resamples and declare the standard deviation of the calculated R values to be the uncertainty, rather than checking for convergence. In this case, what is the right number of resample sets to create? Recommendations vary widely, from 50 to 5000. Given a data set in the thousands, I recommend a set of bootstrap resamplings in the hundreds, with a typical number that leads to good convergence for nearly all distributions being M between 100 and 500.

This ideal number for M comes with the caveat that the distributions of the data sets are relatively well behaved. That is, if the data sets have multiple peaks, or there are clusters of points in the (x, y) space away from the main trend, then the calculated R values will have a large spread. Really, it is a matter of convergence. If the scatterplot of the original data shows a nice, well-formed distribution along both axes, then 50 iterations might be enough for good convergence on an uncertainty for R. If the original scatterplot has a bifurcated distribution in the y values for the upper half of the x-value range, then the distribution of bootstrap R values will most likely not have a Gaussian shape. If the points in one of these two groupings are favored in a particular resampling—with many multiple selections from one branch and many omissions in the other—then the correlation coefficient will be systematically better than the original. However, this is only the case if the sampling along the x-axis values is fairly uniform. If there is clustering in x as well as y, then the resampled distribution could have unstructured scatter and a lower R value. This will also happen when there is a cluster of points far from the rest of the (x_i, y_i) pairs. When pairs from this cluster are oversampled, then the R for that resample will be lower than the original R, but when they are undersampled, the resampled R will be higher than the original R. This means that the average will be close to the original R value, but that the spread of R values will be rather large. Again, $M = 50$ is perhaps not enough for a converged standard deviation value of R. What will result in both of these cases is a slight shift in the peak of R to higher values and the formation of a long tail of low R values. That is, a skewed-left distribution.

If the distribution of R values from the resampling process is not Gaussian (i.e., do the check described in Chapter 5), then a different measure of spread for R can be used. Specifically, the IQR or FWHM of the bootstrapped R values could be used instead, or any of the other values for spread defined in Chapter 4.

6 Correlating Two Data Sets: Analyzing Two Sets of Numbers Together

> **Quick and Easy for Section 6.4.2**
>
> The bootstrap method is an alternative calculation for finding an uncertainty on a statistic, providing the standard deviation from a set of random resamplings of the original data set.

6.4.3 Uncertainty of R for the Ozone-Temperature Example

Let's look at our example of correlating ozone and temperature measurements. Because the daily peak values have a higher correlation and are more physically defensible, these are the ones that will be considered here for an illustrative uncertainty calculation. Note that this is a very small data set, with only 15 values, but works as an illustration of how to implement the techniques.

The jackknife method is a collection of 15 data sets, each with one from the original set missing. The correlation coefficient R can be calculated for each of these sets; the histogram of that number set of R values is shown in Figure 6.9. The shape is somewhat Gaussian but with some gaps. We do not really have to worry about the normality of this set of numbers, though. From these numbers, we can calculate their mean and standard deviation. The mean is 0.89, which is exactly the original value. The standard deviation of these values—using the sample standard deviation formula, not the population standard deviation—is 0.015. Remember that the standard deviation of the statistics determined from the jackknife method has the functional form of standard error. This value then should be multiplied by $\sqrt{15}$ in order to interpret it as a standard deviation on the original R; that is, as its uncertainty spread. This yields 0.058. So, from the jackknife method, the correlation coefficient, with uncertainty, can be reported as 0.89 ± 0.06.

For the bootstrap method, the number of resamplings, M, is not fixed but something that needs to be chosen. Figure 6.10 shows the calculated mean of the R values from a set of

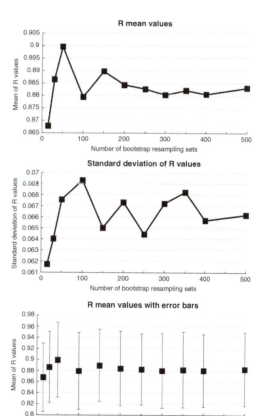

Figure 6.10 The top panel shows mean R values from the bootstrap method, as a function of the number of resampling sets used (M). The second panel shows the standard deviation of these R values. The third panel puts these two elements together, showing the average R values with the standard deviations as error bars on each point. Note that none of the y-axis ranges start from zero.

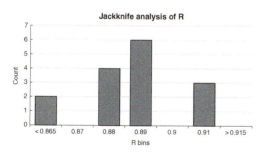

Figure 6.9 The distribution of R from the jackknife method for the daily maximum ozone-versus-temperature analysis at the end of Section 6.3.

bootstrap calculations, ranging from $M = 15$ resampling sets to $M = 500$ sets. Shown in the second panel of Figure 6.10 is the standard deviation of the R values from the bootstrap resamplings. For low values of M, these two quantities vary, but then settle into nearly constant values of 0.88 and 0.07, respectively (rounded to appropriate significant figures). While it looks like the values are still changing substantially even when M is in the hundreds, remember that the y-axis scale is limited to only the region of interest. If plotted from zero to the maximum value, then both of these curves would look like a nearly straight line. To further emphasize that these values are decently stabilized when $M > 100$ or so, the final panel in Figure 6.10 includes the standard deviations as error bars on the mean values of R from each bootstrap analysis. This final panel really shows that the R values from the bootstrap method with $M > 100$ are very similar to each other.

Just to understand the variability for low M, the bootstrap method was run many times for $M = 15$, equal to the number of sets in the jackknife calculation. The mean of these small sets varied considerably, from 0.85 to 0.92, and the standard deviation changed from a low value of 0.03 to a high value of 0.09. This indicates that such a low value of resampling sets is inadequate for obtaining a robust distribution of R values, from which a standard deviation can be determined. Applying a standard deviation to the original R value from such a small bootstrap set would not be wise.

The histogram from the final and largest bootstrap calculation of $M = 500$ resampling sets is shown in Figure 6.11. The R values for each individual resample-set range from 0.62 to 0.99. It is clearly seen that this distribution has a left-skewed tail. This is also revealed by a few statistics of the distribution. For instance, the median is larger than the arithmetic mean by 0.014 and the skewness coefficient γ is -0.99. Both of these facts indicate a left tail-heavy distribution. In fact, as shown in Chapter 5, $|\gamma| \sim 1$ is a severe skew value. Additional tests from Chapter 5 could be conducted on this data set to further determine whether it is normal and quantify the similarity or difference to a Gaussian distribution.

As mentioned in Section 6.4.2, a left-sided skew to the distribution of R indicates a bifurcated point spread or a cluster of outliers in the original scatterplot. It is such a small distribution that this is hard to tell, but examination of Figure 6.8 shows that there is a bit of a

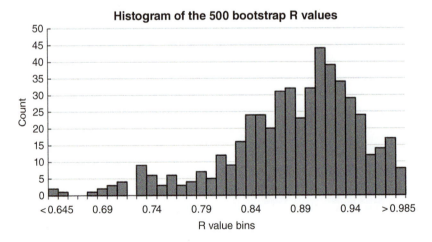

Figure 6.11 The distribution of R from the bootstrap method with 500 resamplings.

bifurcation in the middle x values. The histogram of bootstrap values clearly reveals this feature, as certain points in this x-value range are either selected multiple times or not included in each of the bootstrap resampling sets.

Because of this non-Gaussian shape to the bootstrap R values, it might be good to forego the use of standard deviation as the uncertainty on the original R. It might be good to use interquartile range, which for this data set is $0.929 - 0.849 = 0.080$, or the full width half maximum from the histogram, which, using the bin centers, is $0.95 - 0.87 = 0.08$, the same value. This is slightly bigger than the 0.7 standard deviation from the $M > 100$ bootstrap calculations, which itself was slightly larger than the 0.6 from the jackknife method. Really, any of these uncertainties could be reported for R, but this analysis shows that 0.8 is probably the safest to use because of the issues with the original data set (being small and slightly bifurcated).

> **Quick and Easy for Section 6.4.3**
>
> When the distribution of statistic values from the resampled sets is not Gaussian, then it might be better to report a spread other than standard deviation, like IQR or FWHM.

6.5 Correlation and Causation

A final comment should be made about correlation. You are equipped with the numerical tools to assess it and, assuming a Gaussian, assign a probability to its chances of not being random. You can calculate an uncertainty on R values and use a t test to compare two different R values, again resulting in a probability of significance on their similarity or difference. None of this actually proves that the "driving parameter" assigned to the x axis caused the response in the parameter assigned to the y axis. It must be said out loud: correlation does not equal causation.

This gets back to Chapter 1 and the scientific method. In designing an experiment to test a hypothesis, the test must be devised such that it assesses the physical processes underlying the unexplained feature. One common issue is that the sample set is too small and significant correlations arise despite their being no real connection (for instance, making a conclusion on Sun-climate connections with anything less than several solar cycles of data). Another common problem is that there is an unobserved third parameter that drives both of the number sets being compared in the analysis (for instance, comparing local insect populations with thunderstorms; they are probably both influenced by season). A third common mistake is that the variables being considered are not the proper ones to evaluate the issue (for instance, using total population instead of population density to assess human influence on the local environment). It is important to consider the scientific connection and design an experiment that assesses not only this possible connection but also explores other possible explanations and scenarios.

Correlation analysis is a good first step to comparing two number sets. But it is only that—a beginning. There are myriad ways in which two number sets can be compared, many of which are covered in the rest of the book. Underlying all of these quantitative assessment techniques, however, is the necessity of building the analysis on a solid scientific foundation.

> **Quick and Easy for Section 6.5**
>
> Correlation is a good first step for comparing two number sets, but it does not imply causation between them. A physical connection should be established within the framework of the scientific method.

6.6 Further Reading

The original paper by the anonymous "Student" (W. S. Gosset) about the *t* test:

Student. (1908). The probable error of a mean. *Biometrika*, *6*(1), 1–25. https://doi.org/10.2307/2331554

The original paper by B. Welch updating the *t* test:

Welch, B. L. (1947). The generalization of 'student's' problem when several different population variances are involved. *Biometrika*, *34*(1/2), 28–35. https://doi.org/10.2307/2332510

This is a good website that explains the difference between these two tests: https://www.datanovia.com/en/lessons/types-of-t-test/unpaired-t-test/welch-t-test/

And here is a good blog about statistics, in particular this post on *t* tests: http://daniellakens.blogspot.com/2015/01/always-use-welchs-t-test-instead-of.html

Here is a calculator for the *t* test probabilities: https://stattrek.com/online-calculator/t-distribution.aspx

Covariance and the correlation coefficient are discussed in many statistics textbooks. Daniel Wilks' book has a nice description:

Wilks, D. 2019). *Statistical Methods in the Atmospheric Sciences* (4th edition). Elsevier, Amsterdam. https://www.elsevier.com/books/statistical-methods-in-the-atmospheric-sciences/wilks/978-0-12-815823-4

John Taylor's textbook on error analysis has a short chapter devoted to this topic as well:

Taylor, J. R. (1997). *An Introduction to Error Analysis*. University Science Books, Mill Valley, CA, USA. https://uscibooks.aip.org/books/introduction-to-error-analysis-2nd-ed/

A calculator for the probability from the correlation coefficient: https://www.socscistatistics.com/pvalues/pearsondistribution.aspx

The jackknife and bootstrap methods are thoroughly covered in the book by Efron and Tibshirani:

Efron, B., & Tibshirani, R. J. (1993). *An Introduction to the Bootstrap*. Chapman & Hall, New York. https://www.taylorfrancis.com/books/mono/10.1201/9780429246593/introduction-bootstrap-bradley-efron-tibshirani

The bootstrap method is also described well in a section of this paper by Patricia Reiff:

Reiff, P. H. (1990). The use and misuse of statistics in space physics. *Journal of Geomagnetism and Geoelectricity*, *42*(9), 1145–1174. https://doi.org/10.5636/jgg.42.1145

Papers on the temperature–NOx–ozone connection:

Bloomer, B. J., Stehr, J. W., Piety, C. A., Salawitch, R. J., & Dickerson, R. R. (2009). Observed relationships of ozone air pollution with temperature and emissions. *Geophysical Research Letters*, *36*, L09803. https://agupubs.onlinelibrary.wiley.com/doi/full/10.1029/2009GL037308

Wu, S., Mickley, L. J., Leibensperger, E. M., Jacob, D. J., Rind, D., & Streets, D. G. (2008). Effects of 2000–2050 global change on ozone air quality in the United States. *Journal of Geophysical Research*, *113*, D06302. https://agupubs.onlinelibrary.wiley.com/doi/abs/10.1029/2007JD008917

A paper with a nice plot of the relationship of ozone to atmospheric concentrations of NOx and VOCs:

Sillman, S. (1999). The relation between ozone, Nox and hydrocarbons in urban and polluted rural environments. *Atmospheric Environment*, *33*, 1821–1845. https://www.sciencedirect.com/science/article/abs/pii/S1352231098003458

More on the connection of ozone and temperature:

Jacob, D. J., Logan, J. A., Gardner, G. M., Yevich, R. M., Spivakovsky, C. M., Wofsy, S. C., et al. (1993). Factors regulating ozone over the United States and its export to the global atmosphere. *Journal of Geophysical Research*, *98*(D8), 14817–14826. https://doi.org/10.1029/98JD01224

The published data set for the ozone measurements:

Gunsch, M. J., Pratt, K. A., Ault, A. P., & VanReken, T. M. (2014, July). UMBS: PROPHET Aerosol and Ozone Data, Environmental Data Initiative. https://doi.org/10.6073/pasta/b0943aacce8ad7dea732e2442d22ac53, 2018b.

Two papers thoroughly examining the PROPHET ozone data:

Gunsch, M. J., Schmidt, S., Gardner, D. J., Bondy, A. L., May, N., Bertman, S. B., et al. (2018a). Particle growth in an isoprene-rich forest: influences of urban, wildfire, and biogenic precursors. *Atmospheric Environment*, *178*, 255–264. https://doi.org/10.1016/j.atmosenv.2018.01.058,.

And the other one: Gunsch, M. J. and May, N. W. and Wen, M. and Bottenus, C. L. H. and Gardner, D. J. and VanReken, T. M. and Bertman, S. B. and Hopke, P. K. and Ault, A. P. and Pratt, K. A. (2018). Ubiquitous influence of wildfire emissions and secondary organic aerosol on summertime atmospheric aerosol in the forested Great Lakes region. Atmospheric Chemistry and Physics, 18, 3701–3715. https://doi.org/10.5194/acp-18-3701-2018

The University of Michigan Biological Station has much of the data collected there online and publicly open for use, including the ozone data: https://mfield.umich.edu/dataset/prophet-aerosol-and-ozone-data-july-2014

The Pellston weather station, from Weather Underground: https://www.wunderground.com/dashboard/pws/KMIPELLS3?cm_ven=localwx_pwsdash

And a link to the specific data used: https://www.wunderground.com/history/daily/us/mi/pellston/KPLN/date/2014-7-13

A more permanent location for this data set is the Iowa Environmental Mesonet: https://mesonet.agron.iastate.edu/ASOS/

Here is a good little description of obtaining high correlation from random data sets:

Altman, N., & Krzywinski, M. (2015). Points of significance: Association, correlation and causation. *Nature Methods*, *12*(10), 899–900. doi:10.1038/nmeth.3587

This one has an excellent argument in favor of well-designed experiments that test causation, rather than stopping the analysis as correlation:

Games, P. A. (1990). Correlation and causation. *The Journal of Experimental Education*, *58*(3), 239–246. doi: 10.1080/00220973.1990.10806538

I love not only the title of this paper but also its message:

Matthews, R. (2000). Storks deliver babies ($p= 0.008$). *Teaching Statistics*, *22*, 36–38. https://doi.org/10.1111/1467-9639.00013

6.7 Exercises in the Geosciences

Part 1: By-Hand Calculations

1. Radon is a radioactive element naturally found in air. Two houses in a neighborhood are monitored for radon levels in their basements for a month. For the first house, the 30 measurements yield a mean level of 1.3 picocuries per liter (pCi/L) with a standard deviation of 0.2 pCi/L. The 30 measurements in the second house have a mean of 1.45 pCi/L with a standard deviation of 0.15 pCi/L.
   A Conduct a Student's *t* test to assess whether these two measurements can be considered to be from the same population.

2. A year later, the second house is measured again for radon, but this time for two months (61 days), yielding a mean of 1.52 pCi/L and a standard deviation of 0.13 pCi/L.
   A Conduct a Student's *t* test between this new measurement and the old one for the second house to assess whether the samples came from the same population.
   B Conduct a Welch's *t* test between these two sample sets. Compare these results with those from the Student's *t* test.

Part 2: Two Measures of Solar Activity

These questions compare two values that serve as proxies for the magnetic activity of the Sun. The first is sunspot number. This is a count of the small dark spots on the solar surface and is a number that ranges from zero during solar minimum to values as large as 250 during the most intense solar maximum intervals. Sunspots have been counted since the invention of the telescope in the early 1600s. The other parameter is a radio signal emanating from the solar atmosphere at the very particular wavelength of 10.7 cm. This is a faint signal from the Sun, first noticed by radio astronomers; to them it was noise interfering with the radio signal from far more distant celestial objects. It varies from a low value of about 70 s.f.u. (solar flux units) to a high value of about 250. These questions explore the correlation between these data sets.

3. Download number of sunspots and the F10.7 cm solar flux daily values for the years 1980 to the present from the NASA OMNIweb site: https://omniweb.gsfc.nasa.gov/form/dx1.html

    The main homepage of the OMNI database is here: https://omniweb.gsfc.nasa.gov/ow.html

    The specific format and data products are described here: https://omniweb.gsfc.nasa.gov/html/ow_data.html#1
    A Make a time series plot of these two data sets.
    B Make a scatterplot of these two data sets against each other.
    C Calculate the covariance and the Pearson correlation coefficient between these two data sets; find a probability for the computed coefficient.

4. Write a paragraph answering the following:
    A Qualitatively, what can be said about the linearity of the relationship between these two data sets?
    B Discuss the quantitative nature of the relationship as revealed by the correlation coefficient.

5. Determine an uncertainty on the correlation coefficient.
    A Conduct a bootstrap analysis with 200 resamplings of the paired data set.
    B Calculate R from each of these resampled data sets.
    C Make a histogram of the R values with an appropriate number of bins.
    D Calculate the mean and standard deviation of the resample-set R values.

6. Write a paragraph answering the following:
    A Based on a qualitative judgment about the histogram of the resample-set R values, is this distribution normal?
    B Discuss the similarity of the original R value and mean of the resample-set R values.
    C Discuss the appropriateness of using the standard deviation of the resample-set R values as the uncertainty on the original R.

7

Curve Fitting: Fitting a Line between Two Sets of Numbers

If two data sets are correlated, then the typical next step is to fit a line through the scatterplot. There are several standard approaches to the construction of a curve through a paired data set. This chapter will walk through a few of the common techniques. The first is a straight-line fit. The second is the creation of a curve through a linear combination of functions, each with a leading unknown coefficient. This is done through the derivation of formulas for these coefficients, a derivation that requires several assumptions along the way. The third approach is iterative curve fitting, often used when the unknown coefficients have a nonlinear relationship to the y-axis values and an analytical expression is difficult to derive. In this case, an initial guess at the coefficients is used with a method for nudging these values toward a final value that minimizes the error between the fit and the original data.

This process of curve fitting is often called linear regression or more generally regression analysis. To "regress" in this case means to make the relationship simpler, to identify the key functional form of the relationship, and distill the scatterplot down to a single curve. When you hear the term regression in reference to the statistical analysis of data sets, know that this is synonymous with determining a best-fit curve through the scatterplot.

7.1 Linear Regression

With a statistically significant and relatively high correlation coefficient, it is appropriate to consider a **linear fit** between the two data sets. We write this relationship as such:

$$y = A + B \cdot x \qquad (7.1)$$

In Equation (7.1), A is the y-intercept of the line—the y value when $x = 0$—and B is the slope of the line—that is, $B = dy/dx$. Figure 7.1 shows an example scatterplot of this relationship, annotating the definition of A and B on the linear fit through the points.

In Section 7.1.1, the derivations are given of how A and B are calculated from the x and y data sets, along with uncertainties on these fits (Section 7.1.2), and a few special cases of this linear fit (Sections 7.1.3 and 7.1.4).

7.1.1 Obtaining A and B

As with so much of processing data sets, obtaining A and B from the data sets starts with the assumption of a Gaussian

---

Linear fit: The process of finding a straight line through a scatterplot of data that minimizes a particular error quantity.

Data Analysis for the Geosciences: Essentials of Uncertainty, Comparison, and Visualization, Advanced Textbook 5, First Edition.
Michael W. Liemohn.
© 2024 American Geophysical Union. Published 2024 by John Wiley & Sons, Inc.
Companion website: www.wiley.com/go/liemohn/uncertaintyingeosciences

distribution. In this case, the Gaussian is relative to a mean value that varies with the corresponding x-axis value. That is, the distribution of y values is assumed to be a Gaussian around the fit line given in Equation (7.1). To say it yet another way, instead of X as the centroid of the Gaussian, this location of the peak is replaced by $A+Bx$. At this point, we do not know if our Gaussian assumption is true, because we do not know the A and B coefficients. Once we get these, then we can go back and assess whether our assumption of a Gaussian distribution around the line was a good one.

The following is a quick description of the derivation of the fit coefficients, A and B. To skip this, go directly to Equations (7.19) and (7.18) for A and B, respectively. The derivation is shown here for reference as well as to highlight the assumptions on which the linear fit is based.

Let's begin by defining a new variable, the fitted y values, as $\hat{y}$:

$$\hat{y}_i = A + Bx_i \qquad (7.2)$$

These are, for instance, the y values at each x_i location along the straight line in Figure 7.1.

Furthermore, we can now define the variable relating the original data and fit, $\tilde{y}$:

$$\tilde{y}_i = y_i - \hat{y}_i \qquad (7.3)$$

Using Equation (7.2), this becomes:

$$\tilde{y}_i = y_i - A - Bx_i \qquad (7.4)$$

This quantity is a function of both the y and x data sets.

With these new variable definitions, the probability distribution for the y values as a function of the x values should be written with this form:

$$Prob_{A,B}(y_i) = \frac{1}{\sigma_{\tilde{y}}\sqrt{2\pi}} e^{-\frac{\tilde{y}_i^2}{2\sigma_{\tilde{y}}^2}} \qquad (7.5)$$

Equation (7.5) can be expanded to be written as this:

$$Prob_{A,B}(y_i) = \frac{1}{\sigma_{\tilde{y}}\sqrt{2\pi}} e^{-\frac{(y_i - A - Bx_i)^2}{2\sigma_{\tilde{y}}^2}} \qquad (7.6)$$

In writing Equation (7.5), we have assumed that the y values have a true value that changes with x according to $A+Bx$ and that the spread

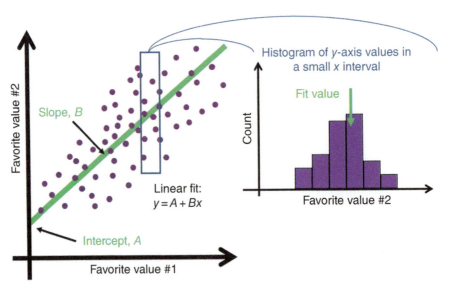

Figure 7.1 A scatterplot of two data sets as (x, y) pairs with a linear fit drawn through it. Also shown is the histogram of the y-axis values within a small interval along the x-axis, to assess whether it is a Gaussian distribution.

around this has an exponential width of $\sigma_{\tilde{y}}$ regardless of x. The tilde is present in the spread term because this is not the spread of the original y values but rather the spread of the y_i values around the linear fit, as given by Equation (7.3). So, with a number set of y values, each is related to a corresponding x value according to Equation (7.5). To describe it another way, at any x_i location along the x-axis data set, the corresponding y_i value could be any number described by a Gaussian probability distribution about a centroid that depends on that x_i location along the x-axis.

Given a list of y measurements at different x values, each one has the probability from Equation (7.5) that each particular y_i value is the one measured for that x_i value. This leads to a formula for the probability with the full set of paired values:

$$Prob_{A,B}(y_1, y_2, \ldots, y_N) = Prob_{A,B}(y_1) \cdot Prob_{A,B}(y_2) \cdot \ldots \cdot Prob_{A,B}(y_N) \quad (7.7)$$

Equation (7.7) states that the probability that the set of N pairs has those particular y_i values is the multiplication of the individual probabilities for each y_i value in the set. This form of the total probability for obtaining a particular data set includes an assumption that the probability for any one y_i value is independent of the others. Applying Equation (7.5) for each of the individual probabilities gets us this:

$$Prob_{A,B}(y_1, y_2, \ldots, y_N) = \frac{1}{\sigma_{\tilde{y}}^N (2\pi)^{N/2}} e^{-g} \quad (7.8)$$

In Equation (7.8), the multiplication of the Gaussian distributions leads to a summation in the argument of the natural exponential, so that the g term is this:

$$g = \frac{1}{2\sigma_{\tilde{y}}^2} \sum_{i=1}^{N}(y_i - A - Bx_i)^2 \quad (7.9)$$

The negative sign was left in Equation (7.8), rather than making it part of g and included in Equation (7.9), to remind us that the argument of the natural exponential is negative, or at most zero.

Being a Gaussian distribution, even the multiplication of many individual Gaussians, Equation (7.8) should have a single peak, the location of the maximum probability in the distribution. Therefore, we can use the principle of maximum likelihood to obtain the values of A and B that match this peak. The process of using this principle is to find the critical values with respect to each unknown—that is, take a derivative of Equation (7.8) with respect to each unknown quantity, in our case A and B, and set that equal to zero—to obtain a best fit for that unknown that maximizes the probability that all of the values came from the distribution given in Equation (7.5).

The derivative of Equation (7.8) with respect to A has this form:

$$\frac{\partial}{\partial A} Prob_{A,B}(y_1, y_2, \ldots, y_N) = \frac{-1}{\sigma_{\tilde{y}}^N (2\pi)^{N/2}} e^{-g} \left[\frac{\partial(g)}{\partial A} \right] \quad (7.10)$$

This derivative should then be set equal to zero. The leading constants cannot be equal to zero, nor can the natural exponential operator, so the only term left in Equation (7.10) that can be zero is the final partial derivative of the g function:

$$\frac{\partial(g)}{\partial A} = \frac{-1}{\sigma_{\tilde{y}}^2} \sum_{i=1}^{N}(y_i - A - Bx_i) \quad (7.11)$$

Equation (7.11) is set to zero. In Equation (7.11), the leading constants cannot be zero, so only the summation terms can be equal to zero. We can break up the summation and rewrite Equation (7.11) as separate summation terms for each component:

$$AN + B\sum_{i=1}^{N} x_i = \sum_{i=1}^{N} y_i \quad (7.12)$$

Equation (7.12) was written in this form, with the A and B coefficients on the left side and terms without A and B on the right side, for a reason, as we will see in a moment.

For now, let us return to the full probability and take a derivative of Equation (7.8) with respect to B:

$$\frac{\partial}{\partial B} Prob_{A,B}(y_1, y_2, \ldots, y_N)$$
$$= \frac{-1}{\sigma_y^N (2\pi)^{N/2}} e^{-g} \left[\frac{\partial(g)}{\partial B} \right] \quad (7.13)$$

This initial step looks very much like our derivative with respect to A in Equation (7.10). Similarly, only the final term, the derivative of g with respect to B, can be zero:

$$\frac{\partial(g)}{\partial B} = \frac{-1}{\sigma_y^2} \sum_{i=1}^{N} x_i \cdot (y_i - A - Bx_i) \quad (7.14)$$

Equation (7.14) is now set to zero. The terms in front of the summation cannot be zero, so we can simplify this a bit more. As done with Equation (7.11), Equation (7.14) can be written as separate summations of each of the original terms:

$$A \sum_{i=1}^{N} x_i + B \sum_{i=1}^{N} x_i^2 = \sum_{i=1}^{N} x_i y_i \quad (7.15)$$

These two equations with the unknown coefficients A and B on the left, Equations (7.12) and (7.15), are known as the **normal equations**. We are not at the end of the calculation, so it might seem rather strange to have a special name for this intermediate equation set. This is because, at this point, we have two equations and two unknowns, and therefore this is the moment in the derivation where we know that we have a solution in our future. In fact, these two equations can be written in matrix form:

$$\begin{bmatrix} N & \sum x_i \\ \sum x_i & \sum x_i^2 \end{bmatrix} \begin{bmatrix} A \\ B \end{bmatrix} = \begin{bmatrix} \sum y_i \\ \sum x_i y_i \end{bmatrix} \quad (7.16)$$

Normal equations: The resulting formulas that remove the Gaussian exponentials and relate the unknown coefficients to summations of the x and y data values.

All of the summations in Equation (7.16) range from $i = 1$ to N. At this point, linear algebra principles can be applied to solve for the unknowns of A and B. For a 2 × 2 matrix, this is straightforward. Equation (7.12) can be easily solved for A with a bit of rearrangement:

$$A = \frac{\sum y_i - B \sum x_i}{N} \quad (7.17)$$

This can be inserted into Equation (7.15) and some algebra yields a formula for B:

$$B = \frac{N \sum x_i y_i - (\sum x_i)(\sum y_i)}{N \sum x_i^2 - (\sum x_i)^2} \quad (7.18)$$

In the numerator of Equation (7.18), the first term has the x_i and y_i values multiplying each other inside of the summation, while in the second term these two values are summed separately and then multiplied together. Similarly, in the denominator, the first term is a summation of the square of x_i, while the second term is the square of the summation of the x_i values.

Finally, with yet more algebra, Equation (7.18) can be put back into Equation (7.17) to get a final form for the function of A:

$$A = \frac{(\sum x_i^2)(\sum y_i) - (\sum x_i y_i)(\sum x_i)}{N \sum x_i^2 - (\sum x_i)^2} \quad (7.19)$$

We now have equations for A and B as functions of the original number sets. While these two equations for A and B look complicated, they are rather simple combinations of summations.

The units of these two terms can be checked, to make sure that it works out to what we expect. For A, we can examine Equation (7.19) and see that there is an x^2 multiplier in all terms. These will all cancel each other out, leaving only the y terms in the numerator. Therefore, A has the dimensions of y, which is exactly what we wanted, as this is the y-intercept value of the fit. For B, the x terms in Equation (7.18) do not cancel out. The numerator has units of x times y, while the denominator has units of x^2. This leaves B having the units of the y-axis data set over those of the x-axis data set, which are the units of slope.

This derivation was presented for several reasons. The first is that it is the basis for more complicated nonlinear curve fits, as we will see in Section 7.4. The second is to reveal the assumptions on which these formulas are based. The first assumption is seen in Equation (7.5); the probability distribution of potential curve fits through the data set is the multiplication of many Gaussian distributions for each y value around the linear fit. This is actually N different assumptions, one about each y value in the number set. The second is that the principle of maximum likelihood was used to obtain the best-fit coefficients, assuming that the best fit is also where this combined probability distribution has its one and only peak. This is only true if the resulting distribution of the discrepancies, the $\tilde{y}_i$ values defined in Equation (7.3), forms a Gaussian distribution. We need to keep these assumptions in mind as we continue through this chapter.

> **Quick and Easy for Section 7.1.1**
>
> Linear fitting is based on the assumption that the distribution of the y values around a straight line is Gaussian. This leads to formulas for calculating the best-fit coefficients defining that straight line.

7.1.2 Uncertainties on A and B

Remember that this came from an assumption of a single Gaussian distribution for all of the y values relative to this linear fit. Now that we know A and B, we can make a plot $\tilde{y}$; that is, of $y_i - \hat{y}$. Figure 7.2 shows this plot for the scatterplot of (x,y) values initially shown in Figure 7.1. In this particular case, it appears to be close to a Gaussian. The closeness of this distribution to a Gaussian can be assessed with the methodology defined in Chapter 5. If it is close to a Gaussian, then continued usage of the linear fit is reasonable. If not, then the linear fit results should be used only with the caveat that the underlying distribution does not coincide with the assumptions underlying the fit.

A standard deviation, $\sigma_{\tilde{y}}$, can be computed from this distribution in Figure 7.2, with the formula we used in Chapter 4:

$$\sigma_{\tilde{y}} = \sqrt{\frac{1}{N}\sum_{i=1}^{N}\tilde{y}_i^2} \qquad (7.20)$$

Expanding this with our definition of $\tilde{y}_i$, we get:

$$\sigma_{\tilde{y}} = \sqrt{\frac{1}{N}\sum_{i=1}^{N}(y_i - A - Bx_i)^2} \qquad (7.21)$$

In Chapter 4, this was corrected by the number of degrees of freedom in this calculation. For the standard deviation of a measurement set, that correction was 1, because the mean of the sample set was used in the calculation of the standard deviation of the sample set. For the spread around a linear fit, we now have 2 degrees of freedom, the A and B coefficients of the fit. So, a corrected formula that includes the degrees of freedom is this:

$$\sigma_{\tilde{y}} = \sqrt{\frac{1}{N-2}\sum_{i=1}^{N}(y_i - A - Bx_i)^2} \qquad (7.22)$$

As discussed in Chapter 4, this influence is small for large N. Because this is a spread, it should only be reported to one significant digit in nearly all cases, so the influence of this extra minus two in the denominator is only important if it causes a change of more than 5% to the value, therefore potentially changing the first digit during rounding. The ratio of Equation (7.20) to Equation (7.22) is $\sqrt{N/(N-2)}$, which will be less than 1.05 for $N \geq 22$. For N less than this cutoff, the degrees of freedom correction should be used; otherwise the spread will be slightly underestimated. For larger N, the influence of the correction is negligible with respect to the first significant digit and either formula can be used.

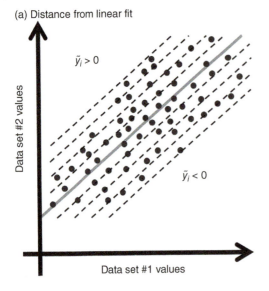

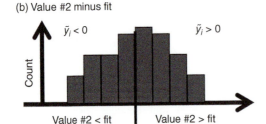

Figure 7.2 Creating a histogram of the $\tilde{y}$ values from the scatterplot in Figure 7.1, the y_i values with the linear fit subtracted. In order to calculate A and B, it was assumed that this distribution was a Gaussian.

We went through this discussion of $\sigma_{\tilde{y}}$ because it is needed for the uncertainties on A and B. Specifically, σ_A and σ_B are both direct multiples of $\sigma_{\tilde{y}}$ with coefficients similar to the formulas for A and B themselves. For the intercept, the uncertainty formula is this:

$$\sigma_A = \sigma_{\tilde{y}} \sqrt{\frac{\left(\sum x_i^2\right)}{N\sum x_i^2 - \left(\sum x_i\right)^2}} \quad (7.23)$$

The formula for the uncertainty of the slope looks similar:

$$\sigma_B = \sigma_{\tilde{y}} \sqrt{\frac{N}{N\sum x_i^2 - \left(\sum x_i\right)^2}} \quad (7.24)$$

In Equation (7.23), the numerator and denominator have the same units, those of the x-axis data set, and therefore the square root operator as a whole is dimensionless. This leaves the units of σ_A as the units of the y-axis data set, whatever that may be. In Equation (7.24), the numerator inside the square root operator is simply the count of (x, y) pairs, which is dimensionless, so the units of the denominator are not canceled out. Therefore, the units of σ_B are those of the y-axis data set divided by those of the x-axis data set—that is, rise over run, the units of slope.

> **Quick and Easy for Section 7.1.2**
>
> There are explicit formulas for the uncertainty associated with the linear fit coefficients.

7.1.3 The Zero-Intercept Special Case

There is a special case of linear fitting that sometimes arises in the natural sciences—those times when you know the relationship is a direct proportionality. That is, when we know that there should not be an offset, it is possible to omit this from the linear fit calculation. In this case, we have

$$y = B \cdot x \quad (7.25)$$

Equation (7.25) assumes that $A = 0$ from the beginning, which simplifies the calculation. In this case, the g function in the natural exponential, given in Equation (7.9), changes to this:

$$g = \frac{1}{2\sigma_{\tilde{y}}^2} \sum_{i=1}^{N} \left(y_i - Bx_i\right)^2 \quad (7.26)$$

The principle of maximum likelihood can still be applied, but now there is only one unknown in the distribution, B, and so only one derivative is needed. As done in Section 7.1.1, it is the derivative of g with respect to B that is the only term in the derivative of the probability distribution that can be zero, so we are left with this:

$$\frac{\partial(g)}{\partial B} = \frac{-1}{\sigma_{\tilde{y}}^2}\sum_{i=1}^{N} x_i \cdot (y_i - Bx_i) \quad (7.27)$$

Equation (7.27) should be set equal to zero, after which it can be solved for B to yield:

$$B = \frac{\sum x_i y_i}{\sum x_i^2} \quad (7.28)$$

If the intercept is naturally zero, then Equation (7.19) would naturally yield a value close to zero. How close is "close" to zero? This is found by comparing A with its standard deviation, and it you calculate that $|A| \ll \sigma_A$, then this indicates that A is a negligible value within its own uncertainty of zero.

> **Quick and Easy for Section 7.1.3**
>
> All of the linear fitting formulas get a bit easier when it is assumed that the best-fit line passes through the origin of the (x, y) coordinate system.

7.1.4 Weighted Linear Fitting

While that was a simplifying case, the more complicated case that will be considered here is where the Gaussian spread of $\tilde{y}$ is not the same for all values in the data set. It could be that the spread changes as a smooth function of x or simply that it is different for the various points in the set, such as values taken by different devices, at different times, or by different investigators. Fortunately, this complication only impacts the formulation of the problem in a small way.

In this section, the derivation of the new A and B coefficients are given. This derivation is included so that the assumptions behind it can be made known. To skip, this, jump to Equations (7.34) and (7.35).

In this case of non-uniform spreads, we need a distinct spread for each of the Gaussian probabilities, rewriting Equation (7.5) as this:

$$Prob_{A,B}(y_i) = \frac{1}{\sigma_{\tilde{y}_i}\sqrt{2\pi}} e^{-\frac{\tilde{y}_i^2}{2\sigma_{\tilde{y}}^2}} \quad (7.29)$$

Expanding to write it in terms of the unknown coefficients, this is:

$$Prob_{A,B}(y_i) \frac{1}{\sigma_{\tilde{y}_i}\sqrt{2\pi}} e^{-\frac{(y_i - A - Bx_i)^2}{2\sigma_{\tilde{y}_i}^2}} \quad (7.30)$$

The full probability of the paired data set, multiplying all of the individual probabilities from Equation (7.30) together, is then this:

$$Prob_{A,B}(y_1, y_2, \ldots, y_N) = \frac{1}{\sigma_{\tilde{y}_i}^N (2\pi)^{N/2}} e^{-g} \quad (7.31)$$

In Equation (7.31), the g function has also changed from its previous form because now the spread must be inside of the summation:

$$g = \frac{1}{2}\sum_{i=1}^{N} \frac{(y_i - A - Bx_i)^2}{\sigma_{\tilde{y}_i}^2} \quad (7.32)$$

As defined in Chapter 5 (specifically, Section 5.10), the inverse of the square of the standard deviation can be defined as a weight:

$$w_i = \frac{1}{\sigma_{\tilde{y}_i}^2} \quad (7.33)$$

For a particular (x_i, y_i) pair, the weight is large when the $\tilde{y}$ spread is small and the weight is small when this spread about the fit value is large. That is, the y values from x locations with smaller spread are given a larger influence on determining the eventual fit values.

Equation (7.33) becomes an extra term within all of the summations in Section 7.1.1, yielding new formulas for A and B that include these weights in every term. The new form of A is this:

$$A = \frac{\left(\sum w_i x_i^2\right)\left(\sum w_i y_i\right) - \left(\sum w_i x_i y_i\right)\left(\sum w_i x_i\right)}{\left(\sum w_i\right)\left(\sum w_i x_i^2\right) - \left(\sum w_i x_i\right)^2}$$

$$(7.34)$$

The revised version of B is this:

$$B = \frac{\left(\sum w_i\right)\left(\sum w_i x_i y_i\right) - \left(\sum w_i x_i\right)\left(\sum w_i y_i\right)}{\left(\sum w_i\right)\left(\sum w_i x_i^2\right) - \left(\sum w_i x_i\right)^2} \tag{7.35}$$

In both equations, the N terms also change to summations of the weights. Because all terms have a "weight squared" multiplier included, this change in the formulas does not influence the units of the final A and B coefficients.

The uncertainties can also be rewritten with these value-specific spreads included in the derivation. For the intercept, the new uncertainty is this:

$$\sigma_A = \sigma_{\bar{y}} \sqrt{\frac{\left(\sum w_i x_i^2\right)}{\left(\sum w_i\right)\left(\sum w_i x_i^2\right) - \left(\sum w_i x_i\right)^2}} \tag{7.36}$$

For the slope, the weighted uncertainty has this form:

$$\sigma_B = \sigma_{\bar{y}} \sqrt{\frac{\sum w_i}{\left(\sum w_i\right)\left(\sum w_i x_i^2\right) - \left(\sum w_i x_i\right)^2}} \tag{7.37}$$

As with A and B, there are now weights multiplying the arguments in every summation and the N terms in Equations (7.36) and (7.37) have been changed to summations of the weights. The inclusion of all of these weights does not influence the final units of the spreads.

Quick and Easy for Section 7.1.4

The method of weighted averaging can also be applied to determining the coefficients for linear fitting, taking into account unequal uncertainties for the y-axis values.

7.2 Testing a Linear Fit

Linear fitting of a distribution of paired values is a form of modeling. It is one of the simplest—the model is based on the data sets themselves—but it is still a model. The relationship between the original data and the linear fit can be quantitatively assessed. The later chapters of this book discuss numerous methods for conducting this assessment, but one should be mentioned here. It is the **analysis of variance**—or ANOVA—table. Although it has general applicability to any fit to the y-axis data set, the ANOVA table was designed specifically for linear fits.

Given our definition of the fit values, $\hat{y}_i$, in Equation (7.2) and our definition of variance from Chapter 4, we can then define some other parameters that are measures of the variation of the different quantities. The first of these is the **sum of squares for regression**, SSR, defined as follows:

$$\text{SSR} = \sum_{i=1}^{N} \left(\hat{y}_i - \bar{y}\right)^2 \tag{7.38}$$

The average of the fitted values is also the average of the original values, so the mean y value in Equation (7.38) does not need a hat accent. Except for the missing $1/N$ multiplier, this is the variance of the fitted values against their mean value. The next new parameter is the **sum of squares for error**, SSE:

$$\text{SSE} = \sum_{i=1}^{N} \left(y_i - \hat{y}_i\right)^2 \tag{7.39}$$

Analysis of variance (ANOVA): A method of assessing the goodness of a curve fit through a scatterplot, it involves creating a small table, leading to the F statistic.

Sum of squares for regression (SSR): The sum of the square of the discrepancy between the fitted y values at each x data set value and the mean of the original y data set.

Sum of squares for error (SSE): The sum of the square of the discrepancy between the original y data set values and the fitted y values.

This is almost the same as $\sigma_{\bar{y}}$ from Section 7.1.2, but again without the averaging denominator. The third new parameter is the **total sum of squares**, SST, defined as this:

$$\text{SST} = \sum_{i=1}^{N}(y_i - \bar{y})^2 \qquad (7.40)$$

This is very close to the variance of the original y data set. The next new parameter is the **mean square for regression**, MSR:

$$\text{MSR} = \frac{\text{SSR}}{d-1} \qquad (7.41)$$

In Equation (7.41), the d in the denominator is the degrees of freedom of the fit, which for a linear regression is 2. This makes this calculation trivial and defines MSR = SSR. This is only the case for linear fitting, however. Because the ANOVA table can be used for any data-model comparison, d does not have to be 2 and can be quite large, depending on the type of model being used. Yet another new parameter is the **mean square for error**, MSE, defined like this:

$$\text{MSE} = \frac{\text{SSE}}{N-d} \qquad (7.42)$$

Total sum of squares (SST): The sum of the square of the discrepancy between the original y data set values and the mean of the original y data set values.

Mean square for regression (MSR): The sum of squares for regression divided by one less than the degrees of freedom of the fit.

Mean square for error (MSE): The sum of squares for error divided by the difference of the numbers in the data set and the degrees of freedom in the fit.

Using the definition of SSE in Equation (7.39), it is seen that MSE is exactly the sample $\sigma_{\bar{y}}$ definition from Equation (7.22). Because N is often quite large, MSE is usually much less than SSE. The final parameter that is part of the ANOVA table is the F **statistic**:

$$F = \frac{\text{MSR}}{\text{MSE}} \qquad (7.43)$$

Because MSR only had degrees of freedom in the denominator and MSE had the number of x–y pairs in its denominator, these two values are usually quite different from each other and, therefore, F is often very large.

The ANOVA table is then defined as given in Table 7.1. It is a quick view of these parameters. It lists the "sources" of the variation in the values, splitting the total variation of the original y data set into two components. The first is the variation that can be attributed to the changing x value, as defined by the fit, regardless of whether it is a linear fit or something more complicated. The second component is the remaining "error" of the values, attributable to the spread of the values around that fit, regardless of whether that spread matches a Gaussian probability distribution or not. The next column, labeled "d.f.," lists the degrees of freedom for that row. The other columns are, in order, the sum of square quantities (found by Equations (7.38), (7.39), and (7.40)), the mean square quantities (from Equations (7.41) and (7.42), dividing by the degrees of freedom), and the F statistic from Equation (7.43), the division of the two mean square values.

A side note about the ANOVA table is that the coefficient of determination, R^2, can be written in terms of the parameters we have just defined:

$$R^2 = \frac{\text{SSR}}{\text{SST}} \qquad (7.44)$$

F statistic: A measure of the goodness of a curve fit, it is defined as the ratio of the mean square for regression to the mean square for error.

Table 7.1 Definition of the analysis of variance, or ANOVA, table.

Source	d.f.	Sum of squares	Mean square	F
Regression	$d-1$	SSR	MSR	F statistic
Error	$N-d$	SSE	MSE	
Total	$N+1-d$	SST		

Table 7.2 The probability for a given F statistic that the score could be from random chance, listed as a function of N for the linear fit value of $d = 2$.

F statistic score	3	5	10	20	30	50
$N = 10$ (d.f. = 8,1)	0.122	0.0558	0.0133	0.0021	0.0006	0.0001
20	0.100	0.0383	0.0054	0.0003	0.0003	$<10^{-5}$
50	0.090	0.0300	0.0027	5×10^{-5}	$<10^{-5}$	
100	0.0864	0.0276	0.0021	2×10^{-5}	$<10^{-5}$	
500	0.0839	0.0268	0.0017	$<10^{-5}$		
5000	0.0833	0.0254	0.0016	$<10^{-5}$		

That is, the coefficient of determination is the ratio of the variance of the fitted values divided by the variance of the original y data set. It is often interpreted as the proportion of how much of the total variance is explained by the fit. This is a slightly different explanation than given earlier in Chapter 6, where R² was defined as the variation in x explaining the variation in y. The difference is that Equation (7.44) could be calculated using any fit, while R² from Chapter 6 is based specifically on the linear fit.

Because of the relationship between the three sum-of-squares terms, another version of the R² formula is this:

$$R^2 = 1 - \frac{SSE}{SST} \qquad (7.45)$$

This is an important version of R², and we will come back to it in Chapter 9.

In addition to a general understanding of how the variation of the y data set is split between what the model can describe and what is left over, the other primary outcome of the ANOVA table is the F statistic. The F statistic can be converted to a probability that the model fit to the data is significant. Table 7.2 lists the one-sided probabilities that a given F statistic could be from random chance (an online calculator is given in Section 7.8). These values are for the linear fit d value of 2. For N = 10, the degrees of freedom for error and regression are 8 and 1, respectively, which are needed for the probability calculation. This probability calculation is nearly always one-sided because the F statistic is positive definite and so it is only the high-end tail of the distribution that matters.

Table 7.2 reveals that the probability calculation is rather insensitive to N. For an F statistic score of 3, the probability slowly drops from 0.12 for N = 10 to 0.08 for N = 5000. For a given F statistic, the probability slowly decreases with increasing N. It is seen that the 5% probability threshold for statistical significance is reached with F = 5 and N between 10 and 20, or for any F > 10 (at least for the N values in the table). Note that these probabilities are based on an assumption that the

Table 7.3 The F statistic needed to reach the 5% and 1% levels that the score could be from random chance, listed as a function of d for an N value of 100.

P score	0.05	0.01
$d = 2$ (d.f. = 98,1)	3.94	6.91
3 (d.f. = 97,2)	3.10	4.84
4 (d.f. = 96,3)	2.70	4.00
5 (d.f. = 95,4)	2.47	3.53
10 (d.f. = 90,9)	1.99	2.62
20 (d.f. = 80,19)	1.72	2.15

spread for the y values around the fit, $\tilde{y}$, is Gaussian. If it is not, then do not trust the probabilities.

The F statistic probabilities are slightly more sensitive to d than to N. Table 7.3 shows F statistic values needed to reach the 5% and 1% significance levels for various settings of d, all calculated with $N = 100$. The dramatic decrease in the F value needed to reach significance with increasing d is readily apparent in this chart. Adding just one coefficient to the fit reduces the F value needed from 3.94 to 3.10 to achieve the 5% significance level, a drop of nearly one. The 1% probability level similarly drops as a function of d, from 6.91 for the linear fit to 4.84 for a fit with just one more coefficient. The F statistic systematically decreases with increasing d, but so does the MSR value in the numerator of the F statistic. For the same SSR value, a change in modeling coefficients from 2 to 3 means a reduction of half in MSR, which is SSR divided by $d - 1$. This means that it is harder to achieve a significant F statistic when the model includes more free parameters.

7.3 Example: Human-Induced Seismicity

Earthquakes have been increasing across the world. Not large ones, but many small earthquakes are now regularly occurring in oil-rich regions of the world. This increase is linked to the recent widespread implementation of **hydraulic fracturing** in these locations, a technique for extracting oil and natural gas from rock, over the last few decades. As the name implies, high-pressure liquids are injected into the ground to break apart the rock formation containing the oil or natural gas, which then comes back up the pipe and is collected. This localized splitting of the rock by high-pressure fluids is not seismicity; this process is too small of scale to directly be felt as a perceptible earthquake without highly sensitive instrumentation. The process that includes injection, fracture, and extraction changes the stress load in the Earth's crust, however, and could eventually lead to an earthquake. This seismicity would not have occurred without the influence of humans.

The process of hydraulic fracturing is actually only one of several main causes of the increase in small-amplitude earthquakes in oil-rich regions. A significant but minority fraction of these earthquakes are due to a different but related human activity—wastewater disposal from oil and natural gas refining. This process generates large quantities of contaminated water that is unfit for not only human consumption but also agricultural usage. Rather than leaving it on the surface to eventually evaporate or seep into the groundwater, this toxic water is sometimes injected deep into the ground, below the local water table. Wastewater disposal is very similar to hydraulic fracturing, except that nothing is removed from the ground—it is a

---

Quick and Easy for Section 7.2

A common assessment of the goodness of a linear fit is to calculate an ANOVA table, culminating in the F statistic, from which a probability can be found that the computed F value could be from random chance.

Hydraulic fracturing: The process of extracting oil and gas from rock by 2injecting pressurized water mixed with certain chemicals to break open the rock formation.

one-way process. This activity has also been rising in the past few decades.

Figure 7.3 shows the relationship between increased hydraulic fracturing and increased small-scale seismicity. The location shown here is the Montney Formation in Western Canada, on the border between the provinces of Alberta and British Columbia. Many new hydraulic fracturing wells have been installed across the region. At the same time, the number of seismic events has also substantially increased. Extending the count within the "background rate" zone of the timeline, the cumulative number of natural seismic events should be less than 100. Instead, over 500 events have occurred in this region with an intensity larger than 2.5 on the Richter scale.

Note that it is not a direct link between more hydraulic fracturing or wastewater disposal wells and more human-induced seismicity. Many wells never cause earthquakes, and some earthquakes could occur far from the well locations.

The linkage critically depends on the local geology and susceptibility of the region to new seismic activity given the changes in the underground stresses. Some places are rather stable, and the extractive industry activity causes little to no additional seismicity. In other places, like the Montney Formation in Canada, the connection is starkly obvious.

To fully use this as an example of linear fitting, data from this region of Canada are shown as a scatterplot in Figure 7.4. Note that the y-axis is actually the base-10 logarithm of the number of earthquakes; because these are all earthquakes, not just those above 2.5 magnitude, this conversion with the log operator makes it a relationship that is closer to linear. Also shown in the plot is the linear fit, as calculated according to the method in Section 7.1.1. The A and B values of the fit are 0.25 and 2.1×10^{-4}, respectively. The fit looks pretty good, and the Pearson linear correlation coefficient between these two number sets is 0.69 (so, R^2 of 0.48), a respectably good value that

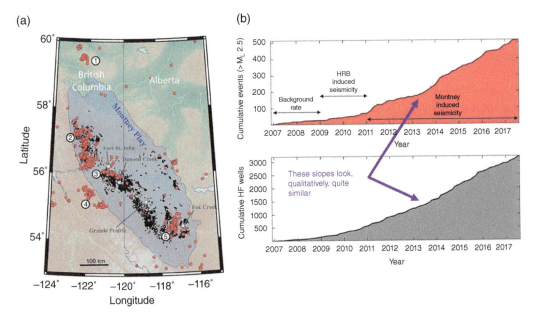

Figure 7.3 The Montney Formation in Western Canada, on the border between British Columbia and Alberta. In the left plot, the black dots are the hydraulic fracturing wells, and the red dots are epicenters of seismicity. The circled numbers indicate regions of seismic clustering. The upper-right line plot shows the cumulative number of seismic events, while the lower-right plot shows the cumulative number of wells. HRB stands for Horn River Basin, a region just to the north of the Montney Formation, in which wells were drilled a few years earlier than in Montney. Adapted from Wozniakowska and Eaton (2020).

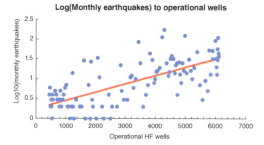

Figure 7.4 The logarithm (base 10) of the number of earthquakes in a given month to the number of operational hydraulic fracturing wells in that region, along with a linear fit (shown as the red line).

is well above the statistical threshold for being considered highly significant.

As just discussed in Section 7.2, this fit can be tested via an ANOVA table. The result of this calculation is shown in Table 7.4. The F statistic has a probability of being from random chance is very small, below 1×10^{-5}. This indicates that this line is a good fit.

A further assessment that should be done is to examine the normality of the distribution of the discrepancies of the y data set values against the fit values. These discrepancies have a range from −0.93 to +1.09, with a standard deviation of 0.4. By definition of the linear fit procedure, the mean of the discrepancies is zero because the average of the linear fit values is always the same as the average of the original y data set. Remembering our process from Chapter 5, the first thing to check is the shape of the histogram. This is shown in Figure 7.5; the top panel shows the histogram and the lower panel includes an overlaid second

Table 7.4 The ANOVA table for our example linear fit for human-induced seismic activity.

Source	d.f.	Sum of squares	Mean square	F
Regression	1	16.75	16.75	99.1
Error	107	18.08	0.169	
Total	108	34.83		

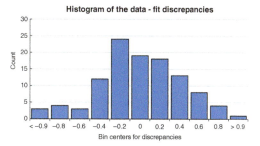

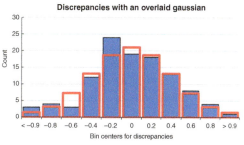

Figure 7.5 Histogram of the discrepancies between the observed values and the fitted values from Figure 7.4 (blue dots minus red line at each x value). The lower panel is this same histogram but with an overlaid histogram of expected count values from a Gaussian distribution of the same mean and standard deviation.

histogram of expected counts from a corresponding Gaussian distribution. Qualitatively, these look close; only two columns (the first and third) look substantially different (by more than 50%). This is below our qualitative threshold of difference.

To be more quantitative, some statistics from this data set can be calculated. First, let's do a comparison of the centroids and spreads. The mean, median, and mode of the distribution are 0, −0.011, and −0.2, respectively. For the mode, the center of the bin was used. These are all relatively close; the median is extremely close to the mean and the mode is only one bin off from zero. The standard error of the data set is 0.05, and the difference of the mean and median is well within this measure of centroid uncertainty. For spreads, we need to compare twice the standard deviation against the 16%–84% quantile range as well as 85% of the full width half maximum. These three values, to a few extra decimals than should be shown, are 0.818,

0.776, and 0.850, respectively. To one significant digit, these are all essentially equal. Another quantitative comparison is to calculate the skewness and kurtosis coefficients and compare them with the ideal values for a Gaussian distribution. For this discrepancy data set, these values are −0.25 and 2.97, respectively. The skew is relatively small and the kurtosis is very close to three. So far, all indications are that this distribution is close to a Gaussian.

The final two assessments from our process are to conduct a chi-squared test and a Kolmogorov–Smirnov test. The first produces a χ^2 value of 6.44, and when divided by the degrees of freedom (11 bins minus 3 constraints makes $d = 8$) it yields $\tilde{\chi}^2 = 0.80$. This is a low number and indicates that the distribution is normal. For the Kolmogorov–Smirnov test, the K-S statistic is $D_{n,m} = 0.030$, while a probability of 0.05 yields $D_{n,m,P}^{\text{crit}} = 0.121$. This also indicates normality of the distribution.

From all of these assessments, the judgment is easy—the use of the linear fit algorithm from Section 7.1.1 is fully justified. If it had not been a Gaussian, then this would have put additional concerns on the use of the probability for the F statistic. Since the distribution of discrepancies is normal, then the probability for the calculated F statistic is reasonable.

There are reasons why we might not have expected a perfect correlation and highly significant linear fit. For one, we are taking the logarithm of the number of earthquakes. There was no physical justification for this operation; it was done because it made the scatterplot look more linearly related. Another is that there are additional possible causes of earthquakes; the operation of hydraulic fracturing extraction wells potentially induces earthquakes, but so might the injection of wastewater from nearby oil and natural gas refining. This could explain the nonlinear relationship and thus the need to take a log of the earthquake count. There could be other factors, too, beyond these two human-induced mechanisms for triggering earthquakes. A final point to make is that this analysis has many more earthquakes included than shown in Figure 7.3. That figure only shows the count of earthquakes above 2.5 magnitude, of which there were roughly 500. The linear fit analysis done here includes all earthquakes in the database, which is a list of over 2000 events.

> **Quick and Easy for Section 7.3**
>
> The underground sequestering of oil and gas refining wastewater has led to an increase in small earthquakes, with a decent linear fit between the two.

7.4 Nonlinear Fitting

What if the functional relationship between the two data sets is not a straight line? What if, instead of a linear fit, it is a quadratic polynomial that should be used for the fit procedure? Some quantities in the natural sciences are expected to have this type of relationship, such as dynamic pressure being a function of the square of wind speed. In this case, the general linear fit formula of Equation (7.1) formula should have another term included, like this:

$$y = A + Bx + Cx^2 \qquad (7.46)$$

This is just one example of a **nonlinear fit** between two number sets. The term nonlinear simply means anything with a functional form of anything other than Equation (7.1). In the following subsections, several common examples of nonlinear fitting are given.

7.4.1 Polynomial Fitting

Polynomial fitting refers to the general relationship of the y variable to terms that contain powers of the x variable. Let's start with the quadratic polynomial case given in

---

Nonlinear fit: The process of finding a curved line through a scatterplot of data that minimizes a particular error quantity.

Equation (7.46). This equation has three unknowns. As done with linear fitting, these coefficients can be written as explicit functions of the original data sets. In fact, the analysis procedure is exactly the same.

In this case, the $\tilde{y}$ variable, the difference between the original values and the fitted curve $\hat{y}$ values, still has this form:

$$\tilde{y}_i = y_i - \hat{y}_i \quad (7.47)$$

The fit now has three unknowns, A, B, and C, so the full form of the y discrepancy variable looks like this:

$$\tilde{y}_i = y_i - A - Bx_i - Cx_i^2 \quad (7.48)$$

We can then assume that the y data set values follow a Gaussian probability around the quadratic curve $\hat{y}$ values with a spread $\sigma_{\tilde{y}}$ that is the same for all of the y values. The combined probability for obtaining the full y data set, with its particular spread of values around the curve, is then exactly the same as in the linear case—Equation (7.8)—but now the g function in the argument of the natural exponential has a different form:

$$g = \frac{1}{2\sigma_{\tilde{y}}^2} \sum_{i=1}^{N} \left(y_i - A - Bx_i - Cx_i^2 \right)^2 \quad (7.49)$$

Following the same procedure as in Section 7.1.1, the principle of maximum likelihood describes a process for obtaining the unknown coefficients. After derivatives are taken and set to zero, we can write the resulting set of normal equations like this:

$$AN + B\sum_{i=1}^{N} x_i + C\sum_{i=1}^{N} x_i^2 = \sum_{i=1}^{N} y_i \quad (7.50)$$

$$A\sum_{i=1}^{N} x_i + B\sum_{i=1}^{N} x_i^2 + C\sum_{i=1}^{N} x_i^3 = \sum_{i=1}^{N} x_i y_i \quad (7.51)$$

$$A\sum_{i=1}^{N} x_i^2 + B\sum_{i=1}^{N} x_i^3 + C\sum_{i=1}^{N} x_i^4 = \sum_{i=1}^{N} x_i^2 y_i \quad (7.52)$$

Equations (7.50), (7.51), and (7.52) come from the derivatives with respect to A, B, and C, respectively, of the full probability distribution equation. As done for the linear version, this can be written in matrix form and a lower-then-upper triangular factoration—typically called the **LU decomposition** method—can be applied to solve for A, B, and C. To briefly describe how to do it by hand, Equation (7.50) can be used to isolate A (but it is still a function of B and C), which can be substituted into Equation (7.51) to isolate B (but this will still be a function of C), and both of these can be substituted into Equation (7.52) to find the form of C that is purely a function of the two original data sets. This expression can then be plugged back into the formula for B, and both can be used in the formula for A. This yields a rather complicated function, so ugly that they will not be listed here. Fortunately, this can be efficiently found computationally.

Examination of the normal equations reveals the diagonal pattern of how the powers of x_i systematically increase moving down through these formulas. This arises from the derivative of g with respect to each of the coefficients. For a particular coefficient, taking the derivative of g results in the polynomial term multiplying that coefficient and becoming an extra multiplier of all of the terms of $\tilde{y}$. To expand the polynomial discussion to an arbitrary power, instead of a quadratic function, let's use a general h-degree polynomial fit, which can be defined like this:

$$y = A + Bx + Cx^2 + \cdots + Hx^h \quad (7.53)$$

LU decomposition: A linear algebra process for solving a matrix, it involves isolating terms by eliminating coefficients in the lower half of the matrix and then solving for the unknowns using the remaining coefficients in the upper half of the matrix.

Polynomial fitting: The process of finding an optimal line through a scatterplot of data based on a function with terms dependent on powers of x.

This will lead to normal equations, written here in matrix form:

$$\begin{bmatrix} N & \sum x_i & \cdots & \sum x_i^h \\ \sum x_i & \sum x_i^2 & & \vdots \\ \vdots & & \ddots & \\ \sum x_i^h & \sum x_i^{h+1} & \cdots & \sum x_i^{2h} \end{bmatrix} \begin{bmatrix} A \\ B \\ \vdots \\ H \end{bmatrix} = \begin{bmatrix} \sum y_i \\ \sum x_i y_i \\ \vdots \\ \sum x_i^h y_i \end{bmatrix} \quad (7.54)$$

Looking at the terms in the leftmost column of the matrix, the first term is a summation over the polynomial multiplying each of the coefficients. This extra multiplier is also seen in the extra x_i polynomial terms in the summations of right-side vector. For A, the polynomial is of power zero, so the multiplier is simply one, and the summation is already conducted, resulting in N.

Equation (7.54) is a set of h equations for h unknowns. It can be solved for explicit relations for all of the coefficients. The results will neither be compact nor look nice, but a closed-form result can be obtained for each of the coefficients.

> **Quick and Easy for Section 7.4.1**
>
> Polynomial fitting follows the same process as linear fitting, resulting in a large matrix to be solved but still a straightforward calculation.

7.4.2 Generalized "Linear Coefficient" Fitting

The pattern noted for the polynomial case can be expanded to an even more generalized function form. First let's consider a relationship like this:

$$y = A \cdot f_1(x) + B \cdot f_2(x) \quad (7.55)$$

The two functions, f_1 and f_2, are assumed to be known, but the linear combination of these is unknown. This is the key point of all of the fitting procedures discussed in this chapter—the dependence on x can be whatever function is necessary and can be very complicated and nonlinear, but the unknown coefficients multiplying these functions must be without power operations or any other mathematical operators acting on them. That is, the coefficients are linear, but the dependence of y on x can be anything.

After the same math that we have applied before, the normal equation matrix looks like this:

$$\begin{bmatrix} A \\ B \end{bmatrix} = \begin{bmatrix} \sum (f_1(x_i))^2 & \sum f_1(x_i) \cdot f_2(x_i) \\ \sum f_1(x_i) \cdot f_2(x_i) & \sum (f_2(x_i))^2 \end{bmatrix}$$

$$= \begin{bmatrix} \sum y_i \cdot f_1(x_i) \\ \sum y_i \cdot f_2(x_i) \end{bmatrix} \quad (7.56)$$

Comparing Equation (7.55) with the linear formula of Equation (7.1), in this case $f_1(x) = 1$ and $f_2(x) = x$. Sure enough, inserting these specific f_1 and f_2 functions into the generalized matrix listed as Equation (7.56) yields the matrix for the linear fit, Equation (7.16).

Equation (7.56) can be expanded to an arbitrary number of functions of x and corresponding unknown coefficients. The first row of the matrix has an extra $f_1(x_i)$ function in each summation and the second row has an extra $f_2(x_i)$ in each summation. Note that each diagonal term of the matrix is a summation of the square of one of the functions. As long as you have a linear combination of these functions, you can have as many functions as you want and make them as complicated as you want.

Just to demonstrate, let's use this technique to write out the normal equations for an example involving trigonometric functions. Remember, the unknown coefficients must be "out front" in each term, not embedded within the trigonometric operators. As an example, let's pick this function:

$$y = A \cdot \sin(x) + B \cdot x \cos(x) + C \cdot \cos^2(2x) \quad (7.57)$$

These functions include both trig and polynomial dependencies on x, powers of the trig functions, and different arguments inside the trig functions. The approach, however, is exactly the same as was done for the quadratic function dependence in Section 7.4.1.

This will yield a 3 × 3 matrix for the normal equations:

$$A\sum_{i=1}^{N}\sin^2(x_i) + B\sum_{i=1}^{N}x_i\sin(x_i)\cos(x_i)$$
$$+ C\sum_{i=1}^{N}\sin(x_i)\cos^2(2x_i) = \sum_{i=1}^{N}y_i\sin(x_i) \quad (7.58)$$

$$A\sum_{i=1}^{N}x_i\sin(x_i)\cos(x_i) + B\sum_{i=1}^{N}x_i^2\cos^2(2x_i)$$
$$+ C\sum_{i=1}^{N}x_i\cos(x_i)\cos^2(2x_i) = \sum_{i=1}^{N}x_iy_i\cos(x_i) \quad (7.59)$$

$$A\sum_{i=1}^{N}\sin(x_i)\cos^2(2x_i) + B\sum_{i=1}^{N}x_i\cos(x_i)\cos^2(2x_i)$$
$$+ C\sum_{i=1}^{N}\cos^4(2x_i) = \sum_{i=1}^{N}y_i\cos^2(2x_i) \quad (7.60)$$

This equation set can be solved using LU decomposition, or long-hand as described in Section 7.4.1.

Quick and Easy for Section 7.4.2

The linear fit process can be generalized to find the best-fit set of coefficients multiplying any set of functions of the x-axis data set that additively combine to model the y-axis values.

7.4.3 Exponential Fitting: Linearizing the Dependence on Coefficients

What if we have an exponential functional dependence? First, let's consider the case where all coefficients remain outside of the operator, like this:

$$y = A + Be^{-x^2} \quad (7.61)$$

In this case, it is exactly like the linear fit problem in Section 7.1.1, but instead of x_i in the summations, the equations will now have $\exp(-x_i^2)$ in its place. Equation (7.61) is still linear with respect to the A and B coefficients, so the procedure described in Section 7.1.1 can still be used.

The problem becomes more challenging when one of the coefficients is embedded within the exponential, as written here:

$$y = Ae^{Bx} \quad (7.62)$$

This is a regular functional form that is often encountered in radiative transfer, for instance, where y is the intensity of a certain wavelength of light being attenuated as it passes through a column of absorbing atmosphere. The trick to finding a data-based fit to this type of functional form is to linearize the equation with respect to the unknown coefficients. In this particular example, we can take the natural logarithm of Equation (7.62) to yield this:

$$\ln(y) = \ln(A) + Bx \quad (7.63)$$

To continue, we must define a new variable, $u = \ln(y)$, and a new coefficient, $C = \ln(A)$. This leaves us with:

$$u = C + Bx \quad (7.64)$$

Equation (7.64) is exactly the linear fit equation from Section 7.1.1 (Equation (7.1)) and we can progress with the regression formulas for this case.

We have one small problem in this analysis, though. Our initial assumption of a Gaussian distribution of the spread of the y values about the fit might no longer be true. That is, if the original y values are equally spread around the true-value exponential curve, then we have a spread for the new dependent variable, u, that can be found through uncertainty propagation and looks like this:

$$\sigma_u = \left|\frac{du}{dy}\right|\sigma_y \quad (7.65)$$

By inserting Equation (7.64) into Equation (7.65), we get:

$$\sigma_u = \frac{1}{y}\sigma_y \qquad (7.66)$$

In this case, because the uncertainty was evenly spread around the exponential curve, the uncertainty for u actually gets smaller with increasing y value.

There are two possible ways around this. The first is to qualitatively check the spread of y and u and determine in which variable the spread is closer to a Gaussian distribution. Note that this should not be an assessment of the histograms of y and u, but rather of $\tilde{y}$ and $\tilde{u}$, the values with the fit subtracted off. The $\tilde{u}$ values use the linear fit from Equation (7.64), while the $\tilde{y}$ values should use the conversion of these linear fit values back into an exponential curve using Equations (7.62) and (7.63). If the spread is Gaussian in u, then the linearized fit of Equation (7.64) is correct. If the spread is Gaussian in y, then σ_u as a function of y should be found with Equation (7.65) and then the weighted least squares fit from Section 7.1.4 should be used.

Quick and Easy for Section 7.4.3

When possible, use math operations to convert a complicated, nonlinear function into one that is linear with respect to the coefficients being fitted, in which case the generalized curve-fitting formula can be used.

7.4.4 Piecewise Linear Fitting

Sometimes, it is useful to have a linear fit but only for a portion of the full (x,y) paired data set. In this case, a **piecewise linear fit** can be done for each x-axis interval. That is, the data should be sorted according to their x-axis values (if they are not already) and then divided into subsets. A linear fit can then be conducted for each of these subsets, ignoring the other (x,y) pairs assigned to the other subsets. The range of validity for that fit, then, is only within the specified x-axis value range for that specific subset.

The easiest way to do this subsetting is to examine the (x, y) scatterplot and manually find the "natural break points" in the data sets. After the fits are calculated, a new plot can be made with the line segments spanning each subset interval overlaid on the scatterplot. Because there is a separate linear fit for each subset interval, however, this can lead to discontinuities in the fitted line at the edges of the subsets. While these linear fits are sometimes plotted beyond the end of the x-axis interval on which they were created, this can lead to confusion, as viewers of the plot might assume that the lines are valid to the point of intersection rather than to the subset interval edge. One way to avoid this is to adjust the subset intervals until all of the intersections occur at the interval interfaces. This is an iterative process of correcting the fits, because once the interval ranges are changed, the pairs in each subset will change and the linear fit will have a new intercept and slope.

This process is seen graphically in Figure 7.6. The first panel shows an illustrative (x,y) paired scatterplot with a visible kink in the values. Also marked on the plot is a guess at a good break point between the two linear intervals. The second panel overlays the two line segments from linear fits of the values. It is seen that a discontinuity of the fit exists at the interface. The boundary location was then iteratively adjusted to find two piecewise linear fits that meet at the interface.

It could be the case that the discontinuities are not a problem. In this case, the boundary adjustment is not needed. This could be the

---

Piecewise linear fit: The process of finding several optimal straight lines through a scatterplot of data, with each line optimized for only part of the x-axis range.

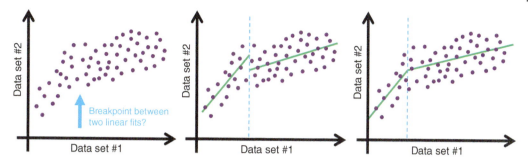

Figure 7.6 Three panels of the same scatterplot with a kink in the y dependence on the x values. The first panel qualitatively identifies the kink location, the second panel has two line segments found on the two subset intervals, and the third panel has the discontinuity at the interface removed.

situation when the change of the values is so dramatic that it does not make sense to adjust the location of the boundary from the logical break point. It could also be the case that the intervals need different functional forms. One interval might be nicely modeled by a straight line, while another might require a more complicated function.

Sometimes, a more rigorous approach than visual inspection of the scatterplot is desired to finding the break points between the piecewise linear fit intervals. This could be done by selecting a fit metric—say, correlation coefficient or the F statistic—and systematically sweeping the break point to find the maximum summed metric value between the two piecewise fits.

> **Quick and Easy for Section 7.4.4**
>
> Sometimes, rather than conducting a nonlinear fit, it is useful to split the data sets according to one or more selected x-axis thresholds and conduct two or more linear fits over "pieces" of the full data set.

7.4.5 Advice about Curve Fitting

There is a trade being made in the choice of using a single, complicated functional fit to the data or using a series of piecewise linear fits. This is a subjective call that can only be made by the analyst for each specific problem. In general, two considerations should dominate your thinking about this issue. The first question to ask yourself is with regard to the physics of the situation. Did something change to require a separate linear fit? If so, then piecewise fitting is appropriate. If not, then perhaps a single functional form for the whole interval is best. The second question to ask yourself is with respect to the usage of the fit. Are you intentionally trying to highlight the break point between the two separate intervals? If so, then a piecewise linear fit might be particularly useful for illustrating the difference. The smooth transition of a single function could soften the feature in the plot, obscuring the change in slope and de-emphasizing the interval interface.

Another point to keep in mind when conducting curve fits between two data sets is that there is a direct trade-off between functional complexity and accuracy. That is, the more complicated the function—single or piecewise—that is used for the fit, the smaller the difference between the data and the fit. The extreme of this is when the free coefficients in the fit are equal to the number of points in the data set, that is, when $d = N$. In this case, the fit could pass through all of the points, yielding $\tilde{y} = 0$ for all y values. To put this quantitatively, it will result in SSE from Equation (7.39) being zero and therefore R^2 will be one and the F statistic will be infinite. Even though the F statistic is infinite, the fit is not a statistically significant one because the degree of freedom for error in Table 7.1, $N-d$, is zero. The simplest illustrative case of this issue is when conducting a linear fit with only two data pairs. The linear fit will connect the dots. This is the trivial case, but the same thing will happen with a polynomial fit of degree $N-1$ to a data set with N pairs.

This is an example of inappropriate fitting. While it is not technically a case of overfitting, it is fitting beyond what should be reasonably done. There are more subtle cases of inappropriate fitting, and this usually is caused by the analyst assuming a functional form for the physics of the situation at hand. If you see a feature in the scatterplot, a more complicated fit might shift the curve to pass through these points, but this comes with an interpretive cost. Now that you have created this fit with a functional form that passes through the feature, this function should be justified with a physical explanation. You should always keep in mind the physics of the scenario and whether a more complicated function is warranted. In general, simple fits are better and more complicated fits should only be used when they can be reasonably defended as appropriate relationships between the two number sets.

> **Quick and Easy for Section 7.4.5**
>
> How complicated to be with a curve fit is a subjective call. It is good to keep it simple and only increase the complexity of the assumed functional form when physically justified and absolutely necessary.

7.5 Example: The Ozone Hole

A good example of the need for a nonlinear fit is found in a concentration of ozone in the middle atmosphere of the Earth. This concentration was steady and stable ever since humans started measuring it, until it started to drop in the early 1970s. The concentration has stopped dropping but has not yet recovered. Let's explain what happened and show the need for a fit to each part of the timeline.

The Earth has a special layer of gas in its middle atmosphere—the ozone layer. Ozone—a molecule of three oxygen atoms—is not found naturally near the surface of the planet. It takes a substantial amount of energy, or just the right chemical reaction, to break the diatomic oxygen molecules and create free atomic oxygen, which can then combine with another diatomic oxygen molecule to form ozone. The primary natural energy source for creating this in Earth's atmosphere is the **near-ultraviolet light** from the sun. This light has more energy per photon than visible light, and photons in the 170–240 nm wavelength range are particularly effective at breaking apart O_2. The solar UV photons in the range 240–300 nm wavelength are effective at breaking apart ozone molecules. There is a trade-off between the availability of O_2 in the atmosphere and the availability of incoming photons from the sun. The density of the atmosphere decreases exponentially with altitude, which favors absorption at lower altitudes. The photon flux is largest at the top of the atmosphere and the flux is attenuated as the photons are absorbed along the downward path, eventually reaching an insignificant level. This trade-off results in what is known as a **Chapman layer**, and for

---

Near-ultraviolet light: Part of the solar photon spectrum that is slightly more energetic than the visible photons, and thus more energetic; they play a key role in the formation of the stratospheric ozone layer.

Chapman layer: A peak of density for some atmospheric species due to the trade-off of incoming solar flux, which is absorbed and weakened as it moves lower in the atmosphere, with the density of the source molecules, which decrease in abundance with increasing altitude.

---

ozone production and loss, this layer is 20–50 km altitude. This absorption of solar UV light by O_2 and O_3 systematically warms the atmosphere in this layer, creating a temperature inversion in this altitude range. This inversion makes the air stable to perturbations and the air, in general, stays at a particular altitude and very slowly mixes with the air above and below it. That is, it is stratified, and thus we call this region of the atmosphere the **stratosphere**. Very little of this UV light makes it to the ground, by the way, which is a good thing because it has enough energy per photon to damage the cells in skin and, especially, eyes.

In the natural cycle of the atmosphere, the density of O_3 in the ozone layer is a function of solar illumination. That is, it varies with latitude and regularly changes with season. In the polar winter regions, there is essentially no ozone layer, because the middle atmosphere is in complete darkness. The air in the stratosphere gets very cold. In the Southern Hemisphere, with the continent of Antarctica helping the wind pattern, a polar vortex forms that isolates the air from mixing with the air at lower latitudes. The air inside this vortex gets so cold that whatever tiny amount of water exists at this altitude, it will condense. This leads to a phenomenon called polar stratospheric clouds. When the sun rises over the horizon in the late winter, ozone production and loss begin again, and these clouds are eventually vaporized. The same process happens in the Arctic winter stratosphere but to a lesser extent and for a shorter amount of time than Antarctica.

In 1928, **chlorofluorocarbons** (CFCs) were developed and introduced to the world as a safe and cheap refrigerant. It quickly found widespread use in homes, businesses, vehicles, and industrial settings. These complex molecules are very stable, nonreactive with the human body, and noncombustible when exposed to flame. The problem is that, as devices using CFCs wore out, this gas was slowly being released into the atmosphere. Harmless to people, this went unchallenged for many decades. Some CFCs migrated upward in the atmosphere and eventually made it to the stratosphere. Here they are still very inert, as even UV sunlight has a difficult time breaking it apart. These photons have enough energy, but the molecule is unwieldy and can absorb some of the photon energy into a kinetic motional channel rather than chemically. If the CFC molecule is held still, then UV photons can split it apart. It is a bit like trying to pop a balloon with a hammer. If the balloon is floating freely in the air, then hitting it with the hammer will rarely pop it, but if the balloon is held against a hard surface, then the hammer swing will usually pop it.

This is the case in the polar springtime stratosphere. During the polar winter, some of the CFC molecules freeze into and onto the water ice crystals in the polar stratospheric clouds. As the sun comes over the horizon, the UV light increases and starts creating ozone. These photons, however, might also find a CFC molecule on the surface of an ice crystal. This combination is sufficient to break apart the CFC, creating a free-floating chlorine atom. This chlorine atom is very destructive to ozone, with one chlorine atom destroying thousands of O_3 molecules before being locked into a more stable form. Eventually, the clouds are dissipated, no more free chlorine is produced, and the existing chlorine is sequestered in stable forms. This catalytic cycle is responsible for

---

Stratosphere: A layer of the atmosphere from roughly 10 to 50 km altitude in which the temperature increases with altitude, making the air very stable to small perturbations.

Chlorofluorocarbons: A chemical developed in the early 1900s used extensively throughout most of that century as a refrigerant, it is very stable, but if it makes it to the stratosphere, UV light can release the chlorine molecule from it, and this atom is effective at destroying ozone molecules.

reducing the density of ozone over the southern pole in the southern springtime.

In the 1970s, it was noticed that the ozone layer over the Antarctic continent in the southern springtime was greatly reduced in density. That is, there was a region of very low concentrations of ozone for a few months, beginning in August and lasting until November or December. Scientists eventually pieced together the physical process of CFCs causing this destruction. The **Montreal Protocol**, adopted in 1987, was the first of several international agreements greatly limiting the production and use of CFCs. The problem was not immediately solved, though, because CFCs were still in the atmosphere, slowly making their way into the southern polar stratosphere and therefore still causing the ozone hole every spring. As the free chlorine atoms move through the stratosphere, the influence of this ozone depletion is seen around the world, reducing stratospheric O_3 concentrations everywhere.

This is seen in the column density ozone concentration observed over Arosa, Switzerland, shown in Figure 7.7. This is an excellent data set because it is nearly continuous over the past 95 years. The trend is nearly flat until the early 1970s and then steadily decreases after this time. This decrease matches satellite measurements from the total ozone mapping spectrometer, an instrument that has been flown on a number of NASA-sponsored low-Earth-orbit spacecraft since the late 1970s. The two linear fits capture the overall trend of the data very well.

Note that Figure 7.7 shows a particular piecewise linear fit through the data, but other fits could have been conducted. First, a different break point could have been chosen; the year 1973 is when there is a substantial dip in the ozone values, but the break point could just have easily been chosen anywhere between 1970 and 1980 with a similar finding of a change from a flat slope to a large negative slope. Second, a different functional form

---

Montreal Protocol: A multinational policy document for reducing the production and use of CFCs around the world.

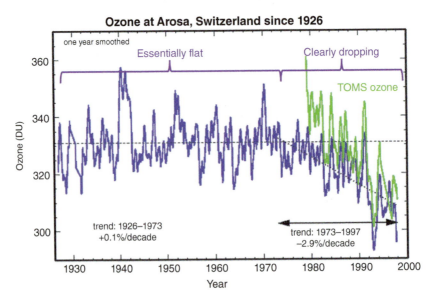

Figure 7.7 Ozone concentration over Switzerland for the last approximately 100 years, with an overlay of satellite data from the Total Ozone Mapping Spectrometer (TOMS) instrument. Adapted from an image courtesy of NASA.

could have been chosen, perhaps a quadratic or cubic polynomial, to capture the large-scale temporal variation with a single function. Third, there is significant short-term variability in the ozone concentration which is not captured by the linear fits. There is clearly another driver of ozone over Arosa, Switzerland that is not represented by a simple function with respect to year. As with any analysis of data, it is good to think about the physical processes involved in the situation. In this case, the physical connection is the release of CFCs and their slow migration from the surface to the stratosphere, the destruction of ozone over the Antarctic continent in the Southern Hemispheric spring, and the slow diffusive redistribution of ozone from other latitudes to compensate for this loss.

> **Quick and Easy for Section 7.5**
>
> Stratospheric ozone was pretty stable from year to year until CFCs worked their way into the middle atmosphere, causing a steady and substantial decline since the early 1970s.

7.6 Iterative Curve Fitting

There are times when the curve fitting routines discussed thus far in this chapter become problematic. This could be the case when the function has many coefficients and therefore it is difficult to get their values from an analytical form from the normal equations. This might arise when the unknown coefficients are embedded in mathematical operators and therefore it is difficult to linearize the equation. Sometimes, the linearization process requires assumptions that would render the fit unphysical and therefore not particularly useful. Whatever the reason, there are cases when it is hard to use the analytical procedures discussed in earlier sections. Fortunately, there are still ways to get a curve fit. One approach is **iterative curve fitting**. Given a functional form with embedded unknown coefficients, the process is as follows.

7.6.1 One-Dimensional Iterative Curve Fitting

Let's start with the simplest case of only a single unknown coefficient in the functional fit. First, make an initial guess at the values of the fit coefficient. This can be done with a completely arbitrary choice, such as setting it equal to one or zero, or by "manually fitting" it by trying numbers and getting a curve that is somewhat close to what you think will be the final fit. Given this initial value, you can then calculate the $\tilde{y}$ error values between y and $\hat{y}$. There are many different error values that could be used, and these are discussed at length in Chapter 9, but a very commonly used one that has already been introduced is SSE, the sum of squares for error, in Equation (7.39).

You then apply a small change to the free-parameter value, recalculate the fit, and find a new error value. If the error decreased, then you changed the coefficient in the correct direction and you can apply another small change in the same direction. If the error increased, then you apply a change to the coefficients in the other direction. This oppositely directed change is usually set to be smaller than the original step size so that you can focus in on the minimum value of the error. The reduction in the step size can be anything from 0.1 to 0.9 of the previous step size, but it is often set around a factor of two decrease. It is good to make the multiplier slightly different from exactly 0.5, otherwise you risk repeating

> **Iterative curve fitting:** The process of finding a curved line through a scatterplot of data that minimizes a particular error quantity by starting from some initial guess at the unknown coefficients and then nudging the values until an optimal setting is reached.

the calculation for a specific coefficient value. You could use the same reduction in step size each time, but this is not the optimal path to convergence. Specifically, the coefficient choices from previous iterations should not be forgotten but used to speed up convergence, sometimes changing the step from a coefficient setting a few iterations back rather than the latest one. This stepping and sometimes back-and-forth changing of the coefficient value should be done until the change in error between steps reaches a very small threshold. At this point, you declare that the minimum in the error has been found and the coefficients producing this minimum error are the "best-fit values" of the free parameters.

Figure 7.8 shows the error value as a function of the coefficient setting for the case of one unknown parameter in the fit. The initial choice of the coefficient is marked with a "1," and each subsequent coefficient setting is labeled similarly with its iteration number. So, "step 1" changed the coefficient from its value at iteration 1 to its new value at iteration 2. For this first step, the error decreased, so a similar step in the same direction was taken. At iteration 3 (i.e., after step 2, the second of the large steps), the error increased, so, to find the next iteration point, a smaller step backward was then used. The reduction factor used was 0.6, so the step went more than halfway back toward the coefficient setting of iteration 2. The error at iteration location 4 was smaller than that of iteration 3 but actually still larger than that at iteration 2, so for step 5, the new coefficient value was based on a small change from the coefficient from iteration 2 rather than that from iteration 4. The error steadily decreased again until iteration 7, when it increased again. The change in coefficient value switched to an increase, with an even smaller step size. Eventually, the process found the minimum error value (within some very small threshold).

One issue with this iterative technique is that you never know if the solution is unique. The process progresses until a minimum error is determined; it is hoped that this minimum is the only minimum in the error curve, or at least the absolute minimum (i.e., the lowest among all relative minima). There might be more than one minimum in the range of the coefficient and you might have found a relative minimum that is not the absolute minimum. In this case, a better coefficient with a smaller overall error value exists.

This is seen in the first panel of Figure 7.9. The step size progresses the coefficient setting slowly, and the process finds the first minimum that it crosses. Unfortunately, this is a relative minimum, and the absolute minimum of the error is a rather different coefficient setting.

This also brings up the issue of iteration step size in how the coefficients are changed. In the left panel of Figure 7.9, it was seen that too small of a step size found the closest minimum to the initial coefficient setting, which may or may not be the absolute minimum. Conversely, too large of a change in the coefficient could cause the process to jump over a region with a minimum, as is the case in the right panel of Figure 7.9.

The solution to this problem of finding the absolute minimum has two steps, at least. The first is to use several different initial settings

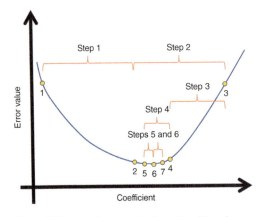

Figure 7.8 Iteration process for a fit with only one unknown coefficient. The y-axis is the error, and the x-axis is the coefficient value, with the specific coefficient settings for each iteration marked and numbered along the curve.

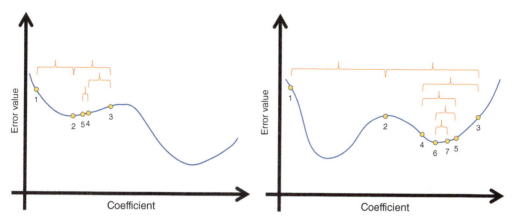

Figure 7.9 Iteration process for the case with multiple relative minima to the error with respect to coefficient setting. In the left panel, the iteration process finds a minimum, but, in this case, it is not the absolute minimum. In the right panel, the iteration process skips over the first minimum and finds the second.

for the coefficient. Conducting this calculation several, perhaps many, times will result in an array of best-fit coefficient values, from which the minimum can be chosen as the absolute minimum and therefore the best coefficient setting. The other half of the solution is to use several different step sizes. This allows the procedure to perhaps find different minima from the same starting locations, especially useful if there are many "localized and shallow" relative minima near each other. The big alternative method is to try a different functional form and see if the error can be minimized even further. However, remember the discussion in Section 7.4.5 about creating inappropriate fits. Making the fit more complicated can often lead to smaller error, but this comes at the expense of physical understanding and solid justification for the functional form of the fit.

> **Quick and Easy for Section 7.6.1**
>
> When the analytical methods mentioned in the earlier text won't work, often because the relationship is too complicated, then another possibility is to iteratively determine the best-fit coefficients.

7.6.2 Multidimensional Iterative Curve Fitting

With two or more coefficients, the problem becomes a bit trickier. One way is to set the value for one coefficient and iterate the second until a minimum is reached, then iterate the value of the first until another minimum is reached. This might not find the true minimum, especially if the "valley" of the minimum error is diagonal through the (x,y) space of the two coefficients. This process then requires an iterative loop around this pair of iterative loops, seeking to minimize the overall error.

A more complicated but usually more efficient process is to combine the coefficients and "vectorize" the change with each iteration. That is, in the multidimensional space of the coefficients, a single **change vector** can be assigned with a certain magnitude and direction. From an initial value set of the coefficients, a vector direction is randomly assigned and the assumed initial magnitude of the step

---

Change vector: In iterative curve fitting, the combination of the iteration step size of all unknown coefficients into a single step with a prescribed direction and magnitude.

---

size is applied. If the error increases, then direction should be reversed and a smaller step size should be taken in the other direction. A small random element should be added to the direction setting for each iteration step change in the coefficient settings in order to allow the coefficients to deviate around the full multidimensional coefficient space. Eventually, a minimum will be found, usually with fewer iterations than in the systematic back-and-forth method of changing only one coefficient at a time.

To vectorize the step change in multidimensional coefficient space, the coefficients should first be normalized. **Coefficient normalization** requires estimating a maximum and a minimum value for each free parameter, which are often only found by examining the scatterplot and making a reasonable guess at these extrema for each coefficient. You then create a normalized version of each coefficient by subtracting the minimum value and dividing by the extrema difference, like this:

$$A' = \frac{A - A_{min}}{A_{max} - A_{min}} \qquad (7.67)$$

In Equation (7.67), A stands for any of the unknown coefficients in the fit. The normalized coefficient, A', has a range of values from 0 to 1. It is okay if you do not get the extrema exactly right, and it is fine for the eventual iteration process to seek error minima beyond the 0–1 range of the normalized coefficient. This normalization simply makes all of the coefficients unitless and on the same order. For two dimensions, this is shown in Figure 7.10. The two axes span the range from 0 to 1, and the initial condition can then be set by two random numbers. The initial error value requires the

---

Coefficient normalization: In iterative curve fitting, the process of converting an unknown coefficient's value into a value between 0 and 1, so all unknown coefficients are of similar scale.

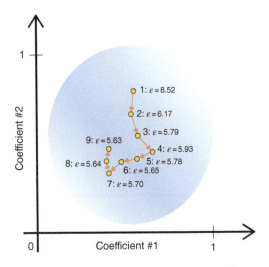

Figure 7.10 Iteration process for the case with multiple coefficients, in this case two. The axes are normalized and therefore range from 0 to 1. The steps are labeled, as shown, with error values listed. The background blue shading represents the error, with darker shades indicating smaller error values.

actual coefficient values, which can be found by rearranging Equation (7.67) for A:

$$A = A_{min} + A'\left(A_{max} - A_{min}\right) \qquad (7.68)$$

Remember that any data-model comparison metric could be used for this error value.

Now the vectorized step size should be found and applied. For any particular iteration i, there will be a magnitude of the step vector, δ_i, and a direction of the step vector, θ_i. The two normalized coefficients are then updated with the x and y components of this vectorized step size:

$$\delta A'_i = A'_{i-1} + \delta_i \cos(\theta_i) \qquad (7.69)$$

$$\delta B'_i = B'_{i-1} + \delta_i \sin(\theta_i) \qquad (7.70)$$

The initial step direction, θ_1, is completely random, and is found by multiplying a random number by 360° and assigning this as an angle counterclockwise from the +x-axis. The initial step size, δ_1, is also a normalized value and depends on how much you are willing to let the iteration process jump around in this normalized coefficient space. It is usually set to a number between 0.01 and 0.1.

Subsequent settings to the vectorized step size are based on the previous value. If the error value decreased from the previous step, then the next step should continue in the same direction. While the magnitude of the step can be the same, choosing exactly the same direction is not good because then a minimum will only be found on the line along the initial direction vector. Therefore, a small random perturbation should be applied to the angular direction of the next step. This perturbation can take many forms. The simplest is a random number across a set width of angles, $\pm \Delta\theta$, around the last iteration's direction. This $\Delta\theta$ setting could be as small as 10°, keeping the new vector almost aligned with the last one, or as large as 90°, allowing the new vector to range anywhere in the semicircle of the "same general direction" as the last step. Another method is to apply a random number to a Gaussian of possible angular deviations, centered on the last iteration's direction but with a Gaussian spread of $\Delta\theta$.

If the error increased, a different process should be used to find the next step vector for the normalized coefficients. As in the one-dimensional case, the magnitude should decrease, which can be any value less than one but usually something around half (but not exactly that value). The direction should be in the opposite direction as the previous iteration, but again a random perturbation should be applied so that the full dimensional space is explored and not just those values along the initial directional line. This perturbation can be different from that used in the "forward" direction when the error decreased. While any width of the random perturbation can be used, it is often the case that this reversed direction has a $\Delta\theta$ spread of $\pm 90°$.

This process of stepping the unknown coefficient set with a change vector should continue until the error changes by less than a small, prescribed amount between iterations. An example of this iterative process is shown in Figure 7.10. Note that the same issue of uniqueness of the best-fit result arises for this case, just as it does in the one-dimensional case. Again, the solution is to use different initial conditions within the normalized coefficient space, and also to use different step size magnitudes. You can even use different random perturbation widths to the direction change between iterations. Conducting the search several, perhaps many, times for the best-fit coefficient values ensures that the absolute minimum error will be found and therefore that the fit is truly the best.

A new problem that arises in the multidimensional case is when the iteration process does not converge. If the step size is too big, the error value could dramatically change between iterations as the location in coefficient space jumps from close to one minimum to close to another. The way around this is to start with a smaller initial step size. This increases the time to convergence, but lowers the risk of nonconvergence.

For more than two coefficients, there is still only one magnitude but now $h - 1$ directional angles, where h is the number of coefficients in the fit. For example, with three unknown coefficients, a "polar angle" that ranges from 0° to 180° should be included in the vector. When reversing the direction of the step, all angles should be reversed, with a random perturbation added to it to allow the iterative process to explore all of the dimensional space.

As with the one-dimensional iterative technique, the error value that is being minimized needs to be selected. Chapter 9 will introduce many possibilities for this error, but a common choice is SSE, Equation (7.39).

Quick and Easy for Section 7.6.2

Multidimensional iterative curve fitting is not an easy task, but there are guidelines for creating a robust method for finding the best-fit coefficients.

7.6.3 Gradient Descent Curve Fitting

A very common technique involving iteration to find the best-fit coefficients is **gradient descent curve fitting**. This is where derivatives of the chosen error function are taken with respect to each unknown coefficient. These derivatives of the error are then used to update the coefficients; the error should decrease fastest in the negative gradient direction. It is more computationally expensive than the iterative process described in the earlier text, but always moves the new iteration error to a lower number.

Let's work out the example of gradient descent curve fitting for linear regression. First, we have to select an error parameter to minimize. A common choice is SSE. For a straight-line fit, SSE from Equation (7.39) becomes:

$$\text{Error} = \sum_{i=1}^{N}\left(y_i - (A + Bx_i)\right)^2 \quad (7.71)$$

The two unknown coefficients are A and B, the intercept and slope of the line, respectively. Derivatives of Equation (7.71) now need to be taken with respect to the coefficients. This yields a gradient formula for A:

$$\frac{\partial(\text{Error})}{\partial A} = -2\sum_{i=1}^{N}\left(y_i - A - Bx_i\right) \quad (7.72)$$

and another for B:

$$\frac{\partial(\text{Error})}{\partial B} = -2\sum_{i=1}^{N} x_i\left(y_i - A - Bx_i\right) \quad (7.73)$$

The two coefficients can then be updated from iteration n to iteration $n+1$ based on these derivatives, for the intercept:

$$A_{n+1} = A_n - \gamma_{A,n}\left[\frac{\partial(\text{Error})}{\partial A}\right]_n \quad (7.74)$$

and for the slope:

$$B_{n+1} = B_n - \gamma_{B,n}\left[\frac{\partial(\text{Error})}{\partial B}\right]_n \quad (7.75)$$

The negative signs are so that the step moves in the "downhill" direction, that is, toward the minimum not the maximum of the error curve.

In Equations (7.74) and (7.75), the γ variable is known as the **learning rate**. This is a scaling factor that sets the relative step size in that coefficient direction. The two learning rates in this example have units of $[A^2/(\text{Error})]$ and $[B^2/(\text{Error})]$ for γ_A and γ_B, respectively. For this example, the selected error formula has units of $[y^2]$, the A coefficient has units of $[y]$, and the B coefficient has units of $[y/x]$. Therefore, in this case, γ_A is unitless and γ_B has units of $[1/x^2]$. The n subscript on the learning rate indicates that it can be, and often is, changing for each iteration (usually decreasing, as in the iterative techniques in Sections 7.6.1 and 7.6.2). The learning rate should not be too large, or the step could go past the local minima, but taking it too small will unnecessarily slow down the calculation. It is often set based on the initial derivative values and then slowly decreased as a function of iteration number.

Note that the gradient descent method has the same issue of nonuniqueness as the process described in Section 7.6.1. This is overcome by conducting several iteration procedures with different starting values and learning rates.

> **Quick and Easy for Section 7.6.3**
>
> Gradient descent curve fitting requires derivatives of the chosen error formula with respect to the unknown coefficients; per step it is more computationally expensive than the method mentioned in the earlier text, but usually results in a decreasing error value with each step.

Gradient descent curve fitting: The process of taking derivatives of the chosen error function to determine the next step in the coefficient values.

Learning rate: The scaling factor that sets the relative step size for each coefficient in the gradient descent curve fitting method.

7.7 Final Thoughts on Curve Fitting

Overall, which method is best, curve fitting with analytical expressions or the iterative approach? For simple formula with linear combinations of the unknown coefficients, certainly the regression approach is easier. It might be inaccurate, though, if the distribution of the y values around the fit is far from a Gaussian. All of the formulas listed in Sections 7.1 and 7.4, as well as the F statistic probabilities in Tables 7.2 and 7.3, are based on this underlying assumption. The coefficients were derived using the principle of maximum likelihood based on the superposition of many Gaussian distributions. The normality of the discrepancies between the y data set values and the fitted curve is something that cannot be checked until after the fit calculation, because the coefficients are needed to subtract the curve and make a histogram of the discrepancies. The assessment tools from Chapter 5 should then be applied to determine if the fit is justified. If not, then an iterative curve fit could be conducted. Alternatively, as most do in this situation, you can continue on with the analytical fit, but explain to others that the coefficients could be off.

For linear coefficients in the curve fit, it might be good to switch from the analytical solution in Section 7.4 to the iterative process in Section 7.6. The reasoning is that the iterative approach is more flexible; it does not have the assumption of a Gaussian distribution for each y value around the fit based on its own uncertainty or that the discrepancies from Equation (7.3) have a Gaussian distribution around the fit. These assumptions become less likely to be true as the complexity of the functional form increases. The cutoff between when to use each method is a matter of preference. One guideline is to use regression for fits with up to three coefficients, and the iterative method for fits with four or more. This is only a suggestion, though.

For more complicated formulas with many unknown free parameters, the iteration method is usually the better choice. When the functional relationship between the x and y is nonlinear, sometimes it can be linearized, but most of the time it cannot without serious limitations, making the iterative approach the only choice. There are many settings to decide when conducting an iterative curve fitting procedure, such as the error function to be minimized, the initial settings for the coefficients, the step sizes to take for each coefficient, and the process of reducing the step size as a function of iteration, just to name a few.

Note that all of these iterative curve fitting procedures are a form of machine learning. There are a number of processes that have been developed over the years for finding the best relationship between a set of input values and a set of output values. That is essentially what is being done here, in this chapter. Most of the techniques within the umbrella of machine learning are methods for optimizing the model coefficients for a best fit to the applied data. The end members of these iterative curve fitting algorithms are artificial neural networks.

The methods described in Section 7.6 are essentially machine learning algorithms where there is one set of coefficients between the input values and the output values. That is, this is a neural network with no "hidden" layers. This does not have to be the case. So-called deep learning algorithms insert hidden layers into the process, at which the inputs are combined, so that intricate combinations of the input values produce the eventual "best-fit" output values.

> **Quick and Easy for Section 7.7**
>
> Many of the techniques that we call machine learning are multidimensional iterative curve fitting procedures for finding the coefficients that minimize some specified error quantity.

7.8 Further Reading

John Taylor's textbook on error analysis has a good chapter on linear and polynomial fittings:
Taylor, J. R. (1997). *An Introduction to Error Analysis*. University Science Books, Mill Valley, CA, USA.
https://uscibooks.aip.org/books/introduction-to-error-analysis-2nd-ed/

The ANOVA table is a standard in most statistics books; I like Alan Chave's straightforward description:
Chave, A. D. (2017). Computational Statistics in the Earth Sciences. Cambridge University Press. Cambridge, UK.
https://www.cambridge.org/core/books/computational-statistics-in-the-earth-sciences/6352412A284512AF4F38766E3717FE38

A good online calculator for F statistic probabilities:
https://www.socscistatistics.com/pvalues/fdistribution.aspx

Here is an excellent review article on the relationship of seismic activity to hydraulic fracturing and refining wastewater disposal, from which Figure 7.3 was taken:
Schultz, R., Skoumal, R. J., Brudzinski, M. R., Eaton, D., Baptie, B., & Ellsworth, W. (2020). Hydraulic fracturing-induced seismicity. *Reviews of Geophysics*, 58, e2019RG000695. https://doi.org/10.1029/2019RG000695

The paper with the data on wells and earthquakes is here:
Wozniakowska, P., & Eaton, D. W. (2020). Machine learning-based analysis of geological susceptibility to induced seismicity in the Montney Formation, Canada. *Geophysical Research Letters*, 47, e2020GL089651. https://doi.org/10.1029/2020GL089651

The Alberta well and seismic data from this paper can be found here:
https://figshare.com/articles/dataset/Wozniakowska_Eaton_2020_GRL_Supporting_Files/12975725/1

Here is a discussion of piecewise linear fitting, from a Penn State University statistics course:
https://online.stat.psu.edu/stat501/lesson/8/8.8

There is a funny xkcd comic about curve fitting:
https://xkcd.com/2048/

The US Environmental Protection Agency has a nice page on "basic ozone layer science":
https://www.epa.gov/ozone-layer-protection/basic-ozone-layer-science

Here is a recent article on an unusually strong Arctic polar winter vortex that created a large Northern Hemisphere ozone hole in March 2020:
Witze, A. (2020). Rare ozone opens over Arctic – and its big. *Nature*, 580, 18. https://www.nature.com/articles/d41586-020-00904-w

The original plot of the Arosa, Switzerland stratospheric ozone data is here:
https://commons.wikimedia.org/wiki/File:Ozone_at_Arosa_Switzerland_1926-1997.png

Here is a paper from quite a while ago that gives a description of a nice algorithm for iterative curve fitting:
van Heeswijk, M., & Fox, C. G. (1988). Iterative method and FORTRAN code for nonlinear curve fitting. *Computers & Geosciences*, 4, 489. https://doi.org/10.1016/0098-3004(88)90031-3

7.9 Exercises in the Geosciences

This set of questions on linear and polynomial curve fittings uses the Arctic sea ice extent. The month of September is chosen because this is typically the minimum of the ice extent, a month that shows the most dramatic change over time. The values are calculated from gridded satellite imagery. There are two numbers in the data set, sea ice area and sea ice extent. The latter is always larger of the former because extent includes regions of partial coverage, while area only includes grid cells with full ice coverage. Only one of these number sets needs to be used for these exercises.

Data for this chapter are taken from this website:

Fetterer, F., Knowles, K., Meier, W., Savoie, M., & Windnagel, A. K. (2017). Updated daily. Sea Ice Index, Version 3. Monthly Time Series. Boulder, Colorado, USA. NSIDC: National Snow and Ice Data Center. http://dx.doi.org/10.7265/N5K072F8.

Get the September Northern Hemisphere data set. This is usually the month of minimum sea ice extent, showing the biggest change due to global warming. The second data set we will use is time along the *x*-axis.

1. Download data of the Arctic sea ice extent from the National Snow and Ice Data Center: http://dx.doi.org/10.7265/N5K072F8
   A Make several time series plots of the sea ice extent data, using lines, symbols, or a combination of the two.
   B Create a linear fit through the full data set. Make a plot that overlays this fit with the data values.
   C Calculate the uncertainty of the two linear fit coefficients.
   D Calculate a Pearson linear correlation coefficient for sea ice extent versus year. Determine a probability value for this coefficient.
   E Calculate an ANOVA table for this linear fit to the data. Determine a probability value for the resulting F statistic.

2. Write a paragraph answering the following:
   A Comment on the qualitative nature of the relationship between the data and the linear fit.
   B Discuss the relative size of the uncertainty values on the linear fit coefficients.
   C Discuss the goodness of the fit relative to the correlation coefficient and F statistic.

3. Calculate a quadratic fit to the sea ice extent yearly values.
   A Create a plot that overlays the quadratic fit with the data.
   B Calculate an ANOVA table for this quadratic fit to the data. Determine a probability value for the F statistic.

4. Write a paragraph answering the following:
   A Comment on the qualitative and quantitative nature of the quadratic fit.
   B Compare the two fits and comment on the similarities and differences.

5. Find the roots of the fits.
   A Find the *x*-axis location (year) where the linear fit crosses $y = 0$ (no sea ice).
   B Find the roots of the quadratic fit. The second root should be comparable to *x*-intercept of the linear fit.

6. Write a paragraph answering the following:
   A Compare and comment on the two projections of when the sea ice will reach zero.
   B Discuss the realism of either of these projections.

8

Data-Model Comparison Basics: Philosophies of Calculating and Categorizing Metrics

We have already encountered modeling. The curve fitting in the previous chapter is a form of modeling based on the data values themselves. We will now explore qualitative data-model comparisons and discuss the various quantitative techniques, going through the fundamental ideas behind the approaches, why so many metrics calculations exist, and some strategies for ensuring that the data-model comparison you are conducting is appropriate and meaningful.

8.1 Example Model: River Flow Rate

A **model** is any set of numbers that tries to replicate another set of numbers. Models exist in a wide variety of accuracy and sophistication levels. Let's consider the example of water flow rate in a river.

In my youth, my school district had all sixth graders spend a week at a camp, sleeping in cabins, eating together in a large dining hall, and doing outdoor learning activities. One day we hiked into the woods, eventually arriving at a stream. Gathering the class on both banks, the instructor asked us to think about how many gallons of water were flowing past us every minute. He said he would give us a few minutes to think about it and then take us to a nearby shelter where we would all reveal our estimates and, more importantly, how we came to that number.

While most of the students took it as a brief interlude to play in the water, others gave it a brief moment of contemplation, while a few took this as a mathematical challenge. When we gathered in the shelter, we first were called on to give our flow-rate values without explanation. Most kids had estimates in the 50–150 range. One student gave an answer of 2000 gallons per minute. On hearing this number, I thought it was very high. She received many strange looks from my classmates—just what a sixth grader wants. Except for one silly answer, this was the highest.

The teacher then called on some of the kids to discuss their method for obtaining their estimate. I do not remember any of their answers, except for one from the girl estimating 2000 gallons per minute. When called on, she described her process to the class. She pictured a gallon jug of milk, placing that volume within the stream. She mapped out the cross-sectional area of the stream in units of milk jugs, then watched a stick flow downstream for 6 seconds, again estimating the distance traveled in units of milk jugs. Multiplying all of these together, she came up with a number of 2000 gallons per minute.

Model: A set of numbers intended to replicate another set of numbers.

Data Analysis for the Geosciences: Essentials of Uncertainty, Comparison, and Visualization, Advanced Textbook 5, First Edition. Michael W. Liemohn.
© 2024 American Geophysical Union. Published 2024 by John Wiley & Sons, Inc.
Companion website: www.wiley.com/go/liemohn/uncertaintyingeosciences

She was very close. The actual flow rate of the stream was around 3000 gallons per minute, even bigger than her value. Her method worked.

The point of this story is that, looking back on it, the instructor was asking us to quickly develop a simple model of calculating the river flow rate. Her method was width times depth times distance, all in units of what she guessed was the size of a gallon jug. This is what we all should have been doing. While there was a large uncertainty on each of these values because of the measurement constraints, being off by less than a factor of two was, in hindsight, amazingly reasonable.

> **Quick and Easy for Section 8.1**
>
> Models are frameworks for explaining phenomena and reproducing measurements from it. They can be very simple, as seen in this example.

8.2 What Is a Model?

For the context of data analysis in the natural sciences, a model attempts to reproduce a complicated number set with a simpler description. There are many different ways to model a number set. A model could be based on the data to which it will be compared. This is called a **fitted model** and is what was done in Chapter 7. It can be based on a similar set of data, usually much larger, so that the model is independent of the values that it is trying to reproduce. This is called an **empirical model** or **reanalysis model**. The model could be found from a physical description of what you think the numbers should adhere to according to some proven analytical law or well-known relationship. This is called a **theoretical model**. The equations themselves can be examined and assessed, contemplating how changes in one term or another will influence the values that you might observe in the natural setting. The model could be a calculation that computationally solves one or more equations. This is a **numerical model**, and allows for the creation of vast number sets for comparison with observations. The model could be an ad hoc tweak to any of the above-mentioned types of model, where some aspect of the model is changed, perhaps based on some physical reasoning but maybe not. Models can be as simple or as complicated as you want. They could be a flat line at some constant value, or it could be a complicated collection of sophisticated numerical codes that use high-order computational algorithms to obtain a solution to an equation set across a multidimensional mesh. If it is an equation or set of numbers that you will then use in comparison to an existing set of numbers, then it fits the definition of a model.

A way to place these four approaches into context is to consider their organization within a framework of the focus of the foundational element of the model and the complexity of the model. This is shown in Table 8.1. While there is a continuum between these boxes, this is a simplistic way to quickly categorize a model within these four basic types.

If we have the data, then why do we need a model of it? There are many reasons for creating models. One reason, especially if you are creating a single-equation fit, is that it could be easier to use than the data itself. In addition,

---

Fitted model: A model based on the data to which it will be compared

---

Empirical model or reanalysis model: A model based on data from field or laboratory experimental observations.

Theoretical model: A model based on fundamental scientific principles.

Numerical model: A model requiring substantial calculational effort to obtain the number set, often a computational solution to a theoretical model.

Table 8.1 Organization of the types of modeling approaches.

	Elegant	Complex
Data-focused	Fitted model	Empirical model
Physics-focused	Theoretical model	Numerical model

an equation-based model can be used to give a y-axis value for any x-axis value, while the data set only has y-axis values at the x-axis locations of the data points themselves. This feature is important for interpolation, within the x-axis domain of the data set, or extrapolation, yielding y-axis values beyond the x-axis domain of the original data set. If you have ever done linear interpolation, finding a y-axis value at an x-axis location between two data points, then you have created a model (the line between the two data points) and applied it for a useful purpose. This is not the only way to get that value between the two data points, though; you might have, for example, found the linear fit through the entire data set, as we did in Chapter 7, to get the y-axis value at that x-axis location. It would almost certainly be different than the first method. Which one is better? That depends on why you needed this value, and it is often a subjective judgment call.

Another reason to create a model is to gain understanding of the factors controlling the features in the data set. As a geophysical example, consider the case of magma flowing under a volcano. The data sets of relevance are all on the surface, including seismometers, temperature sensors, soil sample chemistry, and surveying measurements to track the rise and fall of localized land features. Examining these data sets could yield the development of a scenario of how the magma is flowing underground. While the analysis of the data sets could be highly quantitative, the description of the magma flow is largely qualitative until it is formulated into a model that is consistent with the known physics of how this fluid is supposed to move.

A big reason for creating models is to make predictions on the future state of the system. Given a set of observations or predictions of the known driving conditions that control the system, then a model could be used to determine how that system will change over time. This is how essentially forecasting—for weather, stocks, or sporting events—is done.

One type of model that deserves more explanation is the "fit." In the previous chapter, a model was created by calculating a linear fit to a data set by using the same mean value and then minimizing the mean square error. A few methods for nonlinear fitting of a data set were also presented. As discussed at the end of that chapter, though, this is not the only way to fit a model to the data itself. You can come up with any functional form with any number of free parameters, usually undefined coefficients, in the equations and then use an iterative method to find the "best-fit model" that optimizes a particular metric. A **metric** is just a formula that combines the data and model value pairs in some systematic method to yield a number with a particular meaning about the relationship between the data and the model number sets. We will get to metrics in a bit, and the next few chapters will cover a plethora of metrics equations. While the "testing a fit" methodology and equations shown in Chapter 7 use the mean square error as the metric being optimized, this is not the only way to do it. Any of the metrics to be discussed in the remainder of the book could be used, resulting in a different set of best-fit coefficients in the fit equations.

That is, when basing the model on the data set itself, you might think that there is a single "best fit," but this is not correct. There is no single "best fit" to the data because it depends on the methodology used to create the fit. We see this when we create a quadratic fit to the same data set for which we did for a linear fit. These are two different "best-fit lines" through the

Metric: An equation that assesses a facet of the data-model relationship.

same data set. While the linear regression formulas in Chapter 7 yield a "best-fit straight line" to the data set, this is only one of many lines that we could call the "best-fit straight line" to the data. If we based the derivation of the A and B linear fit coefficients on another assumption besides a Gaussian distribution using the principle of maximum likelihood, then we will most likely create a different best-fit straight line. You might even choose a combination of metrics, in which case you introduce another layer to the methodology, perhaps in the form of weighting factors that combine these different metrics into a single optimization. Determining a best-fit line to a data set often depends on the purpose for creating a fit and the eventual usage of the model or data-model comparison.

Models based on physical equations come with their own issues and can be equally numerous in terms of a "best" comparison. Continuing the volcano example, the magma flows should follow the equations of continuity, momentum, and energy. These equations could be solved, perhaps analytically but most likely numerically. The results could then be compared to the various data sets from the surface. How the model is created from the equation set will influence the resulting model output and the uses for which the model can be considered "good."

It should be noted that assumptions are always built into the use of such equations. In the case of magma flows with fluid equations, the leadoff assumption is that the magma behaves like a fluid. Certain terms in the fluid equations of motion are often omitted from the solution under the assumption that they are small relative to the other terms. The selection of which terms to keep and discard is usually done through the calculation and comparison of certain **dimensionless parameters** that compare terms in the fluid equations, such as the Mach, Reynolds, Froude, Knudsen, Prandtl, Rossby, Péclet, and Argand numbers, to name a few. When these dimensionless parameters are very large or very small compared to one, it means that certain terms in the equations will not contribute much to the solution. The calculation of these parameters, however, usually involves assumptions about the dominant scales—scales of size, time, speed, or temperature—and any simplification of the equations based on these parameter values is dependent on the choices for these assumed scale values. Furthermore, if solving numerically, discretization algorithms have to be applied, each of which comes with its own assumptions about flow relative to the numerical grid. All of these assumptions lead to uncertainties in the model values. That is, while models often provide a number for comparison with a data set, this number has an associated uncertainty, just like the measured value.

No matter how you obtain your model, you can then conduct analysis to determine the **"goodness of fit"** between the model and the data set. This analysis can be simple or complicated, visual or numeric, qualitative or quantitative, tightly focused or multifaceted. Like in creating the model, there are many strategies for determining the goodness of a data-model comparison.

To go back to the river flow rate example of Section 8.1, better models than our simplistic calculations could be developed. Measurements could be taken of the shape of the riverbed, and flow rates could be measured throughout the day and, in fact, over the year, curve fitting through the values to get a prediction of the flow rate for the next year. Even more data could be included, such as

---

Dimensionless parameters: A ratio of the characteristic scales for two terms in a model, used to assess their relative importance to the phenomenon being studied.

Goodness of fit: The ability of a model to reproduce an observed number set, this assessment can be done qualitatively or quantitatively with any number of assessment tools.

rainfall in the area, or just the rain accumulation upstream of the chosen location. There could be temperature and humidity influences on evaporation from the stream surface, or a search could be made for springs along the stream to take into account local water sources. The entire riverbed could be mapped out and a computer model of the flow rate conducted based on the solution of hydrodynamic equations. The flow rate of another stream, somewhere else, could be used as a model, adjusting the values to match the conditions of the selected stream. Any of these choices are valid models for the stream flow rate, and each comes with its own measure of uncertainty.

> **Quick and Easy for Section 8.2**
>
> There are many different types of models out there, most that can be classified as empirical, theoretical, or numerical models. They all have different approaches to how they try to describe the data and they all come with caveats and limitations.

8.3 Visualizing Observed and Modeled Values Together

There are many ways to compare two number sets, with some of those plotting techniques discussed in Chapter 2. All of those still work, but there are a few modifications to them that are specific to data-model comparisons. This is because the model results are attempting to reproduce the observed values. Specifically, the primary difference from one has been done earlier is this: one of the number sets is trying to match the other. In the best-case scenario, the two number sets are identical. This was not the case in the earlier chapters, where the two number sets were possibly related—as quantified by correlations and fits—but one was not defined as trying to match the other.

8.3.1 Scatterplots of Data and Model Values

One of the most straightforward comparisons to make is the scatterplot. If you have model values that directly correspond to the data values, then the plotting procedure is exactly the same as in previous chapters. Sometimes, however, the model is calculated as a function of x, the "input data set" to the model. This is exactly the case we had in Chapter 7, when we calculated linear and nonlinear fits to a y-axis data set given a correlated x-axis data set. More complicated physical models often produce output that similarly connects to specific input values.

In this case, we now have three sets of numbers: the x-axis values at which you have data; the y-axis values of the data set; and the y-axis values of the modeled fit at those same x-axis values. How does one make a scatterplot with three sets of numbers?

First, make two scatterplots, repeating the process in Chapter 2. Make one with the data at their x-axis locations and another plotting the model values at those same x-axis locations. Take a look at these two plots. Do they look similar?

As an example of this, Figure 8.1 shows a linear fit to a data set, both overlaid on the data and then the fitted values as individual points at each of the x-axis locations. This is an extreme case of a simple model; all of the points are arranged in a straight line. Furthermore, it is seen that the y-axis data have a larger range than the model fit values. In general, neither the straight-line nature nor the smaller range needs to be the case, especially for a physics-based numerical model. For a complicated physics-based model with nonlinear feedback embedded in the system, there could be significant y-axis spread in the modeled scatterplot, perhaps more than that in the data.

Next, combine these into the same scatterplot, overlaying the model values on the data. Use a different symbol style, size, or color for the model points so that they are clearly

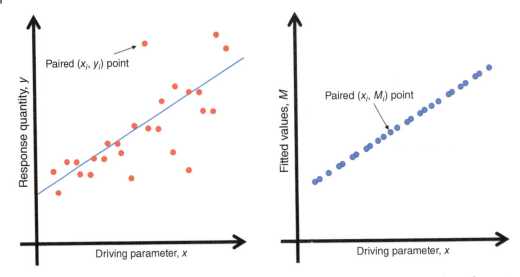

Figure 8.1 An example comparison of two data sets against each other, and then a comparison of output from a modeled linear fit of the y-axis values against the x-axis data set.

different. Take a look at this new plot. Again, do the two point sets look similar? Can you see any systematic differences between the data and the model values?

Figure 8.2 shows this for our example. The model is a straight line of dots, but now with the two sets overlaid, it is easier to see at which x-axis locations the model is close to the data and at which x-axis values it is farther from the observations. There are now two points on the plot for every x-axis data value location. If the red point is higher than the blue point, then the observed value is larger than the model; the model has underestimated the observation. If the red point is below the blue point, then the model is overestimating the measurement. For this plot, the qualitative assessment is that the model is better at lower x-axis values, and seems to worsen with increasing x magnitude.

Note that the trend does not have to be "upward," sloping from lower y values at low x values toward higher y values at higher x values. The trend could be negative, or highly nonlinear. With a plot like that in Figure 8.2, the assessment between the data and model is qualitative.

Why start with the two-plot version instead of simply making the overlaid plot first? The main reasons are that the extra plots serve as a

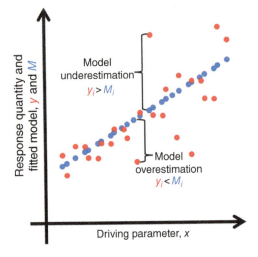

Figure 8.2 The same values but now with the model values overlaid on top of the original y-axis data set.

check to ensure proper plot-making technique and having separate versions allows for better understanding of the overlay plot. Sometimes the differences are easy to pick out, while other times they are subtle. Sometimes they overlap each other so well that it is hard to see the data points. Having the separated version allows you to, at least qualitatively, assess the similarities and differences even better than with only the overlaid scatterplot.

Keep in mind that this is not done yet. Instead of having one $2 \times N$ element array of the (x,y) pairs for the data against their x-axis locations and another same-sized array with the (x,M) pair set for the modeled values at those same x-axis locations, it is useful to remove the x-axis values from the analysis. That is, align the data and model directly in this ordered pair configuration, creating (M,y) pairs of values. Figure 8.3 shows this for our example data set and model output. You now have a scatterplot that has the same form as the (x,y) scatterplot from Chapter 2, implying that all of the analysis calculations done there can be applied to this scatterplot. Now, however, the x-axis values are a model-output number set, which we will label M, and the y-axis values are the observations that the model is attempting to reproduce, which will label the O number set. Comparing Figure 8.3 with Figure 8.2, the points are not spread as wide along x now that the model values are being used for the locations along that axis. The x-axis domain in Figure 8.3 is the blue-point range in Figure 8.2, and the y-axis range in Figure 8.3 is the red-point range in Figure 8.2.

Note that, because the model is supposed to reproduce the data, these points are expected to lie along the one-to-one line of unity slope with no y-intercept. In the case of a perfect model, this linear fit should have the ideal coefficients of $A = 0$ and $B = 1$. In fact, it is encouraged to conduct the linear regression calculation because any differences from these optimal values reveal information about the data-model comparison. Note that these fitted coefficients are numbers, which means that this is an initial quantitative examination of the comparison between the two sets.

Is the example data-model comparison shown in Figure 8.3 a good one? Qualitatively, this can be assessed by the tightness of the spread of the points around the ideal-fit dashed line. In addition, the uncertainty of the measurements needs to be taken into account as well; they could be large and explain some of the variability in the y-axis values. Overall, the trend of the paired (M_i, O_i) points is upward and generally follows the dashed line, but I would probably answer the "is it good?" question with a no. While the spread is small in the left region of the plot at lower model values, it worsens with larger model values. Conversely, the fit is not particularly good at lower observed values; there is significant spread of the corresponding model values for the low y-axis range. At high y values, the spread in the modeled values is smaller, but the model values appear to underestimate the observations.

> **Quick and Easy for Section 8.3.1**
>
> The scatterplot is often the first step to proper data-model visualization.

8.3.2 The 2D Histogram Plot

Sometimes, the density of the point spread within the two-dimensional space of the x and y values is an important feature of the

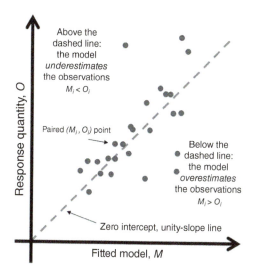

Figure 8.3 The same values but now with the model values as the x-axis values instead of the original second data set. A zero-intercept, unity-slope dashed line is drawn for reference.

comparison that needs to be highlighted. In this case, a 2D histogram plot can be used. This involves binning the values in both x and y directions and then using a grayscale, color scale, or a 3D angled perspective of a raised surface to display the counts within each bin. A couple examples of these were already shown in Chapter 1. As with the running average plot style, this plot is usually only made when the number of points is very high, well into the thousands if not the millions. Anything less and the plot is either very coarse, and therefore featureless and uninformative, or it is a fine grid that is subject to counting uncertainties, perhaps revealing "features" that are really just Poisson noise.

The examples of 2D histograms are shown in Figure 8.4, showing soil organic carbon measurements made across the United States compared against three models that predict this value. The two axes are identical in range, and the bins along them are identical as well. The color in each bin shows the counts in that bin, ranging from 1 (dark purple) to 27 000 (bright yellow). As a scatterplot, the density of points in some regions, especially near the diagonal line—showing a perfect data-model match—would be too great to see any features. Most of the plot would simply be a featureless mass of uniform color as many points overlap. Using the 2D histogram approach, several features of the comparison can be seen and interpreted.

A decision to be made when constructing this type of plot is the number of bins to use. Note that these bins should be of equal value range, not equal point count. Choosing unequal bin widths results in larger bins having seemingly higher counts when really the count rate per axis increment is small. This often creates confusion for the viewer rather than insight. The easiest and most common choice is to use the same number as you would have chosen for the 1D histograms. These 2D histograms, however, are often made with more bins than that choice. This seems counterintuitive because the bins along one axis are now subdivided by the bins in the other direction. This is exactly why this type of plot should

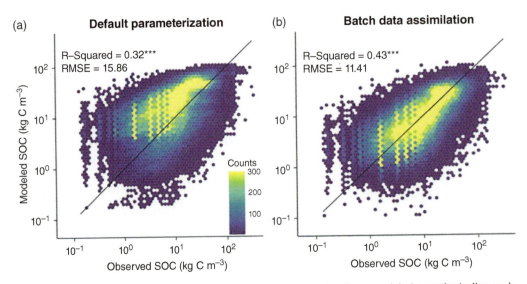

Figure 8.4 Example 2D histograms of data-model comparisons, showing a modeled quantity (soil organic carbon (SOC) at locations across the contiguous USA) against a corresponding observed quantity (Luo et al., 2022). Two model results are shown, along with two metrics (R^2 has been introduced in Section 6.2.2 (Chapter 6) and will be discussed again in Section 9.6.1 (Chapter 9), RMSE will be introduced in Section 9.3.1 (Chapter 9)). The x and y-axes are the same logarithmic scale and the color scale showing the counts in each bin. A diagonal line shows the perfect comparison, for reference.

only be made when the number set count is very large. A good upper limit on the number of bins is 100, because any more than this and the individual bins become tiny and often not well resolved in the plot. The numbers of bins along the two axes do not have to be equal, but, because the numbers of points is equal along the two axes by definition of the (x,y) pairing, it is often the case that the plot looks best with an equal number of bins in each direction. This partly depends on the cluster of the points within the 2D space, which means that it could be necessary to make several versions of this plot to see if physically real features become apparent with different bin-width settings.

Note that it is perfectly acceptable to truncate the x or y extent of this plot. That is, the plot does not have to be made with the full range of the x and y data sets as the upper and lower boundaries in each direction. Because the bin widths must be equal in this type of plot, it is fine to omit outliers from the plot. This allows you to better focus on key features of the histogram, rather than wasting bins to cover the regions out to the extreme endpoints, many of which might be empty. If you do this, though, then state it in the caption or main text description of the plot, noting the difference between the plot ranges and the full data set ranges.

An alternative way to make this plot is to use color contours on top of the scatterplot of points. The contours are constructed from the same method as the 2D histogram and provide the same information, but rather than showing the count value in every cell, it highlights selected levels of the count. It is best to make the contours evenly spaced in count value from each other, otherwise the line spacing will make viewers perceive a sharper gradient for some count values even when no sharp gradient exists. The number of contours to show should be kept relatively small. More contours means that smaller-scale features will be apparent in the (x,y) value relationship, but this comes at the expense of cluttering the graph with many lines. At most 10 contours should be used, and perhaps as few as two.

> **Quick and Easy for Section 8.3.2**
>
> When the scatterplot becomes too crowded with points, then converting it to a 2D histogram helps to clarify the visualization of the relationship.

8.3.3 Overlaid Histogram Plots

Overlaid histograms were used in earlier chapters, in particular Chapter 5 when comparing the histogram of a data set against a Gaussian distribution. Here, the second distribution is not an expected distribution but rather the modeled distribution. An illustrative example is shown in Figure 8.5. This is not the distribution from the previous figures, for which the modeled histogram is a rather uncomplicated square wave with all columns of uniform height. This is a different set of numbers with each distribution resembling a Gaussian (but not quite).

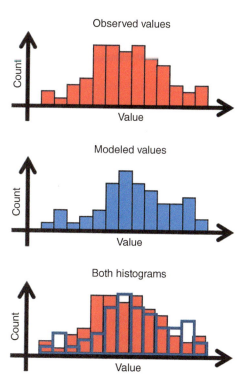

Figure 8.5 Histograms of the observed value number set, the modeled value number set, and the two overlaid on the same chart.

There are many ways to overlay histograms. In Figure 8.5, the modeled values are "on top" but with no color to the actual boxes, just a thick outline to each column. As seen in Section 2.4.1 (Chapter 2), it could also be done with transparency, making the front histogram less-than-fully opaque so that the other can be seen through it. It could also be achieved with striping or some other pattern in the front columns that does not totally block site of the histogram behind it.

> **Quick and Easy for Section 8.3.3**
>
> Because the model values seek to reproduce the data values, an overlaid histogram can qualitatively assess the similarity of the number sets.

8.3.4 Cumulative Probability Distribution Plots

Another way to explore the relationship between a data set and a model trying to reproduce that data set is to examine their cumulative probability distributions. The cumulative probability distribution (CPD) was introduced in Chapter 1 when first discussing the Gaussian distribution, but now it is being used as a comparative tool between two number sets that should match each other.

You are already familiar with the probability distribution. This is the histogram for a discrete set of values, specifically the fractional histogram where the y-axis is not the counts for each bin but this count divided by the total number of values. In this case, adding all of the column heights will result in a summation value of one. For a continuous function, the function could be called a probability distribution if it has the feature that the integral area under the curve is equal to one. This is the reason for the normalization factor in the denominator out in front of the exponential in the Gaussian distribution, as mentioned in Section 1.6 (Chapter 1).

The CPD is simply a plot of the running integral value from one end of the distribution to

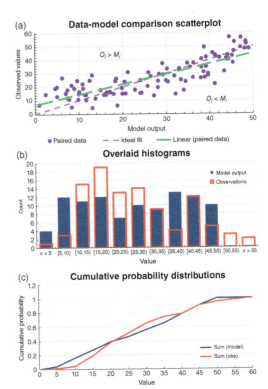

Figure 8.6 (a) A scatterplot of a data set against model values, with a linear fit shown in green and the ideal fit shown in light purple. (b) Overlaid histograms of the data and model values. (c) The cumulative probability distributions of these two data sets.

the other. Let's consider the example data-model comparison from Figure 8.6, which shows an example data set compared with model output, for which there are 100 (M_i, O_i) pairs. A scatterplot of the values is presented in the upper panel (Figure 8.6a), along with a linear fit between the two number sets. It seems that the linear fit is okay, but might not actually capture the true relationship between the model and the data. In the lower panel (Figure 8.6c), it is seen that the connection between the two is not linear, but perhaps better described as a polynomial, at least quadratic and perhaps cubic or quartic. The model overestimates the observations in the middle of the model value range, but underestimates the data at each end.

The histograms for the observed values and model output values are shown in Figure 8.6b. This panel shows histograms, with a y-axis unit of counts. For the fractional histogram, these count values should be divided by the total number of values in the set, which in this case is 100. For the CPD, the value at this subset range is equal to not only the fraction of values in this range but also the total of all fractions from there to the starting bin. This is done through all subset ranges from the fractional histogram until you reach the final nonzero range, where the column height for the CPD reaches the value of one. These full curves for the CPDs from the example histograms are shown in Figure 8.6c.

The CPD lines should monotonically increase from zero to one (or monotonically decrease, if creating the CPD from high to low values). When the CPD is "flat," it indicates that the number set has few values in that range. When the line has a steep slope, it indicates that there are many points in that value range. When using a cumulative probability plot for data-model comparisons, the ideal case is that the two CPD lines lie on top of each other. The specific shape does not matter; the comparison is not against a particular distribution function, but against each other. Note that the exact shape of a CPD curve depends on the histogram on which it is based. A fine-scale histogram will result in a smoother CPD curve compared to a one based on a coarse histogram bin resolution.

Several features can be seen by comparing the curves in Figure 8.6c. In this example case, the data CPD is, at first, below the model CPD, then it is above, and then below it again. This means that there are fewer data values at the lowest range, below 10, but the steeper slope of the red curve compared to the blue curve indicates that there are more data values in the 10–30 range. The data slope then gets more horizontal, allowing the model CPD to catch up, and the two lines are exactly on top of each other for the 40–45 range. Above this, the model CPD quickly rises to 1, indicating that all model values have been counted by a value level of 50. The observational data set includes points with values above 50, so it rises slower toward unity. These could be compared more rigorously with a Kolmogorov–Smirnov test;

we'll get into quantitative data-model comparisons in the next few chapters.

For a continuous function, the CPD is created by conducting a series of integrals of a probability distribution with ever-changing upper limits. You start from the trivial case where the upper and lower limits are the same (therefore, the integral is zero) and continue to the opposite extreme where the entire domain is included (therefore, the integral is equal to one). These integral values are then plotted as a function of that upper limit value.

Note that this summation or integral does not have to move from low values to high values of the number set. The choice of directionality depends on the assessment being conducted. Sometimes it is advantageous to move from high to low in the values, if the interpretation of the results is made more meaningful by such a switch. Usually, it does not matter; the two forms are essentially equal. The primary way in which it might matter is if you would like to highlight one end of the distribution by plotting the CPD with a logarithmic y-axis scale. In this case, you cannot plot all the way to zero, but rather the approach to zero is expanded by the decadal change of the y-axis. This is highlighted in the four panels of Figure 8.7. The panels show the same data-model CPD curves from Figure 8.6, but now done in different ways. The upper panel CPDs were created with the integral ranging from low to high and the lower panels were done with the sum sweeping in the opposite direction. The left panels have a linear y scale, and the right panels have a logarithmic y-axis scale. These plots reveal that the wings of the distribution are indeed different from each other.

> **Quick and Easy for Section 8.3.4**
>
> In addition to histograms, comparisons of cumulative probability distributions can reveal similarities and differences in the data set, especially when plotted with appropriate axes and scales to highlight features of the comparison.

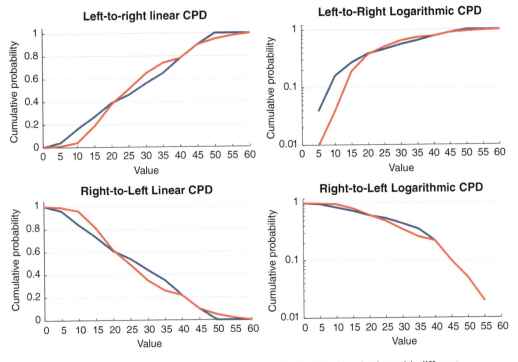

Figure 8.7 Four versions of the same cumulative probability distribution plot but with different formulations. The left-column panels use a linear scale, and the right-column panels have a logarithmic scale. The upper-row panels do the summation from left to right (i.e., small to large values), while the lower panels are created with the summation from right to left.

8.3.5 Quantile–Quantile Plots

Another way to sort, visualize, and understand the relationship is to use the **quantile–quantile plot**. By quantile, you are choosing a specific percentage of the total number of values and calculate the median (or the mean) of that subset. This was discussed in Chapter 4 with respect to using the median and interquartile range as the centroid and spread of a number set. There, it was noted that quantiles are simply cumulative probabilities of the distribution (i.e., summing the histogram from left to right). The quantile–quantile plot is, in many ways, similar to the CPD plot in terms of organizing the two sets of numbers for analysis—both should be rank sorted, separately from each other, so that the original data-model pairings are lost and you are left with the distribution of the values for each set without regard to how they match with the other set.

A typical quantile–quantile plot has bins that are a subset range of 1%, 5%, 10%, or even 20% of the total number of values in the set. After sorting each data set separately and then calculating the mean (or median, if desired) within each percentage subset range of the rank order of the values, you will have 100 (for 1%), 20 (for 5%), 10 (for 10%), or 5 (for 20%) observation-to-model comparison pairs that can be plotted against each other. You can even calculate standard deviations on each of those subset means and put vertical and horizontal error bars on all of the points. It is also useful to draw the zero-intercept, unity-slope line, on such a plot to show the ideal situation

---

Quantile–quantile plot: A figure created by the pairing of the values of equal quantile (i.e., percent) among the rank-ordered listing for each number set.

for which the two histograms exactly match each other.

As an example of this, Figure 8.8 shows two different quantile–quantile plots of the data-model comparison from Figure 8.6. The two quantile–quantile plots follow each other closely, as they should. The (x,y) coordinates of each point are essentially the x-axis values of the two CPD lines in Figure 8.6c, taken at particular y-axis values corresponding to the quantile being plotted. It is seen that neither lies exactly on the unity-slope perfect-match line.

An interesting feature of Figure 8.8 is seen in the final point at each end of the 5% quantile–quantile line. Remember that each of these points is created by finding the median of the values within that quantile range, separately for the model values from the observed values. Table 8.2 lists the model–data pairs used in determining the modeled and observed values for the lowest and highest 5% quantiles in Figure 8.8. It is clearly seen that the five points contributing to the lowest model bin are not the same five points contributing to the lowest data bin. For the latter, the first 5% quantile of the observations include the very low data values at model values of 1, 8, 12, 13, and 19. They have a median observed value of 6.5. The first 5% quantile of the model values, however, are located at observed values of 7, 15, 19, 20,

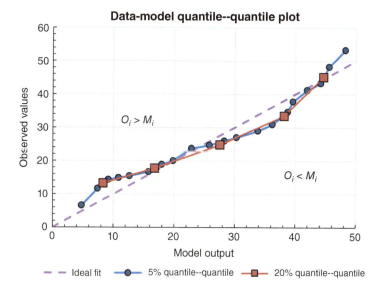

Figure 8.8 A quantile–quantile plot of the data-model comparison in Figure 8.6, using 5% and 20% quantile bins.

Table 8.2 The values used in the lowest and highest 5% quantile bins in Figure 8.8.

	Lowest model bin	Lowest data bin	Highest model bin	Highest data bin
Pair #1	(0.91, 6.8)	(13.1, 4.6)	(47.2, 56.3)	(48.5, 51.8)
Pair #2	(2.4, 14.6)	(8.4, 5.2)	(48.1, 48.1)	(45.7, 53.1)
Pair #3	(4.6, 19.1)	(18.7, 6.3)	(48.3, 53.4)	(48.3, 53.4)
Pair #4	(5.0, 19.9)	(0.92, 6.8)	(48.5, 51.8)	(47.2, 56.3)
Pair #5	(5.0, 13.0)	(11.7, 10.4)	(49.1, 48.2)	(44.5, 56.6)
Median	4.6	6.3	48.3	53.4

and 13. The same is true for the top 5% bins. In this case, however, it turns out that, purely by chance, the medians of these two highest bins select the model and data values from exactly the same (M_i, O_i) pair.

> **Quick and Easy for Section 8.3.5**
>
> Quantile–quantile plots are a way to assess whether the distributions are close to each other at specific subset ranges of the full value sets.

8.4 Example: Total Solar Irradiance

The Sun emits photons across a broad range of wavelengths. Some parts of the spectrum, like the low-wavelength, high-frequency component emitted by the Sun's upper atmosphere (the corona) in the X-ray range of a few nanometers or smaller, are quite variable and can fluctuate by orders of magnitude over timescales of minutes during a solar flare event. The photon flux over most of the spectrum, though, is quite stable, especially in the visible part of the spectrum in the 400–700 nm wavelength range, where the peak of the photon flux is located. Integrating across all wavelengths, the photon energy flux from the Sun can be found. This value is known as **total solar irradiance**, or TSI. At Earth's distance from the Sun, this is in the range of 1361 W/m^2.

Measurements of TSI have been made for hundreds of years. Those from the ground, though, are always too low; the atmosphere absorbs and scatters some of the incoming light. Reliable measurements of TSI were not collected until the Space Age, when such measurements could be made by satellites orbiting far above the atmosphere. Based on a series of satellite missions with appropriately calibrated sensors, a composite record of TSI now exists going back to the mid-1970s. The top panel in Figure 8.9 shows the time series of this value.

There are several key features to note in this plot. First, note the y-axis scale; all of the values, including observed extremes, span a range of only 7 W/m^2. Compared to the average value, this is about half of a percent variation. Most of the values fall in an even smaller range of roughly ± 1 W/m^2, which is less than 0.1% of the base value. This is so narrow of a range that TSI is often called the solar constant. The second key feature to notice in the TSI time series is the sinusoidal nature of the values. This corresponds to the solar cycle, a roughly 11-year oscillation in the Sun's magnetic field, in which it completely flips polarity. The peak of the TSI curve occurs at the interval known as solar maximum, when there are many dark **sunspots** on the solar surface. This sounds counterintuitive; if there are more dark spots on the Sun during solar maximum, why is TSI larger during this part of the cycle? This is because for every sunspot there are hundreds of smaller, brighter regions known as **faculae** that also emerge on the Sun's surface during this time. All of these phenomena are related to the solar magnetic field.

Sunspots: A relatively small dark region on the surface of the Sun; a place where a bundle of magnetic flux crosses the photosphere and the magnetic field contribution to pressure balance means that the particle content is cooler and less dense, emitting fewer photons.

Faculae: Very small bright spots on the Sun surrounding sunspots where some magnetic flux opens the photosphere just a bit, allowing light from the slightly hotter subsurface layers to escape.

Total solar irradiance: The energy flux of photons from the Sun, integrated across all wavelengths, usually reported in Watts per meter squared, also known as the solar constant.

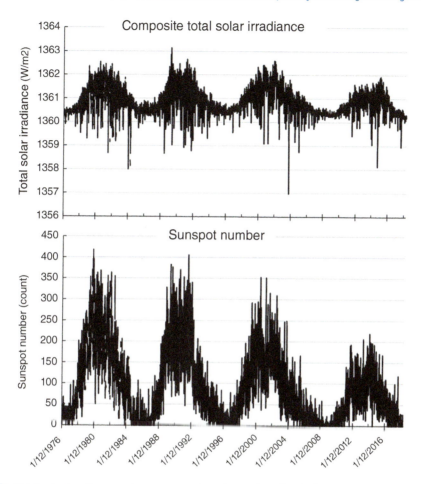

Figure 8.9 Total solar irradiance and sunspot number observations for the last 45 years.

Since Galileo's development of the telescope, in the early 1600s, people have made careful observations of the solar surface and counted the number of sunspots on it. This value is, very creatively, known as **sunspot number**, or SSN. The time series for sunspots that corresponds to the satellite-based TSI values is shown in the lower plot of Figure 8.9. The resemblance between the two curves is very high. Because this sunspot count goes back several centuries instead of only a few decades like the TSI values, it is a potential proxy for TSI over this time span.

Sunspot number: The count of sunspots on the Earth-facing surface of the Sun.

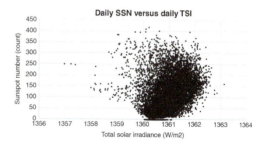

Figure 8.10 Daily sunspot number versus daily total solar irradiance, from January 1976 through May 2018.

Figure 8.10 shows a scatterplot of the two time series of daily TSI and SSN values from Figure 8.9. Qualitatively, the plot does not look particularly good. The Pearson linear correlation coefficient between these two data sets is 0.43, which is statistically highly significant for the over 14 000 values in each number set. The usefulness of SSN to serve as the basis for a model for TSI, however, is limited. No curve fit through that cloud of points would yield a particularly good representation of TSI based on SSN.

The unformed cloud of points in the scatterplot of daily values between TSI and SSN is because of the competing influence of sunspots and faculae on TSI. The impact of a sunspot on TSI is very directional; the sunspot must be near the center of the solar disk (as observed from Earth) for the dark region to have a maximal influence on the emitted solar photon flux reaching our planet. When this happens, TSI decreases. When a large group of sunspots is near disk center, then TSI could drop by several W/m^2, as seen in the top panel of Figure 8.9. SSN, however, does not take into consideration this location of the sunspots on the disk. Complicating the relationship is the existence of many faculae surrounding every sunspot. Faculae have a maximal positive influence on TSI when observed from an angle. For every sunspot moving from left to right across the solar disk as the Sun rotates, there is first a positive contribution to TSI as the faculae dominate the change, then there is a dramatic drop to a negative net influence on TSI as the sunspot moves across disk center, followed by a return to a positive contribution as the sunspot and its associated faculae continue to rotate across the solar disk to the other limb. The Sun rotates once every 27 Earth days, so this progression takes roughly two weeks for each sunspot. Therefore, a comparison of daily values is not appropriate.

A better comparison is made by averaging across a solar rotation. This comparison is shown in Figure 8.11. Because we want to model TSI with SSN, TSI is now shown as the y-axis value and SSN along the x-axis. The comparison is much better; the Pearson linear correlation coefficient for this scatterplot is 0.74, which is a good number above our general rule of thumb about a good R value. Also

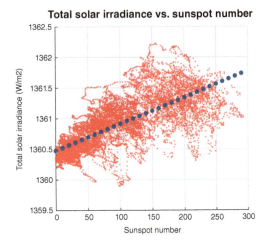

Figure 8.11 Total solar irradiance as a function of sunspot number. The 27-day running average values are in red with a linear fit model in blue.

shown in Figure 8.11 is a linear fit between the two data sets. The linear fit model is this:

$$TSI = 1360.4741 + 0.0043 \cdot SSN \qquad (8.1)$$

It appears to be a good fit through the data, albeit with substantial spread around the fit. The F test for this fit has a probability that is many orders of magnitude beyond the 0.01 threshold for being highly significant.

The plots discussed in Section 8.3 can be made for the observed and modeled TSI number sets. These are shown in Figure 8.12. The data-model comparison scatterplot looks a lot like the scatterplot of Figure 8.11; indeed, the y-axis values of observed TSI are the same. The x-axis values are different, though, with the driving parameter of SSN now replaced by the TSI linear fit model values. The light-shaded dashed line is the unity-slope reference line, demonstrating the ideal fit to the data. Note that the x- and y-axes ranges are slightly different; the model has a smaller range than the observations. This is brought into sharper relief in the other three panels of Figure 8.12. The quantile–quantile plot in the lower-left panel shows that, while the quantiles match the ideal fit in the middle of the range, the model underperforms in the extremes of the observed ranges. The histogram in the upper-left panel

8 Data-Model Comparison Basics: Philosophies of Calculating and Categorizing Metrics | 229

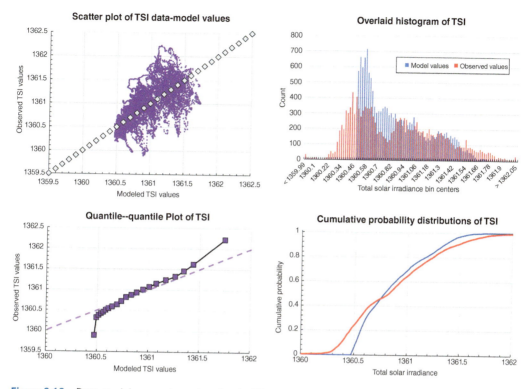

Figure 8.12 Data-model comparison plots for the TSI example. The top-left panel is a scatterplot of the observed versus modeled TSI values, with a light dashed line indicating a perfect fit. The lower-left panel is a quantile–quantile plot comparing the two number sets, again with a light dashed line indicating a perfect fit. The upper-right panel shows overlaid histograms of these two number sets. The lower-right panel contains the cumulative probability distributions of the two number sets.

shows that the model misses all of the very low and very high values of the observed distribution. This is also seen in the cumulative probability distribution curves in the lower-right panel, with the data rising from zero earlier and reaching one later than the model curve. All of this is fully expected for a linear fit; by definition, there is a spread of observed values around the fit and therefore the highest and lowest extremes of the distribution are not included in the model number set.

> **Quick and Easy for Section 8.4**
>
> We now have measurements from satellites of total solar irradiance, but if we want to know this value before the space age, you have to model it from whatever available data are most relevant, like sunspot number counts.

8.5 A Diverse Zoo of Metrics

Now that you have an understanding of some ways to plot and examine data together with a model output, it is useful to consider the many ways to quantitatively examine this relationship. The word we use for these quantitative measures of comparison is metrics. A single metric that fully describes everything about the data-model relationship does not exist. Each metric was designed to test a specific aspect of the data-model relationship, and cannot tell you about other facets of the comparison. Therefore, it takes many metrics to conduct a robust data-model comparison analysis. Chicken Little understands this and therefore they like to visit the zoo of metrics, as seen in Figure 8.13, to consider the options for how to compare a data set to output from a model.

Figure 8.13 Chicken Little likes to go to the zoo—the zoo of metrics—to select a comprehensive and robust set of assessment tools for data-model comparison. Artwork by Asher/Anya Hurst.

Metrics can be classified into several categories, depending on the feature that they are quantifying about the relationship. The major categories are accuracy, bias, precision, association, and extremes; metrics that test subsets of the data or model values fall into the categories of discrimination and reliability; and metrics that test a model's capabilities against a reference model fall into the category of skill. Note that quite a few other metrics categories exist, but this set yields a basic yet robust assessment of a model's ability to reproduce a data set.

8.5.1 The Primary Categories of Metrics

Let's start with the major categories. These are all independent of each other in their calculational approach. Together, they allow for the exploration of many facets of the data-model relationship.

Accuracy: This category includes metrics for the average "goodness of fit" between the models and the observations. It condenses the entire comparison into a single number answering this question: *how close is the model to the data?* There are many ways to assess the accuracy of a data-model comparison, and therefore many metrics have been devised that fit within this category. Each of those measures of accuracy have strengths that focus on a particular aspect of the comparison, so care must be taken when choosing an accuracy metric.

Of the many categories of metrics, this is often considered to be the most important, and therefore sometimes it is the only one discussed. This is definitely a good starting point for data-model comparisons, as an accuracy metric provides an overall assessment of how well the model reproduces the observations. Because they distill the entire number set down to a single value, information about the *way* that the model is similar to or different from the data is lost. To get at this aspect, use of additional metrics is needed. Whenever considering how to interpret a measure of accuracy, it is useful to calculate several more metrics from the other categories to better probe the quality of the model's reproduction of the observations.

Bias: This category encompasses metrics quantifying the average offset between the modeled and observed values. Again, it condenses the entire comparison into a single number answering the question: *what is the overall discrepancy between the model and the data?* It sometimes has an adjective included before it, being called unconditional bias or systematic bias. Bias is often a comparison of some estimate of the "centroid" of the data and model values, perhaps quantified by the mean, median, or mode of each set of numbers. In some sense, it is a measure of accuracy, but it is a specific type of accuracy and thus deserves its own category.

Accuracy: The category of metrics that assess the overall closeness of the two number sets.

Bias: The category of metrics that compare the centroids of the data and model number sets.

Precision: This is a category that is complementary to bias in quantifying a particular aspect of accuracy, providing a measure of the closeness of the model values to the data values with the centroid "bias" removed from the assessment. Metrics in this category seek to answer the question: *do the model and data values have the same clustering?* The formulas for these metrics often have elements of the accuracy and bias metrics, but use them in a manner that yields new information about the comparison.

Association: Sometimes it happens that the model values follow the upward and downward trends in the data values, but the quantities themselves are actually quite far from each other. In this case, the three metrics of accuracy, bias, and precision might not be very good, but the model has some predictive quality. This category includes the metrics that try to assess such relationships, even when the values are not close to each other. That is, they seek to answer this question: *how well do the model and data values rise and fall together?* They examine the local peaks and values in the two number sets, quantifying if they have the similarly located extrema. The absolute magnitudes of the extrema do not matter, only the relative magnitudes when all local extrema are considered.

Extremes: For some models, we do not actually care if it is getting the data exactly correct for most of the values, instead we might only care about the model's ability to capture the maxima and minima of the data range. That is, metrics in this category seek to answer this question: *how well can the model get the outliers in the data?* It often completely ignores the "middle" of the data-model pairings, focusing instead on the tails of the distributions. This category is sometimes called sharpness or refinement.

> **Quick and Easy for Section 8.5.1**
>
> One way to categorize the many different metrics that exist is to cluster them by the aspect of the data-model comparison that they assess. Such categories are defined here as accuracy, bias, precision, association, and extremes.

8.5.2 Skill

There is one more major category of metrics that needs to be discussed. This is a category that is a derivative of all of the other metrics categories listed in Section 8.5.1, plus it is more complicated in its interpretation, so it deserves a special introduction.

Skill: This category seeks to quantify some aspect of the goodness of the model fit to the data, but now with a twist; the metrics in this category do the calculation of the metric relative to some reference model. To distill it down to a question—*how good is the model at reproducing the data relative to my previous model?*—two things need to be chosen for the creation of a skill metric. The first is the comparative metric, which can be any of the metrics from the categories listed in Section 8.5.1. While many skill metrics use one of the accuracy metrics in its assessment, this is not required

---

Precision: The category of metrics that assess the clustering aspect of the data and model number sets.

Association: The category of metrics that assess whether the model values tend to increase and decrease coincidentally with the observational values.

Extremes: The category of metrics that focus on the model's ability to reproduce the outliers or end members of the observed values.

Skill: The category of metrics that assess a model's ability to reproduce the data relative to a reference model.

and any metric could be used. The second thing that is needed is a reference model. This could be an actual model that predicts the data values, but it does not have to be that. It could be derived from the data to which the model is being compared, or from a similar data set, such as comparing accuracy to the variability of the data values or assuming persistence and using the data values collected during some other measurement interval.

Skill is usually calculated with a specific relationship between the metric value of the new model and the reference model. This is known as a **skill score**, and nearly always is defined this way:

$$\text{Skill score} = \frac{\text{Metric (new model)} - \text{Metric (reference model)}}{\text{Metric (perfect value)} - \text{Metric (reference model)}} \quad (8.2)$$

In Equation 8.2, the term "metric" refers to any of the equations fitting into the categories of the previous section, with those specific formulas presented and discussed in the following two chapters. For convenience in this discussion, let's call the value of the chosen metric for the new model as X and the value of the metric for the reference model to be Y.

Regardless of the particular choice of the metric, the skill score formula has some particular properties. If the new model is "perfect" and its value for the chosen metric matches the metric's perfect score, then the numerator and denominator will be identical and the skill score will be one. This is a universal truth for all skill scores based on this formula. Another universal truth of this formula is that if the new model has a value for the chosen

---

Skill score: A particular method of defining a skill metric, it is the difference of metric scores between the new model and reference model divided by the difference between the perfect score of that metric and the score for the reference model.

---

metric that is equal to that of the reference model, then the numerator is zero and therefore the skill score is also zero. Finally, this definition of skill score breaks down if the metric value for the reference model is perfect, yielding zero in the denominator.

For many metrics, a perfect score is either zero or one. Let's explore some of the common features of skill score values with respect to these types of values for the chosen metric. A summary of these skill score attributes is given in Figure 8.14. The left panel is the case when the metric's perfect score is one. In this case, Equation 8.2 can be written as $(X - Y)/(1 - Y)$. The right panel shows the case for the perfect score of the chosen metric being zero. When we have this, Equation 8.2 reduces to $1 - X/Y$. To avoid the issue of dividing by zero, the range for Y is kept to 0.05–0.95. For convenience, the range of the new model metric scores is set to the same interval.

If the perfect score of the chosen metric is one, as is the case for the left panel of Figure 8.14, such metrics usually have a possible range of values that extends less than this, then the denominator, $1 - Y$, is always positive. This is true whether the range of possible values for the metric extends only down to zero or continues down to negative infinity. The numerator could be positive or negative, depending on the values of X and Y. In any case, because the score for both models is less than perfect, the numerator will be less than the denominator and the resulting skill score will be less than one. If the new model is better than the reference model for this metric, such that $X > Y$, then the numerator is also positive and the skill score will be greater than zero. If, however, the new model is worse than the reference model for the chosen metric, then we have $X < Y$ so that the numerator is negative and the resulting skill score will be less than zero.

Now let's consider the case when the chosen metric has a perfect score of zero, with other values above this (extending to either infinity or a finite upper bound). This is shown in

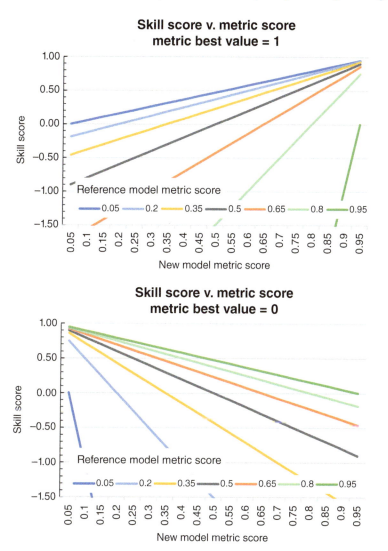

Figure 8.14 Skill score value as a function of both the new model metric score (x-axis values) and the reference model metric score (line colors). The left plot is for the case of an optimal metric score of one and the right plot shows the case for an optimal metric score of zero.

the right panel of Figure 8.14. In this case, the denominator is $0 - Y$ and so it will always be negative. If the new model is better than the reference model for this metric, $X < Y$, then the numerator is also negative and the skill score will be positive. If the new model yields a metric score worse (i.e., larger) than that of the reference model, $X > Y$, then the numerator will be positive and the resulting skill score will be negative. That is, for this case, like the one just discussed, skill scores range from one to minus infinity with skill scores between zero and one showing that the new model is an improvement on the reference model.

One more common case to consider is when the chosen metric's perfect score is zero but the values can be any real number, positive or negative. In this case, the interpretation of skill score gets a little more complicated. First, let's take the case when the metric values for the two models are on the same side of zero (i.e., $X * Y > 0$). If both are above zero, then it is exactly like the case described in the previous paragraph, with skill scores between zero and one when the new model is better than the reference model (i.e., when $X < Y$). If both are below zero, then it is exactly like the case described two paragraphs earlier, again with skill scores between zero and one indicating that the new model is better than the reference model (i.e., when $X > Y$). Now consider the case when the two metrics values are on opposite sides of zero from each other (when $X * Y < 0$). If the absolute values of the two scores are equal, $|X|=|Y|$, then the resulting skill score is equal to two. If $|X|<|Y|$, then the skill score will be between one and two, and if $|X|>|Y|$, then it will be above two. That is, in this special case when the range of a metric is both above and below its perfect value, the skill score can be both above and below its perfect value of one, with a symmetric interpretation of values between zero and two (plus or minus one from a perfect skill score) indicating that the new model is better than the reference model, and skill score values beyond this range, in either direction, indicating that the new model is worse than the reference model.

> **Quick and Easy for Section 8.5.2**
>
> Skill is the assessment of how well a model reproduces the data relative to a reference model. It usually takes the form of a skill score.

8.5.3 Metrics Categories Based on Subsetting

There are additional metric categories that assess the model's ability to predict groupings of the data. These can be very helpful for parsing out the reason that a metric from one of the categories in Section 8.5.1 is yielding a particular value. These are derivative categories because they can use any of the metrics defined in the two previous sections, including skill, but only applied over part of the data-model number set. Subsetting allows for a detailed examination of the data-model relationship in ways that the calculations over the total number set cannot provide.

Discrimination: This category of metrics is essentially the accuracy category defined in Section 8.5.1, but instead of considering all data-model pairs, it only considers those pairs within a particular grouping of the data values. The question being asked is this: *how good is the model for a specific range of the data?* Given a real-valued numerical data and model sets, such as density or temperature, this would mean grouping the data-model pairs according to specific ranges of the data values and then using the metrics for the accuracy category.

Reliability: This category contains metrics which are the inverse of those in discrimination. Instead of taking a grouping of the data and assessing the spread of the model values within this subset, these metrics take a group of the model values and assess the spread of the data within this subset. The question being addressed by this category of metrics is, therefore, this: *how close is the data to the model values for a specific range of the model output?* The calculations will look very similar to those for discrimination but with the observations and model values transposed.

These are also particularly useful when dealing with data that come in discrete events, such as weather—the weather at a particular time and location can be classified into distinct groupings, such as sunny, cloudy, rainy, or snowy. The model, in this case, is not predicting a numeric value but rather a state of a system,

---

Discrimination: The category of metrics that use only a subset of the full number set by selecting values within a specific range of data values.

Reliability: The category of metrics that use only a subset of the full number set by selecting values within a specific range of model values.

and often this is one of multiple states. These are metric categories that specifically address this type of data-model comparison.

There is another type of subsetting when a third number set is used to subset the full data-model paired set. This method does not rely on ranges of either the data values or the model values. Instead, a specific range of values of this other set is defined to selected data-model pairs for the subset of interest. This is usually done when there is a prominent driving parameter that is thought to organize the observations. It could also be applied when there is a related response data set, and you are interested in how well the model works for particular value ranges of this second (unmodeled) response. In either usage, this third number set needs to have a value associated with each of the data-model pairs.

> **Quick and Easy for Section 8.5.3**
>
> You can learn a lot about the data-model relationship by using values from only part of the full paired number set, either by using only a subset of the data range or a subset of the model range.

8.6 The Concept of Model "Goodness of Fit"

As you can see from this list, each of these metrics categories addresses a different aspect of the data-model relationship. Each of the metrics in these categories, presented in the next few chapters, assesses a specific way that the model output is "good" with respect to the observational values. The concept of declaring a model to be a "good fit" against a data set is, therefore, a difficult task, because every "but what about..." question needs to be answered with yet another metric calculation.

When conducting a comparison between a data set and a model, choosing the specific metrics for the analysis should be guided by what you want to highlight about the model's capabilities. Do you care if the values are exactly right everywhere along the timeline or spatial extent? If so, then accuracy and bias might matter most to you. If, however, you only care about predicting the times or places with the most severe data values, then association or extremes might be the metrics categories of most importance. If you want a robust analysis of a model's performance in order to determine its strengths and weaknesses, then conducting metrics from all of these categories would provide this comprehensive assessment of the data-model relationship.

For each of the categories listed in Section 8.5, the resulting value from a metric should be interpreted with respect to both the general definition of the category as well as the specific formulation of the metric itself. That is, metrics equations within a category are not identical in their meaning about the data-model relationship. For example, two measures of bias might both quantify the similarity of the centroids of the model and data values, but the formulas might do this in systematically different ways. One could be based on the arithmetic mean, while another on the median. If the distributions are Gaussian, then these two methods will yield the same values. They will yield different numbers, however, if the distributions are nearly anything else. If the logarithm of the values has a Gaussian distribution, then the mean will be higher than the median. As we have explored earlier, if the number set extends over several decades, the difference between its mean and median could be huge. This needs to be taken into account when considering whether a metric implies a good or a bad fit.

All data-model comparison metrics need to be understood in the context of the uncertainty or spread of the observation and model uncertainties. When a particular metric produces a number of 15, is that a good value or a bad one? This metric needs to be compared against a normalization factor, in fact perhaps more than one. If the value is large compared to the spread of the original data, or against the spread of the model values, then that metric score is probably not particularly good, even though the value might "feel" like a small number to you. There is no "right" way to normalize a particular metric, either. For each metric, there are usually several different normalization factors that could be applied to interpret the value is either large or small. For instance, if you are calculating a bias metric, should it be compared against the full maximum-to-minimum range of the data, its interquartile range, or its standard deviation? There is no correct answer to this question. Because it is susceptible to one value being particularly far removed from the others, the full range could produce a rather small, normalized metric value, which might then be interpreted as meaning that the model is particularly good. It might not be for your purposes, though, if the range is dominated by a single outlier many standard deviations from the mean.

Nature is so complex that there is, essentially, always something that can be improved about the model to make it more realistic and a better predictor of the observations. Measurement techniques continually improve, as does data coverage in space and time, putting more pressure on models to be made better. As these observational and numerical improvements are made, the next level of physical understanding awaits to be discovered. In an operational setting, the specific needs of the users of Earth, atmospheric, and space science forecasts are continually changing, and a model that was considered good for the previous decision-making scenario might no longer be adequate.

8.7 Application Usability Levels

Knowing the end use of a model brings up the concept of the **application usability level** (AUL). The AUL framework is like the technology readiness level convention used for the maturity of specific hardware toward a particular purpose, but is focused on a technique, model, software, or method rather than a physical device. Figure 8.15 shows the progression through the nine levels, from the initial idea of the application toward the validated implementation of the application for making operational decisions. The first three levels—in Phase 1—move a basic research concept toward applied viability by identifying users and establishing requirements. The next three levels—those in Phase 2—are the process of verification and validation of the application for the specific intended purpose of the eventual user. The final three levels—Phase 3—include the steps of transition to operations and full implementation for the end user.

AULs are mentioned here because there are "validation steps" among them, from the very beginning (the basic research usage of the code), through the development stages (especially levels 3 and 6), all the way to operational implementation (not only level 8 but continued assessment while at level 9). Because the eventual use of the application can be rather specific, the methodology of the validation process should be selected with this end goal in mind. This could require emphasizing some metrics categories, or even specific metrics, and minimizing the inclusion of others. If a different usage arises for an application that has already been fully vetted and rising to AUL 9, the usability level will

---

Quick and Easy for Section 8.6

The quality of a data-model comparison depends on its eventual usage and metric values should be interpreted in the context of uncertainties.

Application usability level (AUL): A process for advancing a model from basic research usage to operational usage, involving substantial data-model validation at several steps in the phased journey.

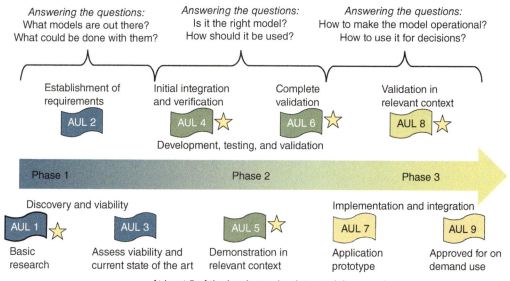

Figure 8.15 The progression from using a model for discovery-level research to using it for operational decision-making, as defined by a series of application usability levels (AULs). Adapted from Halford et al. (2019).

drop back. It could slip just a single level, if the new usage is very close to the old one and the qualities of the model that made it appropriate for the first usage are the same ones needed for the new usage. It could be, though, that it might drop all the way to one of the Phase 1 levels, in order to properly test the application for the new usage.

This concept of "goodness of fit" for a model relative to data is, therefore, a highly subjective judgment call. Assigning a "good" label to a model requires knowledge of how the model output will be used, which dictates the metrics for making an appropriate assessment. The value that the metric produces is subject to interpretation, and might be considered good for one purpose but not good enough for another.

> **Quick and Easy for Section 8.7**
>
> Application usability levels are a framework for validating a scientific model for operational decision-making context; assessing the code with the proper metrics is critical.

8.8 Designing a Meaningful Data-Model Comparison

It is not an easy task to design a good data-model comparison. There are many factors that come into play; here are a few of the topics that need to be considered in order to formulate the comparison in such a way that will truly assess the aspects of the data that are important for the specific needs of the user. Let's discuss three vitally important factors: interval, resolution, and quantity.

The first choice is **selecting the right spatial or temporal interval for the comparison**. Let's consider an example: do you want to test your model of precipitation for Santa Fe, New Mexico? Then do not choose February, when the rainfall is only 0.5 inches for the month, on average. Instead, select a

> **Interval for comparison:** Deciding what interval—spatial, temporal, or otherwise—of the number set is important for the eventual usage of the data-model comparison.

different month when there is a good mix of rainy and dry days, like July or August, or a longer time interval that includes some wetter months, perhaps even a full year. Will I conduct the test at a single station in downtown Santa Fe, or include data from many stations across the city, or even all of Santa Fe County? As another example, what if you want to test your solar flare prediction model? Well, do not choose the year 2020, which is at the trough of the approximately 11-year solar cycle (which has been a little longer than this, lately). Choose a more active year for solar flares, such as 2013 or 2014, the peak of the last cycle where the monthly averaged sunspot number was often over 100 instead of down in the single digits, or even better, 2000 or 2001, the peak of the cycle before, where the number of sunspots peaked at values over 200. This will better test your model. The choice of the interval can dramatically influence the results of the assessment.

Another choice is in **selecting the spatial or temporal resolution** for creating the observational values to be compared against the model output. Using the same two examples, do you want to predict whether it rained on any specific day, or down to the hour, or even down to the minute? Do you care if it rained *anywhere* in Santa Fe County, or did it have to rain over a certain percentage of the county for it to count? Did it have to rain the full time interval or simply any portion of the time interval for it to count as raining in that interval? Similarly for the solar flare prediction model, the question can be asked: do I model every sunspot/active region on the Earth-facing hemisphere of the solar disk, or just those near disk center that are best observed from Earth? Do we treat a successful event prediction as a modeled flare occurring in the same day, or hour, or minute as the observed flare? What if the same active region produces two solar flares within the time interval, but my model only predicts one? As with the interval, the resolution of the values within the interval is a critical choice.

Yet another factor needed for defining the observational values of a data-model comparison is **the nature of the quantity being compared**. Again, discussing the rain in Santa Fe: will the *y*-axis values be an average rainfall across the city, or the peak value anywhere in the city? That is, how will you determine the value in the combined "phase space" of spatial and temporal parameters contributing any one observed value for the analysis? For the solar flare assessment, what quantity is most important? Will it be the escape velocity of ejected material? Will it be the brightness of the flare in a particular wavelength? Will it be the helicity of the magnetic field? As with the rain, are these values averaged or are you finding the maximum value within each selected spatial and temporal window? This definition of the quantity to be considered can also significantly influence the resulting assessment.

Usually, the selection for each of these factors revolves around the question being asked and eventual use of the assessment. For an example, let's say that the rain in Santa Fe is being predicted to decide on how to operate the stormwater drainage system. Specifically, it will be used to decide whether the valves in certain connector pipes should be opened or closed. Let's say that this is a process that requires a worker to go on site, open a hole cover, climb underground, and physically turn a large wheel; a process that could take half an hour. In this case, you might be desiring an assessment of very localized rainfall, down to the stormwater system zones, with a temporal cadence of perhaps 15 min. Being off

---

Resolution for comparison: Deciding what cadence of values within the interval is important for the eventual usage of the data-model comparison.

Nature of the quantity being compared: Deciding what aspects of the number set are important for the eventual usage of the data-model comparison.

by more than this could mean that you missed the chance to open the valve and the surface flooding would begin. Similarly for the solar flare prediction model, let's say that it is for the operation of a space telescope, observing planets and collecting photons in very specific wavelength regimes. Let's say that this telescope is susceptible to signal contamination from solar flares, during which it receives light in the same part of the spectrum as the faint signal of the observation, and too many of these unwanted photons getting through the filters and baffles meant to block it would result in damage to the very sensitive detector. In this case, the prediction should be not on the timescale of days, but less than 8 min. However, it could be that only the largest flares are potential problems, with smaller flares being sufficiently blocked by the design of the instrument. This changes the assessment; only large and highly dynamic active regions need to be accurately modeled in this case, and in fact only those types of active regions that are historically known to release large fluxes of photons in the observational wavelength window. This greatly focuses the problem from the rather difficult one of predicting all solar flares of any size at an accuracy of a few minutes and scales it down to a more tractable one.

When designing a data-model comparison, do not simply go about it with a single randomly selected interval, but rather do a few randomly selected metrics calculations. A thoughtful approach to configuring the comparison to optimize its assessment of what truly matters for the intended usage of the information will go a long way toward good decision-making based on this information.

> **Quick and Easy for Section 8.8**
>
> There are several key factors to consider when designing a good data-model comparison, such as the interval to be considered, the cadence of values within that interval, and the specific nature of the quantity to be compared.

8.9 Further Reading

A great discussion of introductory data visualization and data-model comparisons for geosciences, including an explanation of the quantile–quantile plot and cumulative probability distributions, is this review by Kleiner and Graedel:

Kleiner, B., & Graedel, T. E. (1980). Exploratory data analysis in the geophysical sciences. *Reviews of Geophysics*, 18(3), 699–717. https://agupubs.onlinelibrary.wiley.com/doi/10.1029/RG018i003p00699

The Luo et al. (2022) paper with the 2D histograms:

Luo, Y., Huang, Y., Sierra, C. A., Xia, J., Ahlström, A., Chen, Y., et al. (2022). Matrix approach to land carbon cycle modeling. *Journal of Advances in Modeling Earth Systems*, *14*, e2022MS003008. https://doi.org/10.1029/2022MS003008

Here is a space physics paper by Swiger et al. with another example of 2D histograms:

Swiger, B. M., Liemohn, M. W., & Ganushkina, N. Y. (2020). Improvement of plasma sheet neural network accuracy with inclusion of physical information. *Frontiers in Astronomy and Space Sciences*, *7*, 42. https://doi.org/10.3389/fspas.2020.00042

For additional descriptions of the meaning of the metrics categories, see Chapter 8 of Wilks.

Wilks, D. S. (2019). *Statistical Methods in the Atmospheric Sciences* (4th edition). Academic Press, Oxford. More about it here: https://www.elsevier.com/books/statistical-methods-in-the-atmospheric-sciences/wilks/978-0-12-815823-4

The book by Jolliffe and Stephenson on forecast verification is a treasure trove of useful information on metrics.

Jolliffe, I., & Stephenson, D. B. (2011). *Forecast Verification: A Practitioner's Guide in Atmospheric Science* (2nd edition). John Wiley & Sons, Hoboken. https://doi.org/10.1002/9781119960003

The opening chapter has a nice qualitative description of model goodness of fit:

Jolliffe, I. T., & Stephenson, D. B. (2011). Introduction. In: I. T. Jolliffe & D. B. Stephenson (Eds.), *Forecast Verification*. https://doi.org/10.1002/9781119960003.ch1

The following chapter by Jaqueline Potts is an excellent discussion of categories of metrics, including definitions for some, like the skill score:

Potts, J. M. (2021). Basic concepts. In: I. T. Jolliffe & D. B. Stephenson (Eds.), *Forecast Verification*. https://doi.org/10.1002/9781119960003.ch2

Murphy's 1991 paper on weather forecast verification provides a good discussion of why multiple metrics are needed for robust assessments and discusses these metrics categories:

Murphy, A. H. (1991). Forecast verification: Its complexity and dimensionality. *Monthly Weather Review*, 119(7), 1590–1601. https://doi.org/10.1175/1520-0493(1991)119%3C1590:FVICAD%3E2.0.CO;2

Kubo (2019) has a nice discussion of metrics categories for space weather:

Kubo, Y. (2019). Verification of operational solar flare forecast: Case of Regional Warning Center Japan. *Journal of Space Weather and Space Climate*, 7, A20. https://doi.org/10.1051/swsc/2017018

Another discussion of metrics categories is available in Liemohn et al. (2021):

Liemohn, M. W., Shane, A. D., Azari, A. R., Petersen, A. K., Swiger, B. M., & Mukhopadhyay, A. (2021). RMSE is not enough: guidelines to robust data-model comparisons for magnetospheric physics. *Journal of Atmospheric and Solar-Terrestrial Physics*, 218, 105624. https://doi.org/10.1016/j.jastp.2021.105624

The Halford et al. (2020) paper on application usability levels:

Halford, A., Kellerman, A., Garcia-Sage, K., Klenzing, J., Carter, B., McGranaghan, R., et al. (2019). Application usability levels: A framework for tracking project product progress. *Journal of Space Weather and Space Climate*, 9, A34. https://doi.org/10.1051/swsc/2019030

Here is a book that describes the connection between sunspot number and total solar irradiance for historic assessment of the role of the sun in climate change:

Hoyt, D. V., & Schatten, K. H. (1997). *The Role of the Sun in Climate Change*. Oxford University Press. https://www.ebooks.com/en-us/273187/the-role-of-the-sun-in-climate-change/douglas-v-hoyt-kenneth-h-schatten/

There are several models for solar spectral irradiance back hundreds of years, like this one:

Fligge, M., & Solanki, S. K. (2000). The solar spectral irradiance since 1700. *Geophysical Research Letters*, 27, 2157–2160. https://doi.org/10.1029/2000GL000067

Or even thousands of years, like this one:

Vieira, L. E. A., Solanki, S. K., Krivova, N. A., & Usoskin, I. (2011). Evolution of the solar irradiance during the Holocene. *Astronomy and Astrophysics*, 531, A6. http://dx.doi.org/10.1051/0004-6361/201015843

The composite total solar irradiance used in the earlier text is version 42_65_1805 in file composite_42_65_1805.dat; available from ftp://ftp.pmodwrc.ch/pub/data/irradiance/composite/

The sunspot number data are available here: http://sidc.be/silso/datafiles#total

One of those satellites measuring TSI is the Solar Radiation and Climate Experiment (SORCE): https://earthobservatory.nasa.gov/features/SORCE/sorce_03.php

8.10 Exercises in the Geosciences

In Chapter 2, a comparison was made between paleoclimate atmospheric temperatures and carbon dioxide levels. Let's make some plots of similar data. The premise is that the variations are the same and that one can be used as a good model for the others. Specifically, we can hypothesize that temperature should be a driver of CO_2, as this gas is released from polar tundra permafrost during the entry into an interglacial period.

1. Go to the Climate Data Information site http://www.climatedata.info/proxies/data-downloads/ and download the ice core pack data brick. Note that there is another column in this file for radiation; this number set will not be used in the exercises below.
   A Open and read in the "Vostok-tpt-co2" file.
   B Make a separate time series plot of the temperature and CO_2 data sets.
   C Make an overlaid time series plot, using left and right axes with different scales.
   D Make a scatterplot with CO_2 concentrations on the *y*-axis and temperatures on the *x*-axis.
   E Make a 2D histogram of the paired values, using 100 bins along each axis.

2. Write a paragraph answering the following:
   A What is your initial reaction to the relationship between these two data sets?
   B What new information was learned by making the 2D histogram?

3. Calculate fits and averages.
   A Calculate a Pearson linear correlation coefficient between the two data sets. Determine a probability on this score.
   B Create a linear fit between the CO_2 concentrations and the temperatures.
   C Make a scatterplot with the linear fit overlaid.
   D Create rank-order versions of the paired data, so that the pairings are preserved but the rank has replaced the actual values.
   E Make a scatterplot of the rank-order pairs.
   F Calculate a Spearman rank-order correlation coefficient. Determine a probability associated with this score.

4. Create a running average of the CO_2 data set. There are two ways to make this: the first is with time as the x axis and the other is with temperature as the x axis. While both can be informative, the point here is to make a running average of the CO_2-temperature scatterplot, so use temperature as the x-axis in the running average calculation.
   A Create a 101-point running average of the CO_2 concentrations. Note that you have to sort the paired CO_2-temperature data sets so that the temperatures are in ascending order. Also calculate a standard deviation (i.e., uncertainty) for each of these averages.
   B Make a plot of this running average line, with overlaid uncertainty. If you do not do the temperature sorting correcting Part 4a, then this plot of the CO_2 running averages against temperature will be an incomprehensible tangle of lines. Also overlay the linear fit line from Part 3b.

5. Write a paragraph answering the following:
   A Discuss the linearity between these two data sets. That is, explain what you see in the plot created for Part 3c.
   B Discuss the monotonicity between these two data sets. That is, explain what you see in the plot created for Part 3e.
   C Discuss the similarity of the linear fit to the running average line. That is, explain what you see in the plot created for Part 4b.

6. Assessing the CO_2 data against the linear fit of these data:
   A Create an array of fitted CO_2 concentration values, from the linear fit created in Part 3b, versus the temperatures.

- **B** Make an overlaid histogram plot, with both the observed and fitted CO_2 concentration histograms on the same axis.
- **C** Create a scatterplot of the data against the model values.
- **D** Make a box-and-whisker version of this scatterplot, using 20 bins along the x-axis (median and IQR for the box, and 5%–95% quantiles for the whiskers).
- **E** Create arrays of quantile values separately for both number sets (20 bins).
- **F** Make a quantile–quantile plot from these two arrays.

7. Write a paragraph answering the following:
  - **A** Discuss the similarity of the two histograms.
  - **B** Discuss the quality of the scatterplot and the box-and-whisker plot.
  - **C** Discuss the features of the quantile–quantile plot, and what these might mean.

9

Fit Performance Metrics: Data-Model Comparisons Based on Exact Observed and Modeled Values

This chapter and the next present and discuss equations for the most common metrics in use in the Earth, atmosphere, and space sciences. This is not meant to be a comprehensive listing, but rather an introduction to basic data-model comparisons and, in particular, a discussion of their strengths, limitations, and some practical applications. There are literally hundreds of metrics in existence, and as researchers and model users explore new ways of comparing models with observations, new metrics are developed, tested, and added to our toolbox.

The approach here splits metrics into two major groupings. The first grouping, covered in this chapter, is called **fit performance metrics**. The second grouping, the metrics which are presented in the next chapter, is called **event detection metrics**. The former includes those metrics that use the specific numbers of the data and model values. For this reason, the grouping is also known as "continuous metrics" or, because linear fitting falls within this group, "regression metrics." The latter grouping assigns an event status to each data value as well as to each model value, often based on an event identification threshold value. Metrics in this grouping then ignore the specific number after this conversion to event status. For this reason, that grouping is also known as "categorical metrics" or, specifically in the case when there is only one event status for the data and model values, "dichotomous metrics."

Note that these two groupings are just one way to organize the myriad metrics. Others have grouped them using different clustering schemes based on the characteristics they thought most relevant. In fact, you will most likely see them grouped in different ways throughout your career, especially when venturing into other research fields. You might even see the same groupings with different names, as mentioned in the previous paragraph. Each discipline develops its own nomenclature, even when a name already exists for an identical concept in another field. Be mindful that other names for these formulas exist, some of which will be mentioned in this chapter. Try not to let this distract you. These two groupings are chosen for use here because they offer a convenient organization that allows for a straightforward presentation.

In addition, these two groupings are orthogonal definitions to the categories defined in the

---

Fit performance metrics: All metrics that use the exact values of the two number sets.

Event detection metrics: All metrics that convert the exact values of the two number sets into yes–no event status.

Data Analysis for the Geosciences: Essentials of Uncertainty, Comparison, and Visualization, Advanced Textbook 5, First Edition. Michael W. Liemohn.
© 2024 American Geophysical Union. Published 2024 by John Wiley & Sons, Inc.
Companion website: www.wiley.com/go/liemohn/uncertaintyingeosciences

previous chapter. Each of these two groupings contains metrics in all of those categories. That is, each metric presented in both this chapter and the next one is part of both a grouping *and* a category. The presentation and discussion of the metrics within these two chapters are arranged according to category, and a summary of how all of the metrics within that grouping align with the categories is given at the end of each chapter.

9.1 What Is Fit Performance?

Metrics within the fit performance grouping are defined here as those comparisons that use the exact values of the observations and model output. This usually involves the difference of each data and model-value pair, but not always. The difference is often then processed in some way, with other mathematical operators and functions, and there is regularly a summation over all of the pairs. Sometimes, the formulas do not sum over the pairs, but perform a different kind of assessment, such as finding a median of the number set. There are some metrics that do not include a data–model-value difference at all, instead comparing the distributions of the two number sets. All of these use the exact values as part of the metric determination and therefore they fit into this grouping.

Figure 9.1 shows an illustrative example of a data-model comparison. The purple dots present each of the data-model pairs, and the purple diagonal dashed line shows the unity-slope, zero-intercept line of an ideal match. To the right is a histogram of the observational values and at the top is a histogram of the modeled values. For the *i*th pair, we define the model value to be M_i and the observational value to be O_i. This will be used

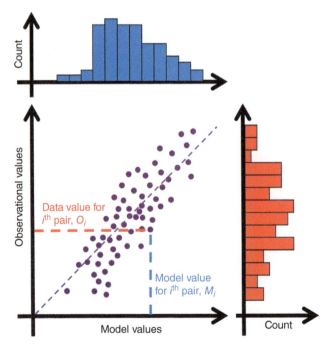

Figure 9.1 Schematic of an example data-model comparison. Fit performance metrics use the exact values of the observations and model output, either as matched pairs or as probability distributions, like the histograms shown for each number set.

throughout the discussion in the rest of this chapter, along with the total number of data-model pairs, N.

A qualitative examination of the scatter plot shows that the model roughly follows the data; it would be a good guess that a linear fit of the data with respect to the modeled values would be quite close to the unity-slope line. Inspecting the two histograms, however, reveals more of the picture. They are not the same, and a mental estimate is that the model values are skewed right, while the observed values are not.

> **Quick and Easy for Section 9.1**
>
> Fit performance is defined here as the collection of metrics, across all metrics categories, that use the exact values of the observations and model output number sets, often but not always taking a difference or ratio of these exact values for each data-model pair.

9.2 Running Example: Dst and the O'Brien Model

To help bring the use and application of these metrics into better focus, a real geophysical example calculation will be employed throughout the chapter. We will examine **Dst, the "disturbance storm time" index** of geomagnetic activity, and a model that seeks to reproduce this time series data. That is a mouthful, so let's step back and spend a bit of time on what this means.

One aspect of Earth's interaction with the Sun is the generation of current systems in near-Earth space. Earth has a strong internal magnetic field that, once you are a bit away from the surface, looks a lot like a magnetic dipole configuration. The solar wind, the supersonic electrified gas streaming from the Sun throughout the entire solar system, carries with it the interplanetary magnetic field. When the solar wind encounters an obstacle such as Earth's magnetic field, it distorts the dipolar shape. We call the region around our planet that is dominated by Earth's magnetic field the magnetosphere, or more generally and including the upper atmosphere, **geospace**. During weak interaction times, the magnetosphere resembles an egg or avocado. During more active times, though, it changes shape into something more like a finger, rounded on the "front" toward the Sun and greatly elongated toward the back, away from the Sun. It also wiggles, rings, flaps, and snaps during these active times. Every distortion away from the internally generated dipolar configuration is associated with an electric current. We have a pretty good idea of the major current systems in geospace and how they vary with geomagnetic activity.

Monitoring these magnetic field distortions can be done with a sensor called a magnetometer, introduced in Section 2.2 (Chapter 2). Nearly every spacecraft launched today has some kind of magnetometer onboard and nearly every smart phone has one inside of it as well. Scientific satellites have sophisticated and well-calibrated magnetometers that can measure field changes at very small magnitude and very fast time resolution. Magnetometers

---

Dst, the disturbance storm time index: A measure of geospace activity based on measurements from several low-latitude magnetometer stations, with quiet time values near zero and strong active time values extending to negative numbers below −100 nT.

Geospace: The region of outer space around Earth dominated by the Earth's magnetic field, it extends from roughly 70 km altitude to about 10 planetary radii toward the Sun and 50–250 planetary radii in the opposite direction.

are also installed on the ground, usually away from buildings, roads, and other equipment to get a clean signal of the perturbations from currents in near-Earth space. There are over 300 ground-based science-grade magnetometer stations located all over the world.

Nearly one hundred years ago, it was conceived of combining individual ground-based magnetometer data into a global measure of geomagnetic activity. These compilations resulted in geomagnetic indices, one of which is Dst. It is derived from four low-latitude stations spread around the world, brought together in a particular methodology and reported on an hourly cadence. This index, measured in units of nanoTesla (nT), is usually near zero, but can rise to perhaps +50 nT and drop as low as −400 nT. Note that this is a very small signal of the Earth's dipolar field, which is roughly 30 000 nT near the magnetic equator. When activity picks up, currents flowing in near-Earth space act against the Earth's dipolar field and suppress the north–south field near the equator. This translates into a negative value for Dst.

One form of exceptionally strong active times is the **geospace storm** (or **magnetic storm**). When a fast-moving structure of charged particles and magnetic field are expelled from the Sun and slam into Earth's magnetic field, the Dst index goes through a characteristic pattern: first quickly rising as the magnetosphere is compressed, then dropping steadily through the "main phase" of the storm toward a minimum Dst value, and then, as the solar wind structure is past and geospace starts to return to its quiescent state, the Dst index rises back toward zero again. This rise is often pretty fast early in the recovery phase and then slows down, perhaps dragging on for days

---

Geospace storm or magnetic storm: When outer space around Earth is highly perturbed, resulting in intense space currents, it is often defined by a characteristic negative excursion in the Dst index.

---

before it settles within noise levels of zero. It can also start dropping negative again, if activity flares up during the recovery process.

For many decades now, people have developed models to reproduce the time series of the Dst index. One well-known and fairly successful version of this is that of O'Brien and McPherron (2000), which we will call the O'Brien model throughout this chapter. They used a differential relationship in a form known in magnetospheric physics as the Burton equation:

$$\frac{d\text{Dst}^*}{dt} = Q - \frac{\text{Dst}^*}{\tau} \tag{9.1}$$

Q and τ are empirically derived source and loss terms dependent on the solar wind electric field. Dst* can then be related to the observed Dst through a simple algebraic expression,

$$\text{Dst} = \text{Dst}^* + b\sqrt{P_{\text{sw,dyn}}} - c \tag{9.2}$$

In Equation (9.2), $P_{\text{sw,dyn}}$ is the solar wind dynamic pressure in nanoPascals (nPa), and b and c are empirically derived coefficients. It is seen that Dst* is a version of Dst with the contributions from a few current systems removed, such as the solar wind dynamic pressure deforming the dayside magnetopause, a quiet-time offset, and a corrective factor due to induced currents within the Earth. For this model, the coefficients were created from 30 years of solar wind and Dst values, yielding $b = 7.26$ nT/nPa$^{1/2}$ and $c = 11$ nT. The source and loss terms in Equation (9.1) have a function dependence on one component of the solar wind motional electric field, $E_y = -v_x \times B_z$, and have the following forms:

$$Q = \begin{cases} a(E_y - E_c) & E_y \geq E_c \\ 0 & E_y < E_c \end{cases}, \tag{9.3}$$

$$\tau = \begin{cases} xe^{y/(z+E_y)} & E_y > 0 \\ xe^{y/z} & E_y \leq 0 \end{cases} \tag{9.4}$$

9 Fit Performance Metrics: Data-Model Comparisons Based on Exact Observed and Modeled Values

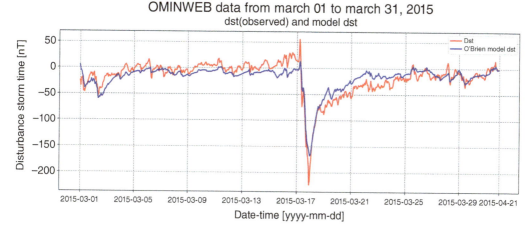

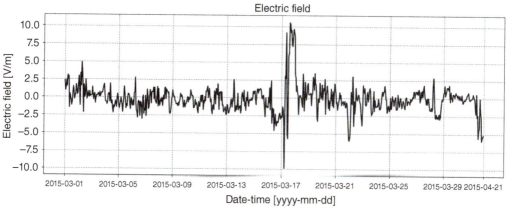

Figure 9.2 Time series of the hourly values of observed and modeled Dst and solar wind input E_y values for the O'Brien model throughout the month of March 2015.

The coefficients in Equations (9.3) and (9.4) are as follows: $a = -4.40$ nT m/(h mV), $E_c = 0.49$ mV/m, $x = 2.40$ h, $y = 9.74$ mV/m, and $z = 4.68$ mV/m.

For our running example through the rest of this chapter, we will use data and model values of Dst for the month of March 2015. This is an interval that provides hundreds of data-model pairs in the number set (744, to be exact). Figure 9.2 shows hourly values of the observed and modeled Dst index along with the primary factor in the model, solar wind electric field E_y. The plot shows that this month contains a rather big geospace storm event, in fact the largest one of solar cycle 24 (December 2008 to May 2020), known as the St. Patrick's Day storm. The plot also reveals quite a bit of lower-level activity both early and late in the month-long interval, where the Dst index is below zero but only by a small amount (remaining positive of -50 nT).

Figure 9.3 shows a comparison of the resulting observed and modeled Dst values, with the modeled Dst* converted into a modeled Dst using a rearranged form of Equation (9.2). The upper panel shows a scatter plot of the observed Dst values with respect to the modeled values. Qualitatively, this looks like a pretty good model, matching the observed values quite well at both quiet times (when Dst is near zero) and disturbed times (when Dst is large and negative). The points are strongly clustered

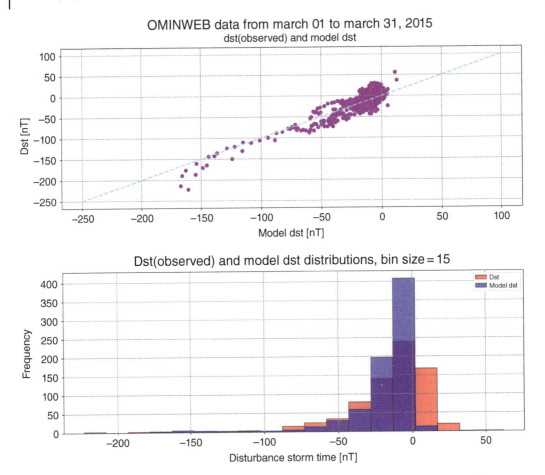

Figure 9.3 The upper panel shows a scatterplot of the observed Dst values against the modeled Dst values. The lower panel displays the histograms of these two data sets, overlaid on top of each other with bin sizes of 15 nT. Red is the observed Dst, blue is the modeled Dst, and purple is the overlap of the two, showing the value of the lower one.

near the dashed light-blue "perfect fit" line. Most of the values appear to be within a few tens of nT from this unity-slope line. The lower plot in Figure 9.3 presents overlaid histograms of the two number sets. Note that because Dst becomes more negative during space storm events, the bulk of the quiet-time values are clustered near zero, with an extended tail of values out to near −200 nT.

To further explore these two numbers, some quantitative calculations about them are listed in Table 9.1. The mean and median values are quite close, with the discrepancy between these values close to or more than a factor of ten smaller than the standard deviation, the mean absolute difference, or interquartile range (IQR). These spread measures are reported to two significant digits as well as to the appropriately rounded value for an uncertainty listing, given in parentheses. Both number sets are highly skewed left, with a long negative tail—the index values during the storm intervals within the month. The kurtosis coefficients are also highly different than the Gaussian value of 3, indicating heavy tails relative to that from a Gaussian with the

Table 9.1 Characteristic quantities of the observed and modeled Dst number sets.

	Observed Dst	Modeled Dst
Mean	−17 nT	−18 nT
Standard deviation	32 nT (30 nT)	24 nT (20 nT)
Mean absolute difference	21 nT (20 nT)	14 nT
Skewness coefficient	−2.6	−3.5
Kurtosis coefficient	10.0	14.8
Median	−9 nT	−11 nT
Interquartile range	27 nT (30 nT)	12 nT (12 nT)
Variance	996 nT^2	591 nT^2
Range	[−223 nT, +56 nT]	[−167 nT, +13 nT]

calculated mean and standard deviation. Therefore, mean and standard deviation may not be the best measures for centroid and spread of these number sets; median and IQR are probably more descriptive. Note that the IQR values are rather different, with the modeled IQR less than half that of the observed number set. This is also seen in the histograms of Figure 9.3 with the model values being more strongly peaked around the centroid than the observations. Note that the significant figure rule dictates that the spread values should be reported to only one digit, unless that digit is a one.

This example will be used throughout the sections in this chapter. As the metrics for a particular category are introduced, they will be applied to this data and model number set, with some interpretive discussion. At the end of the chapter (Section 9.11), this running example is summarized, assessing what is learned by the application of many different metrics, especially the new knowledge beyond a single accuracy measure of the data-model relationship.

Let's briefly address the design considerations for meaningful data-model comparisons brought up at the end of Chapter 8. The first is selecting the correct interval. The choice of the month of March 2015 includes a large magnetic storm, allowing a test of the model for extreme values of the index. There is, however, only one storm interval, so if that one storm is peculiar in some way, then perhaps the comparison is not a true test of the model's capabilities. A longer interval—a full year or even a full solar cycle—would be a better comparison that more robustly assesses the model's quality at reproducing the observed values. The second choice is the selection of the resolution to be used in the comparison. Dst is an hourly index, which is a decent cadence for resolving the main features of geospace storm sequences. There are similar indices available at 1-min time resolution that are better for examining any feature shorter than a few hours, phenomena that would only be represented by one or a few values in the Dst time series. This comparison, therefore, focuses on storms, not on short-timescale activity within geospace. The final choice is the nature of the quantity being compared. Because we are using this as an example of a fit performance assessment, the focus is on exact reproduction of the observed quantities. We could distill the number sets, however, by only focusing on a daily minimum or average Dst value. This might be more appropriate if the comparison was considering a full 11-year solar cycle. Focusing on the daily average would assess the model's ability to match the climatological "storminess" seen in the observed values, while choosing the daily minimum would assess the model's ability to reproduce the worst cases of the observations.

> **Quick and Easy for Section 9.2**
>
> The reproduction of the observed Dst index by the O'Brien model will be used as an example throughout the rest of this chapter. Qualitatively, the comparison seems reasonably good, but the model doesn't span the full range of observed values.

9.3 Accuracy

A common feature of accuracy metrics for the fit performance grouping is that essentially all of them have a perfect score of zero. Because they are usually based on a summation of the difference between the data and model values, if these two numbers are equal, then the contribution to the sum will be zero. If all data-model differences are zero, then the resulting total value will also be zero. What constitutes a mediocre or a bad accuracy score, however, depends completely on the metric and its functional form, as well as the context of the particular number sets and the assessment being conducted.

Figure 9.4 shows a schematic of a data-model comparison, the same illustrative data and model number sets as shown in Figure 9.1. Accuracy metrics assess how closely the points cluster around the diagonal dashed line. Many of the accuracy metrics are based on the difference between the model and data values (i.e., the discrepancy). Because the purple dashed line has a slope of one, the difference between the model and data values for any given (M_i, O_i) pair is equal to the vertical or horizontal distance of the point to this line. It is also directly proportional to the diagonal distance to the purple line, with a

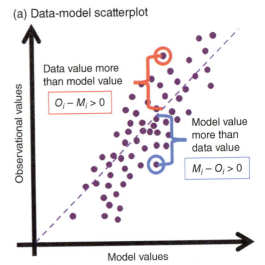

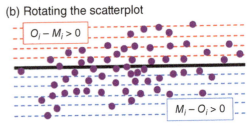

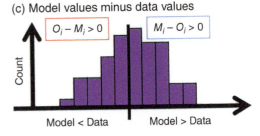

Figure 9.4 Schematic data-model comparison scatterplot with a typical model value minus observed value subtraction as the core operation of the metric. The upper panel shows the scatterplot, the middle panel is a rotated version of the scatterplot with guidelines to show the counting bins, and the lower panel shows a histogram of the discrepancies.

scaling factor of $\sqrt{2}$. In the middle panel of Figure 9.4, the scatterplot is rotated to better see how the values fall within bins of certain intervals away from the ideal comparison line. The lower panel of the figure then shows the counts within each of these intervals,

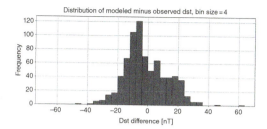

Figure 9.5 Histogram of the subtraction of the modeled Dst minus observed Dst for the running example comparison.

plotted as a histogram. Note that the points within the scatterplot "below and to the right" of the diagonal line are those points for which the model value is larger than the data, and are therefore included in the columns on the right side of the histogram plot, given our definition of $M_i - O_i$ for the value in this histogram. The points "above and to the left" of the diagonal line have observed values bigger than the model output, and contribute to the columns on the left of the histogram.

The histogram of $M_i - O_i$ differences for the running example of Dst is presented in Figure 9.5. The mean of this distribution is −1.4 nT, with a standard deviation of 14 nT, the median is −4.2 nT, and the IQR is 18 nT. From these centroid and spread values, it can be stated that the distribution of discrepancies is slightly offset from zero, but this offset is well within the distribution's spread. This offset indicates that the model systematically underestimates the observed values, but only by a small amount. With Dst, though, the disturbed times are more negative, so this actually indicates that the model values are a bit more "disturbed" than the observations.

> **Quick and Easy for Section 9.3 Introduction**
>
> Accuracy fit performance metrics assess the overall quality of the model. There are many ways to formulate the relationship for this assessment, but most have a perfect score of zero.

9.3.1 The Big Three of Accuracy: MSE, RMSE, and MAE

Accuracy measures abound. In fact, we have already been introduced to one, back in Chapter 7, during the discussion of the ANOVA table—**mean square error** (MSE). This is very similar to variance, but with the model values for each pair replacing the average of the data values:

$$\text{MSE} = \frac{1}{N-d}\sum_{i=1}^{N}(M_i - O_i)^2 \quad (9.5)$$

In Equation (9.5), we have the model degrees of freedom of the model, d. As we saw in Chapter 7, this parameter is 2 for a linear fit, 3 for a cubic fit, and so on. For a more complicated model, d could be not only large but also difficult to determine.

Like variance, MSE includes a squaring of the difference between the data and model values. This emphasizes the larger differences and they will contribute disproportionately more to the summation. As discussed in Chapter 4, this weighting is useful because it is related to certain statistical properties, especially associated with a Gaussian distribution of the underlying values.

What constitutes a good value for MSE? Because of the similarity of form, it should be compared against the variance of either the observation set or the model number set. An even better usage is to compare it against both of these numbers. MSE, however, has the units of the square of those of the model or observation (like variance), making it less intuitive for direct interpretation. That is, because MSE is left as a square, it is not directly comparable to the physical units of either the data or model values. It looks large relative to the other accuracy measures and therefore appears to be

> **Mean square error (MSE):** A fit performance accuracy metric based on the average of the squares of the data-model differences.

"bad," but in reality, the size of the number is purely a feature of its functional form. This is one of the difficult aspects of MSE. As a stand-alone value, it is often considered not particularly useful. That said, it is a common number within other, more complicated metrics and therefore needs to be introduced here so it can be used later.

For our example of O'Brien's Dst calculation, the MSE value is 195 nT^2. To one significant figure, this should be reported at 200 nT^2. Is this good or bad? Let's compare against the variance values in Table 9.1, which are 1000 nT^2 and 600 nT^2, respectively. This metric value is, therefore, quite good, well below the variance for either data set, respectively.

Another very common accuracy measure in the fit performance grouping is the square root of MSE, with the obvious name of **root mean square error** (RMSE):

$$\text{RMSE} = \sqrt{\frac{1}{N-d}\sum_{i=1}^{N}(M_i - O_i)^2} \qquad (9.6)$$

Even though Equation (9.6) is simply the square root of Equation (9.5), it deserves special attention because RMSE is one of the most common metrics in use. This prevalence is because of its formulaic similarity to standard deviation and its units being the same as the original values. As with MSE, the square of the differences gives a preferential weighting to the larger data-model errors, favoring the outliers.

A good RMSE is usually considered one that is less than the standard deviation, σ, of either the data or the model number set. This comparison is only truly useful, however, if the number set has a Gaussian distribution, or something close to it. If the number set distribution is bimodal or has a high kurtosis, for instance, then σ is not a particularly good representation of the true spread of the values, being systematically too large in the former and too small in the latter. So, some thoughtfulness is needed when interpreting RMSE, and additional information beyond σ should be included in this assessment. The most common comparison for RMSE, however, is with σ.

For our example, the O'Brien model has an RMSE value of 14 nT. Again, this is well below either of the standard-deviation values for observed and modeled data sets, seen in Table 9.1 to be 50%–100% larger than this.

Another very common accuracy metric is **mean absolute error** (MAE):

$$\text{MAE} = \frac{1}{N-d}\sum_{i=1}^{N}|M_i - O_i| \qquad (9.7)$$

The absolute value within the summation in Equation (9.7) reveals that this is not simply an average of the differences but a linear combination of the discrepancies regardless of which side of the observation the model value it is located. This is exactly analogous to the mean absolute difference (MAD) of the individual number sets against their arithmetic mean.

Another feature of MAE is that it does not preferentially favor the outlier data-model discrepancies in the summation. Because of the weighting of outliers in RMSE, MAE is typically smaller than RMSE. Neither is inherently better; the two metrics reveal different aspects of the data-model comparison with respect to accuracy. This is illustrated in Figure 9.6, which shows, for the illustrative

---

Root mean square error (RMSE): A fit performance accuracy metric, it is the square root of MSE and therefore has the same units as the numbers in the sets.

Mean absolute error (MAE): A fit performance accuracy metric, it is the average of the absolute values of the data-model differences.

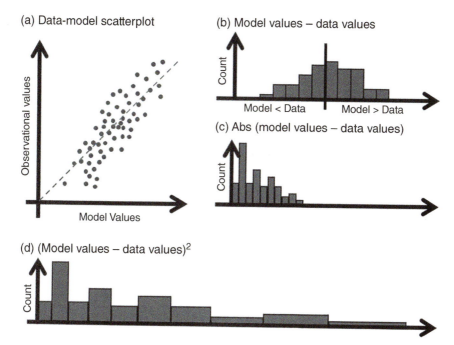

Figure 9.6 Histograms of the subtraction of the modeled and observed values, one as the absolute value of the difference and the other as the square of this value, showing the change in the quantity being summed. MAE uses the former, while both MSE and RMSE use the latter.

scatterplot in the upper-left panel, the histograms for $M_i - O_i$, $|M_i - O_i|$, and $(M_i - O_i)^2$. The absolute value histogram extends over only half of the domain of plot of the signed difference, and the squared values in the final panel stretch across a much larger domain than either of the other two.

To assess whether a calculated value of MAE is good, the typical comparison for MAE is with the MAD of the data set against its mean. Another option is to compare it with standard deviation, but the functional forms of MAE and σ are not identical, so the comparison is not particularly meaningful. MAE should be below either or both MAD values to be considered good; the lower the better.

For the O'Brien Dst calculation, MAE is 12 nT. The MAD scores for the two number sets are 20 and 14 nT. MAE is below both of these MAD scores, indicating that the model has some value for accurately reproducing the observations. As expected, MAE is below the RMSE value of 14 nT, but only by a small amount. The close similarity between the MAE and RMSE metrics is reflected in the kurtosis of the model minus data histogram in Figure 9.5. The kurtosis of this distribution is only 0.25, much smaller than the Gaussian value of 3, indicating a very light tail to the distribution of differences.

> **Quick and Easy for Section 9.3.1**
>
> The three most commonly used accuracy metrics are MSE, RMSE, and MAE.

9.3.2 Neglecting Degrees of Freedom

It is useful to spend a bit more time on model degrees of freedom because this is a challenging aspect of data-model comparisons. The basic definition of model degrees of freedom d, as used in the previous section, is the number of intermediate parameters needed to be calculated to obtain a statistic. Here, intermediate parameters are coefficients found from historical data sources or the data-model pairs themselves and then used to obtain specific model output values. For a linear fit, the two intermediate values needed to obtain the M_i number series were slope and intercept. For our example of the O'Brien prediction for Dst, the intermediate values include b and c in the calculation of Dst*, a and Ec in Q, and x, y, and z in τ, all fit from historical data. In this case, $d = 6$. It could be argued that the degrees of freedom for the O'Brien model is only 3, because all of these fit parameters in the model were obtained from three training number sets: solar wind dynamic pressure, solar wind electric field, and Dst. Because several parameters were extracted from these number sets, though, it is more appropriate to count the coefficients, setting $d = 6$. If the model is a physics-based first-principles code, then d should be found by carefully considering the intermediate values needed to obtain the output from the code, such as initial condition parameters, boundary condition parameters, and flexible-valued coefficients within the physical equations or numerical implementation of the model.

The good news is that it often does not matter because N is usually much bigger than d. This should be the case, even for complex first-principles-based models. As an error calculation, it is often not useful to report more than a few significant digits. Because of this, the d is often completely ignored in the calculation of accuracy metrics and many other fit performance metrics that otherwise should have included it.

As an example of this, let's say that you want to know MSE to two significant digits. We'll do a similar calculation to what we did in Chapter 4 when comparing population versus sample parameters on a single data set. To ignore d in Equation (9.5), this means that the influence of including d in the denominator needs to influence the second significant digit of MSE; if d is small enough that it only influences the third digits of MSE, then it can be ignored. What, then, should be the ratio N/d to reach this desired objective of ignoring d? Remembering our discussion of significant figures in Chapter 1, we can create two parameters, MSEwith and MSEwithout, Equation (9.5) including and omitting d, respectively, for which we want the relative difference between these two values to be smaller than 1%. In fact, smaller than half of 1%, so that rounding influences are also avoided. This criterion for the relative difference, let's call it α, can be generalized for any desired number of significant digits, β, this way:

$$\alpha = \frac{10^{-\beta}}{2} \quad (9.8)$$

The relative difference formula for being able to ignore d is then this:

$$\frac{|MSE_{with} - MSE_{without}|}{MSE_{without}} < \alpha \quad (9.9)$$

This specifies a level of influence below which d can be ignored. Because the summation is exactly the same in all terms on the left side of Equation (9.9), this inequality can be rewritten and reduced to a ratio of N over d:

$$\frac{N}{d} > \frac{1+\alpha}{\alpha} = 2 \cdot 10^{\beta} + 1 \quad (9.10)$$

For our example, we want two significant digits, so $\beta = 2$ and therefore $\alpha = 0.005$. Plugging this into Equation (9.10) reveals that N should be 201 times larger than d to allow us to ignore the influence of d in the second significant digit. For a linear fit with $d = 2$, this means that 402 or more data-model pairs are needed to ignore d in MSE. If only one significant digit is desired, then this minimum data-model pair set required to ignore d in MSE is only 43 ($N/d > 21$, with $d = 2$).

Note that this criterion, Equation (9.10), only informs us about the influence of d on

MSE. It needs to be met in order to ignore d in Equation (9.5) for MSE. The values themselves have uncertainty associated with them and impose their own limit to the significant figures of any data-model comparison metric, which could be more restrictive than this criterion for ignoring d in Equation (9.5). If you choose to include d in Equation (9.5), then the uncertainty on the values determines the number of significant figures for MSE. If you wish to omit d from Equation (9.5), then both the data set uncertainties and Equation (9.10) need to be considered in the significant figure determination, and the more restrictive of these should be applied.

For our example of O'Brien's Dst calculation with 744 hourly data-model values in the month of March 2015 and 6 regression effective degrees of freedom, we have $N/d = 124$, which means that we have a choice—either include d in the calculation of MSE, in which case we can trust MSE to the uncertainty level of the values, or ignore d and then we can only trust MSE to one significant figure. We would need to nearly double our number of data-model pairs in order to reach the 201 threshold, allowing us to add a second significant figure to MSE while ignoring d in Equation (9.5).

RMSE includes d, but, like MSE, this parameter could be omitted from Equation (9.6) if N is sufficiently large relative to d. Because of the square root operation, though, the determination of the threshold for the influence of d on significant figures in RMSE is not equivalent to that of MSE. To determine the N/d relationship for RMSE, an analogous formula to Equation (9.9) can be written with RMSE replacing MSE, yielding this:

$$\frac{N}{d} > \frac{1}{(\alpha+1)^2 - 1} \quad (9.11)$$

While it is slightly more complicated, it yields smaller ratios for a given β value, the number of desired significant figures. Specifically, for one significant digit, applying $\beta = 1$ in α gives an N/d value from Equation (9.11) as 9.8. That is, N should be 10 times larger than d to maintain one significant digit with the omission of d from the definition of RMSE, which is smaller than the value of 21 from Equation (9.10) for MSE for $\beta = 1$. Using $\beta = 2$ yields an N/d that should be greater than 99.8, so N only needs to be 100 times larger than d in order to ignore d in Equation (9.7) and still maintain two significant figures in RMSE. Remember that this is only one part of the limitations on significant figures; there is still the uncertainty of the values themselves, propagated through the calculation following the procedures in Chapter 3. We also need to do the summation of measurement uncertainty, systematic uncertainty, and random uncertainty as described in Section 4.3 (Chapter 4).

Going back to our example, our chosen interval of one month with 744 values, the O'Brien model with 6 regression effective degrees of freedom, means that the resulting N/d value of 124 is above the $\beta = 2$ threshold. As long as our uncertainty propagation (the calculation from Chapter 3) yields a significant figure in the second decimal place of our calculation for RMSE, we are free to omit d from the calculation of RMSE without influence on the result to two significant digits.

Because MAE has a similar form to MSE with regard to d, specifically that it does not contain a square root around the entire formula as does RMSE, the analysis of when d can be omitted from Equation (9.7) is identical to that for MSE. Equation (9.10) applies to this metric with regard to significant figures and dropping d.

> **Quick and Easy for Section 9.3.2**
>
> When N is large, it is usually just fine to omit d from the metrics calculations because its inclusion will only influence the result beyond the last significant digit.

9.3.3 Normalizing the Accuracy Measure

To interpret the metric scores of MSE, RMSE, and MAE, they have been compared with an appropriate parameter of the observational and modeled value sets. This can be formalized with the introduction of a normalization factor Λ in the formulas, like this:

$$n\text{MSE} = \frac{1}{\Lambda^2}\text{MSE}, \quad (9.12)$$

$$n\text{RMSE} = \frac{1}{\Lambda}\text{RMSE}, \quad (9.13)$$

$$n\text{MAE} = \frac{1}{\Lambda}\text{MAE} \quad (9.14)$$

The Λ parameter can be any quantity that serves as a baseline. It is usually defined as an expected range of variability for this quantity, but could be an expected absolute value of the quantity. A common option is to set Λ by a parameter related to the observations or the modeled values. It could be defined from another observational data set, say from a different experiment, or perhaps from some climatological long-baseline data set. It could be from a related theoretical derivation or analytical calculation of an expected range. Because there is no consistent normalization value, when creating a normalized accuracy measure it is absolutely essential to report the choice for Λ. Similarly, when evaluating a normalized accuracy measure created by someone else, always inquire about the setting for Λ. The normalized metric cannot be interpreted without knowing this value.

Setting Λ equal to the standard deviation, σ, of the observed value set is a common normalization factor for RMSE. Value for Equation (9.12) or Equation (9.13) of greater than one indicates that the model-data differences have a larger spread than the observed values about their own mean. This provides a quantitative judgment on the quality of the model, with values much less than one revealing that the model is very good at reproducing the observations. Setting Λ equal to MAD is a good choice for nMAE in Equation (9.14). The functional forms are the same, so any score below one can be interpreted as a good score. You could also define nMAE with σ for Λ, which will most likely yield an MAE value below one. This score of less than one, though, might not indicate that the model is good. While it is perfectly fine to normalize by any value, it should be noted that the functional forms of σ and MAE are different. The preferential weighting of the outliers means that σ is systematically larger than MAE.

There are many other options for Λ. Two other useful options are either the minimum of the two number set standard deviations or the average of the two σ values. This has the advantage of including the model variability when interpreting the quality of the data-model comparison. If the model is unresponsive to the input and the values are quite similar to each other, then inclusion of this small model-value standard deviation in the normalization of the accuracy metrics will reveal that the quality of the fit is not particularly good. Another option is to set Λ equal to the interquartile range of the observed values, the modeled values, or a combination of the two. While this normalization factor is slightly larger than standard deviation, it is particularly useful when the observations or model values are not normally distributed, and therefore a comparison against σ is not fully justified. It does not have to be the interquartile range of the 75% value minus the 25% value, but could be any designated quantile range across the rank ordering of either value set. Specifically, three other choices for Λ are the 0%–100% range, the 5%–95% range, and the 10%–90% range. The first of these is, of course, the full range of the underlying data set. This will always yield a normalized value from Equations (9.12), (9.13), and (9.14) of less than one. It can be hard to interpret, though, because it is susceptible to extreme outliers. This problem becomes especially bad when the data set is large, as the probability of extreme outlier values increases (remember, for example, our calculations with Chauvenet's criterion in Chapter 5). The other two options are nearly the full range, but exclude, to different extents, these outlying values from the definition of Λ.

Finally, it is useful to note that the normalization factor does not have to be the same for different metrics. The choice of Λ should depend on your usage of the normalized metric. For instance, the Λ choice for Equations (9.12) and (9.13) could be the observational σ, while the choice of Λ in Equation (9.14) could be average absolute relative difference of the observations against their mean. In this case, the metrics are normalized by the observational value set with an analysis parameter that has the same function form. Alternatively, Λ for nRMSE could be set to the interquartile range, while it is set to the 5%–95% range for nMAE. This large difference in normalization clearly separates the two metric values so that they are not mistakenly compared against each other or inadvertently swapped. Again, whatever the choice for normalization factor, it is important to state what has been chosen so that others can correctly interpret the meaning of the resulting metric scores.

For the running example, RMSE was 14 nT. As just discussed, its normalized version could be defined with a wide variety of Λ choices. A few of these are shown in Table 9.2. Values for nRMSE go from 0.05 up to 1.2, depending on the Λ selection. None is inherently better or worse than the others; any of them could be reported.

Normalization of an accuracy metric was introduced back in Chapter 7, in Section 7.2 with the introduction of the F statistic. This dimensionless parameter is almost an inverse of normalizing MSE by the observation data set variance, with an extra multiplier. Because MSE is in the denominator of the F statistic, F is best when the value is large. The F statistic is a useful accuracy measure when the number sets are Gaussian because it can be converted into a probability that you would get that data-model relationship from random chance. It is a difficult number to interpret without this conversion to a probability, so other metrics are recommended for most geoscience situations.

Table 9.2 Different possible normalizations for RMSE for the running example.

Λ definition	Λ value	nRMSE
σ_O	32 nT	0.44
σ_M	24 nT	0.58
$\min[\sigma_M, \sigma_O]$	24 nT	0.58
$(\sigma_M + \sigma_O)/2$	28 nT	0.50
Range[O]	279 nT	0.050
Range[M]	180 nT	0.078
$\min$[Range(O),Range(M)]	180 nT	0.078
IQR(O)	27 nT	0.52
IQR(M)	12 nT	1.2
$\min$[IQR(O),IQR(M)]	12 nT	1.2

> **Quick and Easy for Section 9.3.3**
>
> Normalizing the accuracy metric score by some measure of spread provides context to better interpret the calculated value.

9.3.4 Percentage Accuracy Metrics

A section detailing a comprehensive list of all of the possible accuracy metrics would be exceptionally long. In order to make this a robust yet concise description of common metrics, a detailed description of all possible accuracy measures of the quality of a model's fit to a data set is not needed here. Moreover, such a list, however long, would probably still inadvertently omit a few of the less-well-known metrics. It is good, though, to cover several other useful formulas beyond the top three discussed in Section 9.3.1.

Despite their ubiquitous usage, there are several problems with MSE, RMSE, and MAE. One of these is that these formulas for MSE and

RMSE are based on an assumption that the underlying distribution of the data-model difference, $M - O$, should be Gaussian. While they can be applied to any distribution of observed or modeled values, the common interpretations of these metrics might not be fully applicable when the distributions have non-Gaussian tails, when they are bimodal or have several relative peaks, or when they are positive definite, which truncates the left-side tail. In this final example of a positive definite number set, there most likely will be a right-side skew to the distributions of both the observations and modeled values. All three focus on values in the uppermost order of magnitude of the M and O number sets.

To remedy this, convert the summation quantity from an absolute difference to a relative difference. Assuming that the model results will have a larger spread than the observed values, one solution is the division of the model–observation difference by the model value, making the summed values into fractions. One metric that uses this augmented term in the summation is the **mean absolute percentage error**, MAPE:

$$\text{MAPE} = 100 \frac{1}{N-d} \sum_{i=1}^{N} \left| \frac{M_i - O_i}{M_i} \right| \quad (9.15)$$

Most definitions of MAPE will omit the model degrees of freedom term d from the denominator of Equation (9.15). It is kept here for completeness, but can be confidently discarded, for a desired number of significant figures, assuming inequality (9.10) is satisfied.

On comparing Equation (9.15) with the metrics defined in Section 9.3.1, it is seen that MAPE has essentially the same functional form as MAE, using the absolute value of the difference without any squares or square roots in the formula. MAPE is an average of the absolute value of the relative difference between the data and model number sets. The factor of 100 at the front of the definition converts the resulting value into a percentage score.

It is difficult to assign a threshold for a good score for MAPE. Since the perfect score is zero, lower scores are obviously better, but how low is "good"? One threshold that is commonly applied is 100%, indicating that the average ratio of the data-model discrepancy to the model value is one. This type of a spread, however, could be large or small, depending on the distributions of the two number sets. Because the numerator inside the summation looks a lot like MAE, a reasonable choice would be to compare MAPE with the MAD scores for each of the number sets. A simple method for obtaining a quantitative threshold, then, is this:

$$\text{Good MAPE} < 100 \frac{\min\left[\text{MAD}_M, \text{MAD}_O\right]}{\bar{M}} \quad (9.16)$$

To make it consistent with the MAPE formula, only the model number set average appears in the denominator. Equation (9.16) is only a guideline and this judgment of the goodness of the metric score should be made within the context of the particular assessment being conducted.

For our example of O'Brien's Dst prediction model, we obtain a MAPE score of 280%. This indicates that, on average, the modeled and observed values are off by a factor of nearly four. This is substantially larger than the "good" value given by Equation (9.16), which is 82%. This sounds bad, and it is; while the earlier metrics indicated relatively good accuracy, this metric is revealing poor quality of the model. This is because it is an average of ratios and most of the values in the set are during quiet times when the model value is near zero.

Mean absolute percentage error (MAPE): A fit performance accuracy metric, it is the average of the absolute values of the data-model differences relative to the model value, multiplied by 100%.

Dst is a "disturbance" index for which the baseline quiet-time geomagnetic field vector has already been removed from the data before compiling the index. The magnetometer in Hermanus, South Africa, used as an example in Chapter 2, is one of the stations included in the calculation of the global Dst index. The values are typically in the 25 500 nT range, but the typical daily perturbations on this are only in the approximately 10 nT range. If this massive offset were still present in the values, then the operator within the summation would produce values very close to one and the MAPE score would be extremely low. Because it is removed so that the quiet-time Dst index is near zero, the MAPE score divides by this near-zero value and the score is large.

Does this mean that the model is bad? It depends on the usage of the model. If you wanted to use it for accurately reproducing all Dst values within a certain percentage error, including quiet times, then MAPE is the correct metric to consider and it indicates that the O'Brien model is not particularly good at this task. A different model, specifically developed to minimize percent error during quiet times, should be used. If, instead, you cared mostly about predicting the severity of storm intervals, then MAPE is a rather useless metric for this assessment because of the influence of the division by near zero inside the summation, heavily weighting the quiet times.

MAPE uses the model value in the denominator of Equation (9.15). This choice, however, could create a systematic bias to MAPE if the model values are consistently offset from the observed values. In particular, MAPE will underpredict the error when the modeled values are larger than the observed values, and overpredict the error when the modeled values are typically smaller than the observed values. That is, MAPE is not symmetric. To illustrate this asymmetry, consider the example of model value M_i being 5 and a corresponding observed value O_i of 10. Plugging these into Equation (9.15), this yields a contribution of one to the sum. If the quantities are reversed, though, setting M_i to be 10 and O_i as 5, then the contribution to the sum will only be 0.5.

To remedy this issue, another version of this metric has been introduced, SMAPE, the **symmetric mean absolute percentage error**, that uses the average of the local modeled and observed values in the denominator of the summation term:

$$\text{SMAPE} = 100 \frac{1}{N-d} \sum_{i=1}^{N} \left| \frac{M_i - O_i}{(M_i + O_i)/2} \right|$$

(9.17)

As with MAPE, the model degrees of freedom d has been included in the denominator of SMAPE but is often omitted, a simplification that is justified if inequality (9.10) is satisfied.

Applying our illustrative example to Equation (9.17), it is seen that, regardless of whether M_i or O_i is assigned the 5 or 10 value, the contribution to the summation is 2/3. SMAPE is, indeed, symmetric.

The definition of a good SMAPE value is not clear. Like MAPE, some apply a threshold of 100%. A more rigorous approach to this definition follows the thinking behind Equation (9.16), adjusting it for the symmetric correction in the definition of SMAPE:

$$\text{Good SMAPE} < 100 \frac{\min\left[\text{MAD}_M, \text{MAD}_O\right]}{(\bar{M} + \bar{O})/2}$$

(9.18)

The means of the modeled and observed number sets are now averaged in the denominator.

For the example data-model comparison, SMAPE is 740%, a score even worse than that for MAPE. This is because the quiet times values for the observations are fairly evenly

Symmetric mean absolute percentage error (SMAPE): A fit performance accuracy metric, it is the average of the absolute values of the data-model differences relative to the data-model average value, multiplied by 100%.

distributed on either side of zero, while the model values are peaked slightly negative, just below zero, as seen in Figure 9.3. The averaging of an (M_i, O_i) pair where they straddle zero sometimes results in a value very close to zero in the denominator, which then makes a huge contribution to the summation. Equation (9.18) yields a good threshold value of 80%, far smaller than the SMAPE score. As with MAPE, SMAPE is only good for an assessment of this Dst model if the goal of the comparison is to quantify the quiet-time accuracy; the majority of points are near zero so that cluster of points dominates the calculation of MAPE and SMAPE for this model and data pairing.

There is another issue of all of the accuracy metrics mentioned here in Section 9.3—they are susceptible to extreme outliers. This is because all of them incorporate the model–observation difference summed over all of the data-model pairs. For example, let's consider the case of a comparison with 50 pairs of observed and modeled values, with 49 "regular" discrepancy values and one extreme difference. To make it specific, let's say that 99 of the pair discrepancies are exactly 1, but the last one is a value that we will vary for illustrative purposes, let it be 1, 10, 100, or 1000, in four cases, respectively. Simplifying by neglecting d, the three cases yield RMSE values of 1.0, 1.7, 14, and 141, respectively. As you can see, the single extreme difference quickly outweighs the influence of the other 49 pairs and dramatically raises the metric score. This will also be the case for MAPE and SMAPE; they mitigate the issue somewhat but extreme outliers can still wreak havoc on their score, resulting in an assessment that the model is of lower quality than it typically is for nearly all of the data-model pairs. While this example has only one outlier point and therefore it could be justified to simply exclude it as discussed in Chapter 5 (specifically, Section 5.9), this becomes problematic when there is a cluster of points far removed from the central cloud, or when the values naturally spread over several orders of magnitude.

A solution to this problem is the use of median instead of the mean of the model-data comparison. The most common metric of this type is MSA, the **median symmetric accuracy**:

$$\text{MSA} = 100 \left(\exp \left[\text{Median} \left(\left| \ln \left(\frac{M_i}{O_i} \right) \right| \forall i \right) \right] - 1 \right)$$

(9.19)

In Equation (9.19), the absolute value around the natural logarithm operator forces all values over which the median is obtained to be positive (or zero, if the two values match). This absolute value operation ensures that the larger value is in the numerator, flipping the two values if O_i is larger than M_i. The metric is, therefore, identifying the median ratio of all of the data–model-value pairs regardless of whether the modeled or observed quantity is bigger. The exponential operation converts the selected median value back into physical ratio. The subtraction of one makes a perfect MSA score equal to zero, which will only happen if over half of the data-model pairs are identical matches. The final step in the calculation is a multiplication by 100 to convert the resulting fraction into a percentage; if the median difference is a factor of two, then the resulting MSA score will be 100%.

Like the other two metrics in this section, there is no accepted guideline for declaring an MSA score to be good. As discussed in Section 9.3.3, one easy option is to use a threshold of 100%. A more robust threshold that takes into account the number set

---

Median symmetric accuracy (MSA): A fit performance accuracy metric, it is the exponential of the median of the absolute value of the log of the model-to-data ratio, multiplied by 100%.

distributions is to use the medians and interquartile ranges of the two number sets, like this:

Good MSA < 100

$$\left(\exp\left(\min\left[\left| \ln\left(\frac{IQR_M}{M_{median}} \right) \right|, \left| \ln\left(\frac{IQR_O}{O_{median}} \right) \right| \right] \right) - 1 \right)$$

(9.20)

This uses the functional form of MSA but with the IQR and median of individual number sets being compared instead of the model values to the data.

For our example case of the O'Brien model prediction of Dst, MSA cannot be calculated. This is because the values could straddle zero, with the M_i and O_i values having opposite sign, which makes the argument of the natural logarithm operator negative. This reveals a shortcoming of MSA—all of the values in both number sets must have the same sign. Taking the absolute value of the individual numbers is not appropriate, as this would reduce the true distance between the values. That is, if one value were 3 and the other −3, taking the absolute value of each separately would make them contribute identically to the natural log argument, which would artificially lower the median of the number set of natural log values. To apply MSA to the Dst comparison of our running example, it can only be applied to a subset of the full month of data that meet the "same sign" requirement. We'll get to subsetting later, in Sections 9.9 and 9.10.

> **Quick and Easy for Section 9.3.4**
>
> There are additional accuracy metrics that are sometimes better than the common metrics, especially when the number sets have non-Gaussian distributions or span orders of magnitude.

9.3.5 Choosing the Right Accuracy Metric

While there are many more accuracy metrics that could be presented and discussed, those covered in this section provide a substantial set for a multifaceted evaluation of the accuracy of the model at reproducing the observations. But which metrics are most appropriate for a specific assessment? This is entirely up to you; there is no magic formula for this selection. However, there are some guidelines. In general, a good assessment will use at least one accuracy metric. While this seems obvious, as we go through the other categories, it will become clear that the plethora of options means that it is possible to conduct a data-model comparison without choosing one from this category. Because the accuracy category of metrics addresses the overall closeness of the modeled values to the observed ones, it is usually good to select at least one from this category. In fact, a nearly perfect accuracy metric indicates that you can, perhaps, forego other metrics calculations. For example, if RMSE is much less than either value set's standard deviation, by an order of magnitude or more or whatever your desired level of quality for the comparison, then it can be argued that, for your purposes, the model output values are "very close" to the observations everywhere. Further metrics calculation will, most likely, simply confirm this finding.

Which one, or several, to choose depends on the question being asked about the data-model comparison, and also depends on the distribution of data and model values. If the distributions are fairly Gaussian, then RMSE is an excellent choice, especially when normalized against one or both standard deviations to provide context to its quality. This is basically what is done with the F statistic in Chapter 7. If the values are not normally distributed, then MAE might be a better choice, revealing the average data-model difference. If the distributions span several orders of magnitude, then SMAPE is a great choice because it is a relative difference inside the summation. If the distribution is highly non-Gaussian, then MSA is probably an even better choice because it excludes the outliers from the calculation, relying only on the median ratio between the data and model values. Comparing RMSE to MAE is a useful step because the separation between these two metrics reveals the influence of outliers, which are highlighted in RMSE due to the squaring operation. Similarly, comparisons between SMAPE and MSA reveal the same thing, with SMAPE susceptible to the influence of outliers, while MSA is not. However, conducting more than two accuracy metric calculations does not provide more insight into the model performance. Your analysis time is better spent calculating metrics from other categories, which will reveal different strengths and weaknesses about the modeled data.

> **Quick and Easy for Section 9.3.5**
>
> It is important to consider the eventual usage of the accuracy assessment in order to choose the most appropriate metric. You don't have to settle for RMSE.

9.4 Bias

In Chapter 8, it was discussed that the category of bias reveals the difference of the centroids of the two value sets. Figure 9.7 shows an illustrative scatterplot noting the aspect of the data-model relationship being assessed by metrics in the bias category. The centroid value used in a bias metric can be defined in a number of ways, and there are also several ways to define "difference." While there are many options, there are two common metrics in this category for geosciences.

9.4.1 Mean Error

The first metric, and by far the more common, is **mean error**, ME. This is a bit misleading to use the singular here because, as seen in this

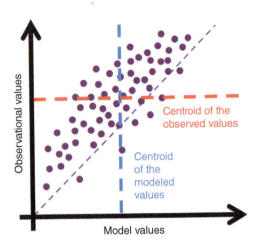

Figure 9.7 Schematic highlighting the property of the data-model comparison being assessed by a bias metric.

section, there are several versions of this metric available. The basic form of ME is simply the difference of the two arithmetic means:

$$\mathrm{ME} = \bar{M} - \bar{O} \qquad (9.21)$$

Equation (9.21) is written with the model mean first and the observational mean second. This order matters, and sometimes ME is defined with the opposite order, causing confusion. With this definition given here, positive values of ME indicate that the model, on average, overestimates the observations, while negative values of ME indicate that the model underestimates the observations. I find this more intuitive than the reverse definition case, where a positive value would indicate underestimation.

Note that Equation (9.21) can be expanded and written in this form:

$$\mathrm{ME} = \frac{1}{N}\sum_{i=1}^{N}(M_i - O_i) \qquad (9.22)$$

Mean error (ME): A fit performance bias metric, it is the difference of the arithmetic averages of the model and data number sets.

There is a striking resemblance of Equation (9.22) to the formula for MAE, Equation (9.7). One difference is the exclusion of the degrees of freedom from the denominator, which is often omitted from MAE anyway. The second difference is the removal of the absolute value operator inside the summation. This is a big change. For MAE, model values M_j and M_k that are equal but opposite in their offset from the corresponding observations O_j and O_k will have equal nonzero contributions to MAE. For ME, those same two data-model pairs will have equal magnitude but oppositely signed contributions to ME, therefore canceling out.

One assumption in Equation (9.21) is the usage of arithmetic mean for the centroids. While this is by far the most common choice, Equation (9.21) could be written with other choices for the centroid. If the value sets are far from Gaussian, then it might be more appropriate to use median instead of arithmetic mean, resulting in **median error**:

$$\text{ME}_{\text{median}} = \text{Median}(M_i \forall i) - \text{Median}(O_i \forall i) \quad (9.23)$$

With the choice of Equation (9.23), outliers play no role in determining bias and only the midpoint values of each set are used. If the value sets span several orders of magnitude, then the geometric mean could be a more informative choice, a metric called the **geometric mean error**:

$$\text{ME}_{\text{geometric}} = \sqrt[N]{\prod_i M_i} - \sqrt[N]{\prod_i O_i} \quad (9.24)$$

Median error (ME$_{\text{median}}$): A fit performance bias metric, it is the difference of the medians of the model and data number sets.

Geometric mean error (ME$_{\text{geometric}}$): A fit performance bias metric, it is the difference of the geometric means of the model and data number sets.

When N gets large, this calculation becomes impossible. Fortunately, Equation (9.24) is equivalent to taking the logarithm of the values, calculating the arithmetic mean of this modified data set, taking the difference of these two means, and then converting back with an exponential operation:

$$\text{ME}_{\text{geometric}} = \exp\left(\frac{1}{N}\sum_{i=1}^{N}(\ln(M_i))\right) \\ - \exp\left(\frac{1}{N}\sum_{i=1}^{N}(\ln(O_i))\right) \quad (9.25)$$

Equation (9.25) is easily calculated, with natural logarithms in the summation rather than trying to calculate N-root operators in Equation (9.24). Because orders of magnitude are reduced to single digit differences by the natural logarithm, the geometric mean error is less susceptible to outliers compared to the arithmetic differences inside the summation in Equation (9.22).

For our example of the O'Brien Dst prediction formula, these equations yield an ME value of -1.4 and an ME$_{\text{median}}$ value of -2.4. A geometric mean error cannot be calculated for this data-model comparison because the numbers within the sets straddle zero and the natural log operator becomes undefined. As discussed qualitatively in Section 9.3 with respect to Figure 9.5, both of these mean errors quantitatively indicate that the model values are systematically lower than the observed values. Perhaps a more descriptive word than "lower" would be "more negative" because the Dst index reveals active times by dropping from near zero to large negative values. This comparison indicates that, for the chosen month of values being compared, the observed values are systematically less disturbed than the model is predicting. This seems counterintuitive from a cursory examination of Figure 9.3, but these measures of bias are dominated by the center of the distribution, not the outliers at the very negative values. That is, bias is sometimes hard to see in the scatterplot, so a calculation of a bias metric is needed to quantify this aspect of the data-model relationship.

ME represents a systematic offset of the model values from the observations. It does not say anything about the spread; we'll get to that next. All of the differences in the ME equations in this section are signed, with underestimations canceling overestimations. Therefore, a perfect score for all of these ME variations is zero.

What constitutes a "good" value for ME? It should be "close to zero," but how close? As with the accuracy metrics, comparing ME with a reference value is useful. Any of the options used for RMSE and MAE from Section 9.3.3 can be applied to ME, resulting in a normalized mean error:

$$n\mathrm{ME} = \frac{\bar{M} - \bar{O}}{\Lambda} \qquad (9.26)$$

In Equation (9.26), Λ can be standard deviation, MAE, interquartile range, full range, or the other choices mentioned in Section 9.3.3. In this case, values within ±1 could be considered good. This, of course, depends on your application, and this criterion for declaring a good bias value might be too lenient.

Applying these normalizations to our running example all yield nME values close to zero. Dividing ME by the smaller of the two standard deviations (24 nT) is nME = −0.06, and use of the smaller of the two MAD scores (14 nT) yields nME = −0.1. For the median error, use of the smaller of the two IQR values (12 nT) gives $n\mathrm{ME}_{median}$ = −0.2. All are close to zero, indicating a small bias between the data and model number sets.

Equation (9.26) is very similar to the t test equations in Section 6.1 (Chapter 6). That is, it has the difference of two number set means in the numerator and a measure of spread in the denominator. Therefore, the t tests are essentially measures of bias, with certain assumptions built into the formulas. Furthermore, there are assumptions built into the interpretation of t tests, especially that the number sets are Gaussian. This is implied in the use of standard deviations in the t test formulas and it is incorporated into the conversion to probabilities that the two means are the same. If the number sets are Gaussian, then any of the t test formulas from Chapter 6 are legitimate bias metrics.

Another option for normalizing mean error is to use the average of the mean values themselves, making it a relative mean error:

$$\mathrm{ME}_{relative} = \frac{\bar{M} - \bar{O}}{0.5(\bar{M} + \bar{O})} \qquad (9.27)$$

This has the advantage of basing the quality of the mean error on the mean values, but this could lead to problems if the means are close to zero. It is even worse if they straddle zero, producing a very large relative mean error. Mean values close to zero are definitely possible in geophysical quantities, such as when considering a single component of a vector quantity such as wind or magnetic field. Therefore, a slight modification to Equation (9.27) is to use the absolute values of the means in the normalizing factor in the denominator:

$$\mathrm{ME}_{relative} = \frac{\bar{M} - \bar{O}}{0.5(|\bar{M}| + |\bar{O}|)} \qquad (9.28)$$

It is best to use Equation (9.28) instead of Equation (9.27), because this new version removes most of the possibility of a zero denominator. It could still happen if both the observations and the model values have a mean of exactly zero, though, in which case this normalization is not needed anyway because ME is very close to zero.

Using Equation (9.28) with the running example, we get $\mathrm{ME}_{relative}$ = −0.08. This number is well within ±1 and indicates that the difference in the biases is small compared to the mean values themselves.

> **Quick and Easy for Section 9.4.1**
>
> Mean error is by far the most common bias metric, but there are several ways to calculate it, so be sure to specify which formula was used.

9.4.2 Percentage Bias

The second most common metric for bias is SSPB, the **symmetric signed percentage bias**:

$$\text{SSPB} = 100 \left(\text{sign}\left[\text{Median}\left(\ln\left(\frac{M_i}{O_i}\right) \forall i \right) \right] \right.$$
$$\left. \left(\exp\left[\left| \text{Median}\left(\ln\left(\frac{M_i}{O_i}\right) \forall i \right) \right| \right] - 1 \right) \right)$$

(9.29)

Taking a close look at the terms in SSPB reveals that Equation (9.29) is quite similar to MSA, Equation (9.19). The key difference is the change in the placement of the absolute value sign within the exponential term. Instead of being within the natural logarithm around the M_i/O_i ratio, it is outside of the median operator. This allows the log values to remain positive and negative, depending on whether the model overestimates or underestimates the observed values, respectively. The ideal situation is that the median of these ratios is at or close to one, yielding a natural log of zero. This would yield an exponential value of one. If the median value is not exactly zero, the absolute value is taken at this point, forcing the value to be positive before the exponentiation. This operation is what gives SSPB its symmetry; and essentially flips the model-over-observation fraction to force the larger to be in the numerator. The resulting exponential value then has one subtracted from it to make the perfect SSPB score equal to zero. The sign operation in front of the exponential retrieves the information about whether the median M_i/O_i ratio was above or below unity. The final operation is a multiplication by 100 to convert the value into a percentage.

SSPB has the same sign directions for overestimation and underestimation as ME in Equation (9.21). That is, a positive SSPB reveals a systematic overestimation of the observations by the model, while a negative SSPB indicates that the model is systematically underestimating the observations. I have never seen it defined the other way with the observational value O_i in the numerator.

For our running example, SSPB cannot be calculated. Like with MSA and the geometric mean error, this is because the Dst values within the two number sets straddle zero, making the natural logarithm operator undefined.

Good values of SSPB are similar to those of MSA. Because it is converted into a percentage, an SSPB of 100 means that the median offset between the value sets is a factor of two. Whether this should be considered good or not depends on the quantity being investigated. For energetic charged particle fluxes, this might be fine. Such values range over orders of magnitude and a model that predicts them within a factor of two would be considered quite good. For temperature in Kelvin (an absolute scale) in a particular location on Earth, this would be terrible, because the record hottest and record coldest temperatures anywhere on the surface of the planet are within a factor of two of each other.

One way to think of bias is that it is a component of accuracy. That is, a modeled value set can have near-perfect bias and poor accuracy. It cannot be the other way around. If the accuracy is excellent, this implies that the bias is also excellent. If the bias is excellent, this does not imply anything about accuracy; the model still could be way off from the observations because the spread of the model around the observed values is bad. This brings us to the next metric category—precision.

Symmetric signed percentage bias (SSPB): A fit performance bias metric, it is the exponential of the absolute value of the median of the log of the model-to-data ratio, multiplied by 100% and the sign of that median value without the absolute value operation.

Quick and Easy for Section 9.4.2

SSPB is a good bias metric that pairs well with MSA as the accuracy metric.

9.5 Precision

Along with bias, the complementary contribution to accuracy is precision. This is a measure of the "spread" or "tightness" of the model values to that of the observed values, all done with the bias removed from the calculation. Like bias, precision measures a component of accuracy, the part specifically excluded in the bias calculation. Therefore, an excellent accuracy metric score implies that precision is also excellent, but the reverse statement is not true. Because the bias could be large, a modeled value set could have a great precision metric score but a poor accuracy. Figure 9.8 illustrates the aspect of the data-model relationship that precision metrics measure. It is the same set of points, simply translated to different offset locations relative to the axes. As the header states, most precision metrics would declare all four of these data-model paired sets to have the same precision. That is, the offset is discarded from the comparison, and the only thing that matters for precision is the relationship between the two number set spreads.

9.5.1 Modeling Yield

There are several ways to construct a comparative measure of spread, but arguably the most common metric in this category is one called **modeling yield**, YI:

$$YI = \frac{\max(M) - \min(M)}{\max(O) - \min(O)} \quad (9.30)$$

Equation (9.30) is simply the ratio of the range of the model output to the range of the observed values. A perfect score is one, where the two ranges are equal. Note that they do not have to span the same number set, but

---

Modeling yield (YI): A fit performance precision metric, it is the ratio of the ranges of the model and data number sets.

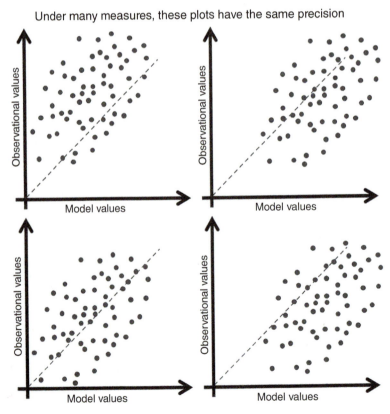

Figure 9.8 The data-model comparison schematic focusing on the quantity being assessed by precision metrics.

could be shifted from each other. That is, even if the centroids of the two value sets are quite different, the yield might be very good if the spread of the modeled value set around its centroid matches the spread of the observational value set around its centroid. By putting the modeled range in the numerator, YI values below one indicate that the model is underestimating the spread of the observations and, conversely, YI values above one reveal that the model overestimates the observed spread.

For our example of predicting Dst with the O'Brien model, we get YI of 0.64. This is not particularly high and indicates that the model range is substantially smaller than the observed range. This is seen in the scatterplot in the upper panel of Figure 9.3; the purple dots of the individual data-model pairs are below the diagonal line at the lower-left extreme of the plot yet above the diagonal line at the upper-right extreme. Being below the diagonal line indicates that the model overestimates the observed values—the observations are more negative than the model output—and the model did not fully reproduce these most active of the observed values. At the other end, the points being above the diagonal line indicate the model underestimates the observations—the observations are more positive than the model output—and again the model did not fully reproduce the extreme of the observed distribution. Even though the model output follows the diagonal line for most of the range, these two excursions away from the diagonal line are highlighted by this metric.

One problem with YI is that it highly favors the maximum value of each number set when the values span more than an order of magnitude. When the maximum values are much larger than the minimum values, then the equation reduces to a ratio of these two numbers. One way around this is to use the logarithms of the extrema instead of the values themselves, resulting in the **logarithmic modeling yield** metric:

$$YI_{log} = \frac{\log[\max(M)] - \log[\min(M)]}{\log[\max(O)] - \log[\min(O)]}$$

(9.31)

Any base will work in Equation (9.31), but the most convenient is log-base-10, so that each integer step represents a decade. The perfect score is still unity, but now the range differences can easily account for order of magnitude differences. In fact, YIlog emphasizes an order of magnitude comparison.

For our running example, YI_{log} cannot be calculated because the minimum numbers for both sets are negative. If all four extreme values were negative, then absolute values could be applied and the calculation could be performed and result in a meaningful metric score. All four extreme values must be on the same side of zero for this metric to work.

Both of these versions of the YI metric use only four values from the two number sets. In fact, it only uses the four most extreme values. Therefore, YI is highly susceptible to outliers and a single point far from the rest will significantly influence this metric. Moreover, the range of a number set increases with more values in the set; it will never get smaller by adding more points but only ratchets upward. As N increases, so will the range, and perhaps in an unpredictable manner. Even if the number set is Gaussian, the probability that the full data set includes a legitimately recorded extreme value from the tail ends of the Gaussian increases with each new value added to the number set.

Logarithmic modeling yield (YI$_{log}$): A fit performance precision metric, it is the ratio of the ranges of the model and data number sets, where all of the extrema values are modified by a logarithm operator.

Quick and Easy for Section 9.5.1

Modeling yield is perhaps the best known precision metric, but it has limitations to its usage.

9.5.2 Definitions of Precision Using Standard Deviation

An alternative precision metric to YI is a comparison of the standard deviations of the modeled and observed number sets. These formulas are a summation over all data-model pairs, therefore including not only the extreme values but also everything else in the two number sets as well. There are two simple ways to formulate this relationship. One option is to write it as the ratio of the standard deviations, giving the **precision ratio**:

$$P_{\sigma,\text{ratio}} = \frac{\sigma_M}{\sigma_O} \qquad (9.32)$$

This version has the same interpretation as YI, with unity being a perfect score, scores above one indicating model overestimation of the observed spread, and scores below one revealing model underestimation of the observed spread. The other way to write it is as a difference of the two standard deviations, yielding the **precision difference** metric:

$$P_{\sigma,\text{diff}} = \sigma_M - \sigma_O \qquad (9.33)$$

This version has the same interpretation as ME, with zero as the perfect score.

The O'Brien Dst model has values of $P_{\sigma,\text{ratio}} = 0.77$ and $P_{\sigma,\text{diff}} = -7.2$ nT. The ratio score is less than one and the difference score is negative; both of these indicate that the model output is not as spread out as the observed number set. This was seen in the overlaid histograms of Figure 9.3 with the narrower peak of the modeled distribution compared to the observed histogram. These metrics quantify that difference in spread in slightly different ways.

Both versions of P_σ are useful in their own way. Equation (9.32) is nice because the metric is unitless and allows for the intercomparison of precision scores across disparate quantities. Like YI, a score of 2 for P_σ, ratio indicates that the spread of the model is twice that of the observations. That could be a relatively good score for a quantity ranging over orders of magnitude, but would be quite poor for other quantities with smaller ranges. Equation (9.33) is good because it has the units of the original quantity and can be compared against most accuracy and bias measurements, such as RMSE and ME. This similarity provides a means of interpreting the quality of the P_σ, diff score.

A final metrics score is simply the standard deviation of the $M_i - O_i$ data-model discrepancies. This can be compared against the standard deviations of each of the two number sets. If it is lower than both of the original number set standard deviations, then it is a good score. If the discrepancies are not a Gaussian distribution, then other measures of spread from Chapter 4 could be used.

Precision ratio ($P_{\sigma,\text{ratio}}$): A fit performance precision metric, it is the ratio of the standard deviations of the model and data number sets.

Precision difference ($P_{\sigma,\text{diff}}$): A fit performance precision metric, it is the difference of the standard deviations of the model and data number sets.

Quick and Easy for Section 9.5.2

There are a few precision metrics based on standard deviations of the observed and modeled number sets, specifically a difference and a ratio version.

9.6 Association

And now let us switch to something rather different—association. As discussed in Chapter 8, metrics in this category are not focusing on how close the model values are to the data but rather to ability of the model to

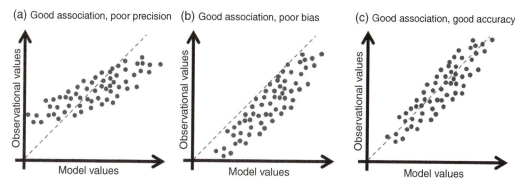

Figure 9.9 Three illustrations of data-model scatterplots with good association but varying quality in other metric categories.

capture local extrema and up–down trends in the observed values. Three scatterplots are shown in Figure 9.9, each with fairly good association but with a different relationship to other metric categories. The first has good bias but poor precision (the observed values have a smaller spread than the modeled values), the second has good precision but poor bias (the observed values have a lower mean than the modeled values). The third panel has good scores for both precision and bias, and therefore good accuracy. All three have good association, though, because the points are fairly closely aligned with a straight-line fit. That is, as the observed values increase, so does the corresponding model output.

9.6.1 Correlation Coefficient

The metric used most widely to understand this has already been introduced in Section 6.2.2 (Chapter 6), the Pearson linear correlation coefficient, or simply the correlation coefficient:

$$R = \frac{\sum (O_i - \bar{O})(M_i - \bar{M})}{\sqrt{\sum (O_i - \bar{O})^2 \sum (M_i - \bar{M})^2}} \quad (9.34)$$

A key feature to note in Equation (9.34) is that the modeled values and observed values are never differenced against each other. This is the important distinction relative to the three categories discussed thus far in this chapter. Instead, the specific O_i and M_i values are differenced against that number set's mean, $\bar{O}$ and $\bar{M}$, respectively. These differences are then multiplied by each other. In the denominator, these differences are squared and the sign of the difference is lost, but not so in the numerator. When the model and data vary away from their mean in the same direction—both above or both below their mean—then the product is positive. If they have opposite differences with their means, then the product will be negative. These products are then summed over all data-model pairs. If the proportion of pairs that trend together (both up or both down relative to their mean) is larger than the proportion of pairs that oppositely trend (one up and the other down relative to their mean), then R will be positive.

Unlike the interpretation of R in Chapter 6, a data-model comparison seeks to maximize R toward unity. Positive values of R indicate that the model is reproducing the up–down trend of the observations, while a negative value indicates that the model is actually predicting the opposite trend from the data. Because we want the model to exactly reproduce the data, a low R score is bad. Although such a model could be useful with a systematic correction applied to the result (i.e., flipping the values about the mean), this type of correction cannot be performed without a theoretical justification if the model is to be of any use. In general, a negative R value is a sign of a poor model.

For our running example of the O'Brien model for Dst, we have R = 0.91. This is a very good correlation coefficient and, from this metric alone, we would declare the model to be reasonable at predicting the up and down trends of the observed number set.

What constitutes a good value for R? Just like in Chapter 6, if the number sets are close to Gaussian, then a significance test can be conducted with a 10%, 5%, or 1% threshold, providing a value of R above which the relationship is considered statistically significant. This threshold depends on the number of data-model pairs N in the assessment, and the threshold for significance decreases as N increases. This can lead to rather low correlation values being deemed statistically significant and some might then ascribe meaning to the correlation. The significance test is really only a guideline. It is a necessary but not sufficient condition for determining a high-quality result (again, only if the sets are close to Gaussian). If R is below this threshold of statistical significance, then it should not be considered particularly good. Values above it, though, require a bit more thought. You must take into account the context of the number of value pairs, the uncertainty of the data and model values, and the situation being assessed. For some applications where excellent models already exist, an R value might have to be 0.9 or above to be considered worthy of attention. Another possible "goodness" threshold could be 0.7, a value for which the coefficient of determination, R^2, is 0.5 (again, see Section 6.2.2, Chapter 6). An R value above 0.7 means that the model is able to explain over half of the variation in the observations. Another good general rule about correlation coefficient significance is to use a threshold of 0.5. Any of these additional criteria on R can be justified, depending on the context.

Because we are discussing R^2, the coefficient of determination, for completeness it is useful to list this as a separate metric:

$$R^2 = \frac{\left[\sum(O_i - \bar{O})(M_i - \bar{M})\right]^2}{\sum(O_i - \bar{O})^2 \sum(M_i - \bar{M})^2} \quad (9.35)$$

Like R, R^2 also has a perfect score of one. Because all terms are squared, its lower bound is zero, and values of R^2 are always closer to zero than the corresponding value of R.

> **Quick and Easy for Section 9.6.1**
>
> The Pearson correlation coefficient, based on covariance and a linear fit between the number sets, is by far the most used association metric.

9.6.2 Nonlinear Association Metrics

There are several cousins to the Pearson linear correlation coefficient; the two to be mentioned here were both introduced back in Chapter 6. One well-known alternate form is the Spearman rank-order correlation coefficient, or just the Spearman coefficient (see Section 6.2.3, Chapter 6). This metric uses the same formula, but instead of including the values and value means, the data and model values are each sorted according from lowest to highest, and the rank-ordering value is used instead of the actual model value:

$$R_S = \frac{\sum\left(\text{rank}(O_i) - \overline{\text{rank}}\right)\left(\text{rank}(M_i) - \overline{\text{rank}}\right)}{\sqrt{\sum\left(\text{rank}(O_i) - \overline{\text{rank}}\right)^2 \sum\left(\text{rank}(M_i) - \overline{\text{rank}}\right)^2}} \quad (9.36)$$

When assigning the ranks, the original pairings must be remembered. It is usually the case that one or the other number set is already in rank order. The data and model values themselves are no longer used, but rather each is replaced by its rank order within its number set. A perfect data-model comparison would have all of the ranks align between O_i and M_i, and so the perfect score is, like R, one.

In Equation (9.36), $\overline{\text{rank}}$ is the average rank value, $(N + 1)/2$, which is the same for both data and model. Because of the transformation to rank order, the two terms in the denominator are equal if there are no repetitions in either value set. The interpretation of this correlation coefficient is exactly the same as that for R—the ideal value is one and a score near zero or even lower indicates a truly poor model at reproducing the variations of the observations.

For our running example, the O'Brien Dst model has a Spearman rank-order correlation coefficient of 0.73. This is not as good as the R = 0.91 Pearson linear correlation coefficient value. The reason is because of the large number of nearly zero quiet-time points; the scatter in this dense cluster is rather random and the rankings can be very different, even for small absolute differences of just a few nT. This is seen in the rank-order scatterplot shown in Figure 9.10. For reference, the bin edges from the histogram in Figure 9.3 are listed on the graph, along with the values for a few specific points. The storm interval is only a small part of the full month, and these values are well aligned in the lower-left corner of this scatterplot. The rest of the plot is a rather unformed cloud of dots. In the "normal" correlation coefficient, the dense cluster near zero contributed rather little to the calculation, but when the values are converted to rank order, all points are now equally spaced (by one) from each other, and dense clusters can take up a disproportionately large chunk of the full range. The correlation coefficient for this scatterplot is still relatively high because the first 80 or so points are so well arranged, but the rest of the number set does not look particularly well correlated at all.

There is one more variation of the linear correlation coefficient that should be mentioned, specifically the version in which the values are replaced with their logarithms, sometimes called the logarithmic correlation coefficient (see Section 6.2.4, Chapter 6):

$$R_{\log} = \frac{\sum \left(\log O_i - \overline{\log O}\right)\left(\log M_i - \overline{\log M}\right)}{\sqrt{\sum \left(\log O_i - \overline{\log O}\right)^2 \sum \left(\log M_i - \overline{\log M}\right)^2}}$$

(9.37)

While this metric has the same meaning as R, with a perfect score being one, it is no longer dominated by those values in the highest order of the number set. Because of this, $R_{\log}$ is often much better than R for data and model number sets with values that span several orders of magnitude and when that full range is what is important for the assessment. If the

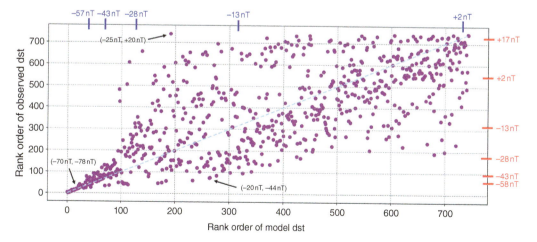

Figure 9.10 Scatterplot of the rank-order values for the observed and modeled Dst values of March 2015. The locations of bin edges from the histogram in Figure 9.3 are indicated along the top and right axes for the model and observed values, respectively. A few specific pairs are indicated within the plot.

observational values of most interest are only those values at the highest end of the range, then R$_{log}$ will misinform you about the quality of the model to reproduce these values because they have been intentionally lowered in significance compared to their influence on R. That is, if the focus of the comparison is on the upper order of magnitude, then it is probably better to use R. R$_{log}$ should be used when values in the lower orders of magnitude are critical to the usage and interpretation of the data-model comparison.

For the example Dst comparison, the R$_{log}$ value is not a valid calculation because the number sets straddle zero.

> **Quick and Easy for Section 9.6.2**
>
> The Spearman rank-order correlation coefficient is an alternative association metric, as is the log version of R, both nonlinearly shifting the locations of the values in a manner that deemphasizes the outliers.

9.7 Extremes

This category assesses the model's ability to reproduce the extremes of the observations. For fit performance, there are several ways to define a metric that does this, because there are several ways to define what is extreme. We'll cover two styles of metric in this category. Additional options for this category are discussed with respect to the discrimination and reliability categories (Sections 9.9 and 9.10), in which only subsets of the model or observation number sets are used. In this case, the subset could be focused on the extreme range of interest; more on this in a bit.

The extremes metrics are not highly used because most assessments focus on getting the bulk of the observed values correct. This metric focuses on the opposite, deemphasizing the data-model pairs near the centroid and highlighting the outliers of the distributions. It can be very useful when the objective of the comparison is to assess rare events. In the fit performance group of metrics, this rare event assessment is done by considering the tails of the distribution of the model output and observational number sets.

9.7.1 Extremes of the Cumulative Probability Distribution

The cumulative probability distribution, or CPD, was introduced back in Chapter 1. It is simply the summation of the histogram of a distribution, starting from one end or the other, normalized by the total number of values, so the scale of CPD ranges from 0 to 1. For a perfect model reproduction of the observations, the two CPD curves would match. We want to focus our attention at the low and high ends of the distribution, that is, the extreme values of the data and model number sets. It is possible to set a threshold value, ε or $1 - \varepsilon$, at the low and high ends of the CPD, respectively, beyond which there is a known number of points, εN. The value at which this threshold is crossed can be easily identified from the CPD curve. At the low end of CPD, it is this:

$$LDE_\varepsilon = M_\varepsilon - O_\varepsilon \qquad (9.38)$$

In Equation (9.38), the quantities of M_ε and O_ε are the modeled and observed values at the ε quantile into that distribution's CPD. The value LDE$_\varepsilon$ is simply the difference of these values. While there is no accepted name for this metric, Equation (9.38) will be called the **low distribution error**. The interpretation of this metric is similar to that of ME—if LDE$_\varepsilon$ is positive, then the model overestimates the value at which this threshold is met. A positive value means that the model has a lighter left-sided

> **Low distribution error (LDE$_\varepsilon$):** A fit performance extremes metric, it is the difference of the model and data values at the given percentage close to the low end of the cumulative probability distribution.

tail relative to the observations, and there are fewer extreme low-end modeled values far from the centroid of the modeled distribution than there are extreme low-end observational values from the centroid of the observed distribution.

At the high end, this metric looks like this:

$$\text{HDE}_\varepsilon = M_{1-\varepsilon} - O_{1-\varepsilon} \qquad (9.39)$$

The $1 - \varepsilon$ value is the quantile of interest for the comparison and is usually set to 0.90, 0.95, or 0.99, depending on the size of the number sets and the specific comparison being conducted. To call it something analogous to its low-extreme partner, Equation (9.39) will be termed the **high distribution error, HDE**.

An example of a case where they do not lie on top of each other is shown in Figure 9.11, progressing from the scatterplot to the histograms to the CPD plot for a representative data-model comparison. In the final panel showing CPD, lines are drawn at the ε and $1 - \varepsilon$ thresholds. It is seen that at the low end of the CPD, the data CPD gets off to a faster rise, indicating that the model is not capturing the full range in the negative end. Similarly, near a CPD of 1, the data CPD reaches unity later than the model curve, indicating that the model is not fully capturing the high end of the observed value range. This type of comparison is common for models; they often miss the extreme values at both ends of the observed range.

The quantities calculated by Equations (9.38) and (9.39) are numbers with the units of the data and model values and therefore directly comparable to other quantities with this unit, such as the means, standard deviations, ME,

---

High distribution error (HDE$_\varepsilon$): A fit performance extremes metric, it is the difference of the model and data values at the given percentage close to the high end of the cumulative probability distribution.

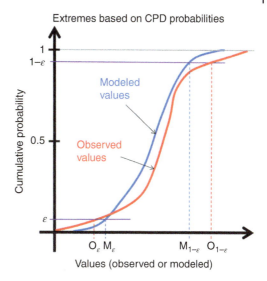

Figure 9.11 Illustrative conversion from a scatterplot to histograms to CPDs for a hypothetical data-model comparison. Only the wings of the distribution are considered for the LDE and HDE extreme metrics, as marked.

MAE, or RMSE. LDE$_\varepsilon$ and HDE$_\varepsilon$ do not compare the differences of the data-model pairs; they assess the tail of the two distributions regardless of how the values are paired together. It could be that the extremes in the model output are not paired with the extremes in the observational number set. That is, the M_i and O_i values within the two number sets could be arranged in any order, but if the two distributions have the same "tails," then the metric scores will be very good. Which one of these two metrics to choose depends on the parameter being considered. For example, for the heat index (based on local temperature and humidity), it should be the high end of the distribution. For wind chill during winter storms, it should be the low end of the temperature scale.

The method used to find quantiles from the CPD, as illustrated in Figure 9.11, is exactly the method used to find values for the quantile-quantile plot format introduced in Chapter 8. In that case, quantile values across the full y-axis range were selected, and the corresponding modeled and observed values at that quantile were paired and plotted against each other as one point on the quantile-quantile plot. Only one of those pairs is needed for LDE or HDE (one near the low or high end of the quantile-quantile curve, respectively), but the process is exactly the same.

To illustrate how CPDs could be compared with these extremes metrics, the CPDs for three Gaussian distributions are shown in Figure 9.12. The three Gaussian distributions have spreads of 0.5, 1, and 2 of some hypothetical baseline standard-deviation value, σ_0, and all three have a mean of zero. In the upper panel of Figure 9.12, it is seen that the CPD curve for $\sigma = 2\sigma_0$ is the "widest" of the three and the curve with $\sigma = 0.5\sigma_0$ is the narrowest. For the extremes metrics of LDE_ε and HDE_ε, most of the CPD curve is ignored; only the very lowest or very highest values matter for this comparison, respectively. As an example, the

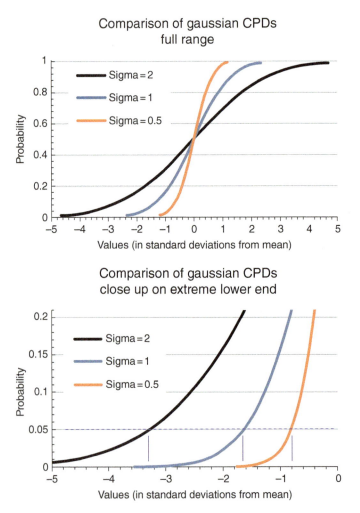

Figure 9.12 CPDs for three Gaussian distributions with different spreads of 0.5, 1, and 2, for orange, blue, and gray lines, respectively. The upper panel shows the full range, while the lower panel is a zoomed-in view near the lower extreme of the probabilities. A LDE_ε line is marked at $\varepsilon = 0.05$ as the horizontal purple dashed line, and the values for the three CPD curves are indicated with the short vertical purple lines.

lower panel of Figure 9.12, therefore, zooms in on the lower range of the probabilities. A LDE$_\varepsilon$ value of 0.05 is marked on this plot, and the values for the three curves are then found by noting the locations of where this threshold occurs for each of the three curves. In this case, these are −3.290, −1.645, and −0.822 for the three curves. The metric score is then the difference between two of these values.

Let's say that the observed data set has a Gaussian distribution. We can then normalize it to have a mean of zero and a standard deviation of one, making it look exactly like the dark blue curve in Figure 9.12. Let's also suppose that a corresponding model number set has the same mean value (i.e., no bias), but is not as variable as the observations (i.e., smaller spread). Let's say it is a Gaussian distribution, but the spread is only half of the observed spread. We can normalize the model output number set with the observational mean and standard deviation and it would look exactly like the orange curve in Figure 9.12. The LDE$_\varepsilon$ score for this case, with $\varepsilon = 0.05$, would then be LDE$_\varepsilon$ = (−0.822) − (−1.645), so +0.823.

A "good" value of LDE$_\varepsilon$ or HDE$_\varepsilon$ is hard to intuitively grasp, so it is useful to consider specific distributions and numbers. Table 9.3 presents values of LDE$_\varepsilon$ for a range of different ε values and different standard-deviation settings of the modeled distribution. Here, the observed distribution is assumed to be Gaussian with standard deviation σ and the modeled distribution is also assumed to be Gaussian but with a standard deviation higher or lower than that of the observed number set. The resulting value of LDE$_\varepsilon$ listed the table is in units of σ. Note that the metric scores in Table 9.3 are listed in units of the σ_O. When the modeled standard deviation is smaller than the observed standard deviation, then LDE$_\varepsilon$ is positive; the modeled left-sided tail is "shorter" than the observed tail. When $\sigma_M > \sigma_O$, then LDE$_\varepsilon$ is negative, indicating that the modeled tail has more numbers at large distances from the median than does the observed

Table 9.3 Values of LDE$_\varepsilon$ in units of the observed data set's standard deviation, as a function of ε setting and modeled standard deviation.

	$\varepsilon = 0.1$	$\varepsilon = 0.05$	$\varepsilon = 0.01$
$\sigma_M = 0.5\,\sigma_O$	0.64	0.82	1.16
$\sigma_M = 0.75\,\sigma_O$	0.32	0.41	0.58
$\sigma_M = 0.9\,\sigma_O$	0.13	0.16	0.23
$\sigma_M = 0.95\,\sigma_O$	0.064	0.082	0.116
$\sigma_M = 1.05\,\sigma_O$	−0.064	−0.082	−0.116
$\sigma_M = 1.1\,\sigma_O$	−0.13	−0.16	−0.23
$\sigma_M = 1.2\,\sigma_O$	−0.26	−0.33	−0.46
$\sigma_M = 1.6\,\sigma_O$	−0.77	−0.99	−1.40
$\sigma_M = 2\,\sigma_O$	−1.28	−1.65	−2.33

distribution. Note that, because of the symmetry of the idealized distributions in this example, the LDE$_\varepsilon$ values in Table 9.3 are equal to corresponding HDE$_\varepsilon$ values, but of opposite sign. By dividing a calculated CPD difference metric score by the observed data set's standard deviation, this table can be used to interpret the calculated CPD differences as if they were two Gaussian distributions of different spreads.

Note that setting $\varepsilon = 0$ selects the minimum values from the two number sets for LDE and maximum value for HDE. These choices are essentially equivalent to the YI metric from association, except that they are assessing only the lower or upper extreme of the number sets. YI, being a difference of maximum and minimum values for each number set (i.e., the full range), is a combination of LDE and HDE with ε set to zero. Because of this similarity, sometimes you will see YI categorized as an extremes metric rather than an association metric.

For our running example of Dst, the important extreme of the distribution is at the low end of the distribution. For the O'Brien model for the chosen month of comparison, Equation (9.38) with $\varepsilon = 0.05$ gives a LDE$_\varepsilon$ value of 15 nT. This positive score means that the model 5% quantile is higher (less negative) than that of the data set; the model does not capture the full extent of the tail of the observed distribution. While this was seen in the scatterplot and histogram comparison of Figure 9.3 and was hinted at in the yield calculation in Section 9.5.1, this is a quantification of that underestimation for a specific quantile of the distribution. Dividing 15 nT by the observed distribution's standard deviation of 32 nT, we get +0.47, which can be compared with the $\varepsilon = 0.05$ column of Table 9.3. It is slightly larger than the value for $\sigma_M = 0.75\ \sigma_O$. Even though the distributions are not particularly Gaussian, this roughly matches our expectation from the two number set standard deviations, which have a ratio of exactly 0.75.

> **Quick and Easy for Section 9.7.1**
>
> One way to calculate an extremes metric is to compare values from the data and model CPDs.

9.7.2 Using Skew and Kurtosis for an Extremes Assessment

An alternative definition for a metric in the extremes category is to use the measures of spread beyond standard deviation, specifically the two have already been introduced in Chapter 5—skew and kurtosis. Skew is a measure of the asymmetry in the outlier values of a distribution, while kurtosis is a measure of the symmetric similarity or difference in the extreme tails of the distribution relative to a Gaussian. Example distributions with different skew and kurtosis coefficients were shown in Sections 5.3 and 5.4 (Chapter 5), respectively, exploring the relationship of these two measures of the tails relative to the core of the distribution.

For skew, one useful metric formula is **skew error**:

$$\gamma_\Delta = \gamma_M - \gamma_O \qquad (9.40)$$

In Equation (9.40), γ_M and γ_O are the modeled and observed skewness coefficients, calculated using the definition of skew for a single data set provided in Chapter 5 (specifically, Equation (5.2)). It has the same properties as ME and other subtractive metrics, with zero being a perfect score, positive values indicating that the model has more extreme outliers in the large-valued tail of the distribution, while a negative score reveals that the model has more extreme outliers in the small-valued tail of the distribution. Note that this metric does not reveal whether the distributions are skewed in the same direction; the two skews could both be highly positive, yet Equation (9.40) could be negative. This indicates that the highly right-sided skew of the observations is underestimated by the model.

For kurtosis, a similar metric can be defined as **kurtosis error**, the difference of the kurtosis coefficients of the modeled and observed number sets:

$$k_\Delta = k_M - k_O \qquad (9.41)$$

The formula for the kurtosis coefficient is given in Equation (5.3) (Chapter 5). This metric has the same meaning, except instead of assessing the model's ability to reproduce the right-sided or left-sided asymmetry of the observed distribution; Equation (9.41) provides an assessment of the model's ability to reproduce the heaviness or lightness of the extreme wings of the observed distribution. The kurtosis coefficient of a Gaussian is 3, but the subtraction removes this offset from the kurtosis metric, and k_Δ can be interpreted just like γ_Δ, with zero being the perfect score.

For our running example of the O'Brien model at predicting the Dst index, the skew and kurtosis error metrics are $\gamma_\Delta = -0.92$ and $k_\Delta = 4.8$. These

---

Skew error (γ_Δ): A fit performance extremes metric, it is the difference of the model and data skewness coefficients.

Kurtosis error (k_Δ): A fit performance extremes metric, it is the difference of the model and data kurtosis coefficients.

---

seem like large numbers; if either of these were an actual skew or kurtosis coefficient, it would be declared to be a large value far from Gaussian. In this case, though, context is important, and the skew and kurtosis error values need to be understood relative to the coefficient values for each data set. Remembering the γ and k values in Table 9.1, both distributions are highly left-skewed with very heavy tails, so it is expected that the difference between the coefficients from the two number sets could also be quite large. The reality is that both of these error metrics are rather small compared to the γ and k values that went into them.

Let's get into the details of the interpretation for the running example. The negative skew error value indicates that the model is more highly left-skewed than the observed values; the positive kurtosis error reveals that the model values have a heavier tail than the observed Dst distribution. These conclusions seem counterintuitive, since the observed values extend to lower negative values during the intense storm event. It should be remembered, though, that the skewness coefficient is defined as the asymmetry of the tails relative to the central peak of that number set, and includes division by the cube of the distribution's standard deviation. Similarly, the kurtosis coefficient includes division by σ^4. The modeled Dst values have a smaller spread than the observed Dst values (see Table 9.1), and this makes all the difference.

To summarize, these two error measures reveal a different aspect of the extremes of the two distributions than that assessed by the CPD difference metrics. The skew and kurtosis coefficients are normalized by the spread of each distribution, so these metrics reveal the asymmetry and heaviness of the tails relative to the each distribution's central peak. The CPD does not take this into account; the central peak can be arranged in any shape, and it will not change the percentile values in either tail. This makes interpretation of the skew and kurtosis metrics more challenging because the value for each number set is not only dependent on the outlier values but also dependent on the core of the distribution.

Interpretation of the quality of these metric scores as good or poor is similar to what was done back in Chapter 5 when skew and kurtosis were introduced. These quantities are already normalized by a standard-deviation term in the denominator. Specifically, since they are already normalized and they are a difference quantity, both metrics can be assessed just like skew, with values within one of zero being relatively good and those beyond this guideline threshold being considered increasingly poor, the farther away from zero the score gets. However, this should be done with respect to the skew and kurtosis coefficients for the two distributions, as we saw in the running example.

A main difference with the CPD-based extremes metrics is that skew and kurtosis use all of the observed and modeled values, rather than only the extreme tails. A key similarity is that the observations and model output are treated independently in calculating their skew and kurtosis values, and then, as with LDE_e and HDE_e, these values are subtracted.

To avoid the issue of unpairing the data and model number sets, another method for finding metrics of this nature is to calculate the γ and k coefficients of the $M_i - O_i$ discrepancy number set. These two γ and k values could then be compared against the γ and k from the data number set and model number set.

A caveat on the interpretation of the skew difference and kurtosis difference metrics is that the normalizations of the modeled γ and k values are different from that of observed γ and k values. That is, the denominators are calculated from the individual number sets, and therefore are scaled relative to its own cluster around its centroid, rather than using the same normalization. So, caution must be taken when using these metrics; context about the two number sets is needed for proper interpretation.

> **Quick and Easy for Section 9.7.2**
>
> Extremes metrics can also be created from the two parameters that assess the tails of the distributions—skew and kurtosis.

9.8 Skill

As noted in Chapter 8, skill is the metrics category that uses the formulas from other metrics and compares it against a reference model score for that metric. Many metrics have been covered so far in this chapter, and any of them could be used in the skill score definition. There are really only two that have received widespread attention and usage in the geosciences, though, so the list of essential formulas that need to be presented here is quite short.

9.8.1 Prediction Efficiency

The first is known as **prediction efficiency**, PE, and is based on MSE as the metric with the variance of the observations as the reference model score. In this case, the perfect score of MSE is zero and the denominator of Equation (8.2) (Chapter 8) is a single summation term, which allows us to rearrange a bit and write PE in its standard form:

$$PE = 1 - \frac{\sum(M_i - O_i)^2}{\sum(O_i - \bar{O})^2} \qquad (9.42)$$

In Equation (9.42), the N in the denominator of variance has been canceled with the $N - d$ denominator of MSE. If you want to keep these terms, you can, but most usages of PE omit them. An exact match of the model and observations for all pairs yields zero in the numerator of the last term, so a perfect score of PE is one with all other PE scores below this. Note that PE can go negative. When the modeled values are systematically farther from the observations than the observations are from their mean, then the last term in Equation (9.42) is greater than one, and the subtraction of this value from one makes PE less than zero. This indicates that the model is worse than the average of the data as a predictor of the observed values.

One reason for the dominance of PE is that it is mathematically equivalent to R^2. This does not look obvious when comparing Equation (9.42) with the R^2 formula, Equation (9.35). A bit of algebra can get you there, though, and the better comparison is with the R^2 formulas in Chapter 7, specifically the second one, Equation (7.45). Indeed, the formulas are identical and it is evident that PE is 1—SSE/SST, as defined in Section 7.2. Note that because PE is written for a general model and not specifically for a fit based on the model, SSE could be larger than SST and therefore PE could be negative.

This raises the question of what is a good value of PE. A score of less than zero is universally considered bad, but this is a rather low threshold for the model. Because it includes terms that resemble correlation coefficient, a general rule is that PE should be at least 0.5 to be considered good, which was the goodness rule of thumb applied to R^2 in Chapters 6 and 7, and in Section 9.6.1. Some argue that PE should perhaps even as high as 0.7 before attaining this descriptor. These general-rule thresholds for declaring the model to be of high quality can be inadequate, though, especially when the observational number set is not Gaussian (and, therefore, probably also the model output trying to reproduce it), in which case variance is not a good measure of the spread of the values. If the values have a strong skew or kurtosis, then the squaring of the differences amplifies the influence of the outliers. This could give a lower PE score than the comparison might otherwise deserve. In this case, it might be useful to apply other accuracy metrics to the general skill score formula (Equation (8.2), Chapter 8) to try to mitigate the influence of outliers on the skill score.

For our running example of the O'Brien Dst model, the PE score for the selected date range

---

Prediction efficiency (PE): A fit performance skill metric, it is a skill score based on mean square error with the observation number set variance as the reference model.

is 0.81. This is a high score and reveals that the model is much better than the climatological average of the observations at reproducing the variations of the index. There is room for improvement, but it is very good.

> **Quick and Easy for Section 9.8.1**
>
> Prediction efficiency, based on mean square error and the observed number set average, is by far the most commonly used fit performance skill metric.

9.8.2 Other Options for Fit Performance Skill

The other common skill score is essentially the same formula as PE, but, instead of using the observational mean as the "reference model," the comparison is done against another model's output. The form of the skill score equation is then this:

$$SS_{MSE} = 1 - \frac{\sum (M_i - O_i)^2}{\sum (M_i^{old} - O_i)^2} \quad (9.43)$$

In Equation (9.43), M_i^{old} are the output values from another model. This could be anything, as discussed in Section 8.5.2. In Equation (9.43), the $N - d$ denominators of MSE for the new and reference models have been omitted, assuming that the degrees of freedom for the two models, d_{new} and d_{old}, are either both small compared to N or similar to each other (as discussed in Section 9.3.2). It has the same interpretation as PE, except that now a score of zero indicates no improvement over the old model, rather than no improvement over the observational average.

For this, "good" is essentially anything above zero, because it shows improvement over the old model. The question to ask, then, is whether the new model introduces more mathematical complexity or computational overhead, in which case a small improvement might not be worth the extra cost of time, resources, or intuitive understanding. Usually, though, the improvement is worth it, and a positive score for SS_{MSE} indicates that the new model should be adopted.

For Dst, a classic prediction algorithm is the predecessor to the O'Brien model, a formula known as the Burton equation. It is a simpler form with a constant restorative term ι and slightly different coefficients. Otherwise, the functional form is the same. A comparison of these two models against the observed index is shown in Figure 9.13. It is seen that the O'Brien

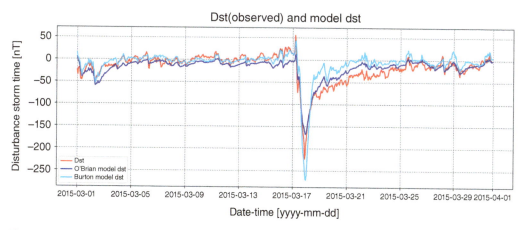

Figure 9.13 Time series of Dst index values with two models compared against the index compiled from observations; the blue line is from the O'Brien model, while the light-blue line is from the Burton equation Dst reproduction model.

model underpredicts the intensity of the storm peak value, while the Burton model overpredicts the depth of the negative Dst excursion. In the recovery phase of the storm, the O'Brien model is closer to the observed index values. During the quiet or low-activity times the rest of the month, the two models appear to be roughly equal in their ability to match the index.

When the O'Brien model is compared against the Burton equation Dst prediction as the reference model, again for the month of March 2015, the S_{SMSE} score is 0.49. This indicates that the O'Brien model is substantially better than the original Burton equation at predicting the Dst index, at least for this interval.

Other formulas that look like Equation (9.43) can be constructed from the metrics discussed in this chapter. Because most metrics have a perfect score of either zero or one, the skill scores have two general forms. As explored in Section 8.5.2 (Chapter 8), the first is when the perfect score is zero, and the formula looks very much like the two skill scores already presented:

$$SS_{perfect=0} = 1 - \frac{\text{Metric}(\text{new model})}{\text{Metric}(\text{reference model})} \quad (9.44)$$

The other common general formula is when the perfect score is equal to one, in which case no reduction of the skill score general formula is possible and the equation becomes this:

$$SS_{perfect=1} = \frac{\text{Metric}(\text{new model}) - \text{Metric}(\text{reference model})}{1 - \text{Metric}(\text{reference model})} \quad (9.45)$$

All of the metrics listed in Sections 9.3–9.7 can be used with the appropriate choice of one of these two formulas.

A final note on using other metrics in the skill score general formula should be made. With MSE, it was possible to use the observational number set mean as the reference model, as is done for PE in Equation (9.42). This reference model choice can be used with most accuracy measures but not with many metrics from other categories. With ME, for instance, using the observational mean for the reference model yields Metric(reference model) = 0, which is also the perfect score value, so the denominator of the skill score equation becomes zero (or, using Equation (9.44), the second term is infinity, yielding the unhelpful skill score of negative infinity). A skill score based on ME cannot be compared against this choice of reference model because this value (observational mean) is built into the metric itself. It works with other reference model choices, just not that one.

A slightly different issue arises with R. In this case, using the observational mean as the reference model yields an R value of 0, because all of the $M_{i,\,ref}$ values are equal to $\bar{M}_{ref}$ and the second half of the numerator of Equation (9.34) is zero for all i indices. For R, however, the perfect score is one, not zero like ME, and so the denominator of Equation (9.45) is one. The problem is that both of the Metric(reference model) terms are gone, including in the numerator, and so the skill score formula reverts to the original metric of R. That is, nothing new is learned with this application. Any other choice of reference model, however, will not have this issue, and a skill score based on R would be a meaningful metric to consider.

Some other metrics have this same issue. So, even though the observational number set mean is applied as the reference model in the most commonly used skill score, PE, this is often not an appropriate choice for most other metrics. If a metric other than MSE matters to your assessment of a model, then it is completely acceptable, and even encouraged, to create a skill score based on that metric of special interest. Care should be taken, however, when constructing that new skill score, to ensure that it will produce reasonable values that you can readily interpret.

> **Quick and Easy for Section 9.8.2**
>
> Skill scores can be created from any of the metrics from the other categories against any reference model choice.

9.9 Discrimination

Discrimination assesses the model's ability to reproduce the observations within a band of observed values. If the metric scores are good for a particular subset, then it means that, when the data are in this selected range, the corresponding model output is most likely trustworthy. That is, the process of discrimination assessment provides a user with the ability to know in what data ranges the model output is trustworthy or not. This is very useful when you are in the field and record a new observation similar to prior data for which a discrimination assessment has been conducted. Based on the new observations you would know whether the model would be able to accurately reproduce this value and therefore be useful to determining the physical, chemical, or biological processes responsible for that observed value.

There are no formulas specifically and uniquely within the discrimination category for fit performance. Remember, as defined in Section 8.5.3 (Chapter 8), this is the category that uses a subset of the data-model pairs, selecting only those pairs that have an observational value within a given range. All of the metrics in this chapter are fully valid, but now the number set is reduced, so the total is no longer N. This also changes any mean or median values embedded in the metric formulas; these should be recalculated for the subset.

In our standard comparison scatterplot with model values along the x axis and observed

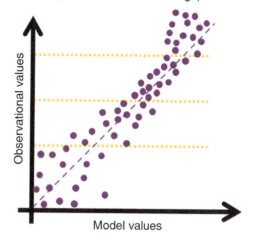

Figure 9.14 Illustrative scatterplot with three horizontal lines indicating discrimination subsetting into quartiles.

values along the y axis, the discrimination category parses the full scatterplot with one or more horizontal break points. This is illustratively shown in Figure 9.14, with three lines drawn dividing the data range into quartiles. The metrics in Sections 9.3–9.8 can then be applied to the points within each horizontal section of the plot. For any one metric calculation, all other points on the scatterplot are ignored; it is as if they did not exist. Because the range is artificially limited for the data but not the model values, the spread in the model values within a discrimination subset will usually be larger than the spread of the data values. Do not be surprised, therefore, when some of the metric scores are worse than those for the full data set. Some metrics, like ME, are linearly additive, and the sum of the metric scores for all of the subsets will be equal to that of the full-range metric score. Others, like MAE, appear to be additive, but this equality is only true for MAE if d is omitted from the denominator; d will be the same for all subsets (it is the same model), so its inclusion makes the sum of the subset MAE scores higher than MAE of the full number set.

The break points can be selected in whatever is most convenient or useful for the assessment. A common method is to split up the data set so that each subset has the same number of values, as is shown in the example plot. It does not have to be quartiles; any fraction can be used. Other common choices are to break the data values in half, into thirds, into quintiles, into deciles, or even into smaller subsets, such as using 20 or 100 bins, partitioning 5% or 1% of the values in each subset bin, respectively. The choice depends on the total number of points, N, and what is reasonable for subsetting bin sizes, as well as the assessment being conducted and the information desired from the comparison. If the number set includes millions of data-model pairs, then 1% bins are perfectly reasonable. For our running example of hourly Dst values over a single month, with $N = 744$, decile bins are about as small as one should go, as anything smaller starts to stretch the data set too thin and could render the subset assessments meaningless because of the uncertainties on the values.

It could be that a specific value is known to be a critical threshold, in which case values above this are important and a separate analysis of only these observations is desirable. For example, this could be rainfall rate, for which there is a value that overtaxes a town's stormwater drain system and localized flooding will occur. It would be useful, then, to assess how well a rainfall rate model is reproducing such observations. For our running example, a possible choice of a critical threshold is Dst = −30 nT, which is a value that has been declared the cutoff for a weak magnetic storm. Dst values above this threshold are considered "nonstorm times," while those below (more negative) this threshold can be given the label of "storm time."

Much can be learned from discrimination subsetting metrics scores. Most importantly, this assessment isolates which parts of the data range contribute the most or least to the full metric score. In the illustrative example shown in Figure 9.14, it is seen that the lower two sections have a lot of spread in the model values, the third section has a much smaller spread, while the top section has a spread somewhere in between. We can qualitatively see this from the plot. Applying metrics to these subsetted regions of the data range, however, quantitatively determines the level of this good or poor reproduction of the data for each of the specific data ranges. While accuracy is often chosen as the sole metric for this analysis, conducting more metrics calculations from the other categories is useful to learn even more about the model's capabilities in each data range.

For our running average, Table 9.4 shows a grid of metrics scores for quartiles of the data range. The ranges are listed in the second row, followed by some selected metrics. Each subset has $N = 186$, which is a large-enough number that the statistics are robust. The cutoff of the first quartile at −25 nT is quite close to the −30 nT threshold that others have defined for a weak storm value, so the first subset is essentially the storm times and the other three are divisions within the quiet-time range of the values. In understanding these subsets, it might be useful to look back at the scatterplot and histogram in Figure 9.3; these subsetting bins were defined by discrete ranges of the observed values along the y axis of the scatterplot.

The comparison of the two accuracy measures, MAE and SMAPE, is quite interesting. For the first quartile, MAE is 17 nT, the largest MAE value of the four subsets, while SMAPE is 48%, the lowest SMAPE value of the subsets. At the other end of the observed data range, in the fourth quadrant, both accuracy values are rather large. This explains that, in absolute terms (shown by MAE), the model is not particularly good at the two ends of the distributions, but in relative terms (shown by SMAPE), the error is lower for more active times. That is, according to this interpretation of SMAPE scores, the model is quite good during storm intervals, but is not that useful at determining specific quiet-time Dst values.

Table 9.4 Discrimination subsetting metrics for the running example of Dst data-model comparison, with the observed Dst range divided into four equal-numbered subsets.

	[−223 nT, −25 nT]	[−25 nT, −9 nT]	[−9 nT, 2 nT]	[2 nT, 56 nT]	Full range
MAE	17 nT	7.9 nT	6.8 nT	15 nT	12 nT
SMAPE	48%	53%	126%	2770%	740%
ME	14 nT	0.84 nT	−5.4 nT	−14.9 nT	1.4 nT
YI	0.83	3.2	2.9	0.88	0.64
R	0.95	0.11	0.30	0.30	0.91
γ_Δ	0.49	−0.71	0.21	−1.4	−0.92
k_Δ	−2.6	1.8	1.24	−1.2	4.8

The centroid and spread metrics of ME and YI complement this analysis. The mean error is worst at the two end quartile subsets, but the spread is best in these two subsets. When the Dst index is near zero, the model values are somewhat randomly sprinkled around the observations, which leaves us with a decent centroid comparison, but a poor spread comparison.

The metrics scores for association are listed next in Table 9.4. It is seen that the index trends in the storm intervals, included in the first discrimination subset, are very well reproduced by the model, but the R scores for the other three subset quartiles are not good. This is yet more evidence that the model is particularly good at times that are highly driven by the solar wind—when E_y is positive and large—while at other times it is not.

The final two rows of Table 9.4 show the values for the extremes metrics of skew error and kurtosis error. Remember that the data values are limited for each subset and are within the specified range, while the model values are those that correspond to these observed times in the series. That is, we might expect the modeled values to have a larger range and therefore more asymmetry and heavier tails. There is only one γ_Δ value that is more than one away from zero—the fourth quadrant in the positive Dst range—and the value is negative, indicating that the model underpredicts the right-sided asymmetry of the observed distribution in this subset. This is further evidence that the model does not do well in this range of the observed index values. For the k_Δ metric, it is positive in the center two subsets, as expected, but negative in the first and last subsets. Again, this is reflective of the fact that the model does not get the extreme values particularly well.

Having conducted a discrimination subsetting assessment, the understanding gleaned from that analysis could be useful given a new set of similar observations. The observed values will fall within one of the subsets, allowing you to judge whether to trust the accompanying model output.

> **Quick and Easy for Section 9.9**
>
> Any of the fit performance metrics can be calculated over a subset of the full number sets. When the subset is defined by a range of the observed values, it is called discrimination.

9.10 Reliability

Reliability is the other common subsetting metric category and is the converse of discrimination. That is, for reliability, subsets are based on selecting data-model pairs within

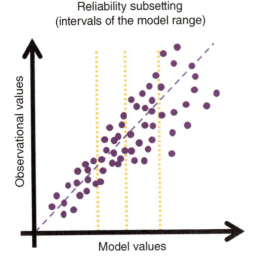

Figure 9.15 Illustrative scatterplot with three vertical lines indicating reliability subsetting into quartiles.

Reliability assesses the model's ability to reproduce data based on the model values. Good metrics scores for a particular subset reveal that when the model is producing a value in that range, then the data will most likely be close to that model value. That is, it determines the model ranges for which the model is useful for data reproduction or prediction. Large spreads or offsets in the observations within a particular subset reveal that the model is untrustworthy for that model-value range.

In the example shown in Figure 9.15, it is seen that first quartile is exceptionally good with the spread of the data in each vertical stripe getting worse with increasing model value. Conducting metrics for these subsets will quantify this change in goodness as a function of model value. For this case, the model is good for low model values but not for high model values.

For our running example of Dst reproduction with the O'Brien model, Table 9.5 shows the reliability metrics for quartile subsetting according to the model output values. The same seven metrics are listed here that were given in Table 9.4 for discrimination subsetting; this time, though, the subsets are "vertical swaths" through the scatterplot in Figure 9.3 at the horizontal model-value ranges given along the top row.

specific model-value ranges. Figure 9.15 shows this "vertical line" subsetting for the same illustrative example as in Figure 9.14 for discrimination. All of the same guidelines apply to reliability that were listed in the previous section for discrimination regarding how to subset and what metrics to calculate. This time, though, the model values are confined to a limited range, while the corresponding observed values can be anything.

Table 9.5 Reliability subsetting metrics for the running example of Dst data-model comparison, with the modeled Dst range divided into four equal-numbered subsets.

	[−167 nT, −18 nT]	[−18 nT, −11 nT]	[−11 nT, −7 nT]	[−7 nT, 13 nT]	Full range
MAE	14 nT	11 nT	11 nT	11 nT	12 nT
SMAPE	37 %	583 %	1570 %	858 %	740%
ME	5.0 nT	−0.97 nT	−4.1 nT	−5.4 nT	1.4 nT
YI	0.69	0.11	0.065	0.19	0.64
R	0.93	0.26	0.21	0.20	0.91
$\gamma\Delta$	−0.24	0.024	0.59	1.4	−0.92
$k\Delta$	−0.025	−0.33	−1.54	1.2	4.8

The overall takeaway from the reliability metrics in Table 9.5 is similar to that from Table 9.4, but with some noteworthy differences. The accuracy metrics follow the same trend; the first subset has the largest MAE, but the smallest SMAPE. The bias scores for these reliability subsets are also similar to the discrimination bias scores, with the first subset being positive—the model values in this subset are systematically higher than the observed values—and the other subsets are near zero and/or negative. Yield and correlation are also similar, with the first quadrant having a relatively good metric score, while the other three subsets have rather poor scores for these two metrics. The extremes metrics are where the difference with discrimination is most visible. For the first two categories, both γ_Δ and k_Δ are exceptionally good, indicating that the model values, even with the truncated nature of the subsetting, capture the observed variation in these model value ranges. For the last subset, both values are positive, indicating the model values have a higher right-sided skew and a heavier tail in this model-value range than does the distribution of the corresponding observed Dst values. This was not expected, but can be easily explained by considering the standard deviations of the observed and modeled number sets within this model-value subset interval. The observed values have a much larger spread, with a standard deviation of 12 nT, compared to the modeled values that have a standard deviation of only 3 nT. The skewness and kurtosis coefficients for each number set are calculated relative to these values, so the values make sense in this context.

Another assessment that can be conducted with subsetting, in particular for our running example of Dst where the values straddle zero, is to perform the assessments for those metrics that require single-signed number sets. A convenient subset to choose is the lowest quartile of the reliability calculations, which has model values from −167 to −18 nT and corresponding observed Dst values from −223 to −6 nT. With all values negative, the ratios within MSA and SSPB will yield positive arguments for the natural logarithm operators. For other metrics with logarithms of individual values, such as ME_{geo}, YI_{log}, and R_{log}, absolute values can be used because all numbers are on the same side of zero. For this subset of the full number sets, these otherwise undefined metrics can be calculated.

For accuracy and bias, MSA is 33% and SSPB is −41%. The MSA score is rather close to the SMAPE score listed in Table 9.5, which indicates that the median error is of similar magnitude as the average relative error recorded by SMAPE. The SSPB score is negative, however, which is opposite of the mean error score for this subset. The problem is that our normal understanding of SSPB, which is identical in interpretation as ME, is reversed when all values are negative. That is, the negative signs cancel within the natural logarithm operator, and so a negative SSPB, for these two completely negative number sets, indicates that the observations are more negative than the model values.

Another bias calculation that can be calculated on this subset is ME_{geo}, which is 0.16 nT. This is very close to zero and much smaller than ME for this subset, which is 5 nT. This is because the geometric means of the subsets are both near −39 nT, calculated by taking the absolute value of the values, calculating the geometric mean, and then multiplying by −1. The model-value geometric mean is ever-so-slightly closer to zero, so the number is positive, like ME, but the large collection of points close to zero in these number sets pulls the geometric means for both subsets closer to zero than their arithmetic means.

A spread metric that requires single-signed number sets is YI_{log}, which has a value of 0.61 for this subset. This is close to the linear yield metric in Table 9.5, which is 0.69. Because the range of the values spans just over one order of magnitude (from 6 to 223), it is expected that these two calculations of yield should be quite similar.

The final assessment that can be considered is R_{log}, which has a score of 0.83 for this subset. That is a bit less than the Pearson correlation coefficient for these number sets (0.93). When all of the values are converted with a logarithm, the values along both axes are brought much closer to zero. Those points that were several times farther along the axis in linear space are now separated by only a value of one or two after the log operator is applied. This "contraction" of the axis can lead to better or worse linearity of the values. In this case, it makes it worse.

A reliability subsetting assessment is especially useful when you are predicting future observations. The new model output will fall within one of the subsets of the analysis, allowing you to judge whether to trust those model values as accurate predictions of the eventual observed state.

> **Quick and Easy for Section 9.10**
>
> When the subset is defined by using only those points with modeled values within a specified range, it is called a reliability metric.

9.11 Summarizing the Running Example

As a brief summary to the running example of Dst modeling and comparison with observations, we can ask the question, what did we learn? More specifically, what did we learn beyond the basic assessment with RMSE and R?

A few of the key metrics for the full number set are summarized in the final columns of both Table 9.4 and Table 9.5. RMSE is 14 nT and R is 0.91, both of which are quite good scores. RMSE is well below the standard deviations of either number set, and R well above our rule of thumb values of 0.5 and 0.7. Perhaps we could have stopped there, declaring victory at using a high-quality model. But the model is not perfect—we can see that in the scatterplot and histograms of Figure 9.3—and our additional assessments have revealed and quantified nuances of this data-model relationship.

In particular, further analysis showed that the model has a smaller range than the observations. It has a difficult time reproducing the very low and very high Dst values. At the low end, although it is still predicting storm-level Dst values, they are not quite as negative as they should be, at least not for the one intense storm in the chosen interval. Additionally, the storm-level subset, in both the observations and the model output, has by far the best correlation of any subset. That is, the model is good at getting the up–down trends of Dst when geospace is highly perturbed and driven by the solar wind. If the exact values at the peak of storm intervals matter to you, then this model is not particularly good. If, however, getting the timing and rough intensity of the storm interval correct is more important, then this model is fantastic.

At the high end of the Dst range, the model simply does not extend to values above about +5 nT, with only two numbers in the whole month higher than this threshold. The observations regularly go beyond this threshold, though, with the right-sided tail of its central peak well formed in this range. If these positive Dst values were a concern for you—perhaps you care about sudden compressions of the dayside magnetosphere, which are often the cause of rapid positive excursions of the Dst time series known as sudden impulses—then this is not the best model. If, however, you only care about storm intervals when Dst is strongly negative, then this shortcoming of the model does not matter.

Another fact revealed by the additional analysis is that a large part of the error arose from precision rather than bias. The two data sets have nearly the same centroids, but the spreads are quite different, with the observed number set having a larger σ than the model output number set. This indicates that there are most likely other drivers of the Dst index beyond those included in the O'Brien model. This is especially true during quiet times, when the relative error between the two number sets was particularly poor. From this, we can conclude that the O'Brien model includes the primary driver of space storms, as measured by the Dst index, but does not include the important drivers during other, smaller activity phenomena within geospace.

For Dst, the main intervals of importance for space weather are when the index plunges to large negative values. These excursions are the defining element of a magnetic storm, and are times when the inner part of near-Earth space is flooded with energetic charged particles that can cause spacecraft surface charging or internal electrical problems, and when strong currents flow in geospace that can induce current flows in long power lines and pipelines, to name just a few of the societal ramifications. Therefore, an assessment of a predictive model for Dst might be best served by focusing on whether the model can reproduce this particular part of the observed values.

To focus that the assessment should be on the extreme lower end of the distribution, we should consider the LDE_e value and the discrimination metrics of the lowest subset of the observed values. If the only concern is capturing the peak of magnetic storms, then all of the other metrics just discussed are essentially irrelevant. For the selected model and the selected interval, the comparison is not particularly good. The best metric in the first column of Table 9.4 is R, which is 0.95, but all of the others indicate that the model struggles to get the actual observed values at storm peak. Of course, only one storm occurred during this interval, so it is an unfair assessment of the true ability of the model to correctly predict storm peaks.

> **Quick and Easy for Section 9.11**
>
> We learn a lot more about the data-model relationship when we conduct a robust suite of assessments using metrics from many categories.

9.12 Summary of Fit Performance Metrics

Many metrics have been covered in this chapter. Let's summarize them in a single place, as shown in Table 9.6. There is no need to use them all for a single data-model comparison; one per category is enough for a robust assessment. Some suggestions for an optimal collection of metrics for particular types of comparisons are discussed in Chapter 12. Keep in mind that the guidelines for declaring the metric scores to be good are purely suggestions. The quality of a particular metric score is highly dependent on the context of the specific data-model comparison being conducted.

> **Quick and Easy for Section 9.12**
>
> All of the fit performance formulas in one place.

Table 9.6 Summary of the fit performance metrics from this chapter.

Metric	Formula	Range	Perfect Score	Guideline for "Good"
Accuracy metrics				
MSE	$\dfrac{1}{N-d}\sum_{i=1}^{N}(M_i - O_i)^2$	$[0, \infty)$	0	$< \min\left[\sigma_M^2, \sigma_O^2\right]$
RMSE	$\sqrt{\dfrac{1}{N-d}\sum_{i=1}^{N}(M_i - O_i)^2}$	$[0, \infty)$	0	$< \min\left[\sigma_M, \sigma_O\right]$
MAE	$\dfrac{1}{N-d}\sum_{i=1}^{N}|M_i - O_i|$	$[0, \infty)$	0	$< \min\left[\text{MAD}_M, \text{MAD}_O\right]$
MAPE	$100\dfrac{1}{N-d}\sum_{i=1}^{N}\left|\dfrac{M_i - O_i}{M_i}\right|$	$[0, \infty)$	0	$< 100\dfrac{\min\left[\text{MAD}_M, \text{MAD}_O\right]}{\bar{M}}$
SMAPE	$100\dfrac{1}{N-d}\sum_{i=1}^{N}\left|\dfrac{M_i - O_i}{(M_i + O_i)/2}\right|$	$[0, \infty)$	0	$< 100\dfrac{\min\left[\text{MAD}_M, \text{MAD}_O\right]}{(\bar{M}+\bar{O})/2}$
MSA	$100\left(\exp\left[\text{Median}\left[\left|\ln\left(\dfrac{M_i}{O_i}\right)\right| \wedge i\right]\right] - 1\right)$	$[0, \infty)$	0	$< 100\left(\exp\left[\min\left[\left|\ln\left(\dfrac{\text{IQR}_M}{M_{\text{median}}}\right)\right|, \left|\ln\left(\dfrac{\text{IQR}_O}{O_{\text{median}}}\right)\right|\right]\right] - 1\right)$
Bias metrics				
ME	$\dfrac{1}{N}\sum_{i=1}^{N}(M_i - O_i)$	$(-\infty, \infty)$	0	$< \min\left[\sigma_M, \sigma_O\right]$
ME$_{\text{median}}$	$\text{Median}(M_i \wedge i) - \text{Median}(O_i \wedge i)$	$(-\infty, \infty)$	0	$< \min\left[\text{IQR}_M, \text{IQR}_O\right]$

ME$_{geom}$	$\exp\left(\frac{1}{N}\sum_{i=1}^{N}\ln(M_i)\right) - \exp\left(\frac{1}{N}\sum_{i=1}^{N}\ln(O_i)\right)$	$(-\infty,\infty)$	0	$< \min[\text{IQR}_M, \text{IQR}_O]$
SSPB	$100\left(\text{sign}\left[\text{Median}\left(\ln\left(\frac{M_i}{O_i}\right)\right)\right]\right)\left(\exp\left[\left\|\text{Median}\left(\ln\left(\frac{M_i}{O_i}\right)\right)\right\|\right]-1\right)$	$(-\infty,\infty)$	0	Same as good MSA
Precision metrics				
YI	$\dfrac{\log[\max(M)] - \log[\min(M)]}{\log[\max(O)] - \log[\min(O)]}$	$[0,\infty)$	1	$[0.75, 1.33]$
$P_{\sigma,\text{ratio}}$	σ_M/σ_O	$[0,\infty)$	1	$[0.75, 1.33]$
$P_{\sigma,\text{diff}}$	$\sigma_M - \sigma_O$	$(-\infty,\infty)$	0	$\ll \min[\sigma_M, \sigma_O]$
Association metrics				
R	$\dfrac{\sum(O_i-\bar{O})(M_i-\bar{M})}{\sqrt{\sum(O_i-\bar{O})^2 \sum(M_i-\bar{M})^2}}$	$[-1,1]$	1	> 0.7
R^2	$\dfrac{\left[\sum(O_i-\bar{O})(M_i-\bar{M})\right]^2}{\sum(O_i-\bar{O})^2 \sum(M_i-\bar{M})^2}$	$[0,1]$	1	> 0.5
R$_S$	$\dfrac{\sum\left(\text{rank}(O_i)-\overline{\text{rank}}\right)\left(\text{rank}(M_i)-\overline{\text{rank}}\right)}{\sqrt{\sum\left(\text{rank}(O_i)-\overline{\text{rank}}\right)^2 \sum\left(\text{rank}(M_i)-\overline{\text{rank}}\right)^2}}$	$[-1,1]$	1	> 0.7

(Continued)

Table 9.6 (Continued)

Metric	Formula	Range	Perfect Score	Guideline for "Good"
$R_{\log}$	$\dfrac{\sum\left(\log O_i - \overline{\log O}\right)\left(\log M_i - \overline{\log M}\right)}{\sqrt{\sum\left(\log O_i - \overline{\log O}\right)^2 \sum\left(\log M_i - \overline{\log M}\right)^2}}$	$[-1,1]$	1	> 0.7
Extremes metrics				
LDE_ε	$M_\varepsilon - O_\varepsilon$	$(-\infty, \infty)$	0	$<\sim \min[\sigma_M, \sigma_O]$
HDE_ε	$M_{1-\varepsilon} - O_{1-\varepsilon}$	$(-\infty, \infty)$	0	$<\sim \min[\sigma_M, \sigma_O]$
γ_Δ	$\gamma_M - \gamma_O$	$(-\infty, \infty)$	0	$\ll \min[|\gamma_M|, |\gamma_O|]$
k_Δ	$k_M - k_O$	$(-\infty, \infty)$	0	$\ll \min[\sigma_M, \sigma_O]$
Skill metrics				
PE	$1 - \dfrac{\sum(M_i - O_i)^2}{\sum(O_i - \overline{O})^2}$	$(-\infty, 1]$	1	$>\sim 0.5$
SS_{MSE}	$1 - \dfrac{\sum(M_i - O_i)^2}{\sum(M_i^{old} - O_i)^2}$	$(-\infty, 1]$	1	> 0

9.13 Further Reading

This is a good description of the Dst index:
Sugiura, M., & Kamei, T. (1991). *Equatorial Dst Index 1957–1986*. ISGI Publications Office.

Here is the original paper on the O'Brien Dst prediction model:
O'Brien, T. P., & McPherron, R. L. (2000). An empirical phase space analysis of ring current dynamics: Solar wind control of injection and decay. *Journal of Geophysical Research*, 105, 7707–7720. doi:10.1029/1998JA000437

This is the original Burton equation model:
Burton, R. K., McPherron, R. L., & Russell, C. T. (1975). An empirical relationship between interplanetary conditions and Dst. *Journal of Geophysical Research*, 80(31), 4204–4214. doi:10.1029/JA080i031p04204.

And here is a comparison of the O'Brien and Burton models for real-time nowcast usage:
O'Brien, T. P., & McPherron, R. L. (2000). Forecasting the ring current index Dst in real time. *Journal of Atmospheric and Solar Terrestrial Physics*, 62, 1295–1299. https://doi.org/10.1016/S1364-6826(00)00072-9

The Dst index values can be downloaded from the NASA OMNIweb data site:
https://omniweb.gsfc.nasa.gov/
Or, more specifically for the user interface to download the data, here:
https://omniweb.gsfc.nasa.gov/form/dx1.html

Here is a detailed article defining geospace storms:
Gonzalez, W. D., Joselyn, J. A., Kamide, Y., Kroehl, H. W., Rostoker, G., Tsurutani, B. T., & Vasyliunas, V. M. (1994). What is a geomagnetic storm?. *Journal of Geophysical Research*, 99(A2), 5771–5792. doi:10.1029/93JA02867.

And one about electric currents in near-Earth space:
Ganushkina, N. Yu., Liemohn, M. W., & Dubyagin, S. (2018). Current systems in the Earth's magnetosphere. *Reviews of Geophysics*, 56(2), 309–332. https://doi.org/10.1002/2017RG000590.

Here is a book with a collection of nice introductory articles about space storms:
Suess, S. T., & Tsurutani, B. T. (2008). *From the Sun: Auroras, Magnetic Storms, Solar Flares, Cosmic Rays*. American Geophysical Union, Washington, DC. ISBN 0-87590-292-8

And this is a good entry-level textbook on space weather:
Moldwin, M. (2008). *An Introduction to Space Weather*. Cambridge University Press, Cambridge. ISBN: 978-0-521-71112-8

Armstrong's book includes many pages devoted to fit performance metrics, including a great introduction to MAPE and SMAPE:
Armstrong, S. J. (1985). *Long Range Forecasting* (2nd edition).Wiley, New York, USA. https://repository.upenn.edu/marketing_papers/211

Yet another good source for a discussion of different fit performance metrics, with a space weather focus, is given by Morley et al:
Morley, S. K., Brito, T. V., & Welling, D. T. (2018). Measures of model performance based on the log accuracy ratio. *Space Weather*, 16, 69–88. https://doi.org/10.1002/2017SW001669

This paper by Murphy is a nice introduction to prediction efficiency and other fit performance skill scores:
Murphy, A. H. (1988). Skill scores based on the mean square error and their relationships to the correlation coefficient. *Monthly Weather Review*, 116, 2417–2424. https://doi.org/10.1175/1520-0493(1988)116<2417:SSBOTM>2.0.CO;2

The book by Wilks contains discussion of a few of these metrics:
Wilks, D. S. (2019). *Statistical methods in the atmospheric sciences* (4th edition). Academic Press, Oxford. More about it here: https://www.elsevier.com/books/statistical-methods-in-the-atmospheric-sciences/wilks/978-0-12-815823-4

The book by Jolliffe and Stephenson also covers aspects of fit performance metrics:

Jolliffe, I., & Stephenson, D. B. (2011). Forecast Verification: A Practioner's Guide in Atmospheric Science (2nd edition). John Wiley & Sons, Hoboken.
https://doi.org/10.1002/9781119960003
Especially of use for fit performance metrics is Chapter 5 by Michel Déqué:
Déqué, M. (2011). Deterministic Forecasts of Continuous Variables. In Forecast Verification (eds I.T. Jolliffe and D.B. Stephenson).
https://doi.org/10.1002/9781119960003.ch5

Another good article that discusses many fit performance metrics is this by Rob Hyndman and Anne Koehler:
Hyndman, R. J., and Koehler, A. B. (2006). Another look at measures of forecast accuracy. *International Journal of Forecasting*, 22, 679–699. https://doi.org/10.1016/j.ijforecast.2006.03.001

A nice article on MAPE and SMAPE was written by Paul Goodwin and Richard Lawton:
Goodwin, P., & Lawton, R. (1999). On the asymmetry of the symmetric MAPE. *International Journal of Forecasting*, 15, 405–408. https://doi.org/10.1016/S0169-2070(99)00007-2

The main highlights of this chapter are summarized by Liemohn et al. (2021):
Liemohn, M. W., Shane, A. D., Azari, A. R., Petersen, A. K., Swiger, B. M., & Mukhopadhyay, A. (2021). RMSE is not enough: Guidelines to robust data-model comparisons for magnetospheric physics. *Journal of Atmospheric and Solar-Terrestrial Physics*, 218, 105624. https://doi.org/10.1016/j.jastp.2021.105624

9.14 Exercises in the Geosciences

Let's keep going with the Dst example. This time, the model will be a slightly newer one than the O'Brien code, that of Temerin and Li. The website for it is here: http://lasp.colorado.edu/space_weather/dsttemerin/dsttemerin.html

Including an archive of result values link at the top of the page. Go here, scan the plots, and select a month with a geospace storm or two (when Dst dropped below −30 nT). The files are easy to read and you could download several consecutive months of Dst values. You could also choose any of the other three geomagnetic indices that this group predicts—AE, AU, and AL.

Next, go to the OMNIweb site and download the observed index values:
https://omniweb.gsfc.nasa.gov/form/dx1.html

Enter in the time range and select "Dst index" from the list of options (AE, AU, and AL are also available here). Download a data set corresponding to the model output you just downloaded.

1. Open and read the two file sets.
   A Make a separate time series plot of each number set.
   B Make an overlaid time series plot of both number sets.
   C Make separate histograms of each number set.
   D Make an overlaid histogram of both number sets.
   E Make a scatterplot of the observed values against the modeled values.
   F Calculate the mean, standard deviation, median, and IQR for each number set.

2. Write a paragraph answering the following:
   A What is your initial reaction to the relationship between the data and the model?
   B Discuss the need to conduct additional assessments of the individual data sets to better understand the functional form of each distribution.

3. Conduct a metrics assessment.
   A Calculate one metric from each of the main categories (accuracy, bias, precision, association, and extremes). Explain your reasoning for your metric choices. Also, calculate

prediction efficiency from the skill category.
B Conduct a reliability subsetting assessment by dividing the modeled values into four bins and repeating part 3a for all subsets.

4. Write several paragraphs answering the following:
   A Based on your chosen accuracy alone, how good is the model? Explain your answer relative to quantities for the individual number sets.
   B What new do we learn about the quality of the model when the scores for the bias and precision values are taken into account?
   C What new do we learn about the quality of the model when the correlation coefficient score is taken into account?
   D Based on the extremes metric, how good is the model at predicting Dst during geospace storm times?
   E What is the skill of the model relative to the observational variance around its mean (as seen in the prediction efficiency score)?
   F Discuss the results of the reliability subsetting. How does this compare to the O'Brien model reliability assessment in Section 9.10?

10

Event Detection Metrics: Comparing Observed and Modeled Number Sets When Only Event Status Matters

Event detection metrics, also called categorical metrics, take a fundamentally different approach to model assessment than the fit performance metrics in the previous chapter. These metrics are based on converting all values into an event state classification. For a single event state and therefore a binary choice of event or nonevent classification, this assessment grouping is sometimes called **dichotomous metrics**. Observed and modeled values are both converted into this event/nonevent status. This conversion can be done independently, decoupling the event status identification in the two value sets. Once the conversion is done, the actual values are no longer used and *only* the event state designation is used in the assessment.

The threshold used to declare this status does not have to be the same for the observations and model output. In fact, it can be completely different for the two, even to the point of having different quantities being compared that have different units. That is, you could compare two quantities that you think are linked, rather than only a model output that is attempting to exactly reproduce the observed values. The specific values of M_i and O_i are discarded after conversion to event/nonevent status, so it does not matter what the units or ranges of the two value sets were before conversion to these categorical states.

Event detection metrics can also be applied to cause-and-effect variables for which the value ranges are completely different. This cause–effect pairing—for which the units do not have to be identical, that is, event detection metrics—can be applied to two observed data sets, rather than an observed quantity and model output trying to exactly reproduce it. This is fine for event detection metrics; the "model" can be an observed quantity that is, hopefully, a good predictor of the measurement being used as the "observed" value. The quality of this causal parameter at predicting the effect quantity is, in fact, what event detection procedures assess. When you have a model that is attempting to exactly reproduce the data, then both fit performance and event detection metrics can be used. If you do not have such a model, then you can still use all of the event detection metrics, but only a handful of the fit performance metrics can be used, specifically those that do not involve direct differencing or ratios of the individual M_i and O_i values. So, R and RS are still valuable and normalized quantities like γ_Δ and k_Δ can be calculated. Most of the other fit performance metrics cannot be applied when the "model" is

---

Dichotomous metrics: Another name for event detection metrics when there is only a binary yes–no choice of event status, as we are doing here.

Data Analysis for the Geosciences: Essentials of Uncertainty, Comparison, and Visualization, Advanced Textbook 5, First Edition. Michael W. Liemohn.
© 2024 American Geophysical Union. Published 2024 by John Wiley & Sons, Inc.
Companion website: www.wiley.com/go/liemohn/uncertaintyingeosciences

no longer trying to exactly match each of the observations.

Consider the running example from the previous chapter—for event detection metrics, we could bypass the model itself and instead use the solar wind parameters that were used as input to the O'Brien model, specifically E_y, as the "model value." Even though Dst has units of nanoTesla and E_y has units of Volts per meter, the two can be directly compared using event detection metrics. Each value set can be classified into events and nonevents separately, and, once converted, these values can be compared with the event detection tool kit of metrics.

The conversion from real numbers to event state designation is part of what makes these two metrics groupings so distinct from each other—the fit performance metrics discussed in Chapter 9 all required that the ideal model output exactly matched the observations. In event detection metrics, the goal of optimizing the "model output" value set is to correctly organize the observations by event state designation, using the event/nonevent designations of the modeled values as a method for aligning all of the observations classified as nonevents to one side and all observations classified as events to the other. A perfect model is one in which all model output values designated as nonevents align with all of the nonevent observations, while all of the model output values classified as events align with all of the observations deemed to be events.

This feature makes event detection metrics rather powerful, because it can be applied to values that are, from the moment of being recorded, already binned into discrete categorical values. The observations could be recorded as the answer to "did it snow today at this location?" with a yes/no designation. The model might be predicting snowfall amounts, or perhaps percentage chance of snowfall. Event detection metrics are often used in the medical field, with the observation being "did the patient recover from the illness?" and the model being the amount of a certain medication that the patient received. The conversion of this latter value into event/nonevent status can then be adjusted to find the optimal dose that minimizes drug usage while maximizing recovery.

The metrics categories discussed in Chapter 8 are all applicable to the event detection grouping. This chapter begins with a discussion of how to define an event and convert this into a contingency table, the essential tool used in all event detection metrics. Metrics within each category are then listed, similar to the format done in Chapter 9.

> **Quick and Easy for Chapter 10 Introduction**
>
> Event detection metrics convert both data and model values into sets of categorical yes–no values, from which many formulas have been created to assess the data-model relationship.

10.1 Defining an Event

The first choice that must be made is to decide what constitutes an event among the observations. This is shown graphically in Figure 10.1, on which a horizontal line is drawn. The observed values above this line are those that will be deemed "events," while those below are going to be labeled "nonevents." When dealing with events and nonevents, the exact values of the observations above this line are not important, it is only the fact that the above-or-below designation matters. The issue faced with the modeler is then this: what model will best predict when the observations are in an event state? In the example shown in Figure 10.1, it is seen that the data-model pairs above and below the horizontal "observational event" line are somewhat randomly spread around the model value space. There are points on the graph above the line that correspond to nearly the full range of model values. That is, there are observed events at both low model values as well as high model values. In short, these

10 Event Detection Metrics: Comparing Observed and Modeled Number Sets When Only Event Status Matters | 297

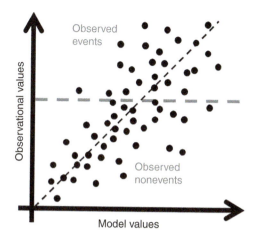

Figure 10.1 Example data-model comparison scatterplot. A horizontal dashed line is drawn to indicate a threshold above which we will define events.

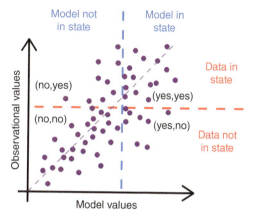

Figure 10.2 Example data-model comparison scatterplot with event identification thresholds indicated as red and blue dashed lines for the data and model values, respectively. In each resulting quadrant, all points have their model and data values converted to yes/no labels, as indicated.

model values on the x-axis are not particularly good at organizing the observed values along the y axis to be above and below the line.

The goal of event detection analysis is to find a threshold along the model output axis that correctly sorts the observed events from the observed nonevents. We'll call this line the **event identification threshold**, and can be defined separately for the observed and modeled number sets. The illustrative model values shown in Figure 10.1 are not the most appropriate model choice for this purpose, but they are not completely without merit. There are, on average, more observed events at high model values than at low model values, so a vertical line could be drawn on this plot to delineate the modeled event and nonevent status of all of the scatterplot points.

Given a scatterplot comparing observed values to model output, two event identification thresholds can be defined to break the observations and modeled values into event/nonevent designations. Figure 10.2 shows an illustrative

example plot of this form. The horizontal red dashed line is the value selected to categorize the observed values. Those data-model pairs above this line have their data value converted from whatever number it is into an event status of "yes." The data-model pairs below this red line have their data values replaced with an event status label of "no." Similarly, the vertical blue dashed line is a designation for the event identification threshold of the model values. To the right of this line, the modeled values of all pairs are replaced with the event status of "yes," and, below this line, the modeled values are replaced with "no." Note that the two thresholds could have different values, and the two axes could even have different units.

In Figure 10.2, the event identification thresholds divide the scatterplot into quadrants. In each quadrant, the conversion of the values is the same, and all data-model pairs in that quadrant now have identical value settings. In the lower-left quadrant, all data-model pairs are now (no, no) pairs, and so forth for the other quadrants. The original values no longer matter, only these yes/no designations will be used in the metric definitions in

---

Event identification threshold: The value beyond which numbers in the set are defined as events.

this chapter. More specifically, only the counts of the data-model pairs in each quadrant will be used.

The selected event identification threshold for the modeled values shown in Figure 10.2 produced nonzero counts of data-model pairs in each quadrant. It could be argued that the modeled event identification threshold should be moved. Whether we should move it to the right or the left depends on what aspect of the prediction you are trying to optimize. The upper-right quadrant is a space where both the model and data are in event state, and therefore this is the region of the chart of "good" data-model pairs. Similarly, the lower-left quadrant has both the data and modeled event status as "no," again indicating this is a region of the chart with "good" data-model pairs. The upper-left and lower-right quadrants, however, are "bad prediction" regions of the chart where the event status of the model does not match the event status of the data. The objective of the modeled event identification threshold is usually to "optimize the prediction," but this is a vague term that can be defined in several ways. What does optimize mean to the user? Does it mean maximizing the upper-right quadrant? This can be done, but at the expense of increasing the pairs in the "bad" quadrants of the scatterplot. This example hints at the nuance in interpreting event detection metrics. More will be discussed about this throughout the sections in this chapter.

Note that the scatterplot could have been created differently, changing the location of the "good" and bad" quadrants. It could be that the "events" have low values, as was the case in Chapter 9 with the running example of the Dst index, in which case the yes/no designations in each quadrant are flipped. We'll do exactly this in Chapter 11. The other change that some will do is to flip the number set being plotted along the two axes, putting the observations along the x-axis and the modeled values on the y axis. This is just fine. You can make this plot and designate events in whatever manner you want, just be clear about it.

There are many possibilities of how event definition might work in the geosciences. Take, for example, the situation of hurricane wind speed and property damage in the region where it made landfall. In this case, both values are observed quantities, neither is actually a model output. One is clearly the cause and the other the effect—the hurricane wind speed should be used as the "modeled value" and the property damage quantity should be the observed number set. Then, we can ask a question of interest that categorizes these number sets into an event status, like this: did the damage exceed a particular threshold, one for which you want to call it a "major damage event?" Those storms that exceed the chosen threshold have their property damage values replaced by a "yes" value, indicating their event status, and all other property damages less than the chosen threshold are converted to a "no" value. For the driving parameter of hurricane wind speed, the threshold setting could be chosen in one of several ways. It could be chosen based on predefined designations of hurricane strength—perhaps you set it to 74 miles per hour, the minimum cutoff for declaring a tropical storm to be a hurricane, or maybe 111 miles per hour, the wind speed for calling a storm a "category 3" hurricane, which is traditionally considered the crossover point to calling it a "major" hurricane. The other option is to look at the scatterplot of values and pick a threshold that you think, qualitatively, would be the optimal threshold for predicting a "major damage event."

This discussion reveals a key step in event detection analysis—make the scatterplot. While this plot is critical to fit performance metrics because, in that case, the model is trying to exactly predict the observations, it is also highly useful for event detection analysis. The calculations in this chapter do not require that this plot is made; those are the quantitative assessments which could be construed as more important than the qualitative assessment of examining a plot. The scatterplot, however, can reveal key information about how to interpret the metric scores. It is a crucial first step to good data-model comparisons.

Another critical step in creating a good event detection is to consider the three aspects raised in Section 8.8 (Chapter 8) regarding the design of a good data-model comparison. Consider the spatial or temporal interval for which the comparison will be made and whether it contains a sufficient number of events. Consider the spatial or temporal resolution so that the right type of event is being investigated. That it, choosing a high cadence could mean that a small-scale fluctuation in the parameter creates many flips between event and nonevent states. If capturing this high-frequency status is important, then fine, but if not, then consider a lower resolution for the number sets to average out the high-frequency signal. Finally, consider the nature of the parameter being assessed. For instance, are you particularly concerned about only extreme events, or do you want to test the model for any slight-above-average value? The choice of the threshold, and the choice of the parameter itself, should be carefully made to best address your specific purpose behind the data-model comparison.

Quick and Easy for Section 10.1

Event detection metrics are useful when the most important parameter is whether the value is beyond some threshold, no matter how near or far away that threshold is.

10.2 Contingency Tables

Continuing the illustrative example from Figure 10.2, the next step in the event detection analysis process is to create the **contingency table**. This is shown schematically in Figure 10.3, where the quadrants of the scatterplot are now assigned labels, the pairs in each quadrant are counted, and the values are entered into the quadrants of a 2 × 2 table. Quadrants are given these names: **hits**, for those data-model pairs that have both the data and the model values marked as being in event state; **correct negatives**, when both data and model values are marked as not in event state; **misses**, when the data value is in event state and the model output has a nonevent status; and **false alarms**, when the data value is in the nonevent state, but the model value is designated as in event state.

The points within each quadrant can then be counted and the totals listed in table form, also shown in Figure 10.3. This table gets a special name—the contingency table. Here, the columns are labeled as the model being in event state and not in event state, while the rows are for the data being in event state and not in event state. Note that the table does not have to be organized in the same quadrant placement

Hits (H): The count of data-model pairs with both the observation and the model values labeled as being in event state; used in the contingency table of event detection metrics.

Correct negatives (C): The count of data-model pairs with both the observation and the model values labeled as being not in event state; used in the contingency table of event detection metrics.

Misses (M): The count of data-model pairs with the observation in event state and the model value not in event state, used in the contingency table of event detection metrics.

False alarms (F): The count of data-model pairs with the observation not in event state and the model value in event state, used in the contingency table of event detection metrics.

Contingency table: The matrix of counts from the quadrants of the scatterplot defined by the observed and modeled event identification thresholds.

300 | *Data Analysis for the Geosciences*

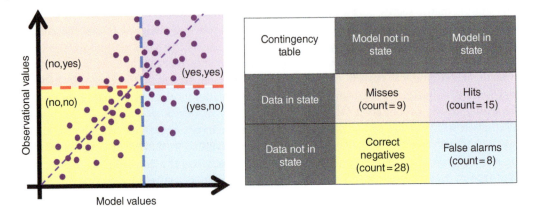

Figure 10.3 A scatterplot with event identification thresholds and the quadrants labeled as hits, misses, false alarms, and correct negatives. The total counts of data-model pairs in each quadrant are then used as the entries in the contingency table.

as the scatterplot; the column and row headings can be arranged in whatever way is convenient for the particular situation, and therefore the hits, misses, false alarms, and correct negatives labels will be shifted around accordingly.

The scatterplot shown in Figure 10.3 looks, qualitatively, like a pretty good match. The contingency table provides a quantitative distillation of the scatterplot. Even though we now have a table of numbers, this is still at a qualitative analysis level. We have not yet introduced the metrics that will use these contingency table values in particular combinations to extract specific information about the data-model relationship.

The name "contingency table" will be used for this matrix of point counts, but it should be noted that there are many names for this table. A list of some common alternative names for the contingency table is given in Figure 10.4. Any of these names is perfectly acceptable, because the name really is not that important, as they are all used the same way toward assessing the data-model relationship.

Another point of clarification should be brought up before moving on—the quadrants in the contingency table have many different names. A few different sets of these names are listed in Figure 10.5. These alternative naming systems for the quadrants extend to the metrics

- Contingency tables are also called:
- Binary table
- Error matrix
- Confusion matrix
- Table of confusion
- Dependency table
- Detection matrix
- Probability of detection matrix
- Hypothesis test matrix

Figure 10.4 Other names for the contingency table.

Alternate names for the cells in a contingency table

Misses	Hits
Type 2 errors	Yes–Yes
False negatives	True positives

Correct negatives	False alarms
No–No	Type 1 errors
True negatives	False positives

Figure 10.5 Other names for the quadrants of the contingency table.

themselves; some metrics have several names in common use. This is often a discipline-specific convention, with each scholarly field settling on a preferred set. Like the name of the table, these labels are essentially irrelevant, because regardless of the label used for a particular quadrant, it will be used the same way and serve the same purpose within the metrics given in this chapter.

From here on in this chapter, the already-defined quadrant names of hits, false alarms, misses, and correct negatives will be used. These are compact yet intuitive names for the quadrants that require no additional explanation for understanding their meaning. In fact, we will use the first-letter labels for them—H, F, M, and C.

> **Quick and Easy for Section 10.2**
>
> Event detection metrics reduce the scatterplot of the comparison into a contingency table, four numbers that summarize the entire data-model relationship. Both the contingency table and the quadrants have many different names in regular use.

10.3 Data-Model Comparisons with Events

The discussion in Sections 10.1 and 10.2 also highlights the purpose of event detection analysis. This type of analysis does not care about the actual values of the observations or model output, the only thing that matters is the event status. Organizing the observed events, all of them and only them, to the right of the modeled event threshold is the only goal. If this is done, then all metrics listed in this chapter will yield perfect scores, regardless of how far off the fit performance metrics are for that model. It means that a model that has less than ideal fit performance metrics might be highly useful for event detection. Choosing the correct threshold can also be done with reliability metrics from the fit performance grouping; knowing which intervals of the model output range are good at matching the observations informs the threshold selection. The conversion to event status also means that it does not matter how close the scatterplot points are to the event identification thresholds. A value could be either right up next to the threshold or very far from it; in both cases, if it is on the same side of the threshold, then it is given the same yes/no label, replacing the original value. That is, a point in one of the "bad" quadrants could be snuggled in very close to the corner of where the two threshold lines cross, where a small change in the value could have sent it into one of the other quadrants. For event detection analysis metrics, this closeness to the other quadrants does not play a role in any of the metrics; they only use the yes/no status. The closeness to the threshold lines will come up near the end of this chapter, when uncertainty of the event detection metrics is discussed.

When contemplating a data-model comparison with event detection metrics, it could be the case that the observed values are already in a binary classification. That is, the data might have been collected as yes/no event state values, rather than real numbers. Such an example is shown in Figure 10.6. The decision about the data "event status" is easy; it is already done. The horizontal "observed event threshold identification line" can be drawn anywhere between "no" and "yes" on the y axis. The model could also be a dichotomous yes/no value set, in which case the modeled event

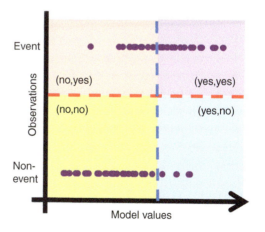

Figure 10.6 Example data-model comparison scatterplot for which the observations were originally recorded with a binary event status designation. The event identification thresholds are indicated as dashed lines.

status is also easily defined. Usually, though, it is a set of real numbers, spread over a range, allowing a choice for the placement of the modeled event identification line. Once this modeled threshold is defined, the analysis proceeds by assigning all model values on the "event" side of the threshold as "yes" and those on the other side as "no," creating four paired possibilities of (model, data) values, with all pairs within a given quadrant having the same designation.

The example shown in Figure 10.6 is fairly well organized. There is a rather small region in the middle of the plot where the no and yes data values mix; at the two ends, nearly all values are no at low x values and nearly all yes at high x values. The number of "bad" (no, yes) and (yes, no) data-model pairs in the down-diagonal quadrants is rather low compared to the number of "good" (no, no) and (yes, yes) pairs. A perfect model would eliminate all bad pairs and the transition from no to yes values would be abruptly sharp. Even a perfect model requires a perfectly chosen event detection threshold; choosing the threshold anywhere other than the abrupt change of observations from nonevents to events would create pairs in the two "bad" quadrants and therefore worsens the scores from the event detection metrics given in this chapter.

A geophysical example of this is an earthquake. Yes, there is the Richter scale to quantify the magnitude, but with earthquakes it is very clear that one either did or did not happen within a particular time interval. The corresponding model could be a calculation of strain rate around the fault line, taking either the average in a localized zone or the peak strain rate in that region. Done for enough fault line regions over enough time intervals, the data-model number set is large enough for an analysis of these yes/no observations against the average or peak strain rate. This addresses the scientific issue of whether local strain rate is the controlling factor in earthquake occurrence. It might yield an excellent distribution, in which the x-axis strain rate values perfectly organize the yes/no occurrences of earthquakes. On the other hand, this might not be the critical factor, in which case the earthquake occurrence data will be mixed along the x-axis.

A final point about the use of contingency tables for data-model comparisons is that the definition of an event does not have to be a binary yes/no designation. For a real-values observational value set, there could be several thresholds defined for events of various intensity levels, such as assigning break points along the Richter scale to have several magnitude levels for the earthquake events. For discretely defined data, this could be different types of events, such as precipitation being designated as rain, snow, sleet, or hail. In this case, the contingency table will no longer be a 2×2 matrix, but a larger $A \times A$ matrix. It could be created with mutually exclusive labeling of the observations, or allowing for multiple labels for the same observation. To continue the two examples, for the earthquakes, this could be a set of categories of earthquakes above a few chosen Richter scale values. In this case, all of the events in the particular category would include not only those observations between one threshold and the next but also all of those above that threshold, even though they are also contributing to other event counts. For the precipitation example, this could be any precipitation within a given region or time interval, in which case there might have been more than one type of precipitation and the observation should contribute to multiple event categories. The model being assessed in the comparison would then need an event identification threshold for each of these observed event categories, thus producing the $A \times A$ sized contingency table.

The metrics presented and discussed in this chapter are only for 2×2 contingency tables. There are extensions for many of the metrics that account for multiple event categories. For any one threshold or group thresholds, though, the $A \times A$ matrix can be reduced to a 2×2 matrix, focusing the analysis only on the event class of interest. In this case, the 2×2 metrics can be used for that reorganization of the larger matrix. This is often a good starting place when assessing a multiple-event observational set, as this can help inform the user about how to target their efforts in calculating the more complicated metric formulas that simultaneously take into account multiple event categories.

> **Quick and Easy for Section 10.3**
>
> The goal with event detection analysis is to align all observed events with all of the modeled events; this will yield perfect scores for all of the event detection metrics.

10.4 Running Example: Will It Rain?

A classic atmospheric science example that cannot be done with fit performance metrics but only with event detection metrics—the question of "will it rain?"—poses a nice dichotomous yes/no classification of events. This is a question to which everyone can relate; we have all looked at the weather, seen the forecast of precipitation over the next few hours, and then decided our plans based on that information. The question to be considered here is this: how good is the model on which those forecasts are based?

The National Oceanic and Atmospheric Administration (NOAA) has a branch called the National Centers for Environmental Prediction (NCEP), of which the National Weather Service (NWS) is one. NOAA also has a branch called the Earth System Research Laboratories, which maintains numerical models used by the NCEP centers for predictions. One of these codes is called the **high-resolution rapid refresh (HRRR) model**, an atmospheric dynamics and cloud physics

---

High-resolution rapid refresh (HRRR) model: An atmospheric dynamics and cloud physics model that produces 18-hour and 48-hour forecasts on a 3-km resolution across the entire United States.

---

model that produces forecasts on a 3-km resolution across the entire United States. It has convective atmospheric dynamics, subgrid processes included as parameterizations, and data assimilation of available records to specify the input conditions as accurately as possible. HRRR is run by the NWS every hour, producing a new 18-hour forecast each run, plus it is run for a 48-hour forecast four times a day. Output from this model includes expected precipitation over the forecast interval.

Output from the HRRR model can be compared against observed precipitation. For this running example throughout the chapter, precipitation data from the weather station located at Detroit Metro Airport in Romulus, Michigan, has been selected (call letters DTW). These data are part of the automated surface observing system (ASOS), which includes stations located across the United States, especially in parks and at airports. The format of this data is in a typical reporting style called METAR (Meteorological Terminal Aviation Routine Weather Report), which includes temperature in degrees Fahrenheit and precipitation in inches, recorded every minute. These data are archived at NOAA's National Climate Data Center.

Before we continue, it is useful to consider the data-model comparison design elements from the end of Chapter 8. The first is selecting an appropriate spatial or temporal interval for the comparison. The HRRR model is continually improved—they are up to what is labeled as version 4 of the main code—and it is not retroactively run for past time intervals, so it is useful to pick as recent a time as possible for the best prediction. For this study, the entire year of 2020 will be used for the assessment.

The second point is the selection of the spatial or temporal resolution of the values within this interval. Because the HRRR model values are produced at a 1-hour cadence, this is a natural resolution for the comparison. Because 2020 was a leap year, there are 8784 hours within that interval, which is a relatively large number for a comparison and serves well for this running example.

The final design choice is the nature of the comparison. For this comparison, it is necessary to choose what aspect of the forecast will be compared against the observed precipitation values. Specifically, the forecast is hourly for the following 18 hours. To make this a bit more reasonable and focused on individuals making decisions about their immediate future, let's consider a 6-hour forecast window. This is long enough that it truly tests the model and not simply the initial conditions, but short enough that it is similar to what people would actually examine when making decisions about outdoor activities. Precipitations from shorter intervals were summed to create this database of hourly-cadence values of the predicted and observed precipitation over the following 6 hours. With these considerations of how to set up the comparison, the result is a data-model pair nearly every hour for the entire year of 2020, for a total of 8555 usable pairs.

A plot of these observed and modeled 6-hour-interval precipitation values is shown in Figure 10.7. The observed values, labeled ASOS/METAR data, are shown as red columns and the HRRR-modeled forecast values are shown in blue. Purple indicates overlap between the columns, with the taller of the two columns protruding above the purple part of the column. Looking at this, it is rather difficult to determine whether the model is good or not at predicting precipitation. There is often substantial overlap, but it is nearly impossible to make a judgment from this plot alone.

Figure 10.8 continues this initial assessment of the two data sets by presenting first the scatterplot of the two number sets against each other and then overlaid histograms of them, both with and without the very tall columns for zero and trace precipitation. The scatterplot is somewhat revealing; it looks like there might be some trend of the points following the ideal

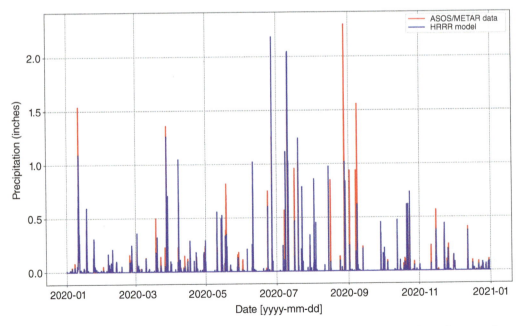

Figure 10.7 The time series of all data and model precipitation values, in inches, for the entire year of 2020 for our running example. The red columns are the observed precipitation values and the blue columns are the modeled forecast values.

10 Event Detection Metrics: Comparing Observed and Modeled Number Sets When Only Event Status Matters

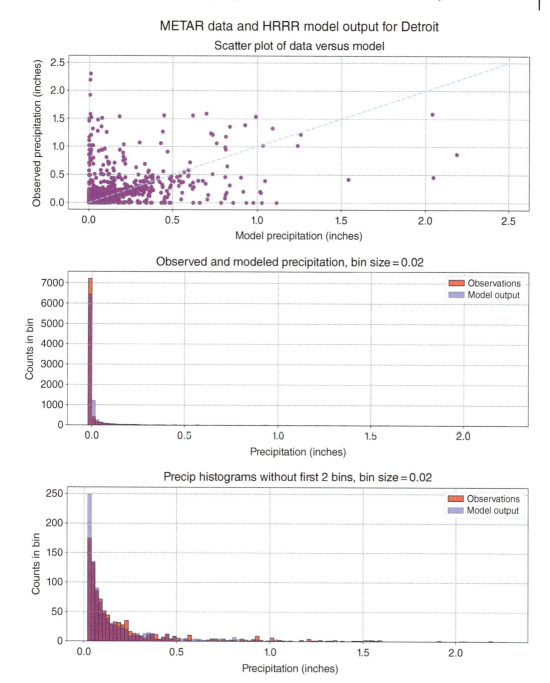

Figure 10.8 The top panel shows a scatterplot of observed versus modeled precipitation values for the running example. The diagonal light-blue dashed line represents a perfect fit. The middle panel is a histogram of the data and model values, with a bin size of 0.02 inches. Red columns are the data and blue columns are the model, with the column overlap shown as purple. The lower panel is the same histogram but without the first two columns, so that the lower-count bins are more visible.

comparison, represented on the plot as a light-blue dashed line, especially below 0.4 inches of precipitation. There is a lot of spread around this, however, and what is not seen is the *very* large number of points crowded into the lower-left corner of this scatterplot. This is better illustrated by the histograms. The middle panel of Figure 10.8 shows the full histograms for the two number sets. The square-root rule from Chapter 2 would have us use 93 bins for this histogram. As mentioned in Section 2.3 (Chapter 2), the various bin number formulas are only guidelines, so to make it a more easily understood histogram, the bin size was set to 0.02 inches, yielding 114 bins up to the highest observed value of 2.3 inches. The lower edge of the lowest bin was set negative, specifically to −0.019 inches, so that the upper edge of this bin is at 0.001 inches. That is, the values in this bin have zero (or essentially zero) precipitation within the 6-hour forecast interval of interest. It is seen that this bin has by far the tallest column, with over 6000 modeled values and over 7000 observed values. From this, we can say that, most of the time, it is not raining at the DTW weather station. The second column is also quite tall, with over 1000 model values having precipitation predictions of 0.001–0.021 inches. The rest of the columns are so close to the *x*-axis that it is hard to see the columns. Therefore, the third panel of Figure 10.8 shows the same histogram but without these first two columns. The tallest column is now only a few hundred counts, rather than many thousands of counts, and it is seen that the histograms of the observations and the model output are rather close to each other across a broad range of precipitation values. The similarity of the distributions does not mean that the forecast was good—the timing of these values could be different—but it is a necessary condition for a consistently good forecast.

There is one more assessment design element that needs to be considered for event detection—the threshold settings for the observed and modeled events. We have chosen a model that is directly predicting the observed quantity—precipitation in inches over the next 6-hour interval—so we could use the same threshold for both number sets. Remember that our model could have been some other parameter with completely different units and a different range; that is completely acceptable for event detection metrics calculations. Because of this similarity of the data and model values, though, let's use the same threshold for event identification. That should be entirely dependent on the assessment we want to conduct. We could set it rather high, in which case we are assessing the model's ability to predict heavy precipitation events—that is, rain or snow storms. These events are very rare and would not be a particularly good example for demonstrating the usefulness of the various statistics to be discussed and computed in the rest of this chapter. Instead, let's choose a very low threshold of 0.001 inches. This is an extremely low number and means that essentially any precipitation puts the value into event status. That is, with this low threshold, we are posing the question, "will it rain?"

With 0.001 inches as the event identification threshold for both the observations and the model values, a contingency table can be constructed. Lines can be drawn on the scatterplot and points counted in each quadrant. The result of this exercise is shown in Figure 10.9. Not only are the counts for each cell listed but also the totals for each row and column are given in the table.

Metrics quantify specific attributes of the data-model relationship. The contingency table and figures in this section qualitatively provide all of the information that the metrics we will learn about will quantify. We see that there are far more correct counts in this table than there are error counts in either of the two bad cells. That is, there are several hundred more counts in hits compared to false alarms or misses, and correct negatives are substantially bigger than all other cell counts. As seen in the following sections, some metrics exclude or minimize the correct negatives

10 Event Detection Metrics: Comparing Observed and Modeled Number Sets When Only Event Status Matters | 307

Quadrant counts for DTW precipitation running example

Running example	Model not in state	Model in state	Totals
Data in state	Misses: 140	Hits: 1207	Observed events: 1347
Data not in state	Correct negatives: 6280	False alarms: 928	Observed nonevents: 7208
Totals	Modeled nonevents: 6420	Modeled events: 2135	$N = 8555$

Figure 10.9 The contingency table for the running example with the event identification threshold set at 0.01 inches for both the observations and the model output.

quadrant; even these metrics will be decently good. We also see that the table is not well balanced, in either the good or the bad cells; false alarms are much more prevalent than misses and correct negatives are much more common than hits. Because of this, we know that this model is a reasonable choice to answer the posed question, even without performing further calculations. In summary, the contingency table looks to be pretty good and without any calculations. In the following sections, we will quantify the model's quality, putting numbers to this assessment.

From the histograms, it is seen that the two distributions are relatively similar, which may indicate that the model performance is good. The scatterplot, however, did not convey this message very strongly because it contains a lot of randomness in the spread of values. This scatter and spread will be assessed in the next chapter, when the event identification thresholds will be systematically varied to explore how well the model captures different levels of precipitation. However, because the event identification threshold is set so low that *all* of the off-axis points in Figure 10.7 are hits, the scatter and spread are less important. After conversion to the yes–no categorical status, event detection metrics do not use the specific values of the data and model; for the analysis in this chapter, most of the clustering of points in Figure 10.7 is ignored.

Quick and Easy for Section 10.4

The example of precipitation at Detroit Metro Airport will be used as a running example throughout the rest of the chapter. Qualitatively, the comparison seems reasonably good, but there is substantial spread in the scatterplot.

10.5 Significance of a Contingency Table

We have already addressed the first significance test for a contingency table: examine the quadrant counts. Also, look at the associated scatterplot. This qualitative assessment is very important, as it provides insight into the expectations for the quantitative metrics later in this chapter.

This initial step of examining the scatterplot and table values addresses a critical question: does the choice of event identification threshold for the data and model look reasonable? From the scatterplot, you can often readily see if the choices of thresholds occur at natural break points or in the midst of a cloud of points. In the latter case, the small changes to the thresholds could lead to large changes in the counts, which puts a caveat on any conclusions drawn from the metrics analysis. From the table, you can see if the counts in the good quadrants are higher than those in the bad quadrants. This is an opportunity to adjust the event identification thresholds, if necessary. Having it the other way around is ghastly and you might even question the need to continue with the quantitative assessment. But, again, this depends on the issues being addressed with the data-model comparison; it could be that you only care about some of the quadrants and not others.

Visual inspection of the scatterplot and count values in the contingency table will help you decide if the design of your comparison, as discussed near the end of Chapter 8, is on target for your particular needs. For example, the good quadrants of the contingency table could be highly imbalanced, with many correct negatives but only a few hits. This might produce some very good metrics scores, but if a property related to the hits quadrants is what is truly desired, then the design of the comparison is not optimal for this assessment. A quick look at the scatterplot and contingency table entries can be useful in deciding whether to proceed with the analysis or whether to make adjustments to the comparison setup or analysis process.

The next step is a quantitative test of significance, and it is one that we have already used for histogram analysis back in Chapter 5—the χ^2 test. This is a test of observed versus expected values in the bins, where the expected values are found either from a theoretical solution or from parameters extracted from the binned values themselves, such as the expectation of a Gaussian distribution using the mean and standard deviation calculated from the value set. For histograms, the bins were created along the x-axis and the number of bins could be set by the user, depending on the level of accuracy desired in the comparison between the observed and expected distributions. For contingency tables, the bins are set by the threshold choices. Here, we consider the 2 × 2 contingency table example, but it is readily extended to an A × A matrix of arbitrarily large size.

To provide a quick recap of the χ^2 process, the general formula, slightly rewritten for use with the contingency table format, looks like this:

$$\chi^2 = \sum_{i=1}^{2}\sum_{j=1}^{2} \frac{(E_{ij} - C_{ij})^2}{E_{ij}} \tag{10.1}$$

In Equation (10.1), the i and j summation indices are for the rows and columns of the contingency table. For a 2 × 2 contingency table, there are only four values in the entire double summation. The C_{ij} values are the counts from the contingency table (i.e., H, F, M, and C, as defined in Section 10.2). For the expected values in each cell, E_{ij}, the formula that is often used is based on the contingency table entries themselves:

$$E_{ij} = \frac{T_i T_j}{N} \tag{10.2}$$

This form of E_{ij} is an estimate of a random table number set that maintains the same row and column totals while redistributing the counts according to Poisson statistics. In Equation (10.2), the T_i and T_j values are the totals for the rows and columns, respectively, of the contingency table values, and N is the total number of data-model pairs in the scatterplot ($N = H + F + M + C$). Note that it is not a "flat" table with all cells of equal counts, but rather what would be expected for a random distribution given the row and column totals, which could be very far from equal. Because N in the denominator is also the total count across the full contingency table, it is sometimes replaced with T in this context to indicate the similarity with the row and column totals. Figure 10.10 shows how these values map to the contingency table entries. Note that it is fine if the rows and columns are transposed in the definition of E_{ij} and χ^2; the resulting χ^2 value is robust to this swap.

The significance of the resulting χ^2 value is then converted to a probability that the table entries could have happened by random

10 Event Detection Metrics: Comparing Observed and Modeled Number Sets When Only Event Status Matters

Contingency table totals	Model not in state	Model in state	Totals
Data in state	Misses ($M = C_{11}$)	Hits ($H = C_{12}$)	$T_{i=1}$
Data not in state	Correct negatives ($C = C_{21}$)	False alarms ($F = C_{22}$)	$T_{i=2}$
Totals	$T_{j=1}$	$T_{j=2}$	N

Figure 10.10 A contingency table with the C_{ij} values labeled and the row and column totals T_i and T_j indicated.

chance. For a model with a significant chance of predicting the event state of the observations, it is desired that the table values be significantly different from random chance. Therefore, in this case, we want a high χ^2 score that implies that the contingency table values are significantly different from a random chance occurrence. For a 2 × 2 contingency table, this should be compared against the $d = 1 \chi^2$ probabilities. If the goal is exceeding the traditional 95% level that the two value sets are different, then the χ^2 value should be larger than 3.8. If a more stringent test is desired and you want the contingency table to meet the 99% confidence level that the value sets are different, then a χ^2 value should be larger than 6.6.

For our running example of precipitation at Detroit Metro Airport, the calculation of the expected values is shown in Figure 10.11. These values can be qualitatively compared against the actual counts for our data-model relationship, shown in the contingency table in Figure 10.9. All of the numbers are off by hundreds, but it is difficult to tell if this difference is significant just by visual inspection. Following the math of Equation (10.1), for this contingency table the χ^2 value is 3570. This is an exceptionally high value and well past any criteria for significance.

The model passes this first test; it is different from the random count values expected for these row and column totals. While this is a good start for quantitative testing, there is much more that we can do with these four numbers of the contingency table. Let's now work through the many categories of metrics defined in Chapter 8.

Expected values for DTW precipitation running example

Expected values	Model not in state	Model in state	Totals
Data in state	Expected misses: 1011	Expected hits: 336	$T_{i=1} = 1347$
Data not in state	Expected correct negatives: 5409	Expected false alarms: 1799	$T_{i=2} = 7208$
Totals	$T_{j=1} = 6420$	$T_{j=2} = 2135$	$N = 8555$

Figure 10.11 The expected values for the running example of precipitation at DTW airport in 2020, showing a contingency table with the E_{ij} values for this case.

> **Quick and Easy for Section 10.5**
>
> A common method for assessing significance of a contingency table is with a chi-squared test, with the expected values being random chance.

10.6 Accuracy

Accuracy metrics for event detection focus on a model's ability to maximize the scatterplot points in the two "good" quadrants. There are three metrics that are most often used for this assessment, with slight differences in what and how the counts from the two good quadrants are included. Let's go through them and then discuss their similarities, differences, features, and limitations.

The first is **proportion correct**, PC, and uses everything from the contingency table:

$$PC = \frac{H+C}{N} \tag{10.3}$$

Remember that N is the total number of data-model pairs in the analysis, and is therefore equal to $H + C + M + F$. Because all terms are positive in Equation (10.3), PC is positive definite. If M and F are both zero and all counts are in the good quadrants of H and C, then PC will be one (its perfect score). If all counts are in the bad quadrants and H and C are both zero, then PC will be zero (its worst score). Some definitions will multiply this value by 100 to make it "percent correct," which is confusing because they are both referred to as PC. So, be sure to double check the formula behind this metric, or at least estimate an expected value so that you know if this extra factor has been included or not.

Proportion correct (PC): An event detection accuracy metric, it is the sum of hits and correct negatives divided by the total data-model pair count.

The second one in the lineup of accuracy metrics is the **critical success index**, CSI, sometimes called the **threat score**:

$$CSI = \frac{H}{H+M+F} \tag{10.4}$$

As you can see, CSI excludes the correct negative cell from its calculation. This is useful if C is large and therefore dominates the numerator. PC could be large even though the model is not particularly good at obtaining hits. It also has a perfect score of one and worst score of zero.

The third metric in this category is the F_1 **score**:

$$F_1 = \frac{2H}{2H+M+F} \tag{10.5}$$

Like the other two accuracy metrics, this one also ranges from zero to one. Equation (10.5) is very similar to Equation (10.4) except that it doubles the weight of H in the numerator and denominator. If, for instance, $H = M = F$, then CSI will have a score of 0.33, while F_1 will have a score of 0.5.

In comparing these three metrics, it is seen that the main difference is with how the hits cell H is emphasized. Specifically, its influence grows to a greater degree as we move from PC to CSI to the F_1 score. In PC, H is added with C, making it an accuracy measure of all "good" event and nonevent categorizations. There are many situations, though, where events are somewhat rare and C is much bigger than any of the other cell values in the contingency table. If correct negatives are not really important to the assessment of the quality of the

---

Critical success index (CSI) or threat score: An event detection accuracy metric, it is the hits divided by the sum of hits, misses, and false alarms.

F_1 score: An event detection accuracy metric, it is twice the hits divided by the sum of misses, false alarms, and twice the hits.

model, then CSI and F_1 are better accuracy metric choices. The difference between these two is subtle, but important—CSI is a true proportion correct, while F_1 is not. CSI, however, is comparing one "good" cell value against two "bad" cell values. The F_1 score, therefore, is considered more equitable because it addresses and alleviates this imbalance by doubling the contribution of H to the formula. The usual ordering of these three accuracy metrics is PC > F_1 > CSI. If, however, C is small and H is large, then F_1 will be larger than PC.

What is a good score for these metrics? This is a subjective call, but a typical value is that the F_1 score is above 0.5 (or CSI of 0.33), indicating that there are more hits than the average of the misses and false alarms. That is, at least half of the points designated as modeled events were also labeled as observed events *and* at least half of the observed events were also labeled as modeled events (technically, an F_1 score of 0.5 means the *average* of these two event state ratios is above 0.5). This is a fairly low bar for the model to achieve; anything lower than this indicates that the model is not reproducing the events very well. It depends on the application; many disciplines, especially medical fields, will put a higher threshold on F_1 before declaring the accuracy of a model to be "good." Figure 10.12 shows the functional relationship of both the F_1 score and CSI with respect to H. Here, the "units" of H are given as the sum of F and M. Demanding that H should be equal to or greater than the *sum* of F and M rather than their average puts a slightly higher threshold on F_1 of 0.67 (or CSI of 0.5) for the model to have its accuracy declared to be good. It is very hard to place a goodness threshold on PC because of the inclusion of C in both the numerator and the denominator. This metric is only useful when C is comparable to H, in which case PC resembles the F_1 score and so a similar minimum threshold of 0.5 can be applied.

For our running example of precipitation at DTW, we have a PC score of 0.88, a CSI score of 0.53, and an F_1 score of 0.69. A PC score of 0.88 is quite high; it indicates that, usually, you can trust the forecast from the HRRR model to give you the correct indication of whether there will be any precipitation during the next 6 hours at DTW. The CSI and F_1 scores are both above not only the nominal goodness level but also the more stringent level of the hits surpassing the sum of the error cell counts. By all of the common metrics and quality judgments, the model is accurate.

> **Quick and Easy for Section 10.6**
>
> Proportion correct is the most common accuracy metric for event detection, but there are others that exclude correct negatives, which are sometimes relatively large.

10.7 Bias

Bias is a measure of whether the model is systematically overestimating or underestimating the observed values. For event detection metrics, this can be represented as a measure of the symmetry of the contingency table.

By far the most common metric that fits this description is the **frequency bias**, FB, defined as the ratio of counts for "model in event state"

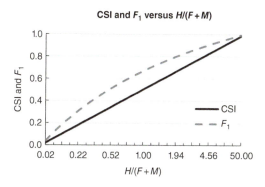

Figure 10.12 Accuracy scores (F_1 and CSI) as a function of H, with H given in units of $F + M$. Note that the x-axis is logarithmic.

Frequency bias (FB): An event detection bias metric, it is the ratio of modeled events to observed events.

to the counts for "observation in event state," like this:

$$FB = \frac{H+F}{H+M} \quad (10.6)$$

Equation (10.6) has a perfect value of one. Note that this does not require F and M to be zero, merely that they be equal. In fact, H could be zero and FB could still have a perfect score of one. It is not assessing the accuracy, but the symmetry of the event state designations. Values greater than one indicate that the model is overpredicting the times or places of being in event state relative to the observations being in event state. FB values less than one reveal that the model underpredicts event state relative to the times that observations were designated as being in event state.

You might be wondering about the necessity of including H in both the numerator and denominator of Equation (10.6). Why not just take the ratio of misses to false alarms? The inclusion of H is because the metric is not assessing the symmetry of the error cells but rather the symmetry of modeled events to observed events. In essence, the inclusion of the H cell count in both places acts as a mitigating factor in the equation. If H is much larger than F and M, then FB asymptotes to one, its perfect score, regardless of the ratio of these two bad-cell values. That is, a simple ratio of F/M could be very far from unity, but if H is large compared to both of these values, then FB will still be close to one.

Note that FB is asymmetric when considered on a linear scale. To make this explicit, let's consider the example of $H = 100$ and then either F or M also being 100, with the other being zero. In the case with $F = 100$, FB would be 2. In the case of $M = 100$, FB would be 0.5. This asymmetry away from unity for the same "error" informs us that FB is better considered on a logarithmic scale, where 0.5 and 2 are equidistant from unity.

To better see this dependence of FB on the relationship of H, F, and M, Figure 10.13 shows a family of curves for FB as a function of H. The units of H are, again, normalized to $F + M$. The many curves on the graph have different ratios of F/M. The curves above unity all have $F/M > 1$, while those below unity all have $F/M < 1$. It is seen that as H gets large, this ratio matters less, with all curves closing in on a value of one.

What is a good score for FB? How close does it have to be to one to be considered "symmetric?" One definition is that F and M are off from each other by a factor of two, and H is only just equal to the larger of these two. As an example of this case, let's set $H = 100$ and then

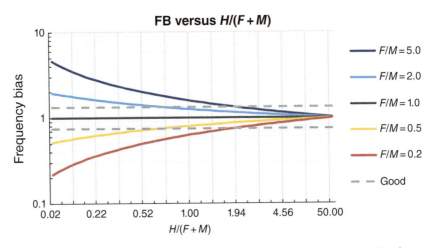

Figure 10.13 Frequency bias as a function of H, with H given in units of $F + M$. The family of curves shows the ratio of F/M. Note that the y axis is a logarithmic scale.

either F or M also be 100 with the other equal to 50. We then have either FB = 1.33 or FB = 0.75. These can be considered minimum thresholds for declaring the symmetry of the contingency table to be good, with the range of good FB scores being between these two values. These are drawn on Figure 10.13 as horizontal gray dashed lines, for reference.

It is seen in Figure 10.13 that it is easier to have an FB score within these two nominal goodness guidelines when H is large relative to F and M. That is, for large x-axis values (>>1), all of the example curves are within the goodness guidelines. The large H score dominates over the imbalance between F and M, making FB very close to unity regardless of the similarity or difference between the two error cell counts. When H is low relative to $F + M$, then it is difficult for FB to be within the goodness guidelines. When H is small, then FB is highly sensitive to differences in F and M. In short, remember that these goodness guidelines are suggestions based on an x-axis value near unity, and the context of your particular data-model assessment should be taken into account.

As you can see, Equation (10.6), it does not include correct negatives, C. FB is unaffected by a large count in this quadrant; it focuses only on the "in event state" cells, comparing "modeled events" to "observed events." It is unconcerned about the times that the model correctly predicted a nonevent. That comes up later.

For the running example, the FB score is 1.59. This is beyond our nominal range of good values; the table is biased in favor of modeled events relative to the observed events. While we knew this from our quick assessment earlier, we now have a specific value attached to this asymmetry of the identified events for our particular case.

> **Quick and Easy for Section 10.7**
>
> Frequency bias, the ratio of modeled events to observed events, is by far the dominant event detection metric for bias.

10.8 Precision

Precision was defined in Chapter 8 as a comparison of the spread of the observed values to the spread of the modeled values. In event detection metrics, we have tossed out the actual values and replaced them with yes/no designations of event status. Therefore, the meaning of "spread" for event detection is fundamentally different than what has been done for fit performance. For event detection, "spread" is defined as a ratio of cell counts rather than variation away from a centroid value, as it was in fit performance.

There is an event detection metric actually called **precision**, formulated as the ratio of hits to all modeled event state designations:

$$\text{Precision} = \frac{H}{H + F} \quad (10.7)$$

Equation (10.7) only uses half of the contingency table—the "model in event state" half. It resembles CSI but without the M in the denominator. Precision ranges from zero to one.

Another metric that is closely related to precision is **recall**, which uses the other half of the contingency table with hits divided by the total number of observed events:

$$\text{Recall} = \frac{H}{H + M} \quad (10.8)$$

Equation (10.8) also looks like CSI but without the F in the denominator. Similarly to precision, it ranges from a worst score of zero to a perfect score of one.

To put Equations (10.7) and (10.8) into the context of spread, both precision and recall

> **Precision (event detection metric):** An event detection precision metric, it is the hits divided by the modeled event count.

> **Recall:** An event detection accuracy metric, it is the hits divided by the observed event count.

measure the converse of spread, **tightness**. Precision is assessing the number of the modeled events that are confined within the hits cell, while recall is assessing the number of observed events that are within the hits cell. By the way, the ratio of recall to precision yields FB.

With only one of the two "bad quadrant" values in the denominator, both precision and recall resemble the F_1 score in terms of how to define a good value from these metrics. That is, for both precision and recall, the curve of the metric score as a function of H follows the F_1 curve in Figure 10.12, with the one change of replacing the normalization of the x-axis to be "F or M," whichever is used in that metric, instead of normalizing H by $F + M$. Like F_1, a nominal good score for precision and recall is 0.5 and a more stringent threshold for a good score is 0.67.

An interesting side note is that the F_1 score was created as the harmonic mean of precision and recall. As discussed back in Chapter 4, the standard formula for harmonic mean uses equal weighting between the values. This equal weighting assumption is used in relating precision and recall to the F_1 score. However, if a different weighting is used, then additional "F_n" scores can be derived. These additional F_n scores are rarely used, so they will not be covered here, but it is trivia regarding the derivation of the F_1 score and its relation to precision metrics. The differently weighted F_n scores could be useful for particular situations in which one of the two errors is more important for the application of the model under assessment.

We can calculate these two scores for our running example. The precision score for that contingency table is 0.56. This beats the nominal score of 0.5 for being called good, but not the more stringent score of 0.67. There are quite a few false alarms relative to the hits, so the model being in event state is not always indicative of actual precipitation coming soon. Recall, however, has a score of 0.90 for this particular contingency table. This is well above both quality levels and reveals that the model is capturing nearly all of the observed precipitation events. The precision of the model is okay, but the recall is excellent.

Quick and Easy for Section 10.8

There is an event detection metric called precision, and its analogous version recall, that assess the model quality within modeled event state and observed event state, respectively.

10.9 Association

Association is an assessment of whether the model reproduces the variation and relative extrema (i.e., up–down trends) of the data. For fit performance, this was best measured by a correlation coefficient. With event detection, the values are replaced with a discrete yes/no designation, or 1 and 0, respectively, to put it into numerical values. Conducting a Pearson linear correlation coefficient on model-data pairs for which all values in both number sets are either 0 or 1 is not particularly informative. That is, because of the discrete jumps from "not in event state" to "in event state," another approach is needed.

To evaluate association of models and data in discrete states, if the model output is systematically offset from the observed value set that yet still reproduces the trends, then the event identification thresholds should be chosen to take this offset into account. That is, the offset should be removed in the conversion to yes/no event status by choosing an appropriate modeled event identification threshold that differs from the observed event threshold. Metrics in the association category are useful when seeking to maximize the good quadrants

---

Tightness: The inverse of spread, tightness has a high score when there are few counts in the bad quadrants of the contingency table.

relative to the bad quadrants of the contingency table.

10.9.1 Odds Ratio

A good choice for association is a metric called the **odds ratio**, θ, defined as the product of the good cell counts to the product of the bad cell counts:

$$\theta = \frac{H \cdot C}{M \cdot F} \tag{10.9}$$

For a perfect data-model comparison in which all events and nonevents are correctly predicted, M and F are both zero and therefore θ is infinity. If all four cells have an equal count, then θ is one.

This metric tests the symmetry of the contingency table in a different way from that of bias. It does not test the symmetry between F and M, it tests the symmetry of good versus bad prediction. This makes the description of association sound like an accuracy metric, but it does not simply count cells. If the sum of the good cells is higher than the sum of the bad calls, then θ will most likely be greater than one, but a high θ score is not a given. This is because the symmetry within the pairs of cells (i.e., diagonal symmetry within the contingency table) is rewarded. This reward for symmetry is what makes the odds ratio a good

---

Odds ratio (θ): An event detection association metric, it is the product of the good cell counts divided by the product of the bad cell counts.

---

metric for the association category rather than the accuracy category.

Let's consider a few examples, with their contingency tables shown in Figure 10.14. The first example has far more hits than correct negatives, while the other two cells have an equal count. The counts in the good cells outnumber the counts in the bad cells, 110 to 70. For this table, the F_1 score is very good—0.74. Similarly, FB is exactly one and precision and recall are both identical to F_1 at 0.74. Based only on these scores, you might declare this model to be good at reproducing events. The θ score, however, is low, at only 0.82. The odds ratio metric rewards the equality between F and M, while penalizing the vast difference between H and C.

The second example in Figure 10.14 has the same summed number for the good and bad cells (100), but the bad cells are lopsided toward F, while the good cells are symmetrically distributed. In this case, the F_1 score is 0.5 and the FB score is 2.3, not particularly good scores for those metrics, but the θ score is 2.8. This is a fairly large θ score, even though the total counts in the good and bad cells were equal.

The final example is a contingency table that is both good and symmetric, with H and C equal to 100 and F and M set to 50. For this table, the F_1 score is 0.67, the FB score is exactly 1, and the θ score is 4; all of these metric scores are good.

What is normally considered a good value for θ? As a bare minimum threshold for goodness, θ being above one indicates that the product of the correct cell counts is bigger than the

---

Lopsided good cells ($\theta = 0.82$)

Contingency table	Model not in state	Model in state
Data in state	35	100
Data not in state	10	35

Lopsided bad cells ($\theta = 2.3$)

Contingency table	Model not in state	Model in state
Data in state	10	50
Data not in state	50	90

Well-balanced table ($\theta = 4$)

Contingency table	Model not in state	Model in state
Data in state	50	100
Data not in state	100	50

Figure 10.14 Three example contingency tables to highlight the relationship of the odds ratio metric to table symmetry.

product of the error cell counts. A more reasonable level for declaring the θ metric score to be of high quality is a threshold of at least $\theta = 2$, and many will set it even higher, at 3 or 4 before declaring the score to be good.

For our running example of precipitation at DTW airport, θ is 58. This is a very high value, well beyond the goodness criteria just discussed. This means that the HRRR model was not getting many more correct negatives than hits at the expense of having many observed events go unpredicted as misses. Rather, it means that there were simply many more observed nonevents than observed events and that the model does very well at separating these. Even if the values within the two error cells were balanced—their sum is 1068, so let's assume $F = M = 534$—the odds ratio would still be 27. The magnitude of the count values in the hits and correct negatives cells in the contingency table in Figure 10.9 far outweighs the error cell counts, no matter how they are distributed.

One drawback of θ occurs when any of the contingency cells is zero. If a good cell is zero, then θ will be zero, regardless of the counts in the other good cell. If either error cell is zero, then θ will be infinite, even if the other bad cell count is large. This can be avoided with a small rearrangement of the values, as discussed in the next section.

> **Quick and Easy for Section 10.9.1**
>
> Odds ratio is an excellent event detection choice for the association metrics category, but it is undefined if any contingency table quadrant is zero.

10.9.2 Odds Ratio Skill Score

There are two extreme end values for the θ score. One of these arises for the case when one of the values in the denominator of Equation (10.9) is zero. In this situation, the θ score goes to infinity. Similarly, if either or both values in the numerator of Equation (10.9) are zero, then the θ score will be zero. Because the best value of θ is infinity, this metric is unlike most others that have a perfect score of either zero or one. An alternative metric has been devised, based on the odds ratio, which transforms the best score from infinity to unity. This new version is called the **odds ratio skill score**, ORSS. It does not use the standard formula for a skill score because the inclusion of the perfect score of θ in the denominator would make the standard formula zero for all values of θ. This nonstandard formula is one reason why it is being listed here, in the discussion about association metrics, rather than later in the section on skill (Section 10.11). It is a convenient conversion of θ into a more understandable metric.

The ORSS metric is defined this way:

$$\text{ORSS} = \frac{\theta - 1}{\theta + 1} \quad (10.10)$$

Applying the formula for θ in Equation (10.9), we get:

$$\text{ORSS} = \frac{(H \cdot C) - (M \cdot F)}{(H \cdot C) + (M \cdot F)} \quad (10.11)$$

Equation (10.11) removes most of the possibility of division by zero. The perfect score for ORSS, when either or both of M and F are zero, is one, while the worst possible score for ORSS, when either or both of H and C are zero, is negative one. If the odds ratio is one, with equal products for the good and bad cells, then ORSS will be zero. Another common name for ORSS is Yule's Q score, and so it is sometimes referred to as Q instead of ORSS.

Nominally, a good score for ORSS is anything above zero. This indicates that the model is better than purely random chance at predicting the event state of the observations. Many

---

Odds ratio skill score (ORSS): An event detection association metric, it is the odds ratio minus one divided by the odds ratio plus one.

use a higher threshold for declaring ORSS to be of high quality. Applying a θ score of 2 to Equation (10.10) yields an ORSS of 0.33. Another threshold is an ORSS of 0.5, which corresponds to a "good" θ score of 3.

For the running example, the ORSS score is 0.97. This is well above nearly any criteria you could reasonably impose as a goodness threshold. According to this metric, the model captures the "trends" of the observations moving between event state and nonevent state exceptionally well.

> **Quick and Easy for Section 10.9.2**
>
> The odds ratio skill score rearranges the odds ratio into a form that is more robust to zero-count quadrants of the contingency table.

10.9.3 Matthews Correlation Coefficient

A final measure of correlation is one originally developed by Pearson but commonly known as the **Matthews correlation coefficient**, MCC, named for a later scholar that popularized its use with event detection assessments. This is directly related to the significance test shown in Section 10.5 and is mathematically equivalent to the χ^2 value divided by the total number of data-model pairs N, all within a square-root operator. The typical form that it takes is this:

$$\text{MCC} = \frac{(H \cdot C) - (F \cdot M)}{\sqrt{(H+F)(H+M)(F+C)(M+C)}} \tag{10.12}$$

Equation (10.12) has exactly the same numerator as ORSS in Equation (10.10), but the denominator has the multiplication of all of the column and row totals from the entire contingency table. Like the Pearson linear correlation coefficient, R, presented in Chapter 6 and listed as an association category metric for fit performance in Chapter 9, MCC ranges from −1 to +1, with a score of zero indicating correlation equivalent to random scatter. However, like the application of R in Chapter 9, a perfect score for MCC is one, not negative one. That is, it is now one-sided, because the model is supposed to be predicting the observed events, and perfect anticorrelation indicates a truly bad reproduction of those observations.

Like ORSS, anything above zero indicates better than random chance and therefore is a skillful model at reproducing the observed events. That is not a satisfying answer, though, because values barely above zero are not particularly good. If H and C are equal and double the values of F and M (meaning that θ is 4 and ORSS is 0.6), then MCC is 0.33. It is harder to score well on MCC than on θ or ORSS.

For our running example, MCC is 0.65. This is a high value for this metric and confirms that the HRRR model quite well captures the shifts of the observations from nonevent to event and back again.

> **Quick and Easy for Section 10.9.3**
>
> The Matthews correlation coefficient is a conversion of the χ^2 significance test into an association metric.

10.10 Extremes

Extremes is another category for which the definition is different between the fit performance and event detection groupings. For event detection, this category focuses on highlighting hits when hits are rare. In the accuracy category (Section 10.6), a similar highlighting was done by simply excluding C from the summations. In the extremes category, the metrics often use other mathematical formulations besides simple summations to still include C in the assessment, but minimize its

---

Matthews correlation coefficient (MCC): An event detection association metric, it is the square root of the ratio of the chi-squared coefficient to the total count.

dominance over H. This is further detailed in the two metrics described in this section.

One well-established metric that fits these criteria is the **extreme dependency score**, EDS, which uses the natural logarithm function to downplay large differences between C and H:

$$EDS = \frac{2 \cdot \ln\left(\frac{H+M}{N}\right)}{\ln\left(\frac{H}{N}\right)} - 1 \quad (10.13)$$

A bit of algebra yields a slightly more intuitive form for EDS:

$$EDS = \frac{2 \cdot \ln(N) - 2 \cdot \ln(H+M)}{\ln(N) - \ln(H)} - 1 \quad (10.14)$$

The EDS formula in Equation (10.13) is listed two ways, one version with ratios of values inside the natural logarithms and the other converting these ratios into subtractions of natural logarithms. In the first version, all of

---

Extreme dependency score (EDS): An event detection extremes metric, it uses logarithms to highlight the differences in the small-count quadrants, especially when hits are rare.

---

the natural logarithms yield negative values. This happens because the ratios are all less than one. The numerator natural logarithm is equal to or less than the denominator natural logarithm, so the ratio of these two negative values yields a positive number between zero and one. The multiplication by two and subtraction of one then makes EDS a metric with a range from −1 to +1. In the second version of EDS, all of the values inside the natural logarithms are counts from the contingency table and therefore positive integers.

The dependency of Equation (10.13) on the cell counts is not intuitive, so let's spend some time understanding it. Figure 10.15 presents the value of EDS as a function of H and M. The H values vary along the x-axis, normalized by N to be values between 0 and 0.2, and the various curves on the plot show different levels of M (again, normalized to N). An example from the plot: if H and M are both equal to 0.1N, then EDS is 0.4. To continue with this, holding $H/N = 0.1$, it is seen that as M/N increases, EDS decreases. For $H/N = 0.1$, an M/N value of 0.05 (that is, M is half of H) yields an EDS of 0.65, while having M/N equal to 0.2 (that is, M is twice H) gives an EDS value of 0.046. This illustrates the very quick change of EDS as a function of M.

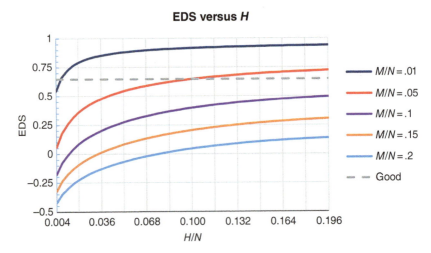

Figure 10.15 The relationship of EDS to H, with H given as a fraction of N. The curves show different proportionalities of M as a fraction of N.

The smaller the values of H and M, the larger the range of EDS. Instead of H being 10% of N, let's set it to 1%, so $H/N = 0.01$. In this case, values of M/N equal to 0.005, 0.01, and 0.02 (half, the same as, and double H, respectively) produce EDS values of 0.82, 0.70, and 0.52, respectively. The range of EDS values is both higher and lower than the example values of EDS for $H/N = 0.1$. When there are very few observed events and $H + M$ is very small compared to N, EDS highlights any successes that the model can produce, rewarding that success more as $H + M$ gets smaller relative to N. To put it another way, we have seen that EDS does not give the same score for the same H/M ratio, but rather EDS increases for the same H/M ratio as $(H + M)/N$ decreases.

A perfect score, which is achieved when $M = 0$ and the natural logarithm terms cancel, is EDS equal to one. This is true regardless of the value of H; it can be any number, large or small (but at least one), and EDS will be one if $M = 0$. In Figure 10.15, the $M/N = 0$ curve would be exactly along the unity line at the top of the plot. In the other direction, EDS can be negative when M is significantly larger than H. As seen in the curves in Figure 10.15, a negative EDS value is easier to achieve when $H + M$ is a significant fraction of N. The lowest score for EDS is -1; EDS approaches this when $M \gg H$, and therefore the denominator logarithm value is much bigger than the numerator and the entire first term goes to zero. If H is zero and $M > 0$, then the denominator of EDS explodes to $-\infty$ and EDS $= -1$. If both H and M are zero, then there were no observations in event state and the metric is undefined because the contingency table was poorly constructed for this metric.

A good EDS score should reflect a contingency table with more hits than error cell counts. One choice is a value of 0.65, which has H at twice the value of M when $H/N = 0.1$. But this value is too relaxed for low H/N and too strict for high H/N. A nonlinear "high-quality threshold" should be adopted. Because the designation of "good" is a subjective call,

a threshold of 0.65 is good enough as a basic guideline.

For our running example of rainy days, we have an EDS value of 0.89. This is a high value. With nearly one in six of the observations being designated as an event, the observed events are not exactly rare, so it is easier to score higher with EDS than if this frequency were lower, but this is still quite a good score.

It is seen that Equation (10.13) includes M in the natural logarithm numerator but not F. EDS focuses on the observations being event state and F only enters the equation along with C as part of N. That is, EDS can be manipulated to yield a perfect score by setting the model to always predict events. In this case, F might be very large, but, because $M = 0$, EDS will be one. This imbalance in how EDS uses the values of the contingency table is addressed in a related metric known as SEDS, the **symmetric extreme dependency score**:

$$\text{SEDS} = \frac{\ln\left(\dfrac{H+M}{N}\right) + \ln\left(\dfrac{H+F}{N}\right)}{\ln\left(\dfrac{H}{N}\right)} - 1 \quad (10.15)$$

As with EDS, this can be rewritten so there are no ratios within the logarithms:

$$\text{SEDS} = \frac{2 \cdot \ln(N) - \ln(H+M) - \ln(H+F)}{\ln(N) - \ln(H)} - 1$$

(10.16)

Comparing Equations (10.13) and (10.15), it is seen that the only difference is the replacement of one of the natural logarithms that use $H + M$ in the numerator with a natural logarithm that instead uses $H + F$ in the

---

Symmetric extreme dependency score (SEDS): An event detection extremes metric, it is a balanced version of EDS that includes both misses and false alarms in the logarithms of its numerator.

---

numerator. This provides balance to SEDS in how it treats the two "bad" quadrants of the contingency table. SEDS has the same range as EDS, with +1 being a perfect score and −1 being the worst possible value.

The similarity of the formulas means that the relationship of SEDS to H is exactly like that of EDS. The relationship of SEDS to M and F is also similar to M in EDS. Figure 10.16 shows several curves relating SEDS to the values of F and M, for a given H/N ratio of 0.01, 0.1, and 0.2 in the three panels. It is seen that the curves are systematically higher in the right-most panel; SEDS scores increase with increasing H. The curves within the panels are also systematically arranged inversely with the $F + M$ proportionality to N; the higher the count in the bad cells, the lower the SEDS

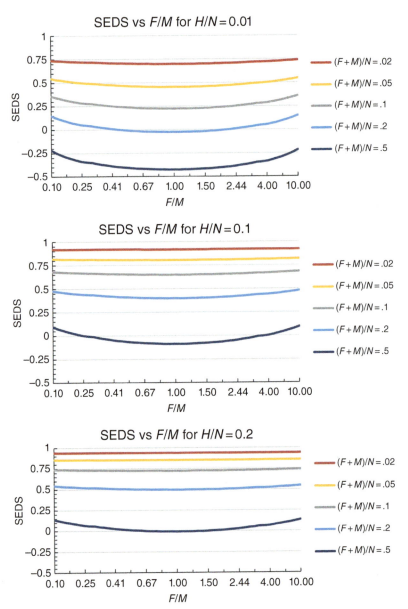

Figure 10.16 The relationship of SEDS to F/M, with H/N set to 0.01 (top), 0.1 (middle), and 0.2 (bottom). The curves show different proportionalities of $M + F$ as a fraction of N.

score. The curves are closer together in the right panel and more spread out in the left panel; SEDS is more sensitive to the bad cell counts when H is small and less sensitive when H is large. Finally, there is a curvature evident in the lines of Figure 10.16, especially for high $F + M$ count values; the SEDS formulation produces a lower score when F and M are equal, and SEDS—as indicated in its name—is symmetric with respect to the asymmetry between F and M in the contingency table.

A single threshold value for declaring a SEDS score to be good is hard to define. Like EDS, it is dependent on the ratio of H to N. It is easier to get high SEDS scores for larger H/N ratios. Like EDS, though, a good general-purpose guideline is 0.65.

For our running example, SEDS is 0.65. The value is lower than the EDS score because of the inclusion of F in the numerator, which factors the number of false positives into the SEDS score. It is still well above zero, so this is a good score, just not quite as outstanding as the EDS score. This tells that if $F > M$, then the model overpredicts events.

Which one to choose, EDS or SEDS? The standard answer to such a question is that, of course, it depends. If the focus of the assessment is on observed event states, then EDS might be the better choice. If the emphasis of the data-model comparison is more general, then SEDS is a more robust measure of hits accuracy when hits are rare.

> **Quick and Easy for Section 10.10**
>
> The best event detection choice for the extremes category is the symmetric extreme dependency score because it is balanced in its usage of M and F.

10.11 Skill

Skill scores are abundant across the event detection analysis world. Several metrics that are skill scores have already been introduced in the two previous sections, such as ORSS, EDS, and SEDS. Here, three more skill scores are provided, with the pros and cons of each discussed.

10.11.1 Heidke Skill Score

The most common metric in this category is known as the **Heidke skill score**, HSS. It is based on PC as the metric with the reference model being random chance, as defined for the expected values in each cell in Section 10.5 when significance of the table was discussed. The final form of HSS is this:

$$\text{HSS} = \frac{2\left[(H \cdot C) - (F \cdot M)\right]}{(H+M)(M+C)+(H+F)(F+C)}$$

(10.17)

Equation (10.17) has several terms, so let's spend some time deciphering its relationship to the contingency table values. For a perfect contingency table, with $F = M = 0$, HSS reduces to $2HC/2HC$; that is, it becomes one. When all cell counts are equal, then the two subtracted terms in the numerator are equal and HSS is zero. For a perfectly bad contingency table, when $H = C = 0$, we see that HSS reduces to $-2FM/2FM$ and therefore HSS $= -1$. So, HSS has a range from -1 to $+1$, with any score greater than zero indicating that the model is better than random chance at predicting observed events.

It is useful to construct a few example contingency tables and discuss the relationship of HSS as the cell counts are changed. To make this easy, let's use the same three contingency tables we considered in Section 10.9, presented in Figure 10.14. For the first table, with a large number of hits but few correct negatives, HSS is -0.037. Yes, even though the model

> **Heidke skill score (HSS):** An event detection skill metric, it is based on correct proportion and uses random chance as the reference model.

produced more correct predictions than incorrect designations, the imbalance between H and C is penalized and HSS is below zero; this model result is ever-so-slightly worse than random chance. For this number of data-model pairs, this difference with zero is probably not significant; uncertainty on metrics scores will be discussed in Chapter 12. For the second table in Figure 10.14, with equal total counts in the good and bad cells but lopsided $F{:}M$ ratio, the HSS is 0.14, slightly above zero but not by very much. In the third table of Figure 10.14, with balanced count levels of 100 in each good cell and 50 in each bad cell, the resulting HSS is 0.33. In this case with $H = C$ and $F = M$, the C and M variables can be removed from Equation (10.17) and the formula for HSS reduces to $(H - F)/(H + F)$. It takes a 3:1 ratio of $H{:}F$ in order to achieve an HSS of 0.5. That is, it is even harder to get a high HSS score than it is to get a high CSI score (which is 0.6 for $H/F = H/M = 3$).

One interesting point to note about HSS is that the numerator of Equation (10.17) is nearly identical to that for ORSS in Equation (10.10). The only difference in the numerators is a factor of two. The denominators, however, are completely different formulations of the cell count values.

Another feature of HSS is that it is robust to imbalances because the formula is symmetric. Take the two examples from Figure 10.14 with a large H relative to C or a large F relative to M. In the first case, swapping the imbalance so that $C = 100$ and $H = 10$ yields exactly the same HSS score. Same thing in the second example—swapping the imbalance so that $F = 10$ and $M = 90$ produces the same HSS score. Both of those examples, however, include balanced values in the other two cells. To further explore this concept, Figure 10.17 includes two more example contingency tables. In these two examples, H and C are imbalanced (in favor of C), as well as F and M being imbalanced. In the first table, F is larger, while the F and M values are reversed in the second table, with M larger. In both cases, HSS = 0.51. This feature of HSS is because it is based on PC, which is a balanced metric.

What HSS value should be considered good? Some put this bar as anything with skill relative to the chosen reference model, which means an HSS value above zero. As seen three paragraphs earlier, though, it is difficult to raise HSS to 0.5; the contingency table has to be strongly in favor of the good cells over the bad cells. Even when it is the case that the sum of the good cells is much higher than the sum of the error cells, HSS could still be negative, as seen in the first example table from Figure 10.14, discussed in Section 10.9.1. If the table is well balanced, then H being a factor of two larger than either M or F, which would normally be considered a good prediction model, yields an HSS value of 0.33. So, a goodness threshold for HSS should be somewhere between 0.0 and 0.3. This can be considered a transition zone for the quality indication from HSS; values above this indicate that there is significant skill for the model.

For our running example of the near-term precipitation forecast at DTW by the HRRR

Table favoring false alarms

Contingency table	Model not in state	Model in state
Data in state	10	40
Data not in state	100	30

Table favoring misses

Contingency table	Model not in state	Model in state
Data in state	30	40
Data not in state	100	10

Figure 10.17 Two contingency tables, one being the transpose of the other. Both would be qualitatively considered good, but have imbalances in both the H-to-C ratio and the F-to-M ratio.

model, the HSS value is calculated to be 0.62. This is well above the goodness criteria just discussed and indicates, yet again, that the model is quite adept at predicting when precipitation is coming.

> **Quick and Easy for Section 10.11.1**
>
> The Heidke skill score, arguably the most well-known of all event detection metrics, is a good measure of skill against random chance.

10.11.2 Peirce and Clayton Skill Scores

The **Peirce skill score**, PSS, is even older than the Heidke skill score in its original derivation and usage. It has this form:

$$\text{PSS} = \frac{(H \cdot C) - (M \cdot F)}{(H + M)(F + C)} \quad (10.18)$$

Considering the three nominal cases of a perfect table with $M = F = 0$, a perfectly "flat" table with all cells equal, and a perfectly bad table with $H = C = 0$, PSS yields scores of +1, 0, and −1, just like HSS. Values above zero indicate that the model is better than a random chance matrix, as defined from the contingency table counts in Section 10.5.

This skill score has several interesting features worth pointing out. The numerator of Equation (10.18) is exactly that of ORSS in Equation (10.10). The denominator is a multiplication of the total counts as sorted by the observations, being the product of observed events and observed nonevents. Also, it has several different names in the literature, sometimes referred to as the true skill statistic or the Hanssen-Kuiper skill score.

Let's again use the three example contingency tables in Figure 10.14 to explore the relationship of PSS to features of the count levels. For the first example, PSS is identical to HSS at −0.037. For the second example, PSS is 0.19, slightly higher than HSS. For the third table, PSS is again identical to HSS at 0.33. Comparing Equation (10.17) with Equation (10.18), it can be seen that, when $M = F$, PSS and HSS become identical in their formulations.

Now consider the two additional example contingency tables introduced in the previous section, in Figure 10.17. Here, the tables are imbalanced in both H-to-C and F-to-M ratios, with the difference between the two tables being that the F and M values are swapped in the second example. For the first table with $F > M$, PSS is 0.57, slightly higher than the corresponding HSS score. For the second table, with $M > F$, PSS drops to 0.48, a little bit below the HSS score. This change in score is because of the denominator of PSS, which is not symmetric to swapping of the values. The two terms that are multiplied together in the denominator of Equation (10.18) are based on observed event status. Specifically, $H + M$ is the total of observed events and $F + C$ is the total of observed nonevents. The multiplication of these two terms is maximized when the totals are equal. In the first example, the totals are 50 and 130 for observed events and nonevents, respectively. These two numbers are not particularly close. For the second example, the multiplied values are 70 and 110, yielding a larger product than for the first example table. This larger product reduces PSS. That is, PSS is systematically higher when there are fewer observed events in the selected data set.

A very similar metric is the **Clayton skill score**, CSS:

$$\text{CSS} = \frac{H \cdot C - F \cdot M}{(H + F)(C + M)} \quad (10.19)$$

Peirce skill score (PSS): An event detection skill metric, it is a skill score based on the split between observed events and observed nonevents.

Clayton skill score: An event detection discrimination or skill metric, it is based on the split between modeled events and modeled nonevents.

Equation (10.19) is formulated very similarly to the PSS in Equation (10.18), except that instead of discrimination metrics being subtracted, it is reliability metrics that are subtracted. That is, for CSS, the denominator is products of the modeled events and nonevents rather than the observed events and nonevents. CSS has essentially the same strengths and limitations as PSS.

For our running example, PSS is 0.77 and CSS is 0.54. The PSS score is even better than the HSS score reported in the previous section. The CSS score is slightly lower because of $F > M$ and the reorganization of the denominator reveals this imbalance. Because, in general, these two skill scores are usually pretty similar to HSS, the criteria for goodness is also essentially the same as discussed in Section 10.11.1. Specifically, both scores are good for our running example, and the PSS value is extraordinarily high because it emphasizes M, the error cell with the relatively small count.

> **Quick and Easy for Section 10.11.2**
>
> The Peirce skill score and Clayton skill score are also nice event detection metrics for skill, but they are susceptible to asymmetries in the contingency table.

10.11.3 Gilbert Skill Score

Let's cover one more event detection skill score that has useful features for some applications, the **Gilbert skill score**, GSS, which uses CSI as the base metric:

$$\text{GSS} = \frac{H - H_{\text{ref}}}{H + F + M - H_{\text{ref}}} \quad (10.20)$$

In Equation (10.20), the "reference model accuracy" $H\text{ref}$ value is the expected value

---

Gilbert skill score (GSS): An event detection skill metric, it is a skill score based on the critical success index and therefore minimizes the influence of correct negatives.

(E_{11}) found earlier in the significance test (Section 10.5):

$$H_{\text{ref}} = \frac{(H+F)(H+M)}{N} \quad (10.21)$$

Once again, let's use our standard three canonical sets of values in Equations (10.20) and (10.21) to conduct an initial examination of the properties of GSS. For a perfect model, with $M = F = 0$, we get GSS = $(H - H_{\text{ref}})/(H - H_{\text{ref}})$, which is one. When all cells have equal value, $H\text{ref}$ reduces to H and GSS becomes zero. For a perfectly bad model, with $H = C = 0$, then $H_{\text{ref}} = (MF)/(M + F)$, and GSS takes the rather complicated form of $(-MF)/(F^2 + MF + M^2)$. If either M or F is zero, then this "worst case" value is zero. If both M and F are nonzero, then it is negative. If M and F are equal, then this extreme value of GSS is $-1/3$.

Let's also calculate GSS for the three example tables in Figure 10.14. For the first table with large H, the GSS value is -0.018, just under the random score of zero. For the second table with large F, the GSS score is 0.074, just barely above zero. For the third table with double the counts in the good cells over the bad cells, GSS is 0.2. All of these are closer to zero than either HSS or PSS.

For the two example contingency tables in Figure 10.17, Equation (10.20) gives the same value for both, GSS = 0.34. Again, this is below the score for either HSS or PSS for these tables. Two things are learned from all of these metrics scores: the first is that the GSS is robust to imbalances in the contingency table, like the Heidke skill score; and the second is that it is difficult to score well with GSS, as expected because it is based on the more restrictive of the accuracy measures, CSI.

The advantage of GSS is that it minimizes the influence of C. If the correct negatives count is large compared to other cells, then this unbalanced value positively distorts the score from the other skill scores described in Sections 10.11.1 and 10.11.2. GSS only includes C in the definition of H_{ref}, and even there only in the denominator. Examining

Equation (10.21), if C is exceptionally large, then H_{ref} will tend toward zero. It removes itself from influence on GSS as its dominance over the contingency table grows.

The GSS score can be calculated for our running example of precipitation forecasting. For this particular case, GSS = 0.45. Because GSS downplays the influence of C, yet C was by far the largest count in the contingency table of Figure 10.9, this score suffers relative to those for HSS, PSS, and CSS. It is still well above our criteria for declaring the model a high-quality predictor of the observed values.

Quick and Easy for Section 10.11.3

The Gilbert skill score minimizes the influence of the correct negatives quadrant of the contingency table.

10.12 Discrimination

Discrimination, as defined in Section 8.5.3 (Chapter 8), is a subsetting category using only a particular part of the data range for the calculation of the metrics. For fit performance, any of the metrics from the other categories could be applied. This is not the case for event detection, where the values have been converted to yes/no settings and split into four quadrants by event identification thresholds for both the observed and modeled value sets. This dramatic change of the number sets, however, can be used to advantage—we already have a natural break point built into the value sets for partitioning the data set into subsets.

Unlike fit performance, metrics in discrimination for event detection have their own names. The most common pair of metrics include the **probability of detection**, POD,

---

Probability of detection (POD): An event detection discrimination metric, it is the ratio of hits to observed events (hits plus misses).

and **probability of false detection**, POFD, defined with cell values exclusively above and below the observed event threshold, respectively. These are arguably the two most well-known and widely used metrics in the event detection grouping. The formula for POD has already been introduced earlier by its different name, recall:

$$\text{POD} = \frac{H}{H+M} \quad (10.22)$$

Equation (10.22) has additional names, such as true positive rate (TPR), power, sensitivity, and recall, depending on the discipline. POD ranges from 0, its worst score, to 1, its best. POFD is defined like this:

$$\text{POFD} = \frac{F}{F+C} \quad (10.23)$$

Equation (10.23) has several names also, such as the significance level or the false alarm rate. While POFD also ranges from 0 to 1, note that POFD has a "bad" cell value in the numerator, while POD has a "good" cell value in the numerator, so the best score for POFD is 0, while its worst value is 1.

These two formulas can be combined as a measure of overall goodness for the contingency table. In fact, this combination has already been given—it is Equation (10.18), the PSS, which can be rewritten as POD − POFD. While it was not originally derived this way, the use of only four values in all of event detection metrics means that some combinations serendipitously arise through other paths.

The reverse of both of these discrimination metrics are also used in some contexts, so it is important to list those as well. Swapping M for

---

Probability of false detection (POFD): An event detection discrimination metric, it is the ratio of false alarms to observed nonevents (false alarms plus correct negatives).

H in the numerator of Equation (10.22) gives the **false negative rate**:

$$\text{FNR} = \frac{M}{H+M} \quad (10.24)$$

This is also known as the miss rate. The reverse formula for POFD is known as **specificity**:

$$\text{Specificity} = \frac{C}{F+C} \quad (10.25)$$

Equation (10.25) is also called the true negative rate.

For all of these discrimination metrics, a minimum level to call it "good" is 0.5. It is important to note the direction of "high quality" away from this: POD and specificity are striving for values close to one for high-quality discrimination, while POFD and FNR have an error cell in the numerator and therefore a low score is desired.

For our running example, POD and POFD are 0.90 and 0.13, respectively. Their converses, FNR and specificity, are 0.10 and 0.87, respectively. These are all very good scores. Among the observed events, most are also modeled events. Similarly, among the observed nonevents, most are also modeled nonevents. These numbers back that up and provide quantitative context, especially when compared against other similar metrics calculations.

> **Quick and Easy for Section 10.12**
>
> Discrimination for event detection data are those metrics that only use the contingency table values from only one side of the observed event definition threshold.

False negative rate (FNR): An event detection discrimination metric, it is the ratio of misses to observed events.

Specificity: An event detection discrimination metric, it is the ratio of correct negatives to observed nonevents.

10.13 Reliability

Reliability is another subsetting technique (Section 8.5.3, Chapter 8), in which the values are considered in only certain ranges as defined along the model value set. Again, because we have a predefined break point along this axis of the modeled event identification threshold, and the real values have been converted to yes/no binary designations, this is the natural subset on which reliability metrics should be based.

The analogous set of four metrics can be defined. Because there is somewhat less usage of reliability metrics in the geosciences, there is no clear order of importance among these metrics. Therefore, the ordering of their presentation will follow that in the last section. The first to list, then, is the metric when the model is in event state with the correct cell, hits, in the numerator. This is one that we have already seen as precision in Section 10.8, but here it will be listed with its alternative name, the **positive predictive value**, PPV:

$$\text{PPV} = \frac{H}{H+F} \quad (10.26)$$

Like POD, PPV ranges from 0 to 1 with a best score of 1.

The analogous reliability metric to POFD has M in the numerator over the sum of times that the model was not in event state. This is known as the **miss ratio**, MR:

$$\text{MR} = \frac{M}{M+C} \quad (10.27)$$

Equation (10.27) has several other names, most notably the false omission rate. The

Positive predictive value (PPV): An event detection reliability metric, it is the ratio of hits to modeled events (hits plus false alarms).

Miss ratio (MR): An event detection discrimination metric, it is the ratio of misses to modeled nonevents.

unfortunate aspect of this metric's name is that it is very close to the alternative name of FNR, from Section 10.12, the miss rate. Like POFD, MR ranges from 0 to 1 with a best score of zero. Putting F in the numerator over the sum of times that the model was declared to be in the event state yields the **false alarm ratio**, FAR:

$$\text{FAR} = \frac{F}{F+H} \quad (10.28)$$

Note that this metric's name is also very close to another—the false alarm rate, which was an alternative name for POFD. When you come across a usage of the FAR acronym, be sure to double check about which of these two metrics it is referring. While both metrics range from 0 to 1 and strive for a perfect score of 0, the change of term in the denominator from H to C could have a rather different score between the two. To avoid this confusion, FAR is sometimes referred to as the **false discovery rate**, FDR.

The final metric in reliability that matches the formulation of the four in discrimination uses the other half of the contingency table, the cells counting the times the model was declared to be not in event state, with C in the numerator. This is usually called the **negative predictive value**, NPV:

$$\text{NPV} = \frac{C}{C+M} \quad (10.29)$$

Equation (10.29) is an analogous formula to the specificity mentioned in the previous section, with 1 as the perfect score and 0 as the worst.

As with the discrimination metrics, the reliability metrics are entirely fractions with one cell's value divided by the sum of that cell with another. Values closer to perfect than worst on each scale, whether perfect is 0 or 1, are nominally considered good.

One more metric can be listed here in the reliability category that has seen occasional use in the geosciences: the **forecast ratio**, FR. This is the ratio of hits to false alarms:

$$\text{FR} = \frac{H}{F} \quad (10.30)$$

This is very similar to PPV, but has a completely different range—from 0 to infinity—with ∞ being the best. It provides a quickly calculated assessment of predictive value, that is, whether you can trust a modeled event to also be an observed event.

These reliability subsetting metrics can be calculated for our running example of precipitation forecasts at DTW. We get PPV = 0.57 and NPV of 0.98, with the converse formulas yielding MR = 0.02 and FAR = 0.43. They are good, especially for the modeled nonevents; the HRRR model has a fantastic NPV indicating that when it predicts no precipitation, there is essentially no chance that there will actually be precipitation in the next 6 hours at DTW. Finally, the running example has a forecast ratio score of FR = 1.3. This is a good but not a great score.

> **Quick and Easy for Section 10.13**
>
> Reliability metrics are those that only use values from one side or the other of the modeled event identification threshold.

10.14 Summarizing the Running Example

Let's recap what has been learned from the calculation of all of these event detection metrics for the case of the HRRR model's 6-hour

---

False alarm ratio (FAR) or false discovery rate (FDR): An event detection discrimination metric, it is the ratio of false alarms to modeled events.

Negative predictive value (NPV): An event detection discrimination metric, it is the ratio of correct negatives to modeled nonevents.

Forecast ratio (FR): An event detection discrimination metric, it is the ratio of hits to false alarms.

Table 10.1 A summary of a few key metric scores for the running example.

Metric	Score
χ^2	3570
PC	0.88
F_1	0.69
FB	1.59
Precision	0.56
ORSS	0.97
SEDS	0.65
HSS	0.62
GSS	0.45

forecast of precipitation as recorded at the Detroit Metro Airport weather station. A few key metrics are summarized in Table 10.1. If we had only calculated one metric, it might have been PC, for example, or its related skill score, HSS. These are all-encompassing metrics that use the entire contingency table in a symmetric and balanced equation. They are very useful at distilling the comparison down to a single value. For this case study, both of these metrics are quite high, with PC = 0.88 and HSS = 0.62. These are both great scores and, like with the Dst data-model comparison of the previous chapter, we could have stopped there and declared victory. Continuing with additional metrics calculations, though, allows us to understand the nuances of the contingency table that are hidden in just these two common metrics.

An important fact that is quantified by a robust analysis is that the table is biased in favor of false alarms over misses. This is highlighted by several of the metric scores. First, it revealed itself the FB score being quite a bit larger than one. It also showed up in the rather large difference in the scores for precision and recall; the first was okay, while the second was excellent. It showed up in a similar way through the reliability metrics of PPV and NPV. This asymmetry is not revealed by any of the accuracy metrics, which are symmetric in their usage of the two error cells.

Another aspect of the data-model relationship that was learned through additional metric usage is the high association scores for θ and ORSS. These scores indicate that the asymmetries in both the good and bad cells—the first pair lopsided toward C and the second pair lopsided toward F – are dwarfed by the substantial advantage in the counts of the good cells over the bad cells.

The conclusion from this assessment is that you can trust the weather report to nearly always provide an accurate answer to the question, "will it rain within the next 6 hours?" That is not a long lead time, but enough for you to plan the day's outing. To trust the forecast 6 days in the future instead of only 6 hours would take a different assessment than the one conducted throughout this chapter.

> **Quick and Easy for Section 10.14**
>
> Nuances of the data-model relationship are brought into focus by the use of many metrics across a number of categories.

10.15 Summary of Event Detection Metrics

The full list of metrics discussed in this chapter is compiled in Table 10.2. While this is not a complete listing of all possible event detection metrics, it is a collection that provides for a robust assessment of a model. The key to a thorough comparison is to choose several metrics that complement each other, examining different aspects of the data-model relationship.

> **Quick and Easy for Section 10.15**
>
> All of the event detection metrics in one place.

Table 10.2 Summary of the fit performance metrics from this chapter.

Metric	Formula	Range	Perfect score	Guideline for "good"
Accuracy metrics				
PC	$(H+C)/N$	$[0, 1]$	1	>0.67
CSI	$H/(H+M+F)$	$[0, 1]$	1	>0.5
F_1	$2H/(2H+M+F)$	$[0, 1]$	1	>0.67
Bias metrics				
FB	$(H+F)/(H+M)$	$[0, \infty)$	1	$[0.75, 1.33]$
Precision metrics				
Precision	$H/(H+F)$	$[0, 1]$	1	>0.67
Recall	$H/(H+M)$	$[0, 1]$	1	>0.67
Association metrics				
θ	$(H \cdot C)/(M \cdot F)$	$[0, \infty)$	1	>3
ORSS	$\dfrac{(H \cdot C)-(M \cdot F)}{(H \cdot C)+(M \cdot F)}$	$[-1, 1]$	1	>0.5
MCC	$\dfrac{(H \cdot C)-(F \cdot M)}{\sqrt{(H+F)(H+M)(F+C)(M+C)}}$	$[1, 1]$	1	>0.33
Extremes metrics				
EDS	$\dfrac{2 \cdot \ln(N) - 2 \cdot \ln(H+M)}{\ln(N) - \ln(H)} - 1$	$[-1, 1]$	1	$>{\sim}0.65$
SEDS	$\dfrac{2 \cdot \ln(N) - \ln(H+M) - \ln(H+F)}{\ln(N) - \ln(H)} - 1$	$[-1, 1]$	1	$>{\sim}0.65$
Skill metrics				
HSS	$\dfrac{2\left[(H \cdot C)-(F \cdot M)\right]}{(H+M)(M+C)+(H+F)(F+C)}$	$[-1, 1]$	1	$>{\sim}0.3$
PSS	$\dfrac{(H \cdot C)-(M \cdot F)}{(H+M)(F+C)}$	$[-1, 1]$	1	$>{\sim}0.3$
CSS	$\dfrac{(H \cdot C)-(M \cdot F)}{(H+F)(M+C)}$	$[-1, 1]$	1	$>{\sim}0.3$
GSS	$\dfrac{H - H_{ref}}{H+F+M-H_{ref}}$	$[-0.33, 1]$	1	$>{\sim}0.2$
Discrimination metrics				
POD	$H/(H+M)$	$[0, 1]$	1	>0.67
POFD	$F/(F+C)$	$[0, 1]$	0	<0.33
FNR	$M/(H+M)$	$[0, 1]$	0	<0.33
Specificity	$C/(F+C)$	$[0, 1]$	1	>0.67
Reliability metrics				
PPV	$H/(H+F)$	$[0, 1]$	1	>0.67
MR	$M/(M+C)$	$[0, 1]$	0	<0.33
FAR	$F/(H+F)$	$[0, 1]$	0	<0.33
NPV	$C/(M+C)$	$[0, 1]$	1	>0.67
FR	H/F	$[0, \infty)$	1	>2

10.16 Further Reading

The paper by Kubo in 2017 calls out a few "best choice" event detection metrics for these same categories:

Kubo, Y., Den, M., & Ishii, M. (2017). Verification of operational solar flare forecast: Case of Regional Warning Center Japan. *Journal of Space Weather and Space Climate*, 7, A20. https://doi.org/10.1051/swsc/2017018

Binary event metrics are thoroughly discussed in this chapter of the Forecast Verification book:

Hogan, R. J., & Mason, I. B. (2012). Deterministic forecasts of binary events. In I. T. Jolliffe & D. B. Stephenson (Eds.), *Forecast Verification: A Practitioner's Guide in Atmospheric Science* (2nd edition, chap. 3, pp. 31–60). John Wiley, Ltd., Chichester, UK. https://doi.org/10.1002/9781119960003.ch3

The Wilks' book also has a lot of discussion of nearly every metric mentioned in this chapter:

Wilks, D. S. (2019). *Statistical Methods in the Atmospheric Sciences* (4th edition). Academic Press, Oxford.

The big paper by Matthews popularizing that contingency-table-based correlation coefficient:

Matthews, B. W. (1975). Comparison of the predicted and observed secondary structure of T4 phage lysozyme. *Biochimica et Biophysica Acta (BBA) – Protein Structure*, 405(2), 442–451. doi:10.1016/0005-2795(75)90109-9

A nice description and application of the extreme dependency score for atmospheric science is presented here:

Stephenson, D. B., Casati, B., Ferro, C. A. T., & Wilson, C. A. (2008). The extreme dependency score: A non-vanishing measure for forecasts of rare events. *Meteorological Applications*, 15, 41–50. https://doi.org/10.1002/met.53

The original derivation of the Heidke skill score by Heidke:

Heidke, P. (1926). Berechnung des Erfolges und der Gute der Windstarkevorhersagen im Sturmwarnungsdienst (Measures of success and goodness of wind force forecasts by the gale-warning service). *Geografiska Annaler*, 8, 301–349.

The original derivation of the Gilbert skill score by Gilbert:

Gilbert, G. K. (1884). *Finley's tornado predictions. American Meteorological Journal*, 1, 166–172. https://www.proquest.com/docview/124374084

The original derivation of the Peirce skill score by Peirce was essentially a half-page commentary about Gilbert's new skill score, published later that same year:

Peirce, C. S. (1884). The numerical measure of the success of predictions. *Science*, 4, 453–454. doi:10.1126/science.ns-4.93.453-a.

The original derivation of the Clayton skill score by Clayton:

Clayton, H. (1927). A method of verifying weather forecasts. *Bulletin of the American Meteorological Society*, 8(10), 144–146. Retrieved June 28, 2021, from http://www.jstor.org/stable/26262138

Another interesting paper on the derivation of skill scores was written by Otto Hyvärinen:

Hyvärinen, O. (2014). A probabilistic derivation of Heidke skill score. *Weather and Forecasting*, 29, 177–181. https://doi.org/10.1175/WAF-D-13-00103.1

The precipitation observations of the ASOS network were obtained from the Iowa Environmental Mesonet in METAR format, with the values retrieved from Iowa State University: https://mesonet.agron.iastate.edu/request/download.phtml

A definition of the METAR format for weather reports is available here: https://catalogue.ceda.ac.uk/uuid/50418e43c3c741618c34e75c22ef43e3

HRRR precipitation model from NOAA:
https://rapidrefresh.noaa.gov/hrrr/
The latest on the model version used in this chapter is summarized in this presentation:
http://rapidrefresh.noaa.gov/pdf/Alexander_AMS_NWP_2020.pdf

The model values for the running example were collected from the University of Utah, who maintain a nice interface to the model output:
Link: http://hrrr.chpc.utah.edu
An article about this website:
Blaylock B., Horel, J., & Liston, S. (2017). Cloud archiving and data mining of high resolution rapid refresh model output. *Computers and Geosciences*, *109*, 43–50. doi: 10.1016/j.cageo.2017.08.005
They are phasing out the maintenance of this database, however, because NOAA is now making this data available through Amazon Web Services:
https://registry.opendata.aws/noaa-hrrr-pds/

The main highlights of this chapter are summarized by Liemohn et al. (2021):
Liemohn, M. W., Shane, A. D., Azari, A. R., Petersen, A. K., Swiger, B. M., & Mukhopadhyay, A. (2021). RMSE is not enough: guidelines to robust data-model comparisons for magnetospheric physics. *Journal of Atmospheric and Solar-Terrestrial Physics*, *218*, 105624. https://doi.org/10.1016/j.jastp.2021.105624

10.17 Exercises in the Geosciences

Again, let's continue exploring precipitation at a location from the NOAA HRRR model. While it is still active, you can go to the University of Utah page to retrieve HRRR output:

http://hrrr.chpc.utah.edu
If not, then go to the AWS website and register to download HRRR values:
https://registry.opendata.aws/noaa-hrrr-pds/

The observations are available at the Iowa Environmental Mesonet site:
https://catalogue.ceda.ac.uk/uuid/50418e43c3c741618c34e75c22ef43e3
Select a "network" (a state or country) and then select a station from the list. The weather parameter can be selected (precipitation in either mm or inches) on the right and a time interval should be set. Get a data set that matches your HRRR output location and interval.
You can also use the precipitation data set from the chapter examples, but choose a specific month from the full number set so that it is a "different" assessment than done in the running example.

1. Open and read the two file sets.
   A Make a separate time series plot of each number set.
   B Make an overlaid time series plot of both number sets.
   C Make separate histograms of each number set.
   D Make an overlaid histogram of both number sets.
   E Make a scatterplot of the observed values against the modeled values.
   F Select an event identification threshold, assign event status to the two number sets, and make a contingency table.

2. Write a paragraph answering the following:
   A What is your initial reaction to the relationship between the data and the model?
   B Examining the contingency table, what is your qualitative assessment of the model's ability to predict this quantity?

3. Conduct a metrics assessment.
   A Calculate one metric from each of the main categories. Explain and justify your choice of metrics.
   B Calculate metrics for the subsets: POD and POFD, PPV and MR.

4. Write several paragraphs answering the following:
   A Based on an accuracy metric alone, how good is the model? Explain your answer relative to quantities for the individual number sets.
   B What new do we learn about the quality of the model when the scores for the bias and precision values are taken into account?
   C What new do we learn about the quality of the model when the association metric score (e.g., ORSS) is taken into account?
   D What does the extremes metric reveal about the quality of the model?
   E What is the skill of the model relative to a random forecast?
   F Discuss the results of the subsetting metrics.

11

Sliding Thresholds: Event Detection Metrics with a Variable Event Identification

In the previous chapter (Chapter 10), all of the metrics assumed a specified event identification threshold setting for both the observations and the model output. This yields a single value for each of the event detection metrics. This is fine if the desired event identification thresholds are selected ahead of time. For the precipitation example used throughout Chapter 10, this was selected to be a very small nonzero value, essentially making it a test of whether the model correctly predicts *any* precipitation in the next 6-hour interval. For the Dst example used in Chapter 9, an appropriate predefined threshold could be the community-accepted cutoff for minor storms at −30 nT. If the situation for the model usage relative to the data is known and a single threshold can be defined as the one and only setting necessary for model evaluation, then this approach is enough.

If the model is a continuously valued number set, though, and the point of conducting the metrics assessment is to determine the best threshold for a particular application, then there is no reason why a single threshold needs to be selected ahead of time. That is, the threshold for defining events in the model output (or driving parameter, or controlling factor, or whatever is being used for the "model" for the reproduction of the observed events) can be varied from low to high values through the entire model value range, creating an array of metrics values that can be analyzed and optimized for the event detection usage. This can be done regardless of the setting for the observed event identification threshold. In fact, the observations could be pre-categorized into yes/no event status; the modeled threshold can still be changed and a list of metric scores created.

Two other methods of sliding event identification thresholds to create an array of metrics values are also discussed. One of these is the option of holding the modeled event identification threshold fixed and sliding the observed event identification threshold. This is essentially the converse of the first sliding threshold method. This technique can only be done if the observations are also continuous-valued numbers and not pre-categorized into yes–no event status. The data and model values, however, can either be the same or different. That is, the model could be trying to reproduce the observed values or it could be a driver parameter with completely different units and range scale. This method is not useful for optimizing the modeled threshold—which is held fixed—but rather for assessing the quality of the model with the chosen event delineation at reproducing the data. That is, this analysis can quantify the regions of the observational data set where the model performs well and where it does not.

Yet another sliding threshold method is to vary both the observed and modeled event

Data Analysis for the Geosciences: Essentials of Uncertainty, Comparison, and Visualization, Advanced Textbook 5, First Edition. Michael W. Liemohn.
© 2024 American Geophysical Union. Published 2024 by John Wiley & Sons, Inc.
Companion website: www.wiley.com/go/liemohn/uncertaintyingeosciences

identification threshold simultaneously. This is only possible if both number sets are continuous and the model is attempting to exactly reproduce the observed values. In this case, the ideal data-model relationship should lie along a diagonal line when the two are plotted against each other. Sweeping both thresholds combines the benefits of the two methods just mentioned, assessing the model for the full possibility of event classifications that are identical in both number sets.

11.1 Sliding the Event Identification Thresholds

Figure 11.1 shows three identical scatterplots of observation–modeled value pairs. On these plots, it is assumed that the definition of an observed event is fixed, but not the definition of a modeled event. In this case, it is possible to try several settings of the modeled event identification threshold, as shown in three of the panels of Figure 11.1. The first panel has a somewhat low setting that would maximize hits over misses at the expense of allowing more false alarms. The lower-left panel is a relatively high setting that minimizes false alarms at the expense of allowing more misses. The upper-right panel is an intermediate choice that attempts to strike a balance between these two competing optimization endpoints. There is no need to stop at three choices for the modeled threshold setting. The fourth (lower-right) panel of Figure 11.1 shows the case of starting the modeled event identification threshold at a very low value, perhaps even below the lowest modeled value, and then sweeping through all possible values, with a reasonably set increment, up to a very high value, perhaps even above the maximum modeled value.

When sweeping the modeled threshold, the "row" totals remain the same. That is, $H + M$ = constant and $C + F$ = constant. There is no crossover of points from one discrimination range setting to another. There are, however, a lot of changes that can occur between the metrics that use the cells above or below the observed event threshold. For the beginning "very low" threshold, most or all of the points are hits and false alarms, but these will be converted to misses and correct negatives, respectively, as the threshold slides. That is, as the modeled threshold moves from low to high values, probability of detection (POD) decreases as the false negative rate (FNR) increases, and, similarly, probability of false detection (POFD) decreases as specificity increases. The denominators of all of these metrics remain the same.

A similar case can be made about sliding the observational threshold. While it is rare that the case would be that the modeled events are pre-categorized and the observed values are continuous data, in event detection analysis any potential controlling factor can be compared against any "effect parameter," even if the latter is binary event data and the former is a set of real-valued numbers. More often, it would be the case that both are continuous-valued number sets but you are interested in exploring the ability of the model, at a particular event threshold, to predict observations at other thresholds. This could be the case if you know the model has a systematic offset from the observations. In this situation, holding the modeled threshold constant would help decide if the model has skill, even though it is not particularly good at predicting the exact observed values.

The left panel of Figure 11.2 shows the case of sliding the observational event identification threshold. In this case, the points stay on either side of the modeled event identification threshold, and therefore the numbers of points in each of the reliability subsets are held constant such that $H + F$ = constant and $C + M$ = constant. For the beginning data threshold at a very low observed value, the points are all misses and hits, but then convert to correct negatives and false alarms, respectively, as the threshold increases. Therefore, as the data event identification

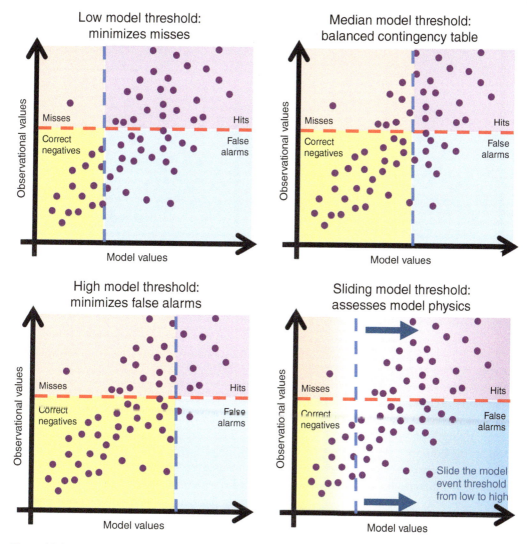

Figure 11.1 An illustrative data-model scatterplot with a given threshold for the observed events, but a sliding threshold for the identification of modeled events. This is shown in the first three panels with different threshold settings, and the final panel shows the process of sweeping the threshold through all possible settings.

threshold is swept from low to high values, positive predictive value (PPV) decreases as the miss ratio (MR) increases, and, on the other side of the modeled threshold, the false alarm ratio (FAR) decreases as negative predictive value (NPV) increases.

Yet another interesting assessment can be made when the model is attempting to exactly reproduce the values of the observed number set. In this case, both thresholds can be swept simultaneously. This can only be done if the real-valued set of model output is attempting to exactly match the observed values. Both thresholds can now be moved together, assessing the model's ability to predict events at every possible event identification setting.

The right panel of Figure 11.2 shows an example scatterplot for which both thresholds are simultaneously swept from low to high values. In this case, for the beginning

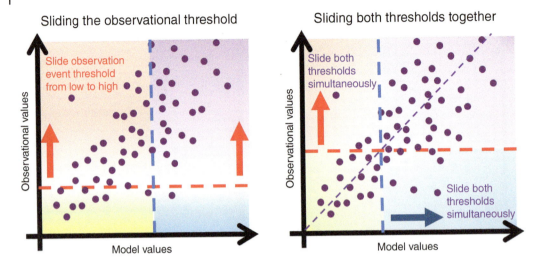

Figure 11.2 Two additional options for sliding the event identification threshold. The left panel shows the case with a given threshold for the modeled events, but a sliding threshold for the identification of observed events. The right panel shows the case of sliding both event identification thresholds simultaneously.

very low threshold setting, all of the points are hits. As they move up, these will be converted from H to one of the other cells. The usual path is to move from H to either F or M as one or the other of the thresholds moves past that point's observed or modeled value. Then, as the other threshold sweeps past the point, it is converted to C. It can be that points are converted directly from H to C, either if they lie perfectly on the unity-slope, zero-intercept diagonal line or if the threshold step size is large enough for this conversion to happen. No matter how the points move through the cells, H will steadily decrease, as this cell only gets smaller and no points are converted into this designation. Similarly, C will steadily increase, as this quadrant only gets bigger, eventually taking over the entire scatterplot. The other two cells, however, will increase and decrease, because the F and M quadrants initially increase in size, reach a maximum, and then decrease. Points might enter these quadrants but eventually leave them. These two cell counts will, in general, increase and then decrease with the change in area, but they could have relative maxima and minima in there count values along the

way through the threshold sweep. Such variability is an exciting feature of sweeping both thresholds together, as it reveals localized clusters of points away from the diagonal best-fit line.

There are two issues that need to be addressed with this simultaneous sweep of both metrics. The first occurs at low thresholds and is a problem for all metrics for which H is not in the denominator—they will have zero in the denominator and become undefined. This can happen when there are still points being converted between two of the metrics, say M and H, but there are no more points in the lower quadrants of the scatterplot and therefore the C and F cell counts are zero. We can think about this in terms of what will happen to the metric value once the threshold is increased and the denominator is no longer zero. What will be that first metric score? For a metric such as POFD, there will be a threshold for which one point is now in either the F or C quadrant, as either the model or data event identification threshold passes its first point in the scatterplot. A logical work-around to the zero in the denominator is to set POFD to either zero or one, whichever value it will become when that first point is crossed by one of the thresholds.

Similarly, there is an issue at the high end of the threshold setting, when all points are correct negatives. Again, there can be metrics for which the denominator is zero and the value explodes to infinity. Take CSI, for example, a metric that does not use C in its formulation. Right before this happens, though, the thresholds will be set so that there is a nonzero value in one of the cells. It will probably be F or M, but it could be H, with the point hopping directly to C with the final threshold increment. In any case, CSI will have a value of either zero or one. This value should simply be continued as the threshold passes the last point.

Note that these same division-by-zero issues can arise when sweeping only one of the thresholds. When sweeping the modeled threshold, calculations of the reliability metrics could yield this error. Conversely, when sweeping the data threshold, calculation of discrimination metrics could leave you with zero in the denominator. This is only an issue at the extreme ends of the sweep, and the same method works in all of these cases—hold the metric score constant from its last defined value.

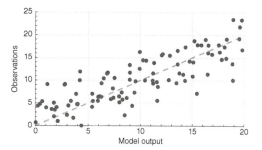

Figure 11.3 Scatterplot of model output attempting to reproduce a set of observations, for use in the later analysis sweeping the event identification thresholds. Specific values are given for both number sets, but, being illustrative, no units are given to keep it generic. The diagonal dashed line shows the ideal data-model comparison.

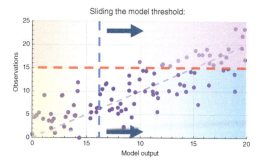

Figure 11.4 Using the illustrative example data set of Figure 11.3, this now shows a fixed observed event identification threshold and a sliding modeled event identification threshold. The shading indicates the contingency table quadrants, which change size as the modeled threshold is swept from left to right.

> **Quick and Easy for Section 11.1**
>
> Sliding the threshold produces metrics scores that span the range of what is expected to be good and bad threshold settings, allowing the exploration of metric score changes as a function of metric setting.

11.2 Sweeping the Modeled Threshold

This sweeping of the threshold will produce a list of scores for each metric, which can then be plotted or otherwise processed. An illustrative example data-model comparison is shown in Figure 11.3. These paired number sets will be used in the next few subsections to explore metrics values when sweeping the thresholds. That is, both are sets of real numbers rather than categorical assignments. Here, in this section, only the modeled threshold is swept from low to high values.

This sweeping of the modeled event identification threshold is shown in Figure 11.4. The event identification threshold for the observations is set to 15. Of the 100 data-model paired points in the scatterplot, this labels 22 of them as observed events and 78 of them as observed nonevents. The breakdown of the points among the four quadrants of the contingency table, however, is a function of the modeled event identification threshold, which will be swept from 0 to 20. When this threshold is small, there are no misses (i.e., no points in the red region) but many false alarms (points in the blue region). As the threshold moves to high values, this situation reverses, with hits converted to misses and false alarms converted to correct negatives. All of the contingency table cell counts, therefore, change as a function of the modeled threshold setting.

Figure 11.5 shows several key metrics as a function of a sliding modeled threshold, for the example scatterplot shown in Figure 11.3. Just to pick a few from the many metrics listed in Chapter 10, shown in Figure 11.5 are the F_1 score, frequency bias (FB), odds ratio skill score (ORSS), and Heidke skill score (HSS). For F_1, in the top panel, the curve starts at a value of 0.36; while there are no misses for this modeled threshold setting of zero, this means that there are also no correct negatives, so all of the points below the data event threshold are false alarms. The F_1 score rises as these F counts are converted to C counts, which is occurring at a rate faster than H counts are moved over to M designations. As the modeled threshold approaches the upper limit, there are very few points left in the false alarm quadrant, so the conversion of H to M dominates and the F_1 score drops. Eventually, all of the points above the data threshold are now misses and the F_1 score plunges to zero.

The next panel in Figure 11.5 shows FB, which monotonically decreases from a very high value all the way down to zero. When the modeled threshold is low, the contingency table is skewed toward false alarms, so the FB score is well above one. For large modeled threshold settings, the reverse is true, with the F value low and M dominating the bad cell counts. For this particular scatterplot, FB crosses unity at a modeled threshold setting of 15.1, which happens to be very close to the

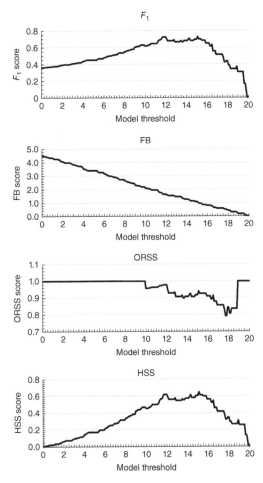

Figure 11.5 Plots of various event detection metrics as a function of modeled event identification threshold.

data event identification threshold setting. This did not have to be the case; it is a coincidence that they are this similar when FB crosses unity.

The third panel of Figure 11.5 shows ORSS, which is a comparison of the symmetry and magnitude of the good cell counts against the symmetry and magnitude of the bad cell counts. It is actually undefined at each end when the threshold surpasses the last point in the scatterplot. When the modeled threshold is very low, the first point encountered is a correct negative, so the initial score for ORSS is one, as the multiplication of the bad cell counts within the ORSS formula still yields zero. This did not have to be the case; if the first point encountered was a miss, then ORSS would have started at negative one instead of one. Similarly, at the high end of the modeled threshold setting, the first point encountered is a hit, so ORSS starts at a score of one here, too. If the first point encountered was a false alarm, though, this end of the ORSS curve would have been negative one. Because there are no misses until the modeled threshold is above 10, the ORSS score remains at one until M becomes nonzero. The fact that the score remains relatively high—above 0.7 everywhere—indicates that the model is rather good at capturing the trend of the observed events.

The final plot in Figure 11.5 is the Heidke skill score. This value has the same numerator as ORSS (with an extra factor of two) but a different denominator, resulting in a score of zero at each end of the modeled threshold setting. In between these two endpoint values, it is above zero, reaching a peak of 0.65 at a modeled threshold of 15. This did not have to be the case; it could have been negative, but the model reproduction of the event state is relatively good and so the HSS score rises to a respectable value.

The assessment of this scatterplot and the relationship of metrics to modeled threshold setting is further illustrated by considering the metrics in the discrimination and reliability categories. The metrics of POD, POFD, PPV, and NPV are shown in Figure 11.6. The first two are measures of discrimination, using contingency table cells exclusively above or below the data event identification threshold. This threshold is static—it is the modeled threshold that is being swept from low to high values—so it is seen that these two quantities are monotonic. Both POD and POFD start at one for low modeled thresholds and end at zero for the highest modeled threshold setting.

The other two quantities in Figure 11.6 are reliability measures, using only cells on one

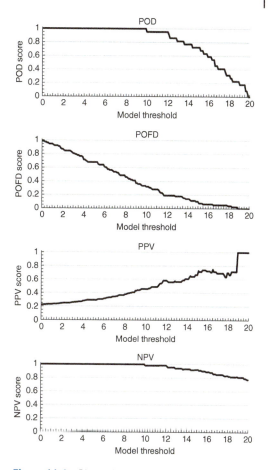

Figure 11.6 Plots of the key discrimination and reliability event detection metrics as a function of modeled event identification threshold.

side or the other of the modeled event identification threshold. Because this is the threshold being moved, these quantities do not have to be monotonic; it depends on the exact location of the points and how many are being added or removed from the cells as the modeled threshold increments to higher values. PPV uses the cells in the modeled event state, so it starts with a score of 0.22, as all points are within these two cells of the contingency table. At the high end, only a few hits remain, so the final PPV score is one. At the very uppermost threshold settings, even H becomes zero and PPV is undefined. The case is similar but exactly the opposite for NPV, which uses the cells for model nonevents. At the highest modeled threshold, all points are within these two cells and NPV = 0.78. At low modeled threshold settings, our choice of a relatively high data event threshold leaves no points within the M quadrant and NPV rises to one. At the very lowest modeled threshold, however, even $C = 0$ and NPV is undefined.

A final example is to consider the metrics that highlight rare events. The extreme dependency score (EDS) and symmetric extreme dependency score (SEDS) metrics for the scatterplot in Figure 11.3 are shown in Figure 11.7. First, let's consider the EDS score. At the low threshold values, there are no misses, and so the numerator and denominator of EDS are identical and the metric score is one. As the modeled threshold increases and hits are converted into misses, EDS drops, eventually going negative. Next, consider the SEDS score. For low threshold, while there are no misses, there are many false alarms, which lowers the SEDS score. In fact, for the lowest threshold for which there are no correct negatives, the second term in the numerator of SEDS is $\ln(N/N) = 0$, and the SEDS formula yields zero. As false alarms are converted to correct negatives, that term becomes nonzero and SEDS slowly rises. It reaches a point, however, when the conversion of F into C is smaller than the conversion of H in M, and the SEDS score starts to drop again. The SEDS score can actually be nonmonotonic because of such dynamics between the conversion of points above and below the observational event identification threshold.

> **Quick and Easy for Section 11.2**
>
> Sliding the modeled threshold reveals how the metrics change as this value is varied, allowing users to choose a setting that optimizes the metric most important to their needs.

11.3 Sweeping the Data Threshold

Continuing the illustrative example, Figure 11.8 shows the scatterplot from Figure 11.3 but now with a fixed modeled event identification threshold and a sliding observed event identification threshold. This is a means of testing the physics of the model against a varying definition of the observed event, allowing you to decipher how well the model captures observed events for this situation of a constant modeled threshold but changing data threshold. This method keeps the reliability subsetting constant—there are 76 of the 100 points to the left of the modeled event identification threshold (labeled modeled nonevents) and 24 points labeled as modeled events. The breakdown of how these are labeled as observed events and nonevents changes as the data threshold is moved. When the data threshold is low, there are no correct negatives but many hits. When it

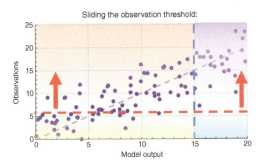

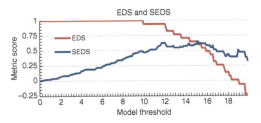

Figure 11.7 EDS and SEDS as a function of modeled event identification threshold for two illustrative examples with few observed events.

Figure 11.8 Using the illustrative example data set of Figure 11.3, this now shows a fixed modeled event identification threshold and a sliding observed event identification threshold. The shading indicates the contingency table quadrants, which change size as the observation threshold is swept from bottom to top.

is high, the hits go to zero and the correct negatives are numerous.

Optimal values of various metrics can be determined from the metrics value curves created by this threshold sweep. A few such curves are shown in the panels of Figure 11.9. The F_1 score starts out at 0.39, just a little bit higher than it did for the modeled threshold sweep in the earlier section. It peaks at a value of 0.75 when the threshold is at 14.5, and then drops again to a value of 0.22 at the highest threshold applied. Note that it does not go all of the way to zero because there are still hits at this data threshold setting; some of the pairs remain in the green quadrant. Note that the F_1 score not only has relative minima at each end of the threshold sweep but there are also many smaller relative extrema throughout the threshold range. This is because the F_1 score changes as any of the three metrics that are included in its definition changes. Misses and hits are dropping as the threshold is increased, but the false alarm count is increasing. These changes do not have to be steady and local maxima and minima can develop in the curve.

The second panel of Figure 11.9 shows the frequency bias. This curve starts small—well below one at low threshold settings—and ends large, with values approaching 10. This increasing trend is opposite than from the modeled threshold sweep seen in Figure 11.5. Misses are in the numerator of FB and they dominate the error counts when the threshold is small. The unity crossover for FB occurs at a threshold of 14.5, coincidentally the same threshold as the peak F_1 score.

The odds ratio skill score is presented in the third panel of Figure 11.9. As with the modeled threshold sweep, it starts and ends with a score of one, but it did not have to; if the first point was an error instead of a correct event status determination, then the start or end score for FB would have been 1. The ORSS values are all greater than 0.87, with this lowest value occurring at a threshold of 11.4.

The final panel of Figure 11.9 shows the Heidke skill score, as calculated with the modeled threshold pegged at 15 and the observation threshold sliding from 0 to 20. Its peak value of 0.67 occurs at a threshold of 14.5, as it did with a couple of the other metrics in this chapter. The HSS value starts at zero for the lowest threshold and ends at zero, but this is beyond the last observed value, which is above our ending threshold of 20, so it has not quite arrived back down to zero by the end of the plotted curve.

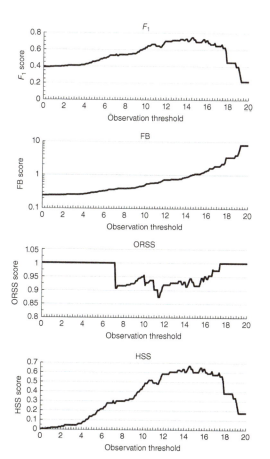

Figure 11.9 Metrics as a function of observed event identification threshold for the illustrative example.

> **Quick and Easy for Section 11.3**
>
> Sliding the observation threshold explores how well the model performs with different settings of this value, illuminating ranges of threshold settings where the model is better or worse at particular aspects of reproducing the observations.

11.4 Sweeping Both Thresholds Simultaneously

The final method of sliding thresholds is to move both the observed and modeled event identification thresholds simultaneously. If the data are a set of real-valued numbers and the model is seeking to exactly reproduce these values, then this type of sweep of the thresholds is possible. Usually, when this relationship between the number sets exists, the common choice is to use fit performance metrics of Chapter 9. Sliding both thresholds simultaneously and calculating arrays of event detection metrics constitute an analysis tool that supplements the fit performance techniques, exploring the quality of the model to reproduce the observations in specific ranges without regard to the exact values of either the data or model output within those ranges.

On the topic of implementation, it is essentially the same as the other methods. The thresholds are moved from a very low to a very high value in small increments. This time, neither is held fixed, but rather set equal to each other at each step. Continuing with the same illustrative example, Figure 11.10 shows this process of sliding both thresholds simultaneously through the scatterplot of data-model pairs.

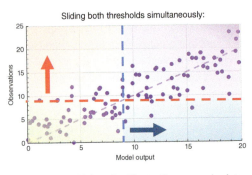

Figure 11.10 Using the illustrative example data set of Figure 11.3, this now shows a fixed modeled event identification threshold and a sliding observed event identification threshold. The shading indicates the contingency table quadrants, which change size as the observation threshold is swept from bottom to top.

The results and the information it reveals, however, are fundamentally different from the two sliding threshold techniques just described in the earlier sections. This is because the data-model pairs move from being all hits to all correct negatives, usually with a stop along the way in one of the error cells of misses or false alarms. This transfer from H to C was not possible in the earlier sliding threshold techniques.

Figure 11.11 presents the curves of several metrics as a function of the threshold setting. All of the metrics are at their perfect score at the lowest threshold, because all of the points are within the hits quadrant. As the thresholds

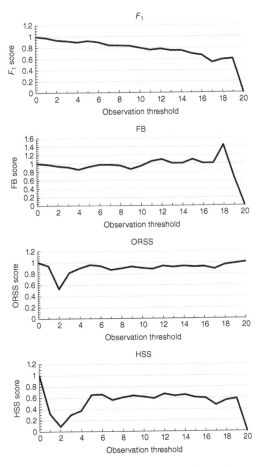

Figure 11.11 Metrics as a function of sweeping both event identification thresholds simultaneously for the illustrative example.

increase, the error cell counts increase and the metrics scores worsen as H counts convert to M and F counts. This drop in quality is clearly seen in ORSS and HSS for the illustrative example. As the thresholds continue to increase, though, the correct negatives count increases, and some of the metrics recover to a better score for a small interval of threshold settings. At the highest threshold settings, when there are very few hits left, the metrics become erratic as individual points move out of not only the H count but also the M and F counts. For this example, the F_1 score steadily decreases through the threshold sweep, with only a few small relative maxima interrupting the downward trend. The downward trend is because points are leaving H and moving to F and M, the upward bumps are when a cluster of points quickly leaves F or M, creating a temporary increase in the F_1 metric score.

Insights about the data-model relationship can be learned from these curves in Figure 11.11. Starting with the F_1 score, it is above 0.5 everywhere except the highest threshold of 20 (when there are more model values above this) and above 0.67 (the more stringent quality criteria) all the way to a threshold of 15. The model is accurate, not only in an overall sense but in specific intervals, in this case from low thresholds to nearly the highest setting.

For frequency bias, it is well within the [0.75,1.33] range designated as a high-quality score for this metric for much of the curve. It hovers near unity for the entire threshold range except the top few settings. This indicates that the model is not systematically underestimating or overestimating the observed values across. The fact that the FB curve is so close to unity for most of the threshold settings implies that this unbiased declaration is true regardless of the discrimination or reliability subset considered.

The third panel of Figure 11.11 reveals that ORSS is close to unity for much of the threshold setting range, except for a brief dip at low thresholds. ORSS includes multiplication operators among the cell counts, specific good-times-good ($H \cdot C$) as well as bad-times-bad ($F \cdot M$), and the product is higher when the two values being multiplied are nearly equal. At low thresholds, C is very small, and so even though H is large, a balanced count between F and M could make $F \cdot M$ rival the size of $H \cdot C$, reducing the ORSS score. The ORSS value, however, stays above 0.5 everywhere, which was the cutoff for declaring this score to be of high quality.

Finally, HSS is shown in the bottom panel of Figure 11.11. It drops very quickly, just as ORSS does. While it stays above zero, meaning that it is always better than random chance, HSS briefly drops below the high-quality mark of 0.3. As C increases in counts, HSS rebounds and hovers near 0.6 for most of the threshold sweep range (until the very last setting). That is, against random chance, this model number set is only marginally better at low thresholds but easily outperforms that "reference model" at all other threshold settings.

To summarize, this technique allows you to assess the quality of the model at separating the observations above and below a threshold value, and because that threshold is varied, you can make this assessment across the full range of possibilities. Depending on the type of assessment you are conducting and which of these metrics is most important, you could then decide whether or not to adopt this model for your application.

> **Quick and Easy for Section 11.4**
>
> Sliding both the event identification thresholds simultaneously is possible when the model is exactly trying to reproduce a real-valued data set. This reveals value intervals where the model performs well or poorly at sorting the observations into events and nonevents.

11.5 Metric-Versus-Metric Curves

As a final section of introducing new event detection assessments, let's consider the case of sweeping one or both of the thresholds and then, instead of plotting the metric against the threshold value, plot a metric against another metric. This is done in a few specific cases resulting in special curves with fun names. There are three that will be discussed here. The first, the **relative operating characteristic curve**, or ROC curve, is by far the most common. The other two are less well-known but provide unique contributions to the data-model comparison, one sweeping the observation event identification threshold and the other simultaneously sweeping both thresholds.

11.5.1 ROC Curves

The ROC curve started out in World War II as a method of interpreting radar signals and assessing the ability of an algorithm to detect an incoming airplane, as opposed to false alarms of flocks of birds or other "events" in the backscattered radio waves. Based on this radar signal processing history, the original name of the metric was the receiver–operator characteristic curve, and it has since morphed through a number of names, such as receiver operating characteristic. The latest words, and perhaps most common in the geosciences, associated with this acronym are relative operating characteristic, but others, including the original, are still regularly used by some disciplines.

The ROC curve is created by sweeping the modeled threshold while holding the observed event identification threshold constant and then plotting the POD scores against the POFD scores. Because the threshold for observed events is held fixed, the ROC curve works for data that are pre-sorted into events and nonevents but require the model, or whatever driving parameter is used, to be multivalued and allow for a sweep of the modeled event identification threshold through the full range of model values. Using the contingency tables at each step through this sweep, the two prominent discrimination metrics, POD and POFD, are calculated and plotted against each other. At the beginning, the modeled threshold is very low and all points are within either the H or F quadrant. This means that both POD and POFD are one. As the threshold is increased, both values will drop. Eventually, for a very high threshold setting, all points will be in either the M or C quadrant, and both POD and POFD will be zero. During the sweep of the threshold, both sets of points—those in the observed event state and those in the observed nonevent state—are unidirectionally converted from the initial quadrants (H and F) to the final quadrants (M and C). Plotting POD along the y axis and POFD along the x axis, the ROC curve extends from the upper-right corner of (1,1) to the lower-left corner of (0,0).

A perfect ROC curve is one in which all observed events are perfectly sorted by the model values and therefore a clean abrupt transition occurs for the data from nonevents to events at a particular threshold. As the modeled event identification threshold is increased, the nonevents are converted from false alarms to correct negatives, decreasing POFD from one toward zero. POD, however, remains one throughout this change of POFD, as all events remain in the hits quadrant. POFD will eventually reach zero, with all observational nonevents converted to correct negatives, while still having POD at one, with all observed events still designated as hits. The modeled threshold has reached the abrupt transition, and from here on POFD will remain zero (all observed nonevents are already counted in C with $F = 0$)

Relative operating characteristic (ROC) curve: A diagnostic produced by plotting POD versus POFD calculated with a constant event status for the observations but a sliding event identification threshold for the model values.

---

and the hits will slowly be converted to misses. At the far upper end of the model range, the last hit will be converted to a miss and POD will reach zero. The ROC curve from this perfect scenario is a curve that follows the top axis and then follows the left-side y axis.

The perfect data-model comparison is rarely the case, though. There is often some overlap in the middle, with observed events paired with model values lower than the highest model value of an observed nonevent. When this happens, POD will start to decrease before POFD reaches zero. This means that the resulting ROC curve will not hug the top $y = 1$ axis but will drift more directly toward the lower left (0,0) corner of the plot. If you have several models and they have ROC curves start away from the upper-left corner and progressively move closer to this corner, this indicates that the models are getting better.

That is, the ROC curve is an overall assessment of the quality of the model to predict the event status of the observations. A perfectly random connection between observations and model output will result in a uniformly distributed scatterplot. In this case, hits will be converted to misses as soon as the modeled event identification threshold starts its sweep. Similarly, the last false alarm will not be converted to a correct negative until the end of the sweep. In this case, the ROC curve will be a straight line diagonally cutting across the space from (1,1) to (0,0). Should the ROC curve drop below this diagonal random distribution benchmark, it implies that the model is particularly bad at sorting the observed events from the observed nonevents.

To illustrate this, the ROC curve for the example scatterplot in Figure 11.3 is given in Figure 11.12. It is seen that the model does a very good job at sorting the data and the calculated ROC curve is close to the upper-left corner, moving along the top axis (POD $= 1$) and then along the left axis (POFD $= 0$) for many of the threshold settings, and only venturing away from these axes for an interval in the

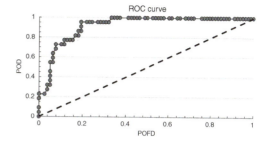

Figure 11.12 Example ROC curve from the illustrative scatterplot in Figure 11.3.

middle of the threshold settings. The diagonal unity-slope dashed line is a reference representing a data-model comparison equivalent to random chance. An ROC curve to the left and above this diagonal line is "good" and indicates the model has some skill at reproducing events, while an ROC curve below this line reveals that the model is not particularly useful for describing the observed events.

Several metrics have been created to distill the information from the ROC curve into a single value. The first is the "**area under the curve**"—widely known as AUC—which is calculated exactly as it sounds. If the step resolution of the modeled threshold sweep is fine enough, then this calculation is often simplified to a Riemann sum:

$$\text{AUC} = \sum_{i=1}^{N_s-1} \text{POD}_i \cdot \left(\text{POFD}_{i+1} - \text{POFD}_i\right) \quad (11.1)$$

In Equation (11.1), N_s is the number of steps in the threshold sweep. This will systematically underestimate AUC, though, especially for large threshold step sizes, so a trapezoid rule is often applied instead:

$$\text{AUC} = \sum_{i=1}^{N_s-1} \frac{\left(\text{POD}_i + \text{POD}_{i+1}\right)}{2} \cdot \left(\text{POFD}_{i+1} - \text{POFD}_i\right)$$

(11.2)

Area under the curve (AUC): The integration of the area between the ROC curve and the x axis, often interpreted as an overall measure of the goodness of the model.

A more sophisticated numerical integration can, of course, be used as well, depending on the accuracy desired for the AUC score. Typically, one of these two formulas is sufficient. One reason to do something more complicated is if you are comparing two ROC curves that are both quite good. In this case, the subtle differences of the deviations from perfection in the upper-left corner of the plot might make the difference between which model is declared the better of the two.

A perfect AUC score is 1. This corresponds to an ROC curve that touches the upper-left (0,1) corner of POFD–POD space. The AUC score decreases with worsening ROC curves, down to a minimum value of 0.

Sometimes AUC is processed a bit more to obtain a metric for comparison with other ROC curves. For example, because a random distribution will result in a diagonal curve, some will remove this part of the graph. In fact, this type of formulation makes it exactly analogous to a skill score, and is often written as such, calling it the **ROC score**:

$$\text{ROC score} = 2 \cdot \text{AUC} - 1 \quad (11.3)$$

The ROC score will sometimes be called the AUC skill score (SS_{AUC}). Equation (11.3) has an upper value of 1, a perfect ROC curve and AUC, and a lowest possible value −1. A score of SSAUC of 0 is achieved if the AUC is 0.5, the score of a random guess.

Another metric from the ROC curve is the **J score**. This metric is calculated as a rather simple linear combination of the x and y values along the ROC curve:

$$J = \max\left[\left(\text{POD}_i + 1 - \text{POFD}_i\right) \forall i\right] \quad (11.4)$$

In Equation (11.4), the i index is any location along the modeled event identification threshold sweep. The J score is a measure of how close the ROC curve comes to the upper-left corner. At the beginning of the sweep, POD = POFD = 1 and the J score will be 1 as well. Similarly, at the end of the sweep, POD = POFD = 0 and J will again be one. In between, however, it is expected and hoped that POD will still be near one, while POFD has dropped to nearly zero. In this case, J will rise toward a value of two. For a perfect sorting, the ROC curve will touch the (0,1) upper-left corner and therefore we get J = 2.

Quick and Easy for Section 11.5.1

ROC curves, based on holding the observed event threshold constant and sweeping the modeled threshold, are used to assess the overall quality of the model at reproducing the underlying processes governing the observed events.

11.5.2 Alt-ROC Curves

The counterpart to the modeled threshold sweep and resulting metric set just described is the **alternative ROC curve**, or Alt-ROC curve, in which the modeled event identification threshold is held steady and the event threshold for the observations is swept from low to high values. The metrics used in this case should not be from the discrimination

---

ROC score or AUC skill score (SS_{AUC}): A measure of the overall quality of a model against a categorized data set, it is twice the AUC score minus one.

J score: A measure of the overall quality of a model against a categorized data set, it is one more than the maximum of the difference of the POD and POFD scores along an ROC curve.

Alternative relative operating characteristic (Alt-ROC) curve: A diagnostic produced by plotting PPV versus MR calculated with a constant event status for the model, but a sliding event identification threshold for the observed values.

subset but rather the reliability subset. To make it perfectly analogous to the ROC curve, the Alt-ROC is defined as a plot of PPV versus MR. With these two metrics, the Alt-ROC curve varies exactly like the ROC curve, starting at (1,1) with a low data threshold setting and ending at (0,0) when the threshold is at the top of the data range. Note that there is no preferred name for this curve in the published literature; the name of Alt-ROC curve is being defined here.

The interpretation of the Alt-ROC curve is exactly like the ROC curve. A perfect prediction will result in an Alt-ROC curve along the top $y = 1$ axis touching the (0,1) upper-left corner, and then moving down along the left $x = 0$ axis. For other models that have a less-than-perfect sorting of the observed values, the Alt-ROC curve will pull away from the two axes and no longer reach the upper-left corner. A random distribution of observed values will result in a diagonal Alt-ROC curve, and an Alt-ROC curve below the diagonal indicates that the model is worse than random chance.

Using the example distributions of Figure 11.3, their curves of plotting various metrics against each other are shown in Figure 11.13. The top panel shows the Alt-ROC curve, revealing that it is indeed monotonic, extending from (0,0) to (1,1). Also, for this data-model comparison, the Alt-ROC curve is well above the diagonal unity-slope line, indicating that the model does well for this data set. However, it is different than the ROC curve in Figure 11.12. While it has many of the same features as the ROC curve, those differences reveal features of the data-model relationship.

Just to include it for comparison, the second scatterplot shows POD versus POFD, the two variables used in the ROC curve definition. As you can see, it is not monotonic because points are converted in and out of these discrimination variables as the data event identification threshold is swept upward from low to high values.

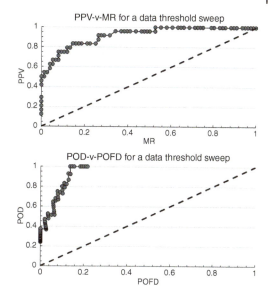

Figure 11.13 The curves creating by sweeping the data event identification threshold: the Alt-ROC curve of PPV versus MR in the upper panel (using the reliability metrics so the curve is monotonic); and POD versus POFD in the lower panel (same variables as the ROC curve).

Quick and Easy for Section 11.5.2

Alt-ROC curves are based on holding the modeled threshold constant and sweeping the observed threshold, which reveals the quality of the model at capturing the governing processes for certain ranges of observed values.

11.5.3 STONE Curves

The final metric-versus-metric plot to present is the **sliding threshold of observation for numeric evaluation curve**, the STONE curve. This is created from the ROC curve and

Sliding threshold of observation for numeric evaluation (STONE) curve: A diagnostic produced by plotting POD versus POFD calculated by sliding event identification thresholds simultaneously for both the observed and model values.

thus uses the discrimination metrics of POD and POFD, just like the ROC curve. In sweeping both event identification thresholds simultaneously, the STONE curve might look like the ROC curve, or might not, depending on the features of the data-model relationship. It shares some characteristics, such as starting from the (1,1) corner and progressing to the (0,0) corner of a POD versus POFD plot, with a perfect STONE curve also lying along the top and left axes of this plot. The biggest difference with the ROC curve is that the STONE curve can be nonmonotonic, either in x or in y, doubling back on itself as points enter and leave the F and M cells at different rates.

These features are clearly seen in the STONE curve shown in Figure 11.14, for our illustrative example scatterplot of Figure 11.3. The actual STONE curve—POD versus POFD—is on the left, while the right panel shows PPV versus MR, the quantities used in the Alt-ROC curve in the previous section. The places where the STONE curve doubles back on itself indicate that there are clusters of points above or below the ideal data-model fit (the light purple dashed line in Figure 11.3). A nonmonotonic left-to-right "wiggle" is an increasing POFD along the x axis, caused by F increasing much faster than C. Such a wiggle—increasing POFD—means that, as the thresholds move to larger values, more hits are transitioning to false alarms than false alarms transitioning to correct negatives. This is evidence of a cluster *below* the unity-slope line; a cluster with model values larger than their corresponding data values (i.e., a cluster of overestimations).

An up-and-down "ripple" is an increasing POD along the y axis, caused by M decreasing much faster than H. A ripple, therefore, happens when there are more misses transitioning to correct negatives than points leaving the hits quadrant. This reveals a cluster above the ideal fit line for which the observed values are larger than the model values (i.e., a cluster of underestimations).

The size and location of these nonmontonicities provide quantitative information about the clustering of the data-model values. The STONE curve, therefore, is useful for extracting additional information not available in a ROC curve. The big limitation, though, is that a STONE curve can only be calculated if both number sets are real values and the model is attempting to exactly reproduce the observed values. That is, it cannot be calculated for categorical observations or for those comparisons when the model is not seeking to reproduce the data.

The second panel of Figure 11.14 shows the PPV versus MR curve for the illustrative example. Because both thresholds are moving simultaneously, it also can exhibit nonmonotonic behavior. Left-to-right wiggles in this curve are times of increasing MR, indicating that hits are becoming misses faster than misses are becoming correct negatives. That is, this also reveals a cluster of underestimated observed values above the ideal fit

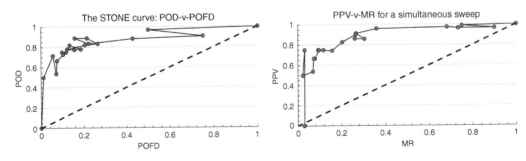

Figure 11.14 The STONE curve for the illustrative example scatterplot in Figure 11.3, along with PPV versus MR for the STONE curve methodology of sweeping both thresholds together.

line. Up-and-down ripples are intervals of increasing PPV, when more points are leaving the false alarm quadrant than are leaving the hits quadrant. This is a sign of an overestimation cluster.

Studies have quantified this relationship, to some degree. One has shown that either a local RMSE of 0.2 the full value domain or a local shift from the unity-slope line of 0.5 of the local RMSE will result in a nonmonotonic feature. These values can be used as a guideline for understanding the appearance of wiggles and ripples in STONE curves relative to the underlying distribution of points.

> **Quick and Easy for Section 11.5.3**
>
> STONE curves, created by sweeping both thresholds simultaneously and plotting POD versus POFD, reveal clustering in the data-model scatterplot that is not easily seen in ROC and Alt-ROC curves.

11.6 Application of Sliding Thresholds to the Geophysical Running Examples

Let's apply this technique of sweeping the event identification thresholds for the two example cases discussed in Chapters 9 and 10. From Chapter 9, the Dst index for March 2015 was modeled using a rather simple differential equation for updating its time derivative. In Chapter 10, the precipitation in 6-hour running intervals at Detroit Metro Airport for the entire year of 2020 was assessed against the 6-hour prediction of precipitation amount from the HRRR model, run every hour by NOAA. For the Dst comparison, 744 data-model pairs were included in the analysis, while for the precipitation assessment, 8555 pairs were used. Both models performed rather well in reproducing the observations using the fit performance and event detection metrics, respectively.

11.6.1 Event Definitions for the Running Examples

The scatterplots of the observed and modeled values for both the Dst and precipitation assessments are shown in Figure 11.15. These are the same plots shown early in Chapters 9 and 10, included here for reference as we go through the process of sweeping the event identification thresholds. Example settings for observed and modeled event identification thresholds are shown on each scatterplot. It is important to remember here that, for the Dst index, the times of important space weather activity are when this index is large and negative. So, for this data-model comparison, events will be defined as those values below the specified threshold. For the precipitation assessment, events are defined in the "standard" way of values larger than the threshold. Therefore, the arrangement of the four quadrants of the contingency table—hits, misses, false alarms, and correct negatives—is rotated 180° between the two scatterplots.

There are a few threshold settings that should be defined before proceeding. When sweeping the modeled threshold, the observation threshold must be set to a constant value. The same is true for the opposite case of sweeping the data threshold and holding the modeled threshold constant. For Dst, the convenient choice is −30 nT, which is the classical definition of a storm event (albeit a weak storm). For the precipitation example, let's choose a value slightly larger than what was used in the previous chapter, now 0.015 inches, so there has to be a bit of a drizzle of precipitation for it to count as an event.

With these definitions, the thresholds can be systematically varied across the full range of possibilities for the two real case examples of geophysical data-model comparisons. First, let's examine the case with a constant observation threshold and a changing modeled threshold. The counts in the contingency table as a function of modeled threshold are shown in Figure 11.16. The first thing to note is that all of

350 | Data Analysis for the Geosciences

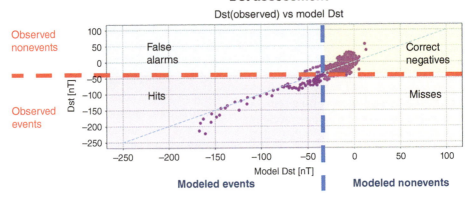

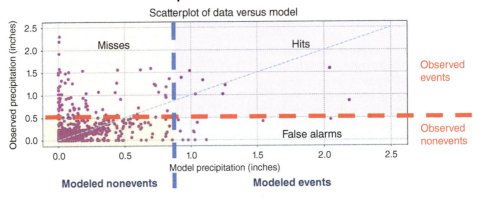

Figure 11.15 Scatterplots for the two running examples of Dst and precipitation, used in Chapters 9 and 10, to be examined again here with sliding event identification thresholds.

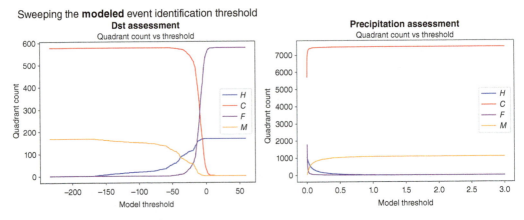

Figure 11.16 Counts in the contingency table cells as a function of modeled event identification threshold for the running examples.

Sweeping the **observed** event identification threshold

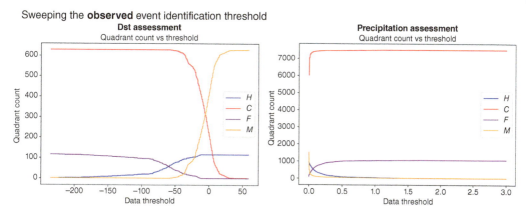

Figure 11.17 Counts in the contingency table cells as a function of observed event identification threshold for the running examples.

the curves are monotonic, either increasing or decreasing as a function of threshold setting. Sliding one threshold and keeping the other constant move points between two cells but not between all four. Remember that the definition of an event is reversed between the two cases. So, the H count, for example, is large when the threshold is "low"—near zero—and the H count drops as the thresholds move to larger values (to large negative values in the Dst case and large positive values in the precipitation case).

For the Dst assessment, there are 577 observed nonevents and 167 observed events. These numbers are then split, the 577 observed nonevents are divided between C and F, while the 167 observed events are in either M or H cells. When the modeled threshold is at +60 nT, all counts are either H or F. At the other end, when the modeled threshold is −240 nT, all counts are either C or M. Most of the variation occurs in the range from roughly +20 nT to −70 nT.

For the precipitation assessment, the observed nonevents total 7473 and the observed events add up to 1082. Most of the change occurs quickly, when the modeled threshold is still quite close to zero, before it reaches 0.5 inches. There are very few observed or modeled values larger than this in either number set.

A similar set of plots can be made for the case of holding the modeled event identification threshold constant and sweeping the observation threshold. Plots of the counts for each contingency table cell are shown in Figure 11.17. Again, the curves are monotonic, because the modeled event status does not change, so the points shift between two cells but not all four. This time, it is hits and false alarms exchanging points, as well as misses and correct negatives.

The plots in Figure 11.17 look like those in Figure 11.16, but the values are different. For the Dst assessment, there are 628 modeled nonevents and 116 modeled events. For the precipitation assessment, the modeled nonevents and events add to 7499 and 1056, respectively. The variation within the plots occurs at roughly the same threshold values, but the ranges are slightly larger. This difference between Figure 11.17 and Figure 11.16 indicates that there is more variation in the observed values than in the modeled values. We knew this from our earlier analysis of these data set, but this is another way to arrive at that same conclusion without having to go through the other assessment calculations.

Figure 11.18 shows the counts in the contingency table cells for the final technique of sweeping both event identification thresholds simultaneously. Remember that this is only possible if both number sets are continuous real values and that the model is attempting to exactly reproduce the observed values. Right

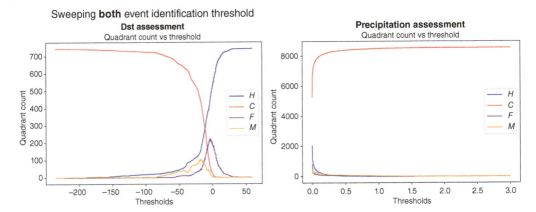

Figure 11.18 Counts in the contingency table cells as a function of simultaneously sweeping both the observed and modeled event identification thresholds for the running examples.

away, something different is seen here—the curves for the error cells are not monotonic. This is especially apparent in the Dst assessment, where the count values are large enough to see the variation. When the thresholds are set at large positive values, all 744 pairs are within the H cell. As the threshold moves down, these correctly identified events shift to one of the other cells. Most will actually shift through two other cells, first joining one of the error cells and then eventually becoming a correct negative. At very low threshold settings, all of the points are in the C cell. The fact that the count for H remains higher than either F or M for all threshold settings is a good sign; this is a qualitative indication that the model is good at determining observed events regardless of the event identification threshold setting.

For the precipitation assessment shown in Figure 11.18, the C curve rises rapidly and the other cell count curves drop to nearly zero faster than seen in Figures 11.16 and 11.17. This is because both thresholds are increasing simultaneously. There are so many values—in both number sets—that are near but just above zero that when both thresholds increment to a slightly larger value, a substantial fraction of the remaining points in H, F, and M are converted to C.

With these curves of contingency table counts, the event detection metrics discussed in Chapter 10 can be calculated as a function of threshold setting. Let's go through the same type of analysis performed in the earlier sections for the illustrative data set, but now for the real examples of Dst and precipitation.

> **Quick and Easy for Section 11.6.1**
>
> The running examples from the previous two chapters are brought back for an analysis using sliding event identification thresholds.

11.6.2 Metric-Versus-Modeled Threshold Curves for the Running Examples

Let's apply the technique of sweeping the modeled event identification threshold for the two geophysical example cases. Using the contingency table count curves in the previous section and the metrics presented in Chapter 10, this is a straightforward task. The resulting curves for a few metrics are shown in Figure 11.19. The left column presents metrics for the Dst assessment, while the right column gives the same metrics for the precipitation data-model comparison. The metrics chosen are the same as in the examples earlier in this chapter: F_1, FB, ORSS, and HSS.

The two sets of plots are quite similar. In the F_1 score, there is a quick rise to a peak and then

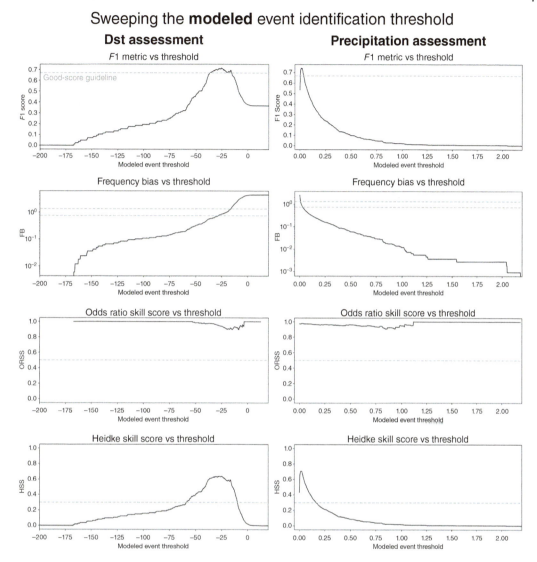

Figure 11.19 Metrics as a function of modeled event identification threshold for the running examples. Shown are the F_1 metric, the frequency bias, odds ratio skill score, and Heidke skill score. The horizontal dashed lines are drawn at the guidelines for declaring the metric score to be of high quality.

a gradual decline toward zero at the very high modeled threshold settings. The FB score starts high and then monotonically decreases, passing through unity and then to very small values at high threshold settings. The ORSS scores for both models are very high, hovering near unity and never dropping below 0.8. The HSS profiles look very similar to the F_1 scores, rising to a peak and then gradually dropping toward zero at large modeled threshold settings.

It would be ideal for all of the metrics to peak at a modeled event identification threshold that is at or near the imposed observed event identification threshold. This appears to be the case for F_1, which is measuring accuracy. As stated in Section 10.6 (Chapter 10), one estimate of a good F_1 score is 0.5, or even 0.67 for a more stringent judgment on quality. The blue dashed lines in the F_1 plots of Figure 11.19 are drawn at the higher of these two thresholds (at 0.67). Both of the models reach this higher value of F_1 near the observed event identification threshold. The models not only have good accuracy, but they reach this near the observation threshold.

A slightly different conclusion is reached when considering frequency bias. The Dst assessment shows that FB crosses unity at a modeled threshold of −20 nT. At the observation threshold of −30 nT, FB is 0.7, which is outside of the "good" range of FB (0.75–1.33). This quantifies that the O'Brien model is systematically underestimating the observed Dst values, not by much but enough that it leads to substantially more misses than false alarms. The precipitation data-model comparison has a better FB result, with the value at a modeled threshold of 0.015 inches (the same as the observation threshold) of 0.98, which is well within the good range for FB. At the particular threshold of interest, the HRRR model is well balanced regarding overestimation and underestimation errors of the precipitation at DTW airport. The FB score drops very quickly, though, as the number of misses only increases (M is in the denominator), while the number of false alarms only decreases (and F is in the numerator).

The odds ratio skill score is always very good for both models because it often correctly determines the numerous observed nonevents. The enormity of the C term for most of the range leads to an ORSS score close to one. The threshold settings where ORSS deviates from unity reveals that there are a few points in the F and M cells, but they never dominate the odds ratio. The ORSS scores for both models are always in the "good" quality range, well above the 0.5 criterion.

The curves for HSS resemble those of the F_1 metric. At low thresholds, they start with a quick rise, reach a peak value, and then gradually decline to zero at high thresholds. The previous chapter defined a rigorously good HSS value as being 0.3 or above, and both models easily clear this bar. Furthermore, in both curves, HSS is above this criterion at the setting for the observed event specification.

To further quantify the details of these metrics, Table 11.1 lists the modeled thresholds and metrics values of some key features of the curves in Figure 11.19. The top rows list the best or worst values for the four metrics, while the bottom four rows list the metrics scores when the two thresholds are equal (−30 nT for the Dst assessment, 0.015 inches for the precipitation assessment). It is expected and hoped that the

Table 11.1 Key values for the metrics curves of the running examples for the case of sliding the modeled event identification threshold.

Feature	Dst modeled threshold	Dst metric value	Precipitation modeled threshold	Precipitation metric value
F_1 maximum value	−25.4 nT	$F_1 = 0.71$	0.012 in.	$F_1 = 0.74$
Threshold where FB = 1	−20.4 nT	FB = 1.01	0.015 in.	FB = 0.98
Threshold of ORSS minimum	−8.3 nT	ORSS = 0.90	0.93 in.	ORSS = 0.91
HSS maximum value	−25.4 nT	HSS = 0.64	0.021 in.	HSS = 0.71
F_1 @ data threshold	−30 nT	$F_1 = 0.70$	0.015 in.	$F_1 = 0.74$
FB @ data threshold	−30 nT	FB = 0.70	0.015 in.	FB = 0.98
ORSS @ data threshold	−30 nT	ORSS = 0.96	0.015 in.	ORSS = 0.97
HSS @ data threshold	−30 nT	HSS = 0.63	0.015 in.	HSS = 0.71

"best" values are at or near the equal threshold setting. This is indeed the case. However, they are rarely at exactly the equal threshold setting. This offset can be taken into account when using the model for event prediction.

By sliding the modeled threshold, we are finding the optimal setting of this value to maximize particular metrics. Depending on which metric matters most for your particular usage of the model, you could then use this other modeled threshold setting for your operational implementation of the model. Considering the F_1 scores, it is seen that the O'Brien model has a slight underprediction of the Dst values, and therefore the prediction of storms reaching −30 nT or greater intensity is most accurately achieved with a modeled event identification setting of −25 nT. For this case, the difference is not large. Similarly, the prediction of 0.015 inches or more of precipitation at DTW airport by the HRRR model is most accurately achieved with a modeled threshold one bin over, set to 0.012 inches. Perhaps accuracy is not the most important factor for you; perhaps you want to balance the two types of error in your predictions so that you are not reporting excessive false alarms or misses. In this case, you might want to set the modeled threshold to the value that gets frequency bias closest to unity. For the Dst model, this is −20 nT, while for the precipitation model, this occurs at the equal threshold setting of 0.015 inches. When making determinations of anticipated observed events from the model output, you do not have to have the modeled event threshold set to the same thresholds as the events being predicted in the data.

> **Quick and Easy for Section 11.6.2**
>
> When sliding the modeled threshold, both geophysical assessments show that the chosen models are highly accurate, but are slightly more accurate with a modeled threshold offset a bit from the observed event identification threshold.

11.6.3 Metric-Versus-Observed Threshold Curves for the Running Examples

A similar set of calculations can be made when sliding the observed event identification threshold. The results of this are shown in Figure 11.20. The same four metrics are presented: F_1, FB, ORSS, and HSS; only the x axis has changed (from the modeled threshold to now being the observed threshold). The modeled event identification thresholds are held constant at −30 nT for the Dst assessment and 0.015 inches for the precipitation comparison.

These plots in Figure 11.20 resemble those in Figure 11.19, but with some key differences that reveal new insights about the data-model relationship. One striking difference is that the trend of the FB curve is reversed; it is now very small at low threshold settings and very large at high threshold settings. At a low observed event identification threshold, there are many misses and few false alarms, making the denominator of FB big and the denominator small. As the threshold steps to larger values, misses are converted to correct negatives and hits are converted to false alarms.

Some key values from these curves are noted in Table 11.2. It is a smaller table than that in the previous section because the last four rows would be identical; equal modeled and observed event identification threshold settings give the same metric scores, regardless of which is being swept and which is held fixed.

For the Dst assessment, sweeping the observed event identification threshold leads to the metrics peaking later, at more negative threshold settings than they did when sweeping the modeled threshold. The peak F_1 score and HSS for Dst is at an observed threshold setting of −40 nT, and FB reaches unity at −37 nT. This is consistent with the model underestimating the observed values, and this increase from −30 nT (the setting of the

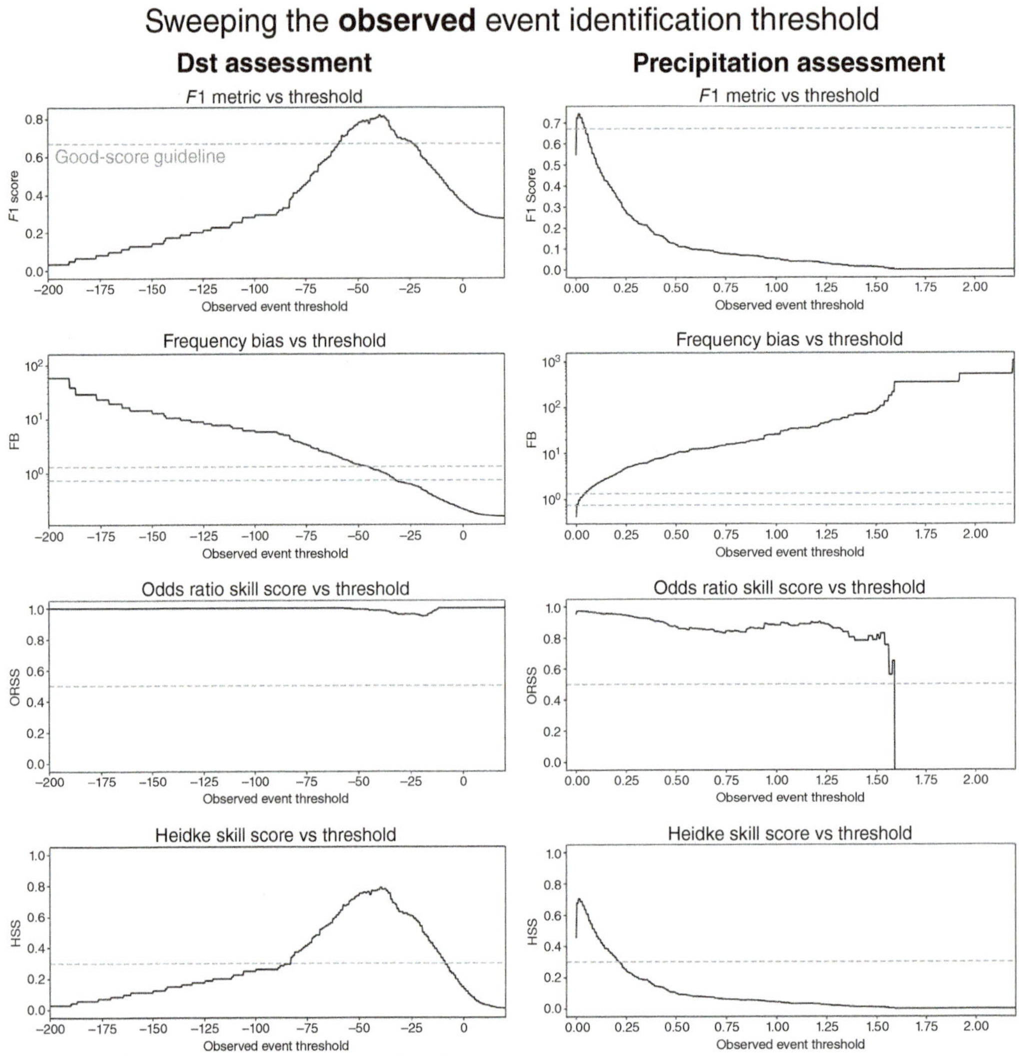

Sweeping the **observed** event identification threshold

Figure 11.20 Metrics as a function of observed event identification threshold for the running examples. Shown are the F_1 metric, the frequency bias, odds ratio skill score, and Heidke skill score. The horizontal dashed lines show the guidelines for declaring the metric score to be of high quality.

Table 11.2 Key values for the metrics curves of the running examples for the case of sliding the observed event identification threshold.

Feature	Dst observed threshold	Dst metric value	Precipitation observed threshold	Precipitation metric value
F_1 maximum value	−39.8 nT	$F_1 = 0.82$	0.012 in.	$F_1 = 0.74$
Threshold where FB = 1	−37.1 nT	FB = 1.01	0.018 in.	FB = 0.98
Threshold of ORSS minimum	−19.8 nT	ORSS = 0.95	1.59 in.	ORSS = −1.0
HSS maximum value	−39.8 nT	HSS = 0.79	0.012 in.	HSS = 0.71

modeled threshold for this assessment) quantifies that underestimation.

For the precipitation data-model comparison, the observation event thresholds for peak metric values are all within one increment of the fixed modeled threshold of 0.015 inches. This indicates that this model is quite well balanced relative to the observed precipitation amounts and is quite accurate at determining the chance of precipitation at this location.

One new aspect of the data-model relationship that is learned from this analysis is the range of observed threshold settings that are well modeled with a given modeled threshold setting. Table 11.3 lists these intervals for the observed event identification threshold yielding a metric score higher than the "high-quality" designation for that metric, as defined in Chapter 10. It is seen that the ranges for the Dst assessment are quite broad. While there are specific threshold settings that optimize certain metrics, this table allows you to see that there are ranges of threshold setting that yield high-quality scores for several metrics. Conducting threshold sweeps for a robust set of event detection metrics allows a user to select a final threshold that satisfies multiple criteria about the goodness of the model relative to the data.

> **Quick and Easy for Section 11.6.3**
>
> When sliding the observed event identification threshold while holding the modeled threshold fixed, the two running examples reveal observed threshold intervals of high-quality metrics scores surrounding the modeled threshold setting.

11.6.4 Metric-Versus-Simultaneous Threshold Sweep Curves for the Running Examples

Because these two running examples are real-valued data with the model trying to exactly reproduce those values, both thresholds can be moved simultaneously. This assesses whether the model captures the observed events at any threshold level. In the previous two cases, the number of events in one of the sets was constant because the event identification threshold was held constant. In this technique, both thresholds are moved, so the number of events for both the observations and the model output begins as the full number set and ends as none of the set. It is somewhat like fit performance analysis because the thresholds are kept equal, but it is still event detection analysis because, once the quadrants are defined for a particular threshold setting, the exact values do not matter (only the counts in the quadrants are important).

The curves for several metrics for the two geophysical running examples are shown in Figure 11.21. The blue dashed lines indicate the regions of the plots where that metric is considered good, per our definitions in the previous chapter. The obvious difference with the two previous figures is that the metrics scores are, in general, much better across a larger range of threshold settings. In addition, the metrics fluctuate and some of them have multiple threshold intervals when they are within the good value range. It is seen that for the Dst assessment, the F_1 and HSS metrics actually reach a value of one—a perfect score—at

Table 11.3 Intervals for which the metrics curves are within the "high-quality" range for the case of sliding the observed event identification threshold.

Feature	Dst threshold interval	Precip. threshold interval
Thresholds of $F_1 > 0.67$	−60 to −23 nT	0.003 to 0.048 inches
Thresholds of FB ∈ [.75,1.33]	−45 to −32 nT	0.003 to 0.039 inches
Thresholds of ORSS > 0.5	−223 to +56 nT	0 to 1.6 inches
Thresholds of HSS > 0.3	−83 to −9 nT	0 to 0.21 inches

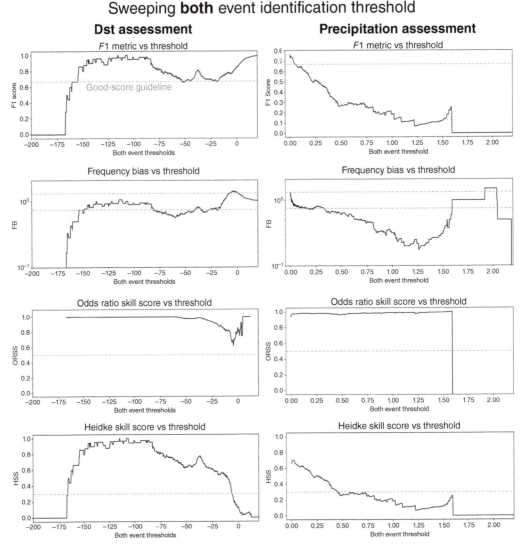

Figure 11.21 Metrics as a function of simultaneously sweeping both the observed and modeled event identification thresholds for the running examples. Shown are the F_1 metric, the frequency bias, odds ratio skill score, and Heidke skill score. The horizontal dashed lines show the guideline values for declaring that metric to be of high quality.

several instances along the threshold sweep. This reveals that there are some threshold settings where the model exactly sorts the observed values into events and nonevents, with no misses or false alarms in the contingency table.

The ranges where the metrics are within the designated high-quality range are listed in Table 11.4. The O'Brien model is quite accurate (F_1 score) at reproducing the Dst index across nearly the entire threshold sweep. It is also balanced, as indicated by the large intervals of good FB and ORSS scores. Interestingly, HSS is only good for negative Dst values. It drops below the high-quality value for Dst thresholds that are near zero or positive.

Figure 11.21 and Table 11.4 show that the precipitation model is not quite as robustly good as the Dst model. Its F_1 score is still only

Table 11.4 Intervals for which the metrics curves are within the "high-quality" range for the case of sliding both observed event identification thresholds simultaneously.

Feature	Dst threshold interval	Precipitation threshold interval
Thresholds of $F_1 > 0.67$	−160 to −52 nT −50 to +59 nT	0 to 0.06 inches
Thresholds of FB ∈ [0.75,1.33]	−154 to −150 nT −149 to −81 nT −40 to −35 nT −26 to −20 nT −18 to −5 nT +1 to +59 nT	0 to 0.15 inches 0.23 to 0.32 inches 1.6 to 1.9 inches
Thresholds of ORSS > 0.5	−167 to +13 nT	0 to 1.6 inches
Thresholds of HSS > 0.3	−166 to −6 nT	0 to 0.46 inches

above the "good cutoff" at very low precipitation levels. FB has a good score across a larger range of thresholds, with an extended interval at low precipitation values as well as another at very high precipitation values. ORSS is good throughout much of the threshold range, all the way up to 1.6 inches. HSS is above the high-quality mark up to nearly half an inch of precipitation. What this says is that the model reproduces the observed precipitation when the levels are low, but for heavy events, the model loses validity at specifying the exact amount of precipitation within the next 6-hour interval.

The fact that these intervals of high-quality metrics are expanded compared to the intervals in the previous two subsections of this chapter is not surprising. In those earlier techniques, one of the two metrics was held fixed, and therefore one of the two number sets was divided into events and nonevents and then these count values were held fixed, while the threshold for the other number set was swept through the range of possibilities. It is therefore expected that the model should not have great metrics scores when the sliding threshold

is far away from the fixed threshold. Those earlier two techniques are useful for identifying threshold settings that optimize particular metrics, or a collection of metrics, of interest to the model user. They are also applicable when the two number sets are not of the same quantity; that is, when the "model" is actually a "driving parameter" or function that does not match the units of the observations. They could have very different value ranges and the sliding threshold technique still works. For the simultaneous sliding threshold method, the extra restriction is placed that the model must be trying to reproduce the exact values of the observations. If this is the case, though, then the simultaneous sweep is a powerful tool for assessing the model's ability to detect events at any and all threshold settings.

> **Quick and Easy for Section 11.6.4**
>
> Because both running examples are real-valued number sets, the technique of sliding both thresholds simultaneously can be applied, revealing threshold intervals where each model does well for one or several metrics.

11.6.5 Metric-Versus-Metric Analysis for the Running Examples

The final assessment to conduct with the two running examples of Dst and precipitation prediction is the technique of plotting a metric score against another metric. Let's apply the three methods discussed earlier in Section 11.5: the ROC curve, the Alt-ROC curve, and the STONE curve. These plots are shown in Figure 11.22, with the top row displaying the curves for the Dst assessment and the bottom row presenting them for the precipitation assessment. Remember that the ROC curve is calculated with a fixed observed event identification threshold and the Alt-ROC curve is calculated with a fixed modeled event identification threshold. As with the other plots in

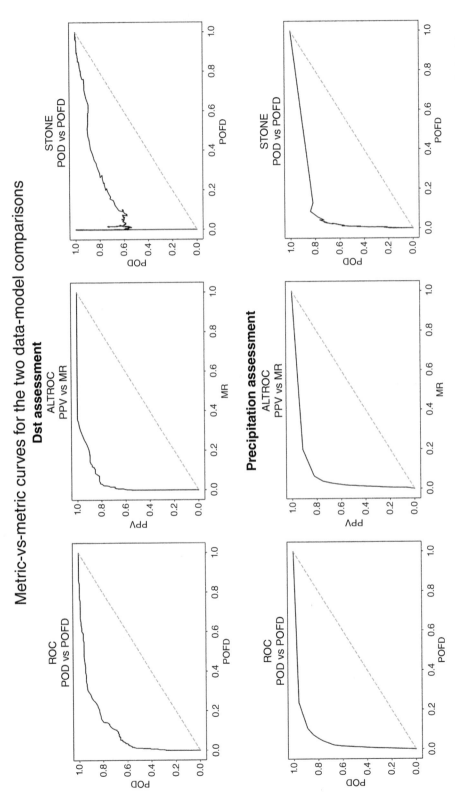

Figure 11.22 ROC curve for the running example of rainy day prediction. The diagonal dashed curve shows the unity slope of a random chance relationship.

this section, these thresholds are fixed at −30 nT and 0.015 inches for the two examples, respectively.

We see in Figure 11.22 that all six curves are well above the "random chance" relationship line in each. Overall, the two models are quite good at capturing the physics of the quantity to which they are being compared. For both models, the ROC and Alt-ROC curves are quite close to the upper-left corner of the plots. The STONE curves are a bit lower and farther away from the upper-left "perfect" corner. All three curves should be equal at the fixed threshold setting. For the Dst comparison, this occurs at a POD value of 0.59 and a POFD value of 0.029. For the precipitation assessment, it happens at POD = 0.73 and a POFD of 0.035. The lines could cross several other times, but this is neither required nor given for any particular comparison. The ROC and Alt-ROC curves are monotonic, as expected and required by their formulation, but the two STONE curves include nonmonotonic features, in particular up-and-down ripples at high threshold settings.

Several metrics can be calculated from these curves, in particular the AUC, SS_{ROC}, and J score, as defined earlier in Section 11.5.1. These are listed in Table 11.5. In addition to the J score itself, the threshold at which the J score occurs is also listed, since this is a score found as the maximum of a formula. Remember that the SS_{ROC} value is direct derivative of the AUC score; while AUC has a range from 0 to 1, SS_{ROC} is the value mapped to a range of −1 to +1. The J score ranges from 1 to 2, with 2 being a perfect score.

Table 11.5 shows that all of these metrics are quite good for both models. In fact, the J score of Dst is a perfect score of 2. This actually occurs at two different threshold settings, for −115 nT and again at −107 nT. There is another "upward spike" of the STONE curve at a slightly lower threshold, up to 1.95 at −87 nT. Removing these "spikes" from consideration, the optimal J score is 1.72 at a threshold of −37 nT. This is more directly comparable

Table 11.5 Distillation metrics of the ROC, Alt-ROC, and STONE curves for the two running examples.

Distillation metric	Dst assessment	Precip. assessment
AUC of the ROC curve	0.91	0.95
SS_{ROC} of the ROC curve	0.82	0.89
J score of the ROC curve	1.66	1.78
Threshold of J score	−16 nT	0.003 inches
AUC of the Alt-ROC curve	0.96	0.92
SS_{ROC} of the Alt-ROC curve	0.92	0.85
J score of the Alt-ROC curve	1.78	1.76
Threshold of J score for Alt-ROC curve	−38 nT	0.003 inches
AUC of the STONE curve	0.85	0.89
SS_{ROC} of the STONE curve	0.70	0.77
J score of the STONE curve	2.00	1.75
Threshold of J score for STONE curve	−115 and −107 nT	0.003 inches

with the other J scores from the ROC and Alt-ROC curves.

Another interesting point from Table 11.5 is that the thresholds at which the J score occurs are identical for the three curves assessing the precipitation data-model comparison, but are quite spread out from each other for the Dst curves. For the precipitation model, this peak occurs very early, at the first nonzero threshold in the sweep. The very quick drop of POFD and MR for this model is seen in the curves as kinks between straight-line segments. This is seen in all three curves for this model. This is because there are so many zero values in both of the precipitation number sets, as soon as the threshold moves to a nonzero value, over half of the points convert from false alarms to correct negatives. This indicates that the HRRR model is really good at addressing the question of "will there be any precipitation in the next 6 hours?" At higher thresholds of precipitation, the model is still quite good, but its ability to predict the specific amount of precipitation is not as good as its ability to predict the yes/no status of any precipitation at all.

For Dst, the J scores peak at thresholds that are quite different from each other. The thresholds for the J scores for the ROC and Alt-ROC curves reflect the slight underestimation of the model values relative to the observed Dst values. For the STONE curve, the uncertainties associated with very small number of points at extreme negative Dst values lead to the STONE curve jumping rapidly from high to low values. The fact that the STONE curve J score is set for a large and negative threshold indicates that the O'Brien model is good at predicting Dst during active times, but less good at predicting the specific Dst value during quiet times.

Both of the STONE curves are relatively monotonic but with a few small vertical non-monotonic features. It is good to remind ourselves what this means: a temporary increase in POD as the threshold is swept upward. Remember that POD is $H/(H + M)$, and H is always decreasing as the threshold moves. POD can only increase if M undergoes a substantial drop in a short threshold interval. Each upward spike implies that a cluster of points in the misses quadrant left the M count as the threshold rapidly swept over them. To be in the M quadrant means that these points were "below and to the right" of the unity-slope perfect-match line in the left panel scatterplot shown in Figure 11.15. That is, the model values are smaller than the corresponding observed values; a vertical ripple represents an underestimation by the model. Every up–down ripple is another small cluster of points that were underestimated by the model. The thresholds where these ripples occur could then be examined in more detail with a fit performance subset analysis.

> **Quick and Easy for Section 11.6.5**
>
> The ROC, Alt-ROC, and STONE curves reveal the high quality of these two running examples, as well as certain aspects of the data-model relationship.

11.7 The Power of Sliding Thresholds

To summarize, there is much that can be learned by sliding the modeled threshold and exploring the resulting metric scores for interesting features. Some will exhibit relative maxima and minima, while others are monotonic with gradual or sharp changes in slope. Some do not change much for long stretches of threshold setting and then reveal a dramatic change over a short interval of threshold settings. Each of these features in the metric curves reveals underlying features of the data-model scatterplot. Sometimes, these features of the scatterplot are easily seen simply by qualitative examination, but many times they are subtle and lost within the mass of points, but an examining of several different metric scores versus threshold can reveal these features. Different qualities about the data-model relationship are learned by sweeping the two thresholds, either independently (holding the other fixed) or both together.

This was seen in our analysis of the Dst and precipitation forecast models and sweeping the thresholds in various ways. For instance, with Dst, the systematic bias showed an underestimation of the observed events. This is not an average underestimation but rather only an underestimation of events beyond that particular threshold. The O'Brien model was then very good at other thresholds, even reaching perfect F_1 and HSS metric scores at some storm-level thresholds. With the HRRR model for precipitation, it was seen that it does very well for low thresholds, predicting the chance of any precipitation in the next 6 hours at

Detroit Metro Airport with high accuracy and essentially no bias in the contingency table.

The processes demonstrated in this chapter are reminiscent of the discrimination and reliability subsetting techniques of the fit performance metrics. Depending on the step size in the sweep, it can be a coarse or very-fine-scale measure of the model's ability to reproduce observed events within certain ranges of the values. Note that the intervals of high-quality metric scores for these techniques are not the same as discrimination and reliability checks with fit performance metrics. In the fit performance case, the data are culled to only those with a value within a particular range of either the observed values (discrimination) or modeled values (reliability). This is not the case here; all points from the data-model sets are used to fill the contingency table. Some metrics ignore one or two of the cell count values, but many metrics use all four quadrants. This analysis is not assessing the same features as fit performance, either, because these are event detection metrics. That is, the only thing that matters is whether the value is above or below the event identification threshold; after conversion to a contingency table count, the exact value of both of the observation and model output is forgotten for the actual metric calculations. While these sliding threshold techniques can be used to identify intervals of interest and inform the break points for fit performance discrimination and reliability subsetting analysis, it reveals a different aspect of the data-model relationship.

One point that should be discussed is the step size for the sweep. A good step size for sliding the threshold is a bit like setting the number of bins in a histogram, as done in Chapter 2—it is highly subjective and depends on the level of information that you want to extract from the relationship. There is an upper limit to the number of thresholds, and that is the number of points in the data-model comparison. If the threshold step count is higher than the total number of points, then the sweep will produce fluctuations in the metrics as individual points change from one quadrant to another. An argument could also be made for a lower limit of 10 steps in the threshold sweep (assuming there are more than 10 points in the scatterplot). Anything less than this means that the step size between threshold settings is rather large and features of the metric score curves could be skipped over. Choosing a finer scale is perfectly acceptable if high resolution is desired for the assessment being conducted, but it runs the risk of the metric score curves becoming noisy, especially near the ends of the sweep. It is better to choose a value that smooths out statistical fluctuations yet resolves large-scale features, such as gradients and relative peaks. The sweep step size is something that has to be tuned to optimize the plots for the particular assessment being conducted.

There is an interesting twist to event detection: events could be "nonevents," at the low end of the y value range. To examine this, simply reverse the $>$ to a $<$ sign and do the same analysis (including sliding thresholds) to see how well the model captures quiet times. This might be very useful for particular users: for example, astronauts taking an extra-vehicular activity (EVA) excursion outside of the safety of the space station, or a satellite operator wanting to turn on an expensive yet very sensitive satellite subsystem. These people do not need to know the size of the energetic particle event that could happen; they are more interested in predicting quiet times to know that it is okay to do the EVA.

This is the same approach as was used with the Dst event analysis in the earlier sections, but with a different emphasis. In the Dst case, the index is designed such that low values (i.e., large but negative) are times of high activity. In searching for quiet times at the low end of a positive definite value, such as precipitation, a reversal of the event status identification (defining events as those values *below* the threshold), would "detect" quiet times of little-to-no precipitation, the events of interest in this case.

> **Quick and Easy for Section 11.7**
>
> Sliding the event identification threshold, for either the observations or the model, allows for an even richer diagnosis of the data-model relationship than with a single threshold choice.

11.8 Further Reading

Many articles exist discussing the process of sliding thresholds for optimizing event detection metrics for various purposes.

Here is a good one for a climate/agriculture-related application:

Mathieu, J. A., & Aires, F. (2018). Using neural network classifier approach for statistically forecasting extreme corn yield losses in Eastern United States. *Earth and Space Science*, 5, 622–639. https://doi.org/10.1029/2017EA000343

And one for weather forecasting:

Manzato, A. (2005). An odds ratio parameterization for ROC diagram and skill score indices. *Weather and Forecasting*, 20, 918–930. https://doi.org/10.1175/WAF899.1

Here is one for a solar physics usage:

Bobra, M. G., & Couvidat, S. (2015). Solar flare prediction using SDO/HMI vector magnetic field data with a machine learning algorithm. *The Astrophysical Journal*, 789, 135. http://dx.doi.org/10.1088/0004-637X/798/2/135

And another for a planetary physics application:

Azari, A. R., Liemohn, M. W., Jia, X., Thomsen, M. F., Mitchell, D. G., Sergis, N., et al. (2018). Interchange injections at Saturn: Statistical survey of energetic H$^+$ sudden flux intensifications. *Journal of Geophysical Research: Space Physics*, 123, 4692–4711. https://doi.org/10.1029/2018JA025391

The history of the ROC curve is summarized here:

Carter, J. V., Pan, J., Rai, S. N., & Galandiuk, S. (2016). ROC-ing along: Evaluation and interpretation of receiver operating characteristic curves. *Surgery*, 159(6), 1638–1645, https://doi.org/10.1016/j.surg.2015.12.029

The ROC curve is discussed in this chapter of the Forecast Verification book:

Hogan, R. J., & Mason, I. B. (2012). Deterministic forecasts of binary events. In I. T. Jolliffe & D. B. Stephenson (Eds.), *Forecast Verification: A Practitioner's Guide in Atmospheric Science* (2nd edition, chap. 3, pp. 31–60). John Wiley, Ltd., Chichester, UK. https://doi.org/10.1002/9781119960003.ch3

Chapter 8 of the book by Wilks also discusses the ROC curve:

Wilks, D. S. (2019). *Statistical Methods in the Atmospheric Sciences* (4th edition). Academic Press, Oxford.

A good discussion of ROC curve for earthquake analysis is given here:

Meade, B. J., DeVries, P. M. R., Faller, J., Viegas, F., & Wattenberg, M. (2017). What is better than Coulomb failure stress? A ranking of scalar static stress triggering mechanisms from 10^5 mainshock-aftershock pairs. *Geophysical Research Letters*, 44, 11,409–11,416. https://doi.org/10.1002/2017GL075875

And another for ROC curves used with volcanism:

Stefanescu, E. R., et al. (2014). Temporal, probabilistic mapping of ash clouds using wind field stochastic variability and uncertain eruption source parameters: Example of the 14 April 2010 Eyjafjallajökull eruption. *Journal of Advances in Modeling Earth Systems*, 6, 1173–1184. https://doi.org/10.1002/2014MS000332

A couple of space physics examples for the ROC and STONE curves are provided by Liemohn et al. (2020):

Liemohn, M. W., Azari, A. R., Ganushkina, N. Y., & Rastätter, L. (2020). The STONE curve: A ROC-based model performance assessment tool. *Earth and Space Science*, 6, e2020EA001106. https://doi.org/10.1029/2020EA001106

A few more space physics examples with the ROC and STONE curves are presented in Liemohn et al. (2021):

Liemohn, M. W., Shane, A. D., Azari, A. R., Petersen, A. K., Swiger, B. M., & Mukhopadhyay, A. (2021). RMSE is not enough: Guidelines to robust data-model comparisons for magnetospheric physics. *Journal of Atmospheric and Solar-Terrestrial Physics*, *218*, 105624. https://doi.org/10.1016/j.jastp.2021.105624

And an analysis of the relationship of scatterplot features to STONE curve nonmonotonicities is presented in Liemohn et al. (2022):

Liemohn, M. W., Adam, J., & Ganushkina, N. Y. (2022). Analysis of features in a sliding threshold of observation for numeric evaluation (STONE) curve. *Space Weather*, *20*, e2022SW003102. https://doi.org/10.1029/2022SW003102

11.9 Exercises in the Geosciences

There are many communications (scientific) and other spacecraft in orbit around the Earth at the special distance of 6.62 Earth radii from the center of the planet. At this distance, their orbital period is exactly one day. If positioned over the equator, they remain fixed in the sky as they orbit with a perfectly matching velocity to Earth's rotation. This location, however, is continually bathed in energetic particles accelerated in Earth's magnetic field. These particles can damage the surface or even the interior of the satellites. Models have been developed to simulate this near-Earth space environment, including one called the Inner Magnetosphere Particle Transport and Acceleration Model (IMPTAM). It can be compared with data from the NOAA Geosynchronous Orbiting Environmental Satellite (GOES), which includes sensors to monitor the intensity of these energetic particles.

One such data set and model output combination is available here:

https://deepblue.lib.umich.edu/data/concern/data_sets/02870v99r?locale=en

1. Open and read in the GOES data and IMPTAM output files. Make all of these plots with logarithmic axes for the flux values, which span several orders of magnitude. Each of these files contains over two years of data for three energy channels. For the following analysis, please choose only *one month* of the full data set and select only *one of the energy channels*.
   A Make a separate time series plot of each number set (again, one month for one energy channel, but for both the GOES data and IMPTAM output).
   B Make an overlaid time series plot of both number sets.
   C Make separate histograms of each number set.
   D Make an overlaid histogram of both number sets.
   E Make a scatterplot of the observed values against the modeled values.

2. Write a paragraph answering the following:
   A What is your initial reaction to the relationship between the data and the model?

3. Conduct a metrics assessment with sweeping thresholds.
   A Implement a sweeping threshold routine that slides both the model and observed event identification thresholds simultaneously. Calculate contingency table cell counts as a function of threshold setting. Note that starting the sweep before the first point (in either the model or the observed number set) will result in some metrics being undefined. A similar situation results at the other end of the sweeps. This is achieved with an "if statement" that manually sets the metric score to the first defined value (or last defined value, at the other end of the sweep).
   B Calculate at least four metrics from various categories for these cell count arrays. A possible good set would be what was done in this chapter: F_1, FB, ORSS, and HSS.

C Make plots of these metrics scores against the threshold.
D Calculate POD and POFD along the same sliding thresholds.
E Create a STONE curve.

4. Write several paragraphs answering the following:
   A What are the threshold intervals where each of the metrics curves are in the "high-quality" value range?
   B Discuss what can be learned about the model from these metrics versus threshold curves.
   C Describe the features of the STONE curve.
   D Discuss these line-plot features in terms of any features that can be qualitatively seen in the various plots from part 1.

12

Applications of Metrics and Uncertainty: Final Advice and Introductions to Advanced Topics

Everything in the preceding chapters is the basic set of data calculations, testing, and comparisons. You are now well equipped for tackling intensive data analysis and modeling assessment tasks. This final chapter summarizes and synthesizes the material covered in the earlier chapters, offering additional advice on how to combine these many tools for conducting robust examinations of number sets. In addition, the last three chapters on data-model comparisons have not included uncertainty calculations for the metrics. A section is included in this chapter on how to calculate uncertainties on metrics values and ways to include these uncertainties in your usage of the metrics. Another topic introduced in this chapter is the process of using the comparison of two number sets for making decisions.

The set of tools presented in this book, however, is just the beginning. This chapter also introduces some additional concepts, many of which are quite important for data analysis in the natural sciences. This is not meant to be an exhaustive list of topics, nor is it a full explanation of the ones that are mentioned. This is a brief overview so that you know the terms and have some idea of when to turn to these other tools.

The final section of this chapter discusses uncertainty within individuals conducting scientific research. Remembering the scientific method presented in Chapter 1, there are many times in that process where one can be wrong. This can lead to feelings of uncertainty and inadequacy, which can trigger doubt in your self-confidence, anxiety about your work life, and perhaps lead you all the way to depression. People working in scientific fields should keep in mind that they are doing something hard and that these feelings are common. You are not alone in having these thoughts and we should embrace the uncertainty before us and persevere together.

12.1 Choosing the Right Set of Metrics

In order to select the most appropriate metrics for your particular application, it is necessary to understand the features of the individual number sets that will be compared. First, make some plots, as discussed in Chapter 2, to get a general, qualitative impression of the relationship. Next, if there are equations and processing involved, do the uncertainty propagation from Chapter 3. Next, quantitatively understand each number set separately with the calculations in Chapter 4, and perhaps also step through the normality check process outlined in Chapter 5. If it makes sense, then do a linear comparison with a correlation from Chapter 6 and a fit from Chapter 7. After all of these

---

Data Analysis for the Geosciences: Essentials of Uncertainty, Comparison, and Visualization, Advanced Textbook 5, First Edition. Michael W. Liemohn.
© 2024 American Geophysical Union. Published 2024 by John Wiley & Sons, Inc.
Companion website: www.wiley.com/go/liemohn/uncertaintyingeosciences

initial steps, you are now ready to jump into the metrics evaluations.

After many chapters of metrics formulas, where to start for a specific data-model comparison can seem overwhelming. Usually, the comparison will fall into one of three main types. The first decision is whether the focus of the comparison is on matching exact values or on sorting the data into events and nonevents. This is the split between fit performance and event detection. If exact values are important, then there is a second criterion to consider—whether the number sets are Gaussian or not. If they are close to Gaussian and span only one order of magnitude (or at most two), then the metrics that are commonly interpreted with an assumption of an underlying Gaussian are applicable and appropriate. If, however, the distributions of the number sets are far from Gaussian, then other metrics are better for assessing the sets against each other.

> **Quick and Easy for Section 12.1 Introduction**
>
> The best set of metrics to use depends on both the type of assessment being conducted as well as the distributions of the two number sets.

12.1.1 Metrics for Fit Performance Assessment on Gaussian Distributions

For the case of a fit performance assessment on near-Gaussian distributions, a list of good metrics to use is given in Table 12.1. For each metrics category (listed in the first column) assessing a particular aspect of the relationship (given in the second column), there are one or two metrics that best quantify that facet (the third column). Let's briefly explain the categories for which two metrics are listed. For accuracy, there are two good

Table 12.1 Summary of the best-fit performance metrics for Gaussian number sets.

Category	Feature assessed	Best choice	Notes
Accuracy	Overall similarity	RMSE	Equivalent to σ
		MAE	Average error instead of σ
Bias	Centroids	ME	Pairs well with RMSE or MAE
Precision	Spreads	$P_{\sigma,\text{diff}}$	Same interpretation as ME
Association	Trends	R	Assesses linearity very well
Extremes	Outliers	γ_Δ	Same interpretation as ME
		k_Δ	Same interpretation as ME
Skill	Improvement relative to a reference model	PE	Skill relative to data variance
		SS_{MSE}	Skill relative to reference model variance, same form as PE
Discrimination	Data–range subset	Any of the above	Assesses what data ranges contribute most to metric scores
Reliability	Model–range subset	Any of the above	Assesses what model ranges contribute most to metric scores
Sliding threshold	Events at all threshold settings	STONE curve and other metrics	Identifies value intervals where model performs well

choices, RMSE resembles standard deviation, while MAE is based on average error, which is usually slightly smaller. Normalizing these by the data set standard deviation is a good practice to provide context about their magnitude. The two metrics of skew error and kurtosis error are listed for the extremes category. Both are useful because they assess different aspects of the outliers, the former testing the similarity of the left–right symmetry of the two-number-set distributions and the latter testing the similarity of the heaviness of the tails of the two number sets. Skill also has two metrics listed. The first, PE, is so common that it is useful to calculate just for comparison with other assessments. The reference model for it, though, is the variance of the observations. If another model is available, then it is perhaps useful to compute a skill score directly against that prior model, rather than conducting a calculation with the intermediate step of the data set as the reference model in a PE score for each model separately.

It is strongly advocated that some form of subsetting analysis be conducted. This allows for a fine-scale assessment of which specific ranges of the data set or model set contribute the most to the metric scores. That is, subsetting provides a much richer evaluation of the data-model relationship. This is often done with a traditional subsetting approach of either the data set (discrimination) or model set (reliability). The ranges of the subsetting, however, could also be more quantitatively determined with a sliding threshold analysis in which both event identification thresholds are moved simultaneously.

12.1.2 Metrics for Fit Performance Assessment on Non-Gaussian Distributions

If the values span more than a few orders of magnitude, then some of the usual selections for metrics become less trustworthy in assessing model fit performance. The data-model pairs in the lower orders of magnitude will be crowded near the origin and only the pairs in the highest order will be clearly seen. This can lead to qualitative misinterpretations when examining such plots, as well as quantitative partialities. This is because many of the metrics are designed with a Gaussian, or near-Gaussian, distribution in mind. The issue is often centered on the use of differencing the modeled and observed values, that is, metrics that include $M_i - O_i$. The issue persists when the metric includes a subtraction of the individual values from the mean of the set, either $M_i - \bar{M}$ or $O_i - \bar{O}$. When the values span several orders of magnitude, then the values in the highest order will contribute the most to the summations over these differences. For non-Gaussian and highly variable number sets, it is best to avoid metrics based on linear or even squared combinations of the values.

There are plenty of examples of highly variable data in Earth, atmospheric, space, and planetary processes. One of the most famous from space physics are energetic charged particle fluxes. For example, the Sun occasionally emits **solar energetic particles**, SEPs, generated by the explosive magnetic reconfigurations in active regions of the solar atmosphere as well as additional acceleration processes at shock fronts when a fast plasma structure streams away from the Sun through the solar system. SEP fluxes are usually very low, often below the one-count "noise level" of the

---

Quick and Easy for Section 12.1.1

For near-Gaussian distributions, metrics based on data-model differences work very well at assessing the relationship between the number sets.

Solar energetic particles (SEPs): Very-high-energy electrically charged particles from the Sun, emitted during solar flares and by high-speed shocks, which pose a radiation hazard to spacecraft and astronauts.

instrumentation designed to detect them. When an active region explodes as a solar flare, or a supersonically fast shock front is created in the inner solar system, then the flux of SEPs rises dramatically, rising over the background levels by a factor of a thousand or even a million. These particles can reach relativistic speeds, spiraling along the interplanetary magnetic field and traversing the distance from the Sun to the Earth in as little as 30 minutes (remember, light, which travels in a straight line rather the curved spiral trajectory that the charged particles must take, reaches Earth in 8 minutes, so these ions and electrons are very fast). SEP flux is a highly variable data set with rare but very intense "events" that requires special care for proper model assessment.

The examples of a solar flare and consequent SEPs are shown in Figure 12.1. The plots show a 3-day window of measurements in 2011 from the Geostationary Operational Environmental Satellite (GOES), specifically GOES-13. This satellite has two sensors measuring energetic radiation from the Sun, the first is an X-ray detector, observing highly energetic photons, and the other is a particle detector, which measures highly energetic protons, which are nearly all from the Sun. Two strong solar flares occurred on August 9. Flares emit their photons in all directions, so every flare occurring on the same side of the Sun as the Earth will be seen by the GOES spacecraft sensors. The second of these flares happened at a spot on the solar surface that was magnetically connected to Earth through the interplanetary magnetic field. The SEPs that were generated and released during the flare soon arrived at Earth and were detected. There was a very sharp rise in SEP flux followed by a very gradual decrease over the next day or two down to background levels.

Another example of highly variable data in nature are insect populations throughout the year, especially in off-equatorial latitudes such as the northern United States. Many insect species are dormant in the winter, with very low counts through those months, and then populations rapidly grow in the spring to reach peak values at some point in the summer. The change in value can be dramatic. While there is a fairly constant 10 quintillion (10^{19}) insects on Earth at any one time, the number of any particular insect in a specific region can vary by many orders of magnitude, from zero to a trillion (10^{12}) in a short time span.

If the number sets are deemed highly variable, then caution should be used when interpreting the metrics in Table 12.1, all of which are based on differencing. Instead, rely on the metrics that use the logarithmic values, ratios of the values, or medians of the values. These metrics minimize the influence of values that span over several orders of magnitude, treating the orders of magnitude more equally in the metric score rather than favoring the values in the highest order.

Fortunately, there are metrics in all of the categories that work in this situation. A list of metrics that work well for such number sets is given in Table 12.2. All of the metrics are distinct from those listed in Table 12.1. These are based on either relative comparisons (i.e., formulas involving M_i-to-O_i ratios) or include logarithmic operations, both of which act to normalize order of magnitude differences. Note that this excludes many famous and commonly used metrics, like RMSE, MAE, R, and PE. That is right; do not use the well-known metrics because these include absolute differences that favor the highest order and therefore are not ideal when the values span two or more orders of magnitude.

In Table 12.2, there are a few categories with multiple listings of "best" metrics. For accuracy, SMAPE uses the relative difference, while MSA is a median of the log of the ratio. The first includes an influence from all data-model pairs, while the second only reports a single value among all data-model pairs. If reporting SMAPE for accuracy, then either ME$_{median}$ or ME$_{geometric}$ is the best choice for the bias category. If using MSA, then SSPB is best because the functional forms of these two metrics are nearly identical. Association also has two options, R_S and R_{log}. The Spearman coefficient

12 Applications of Metrics and Uncertainty: Final Advice and Introductions to Advanced Topics | 371

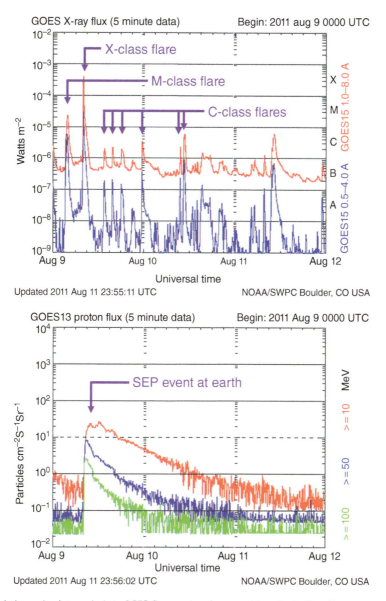

Figure 12.1 A three-day interval plot of SEP fluxes, showing a sudden rise in flux. The top panel shows solar X-ray fluxes, while the lower panel shows solar energetic particle fluxes in three proton energy channels. Adapted from the NOAA Space Weather Prediction Center.

converts all values into their rank order, which removes the actual values of the two number sets. The second approach takes the log of all values in the correlation coefficient formula. This reduces orders of magnitude to integer differences. Both are effective at de-emphasizing the values in the highest order of the range, but do so in different ways. For skill, which is often based on an accuracy metric, it is appropriate to use either SMAPE or MSA in the skill score formula. The reference model can be either the data set against its median or another model. Subsetting for highly variable data sets works exactly the same as with near-Gaussian sets, but the ranges of the subsets will most likely be very different size (or,

Table 12.2 Summary of the best-fit performance metrics for non-Gaussian or highly variable number sets.

Category	Feature assessed	Best choice	Notes
Accuracy	Overall similarity	SMAPE	Individually normalized MAE
		MSA	Ignores the outliers
Bias	Centroids	ME_{median}	Ignores outliers
		$ME_{geometric}$	Takes a log, then averages
		SSPB	Pairs well with MSA
Precision	Spreads	YI_{log}	Takes a log first
Association	Trends	R_S	Minimizes outliers to their rank
		R_{log}	Takes a log first
Extremes	Outliers	$CPD_{\Delta,\varepsilon}$ & $CPD_{\Delta,1-\varepsilon}$	Based on quantiles, not spreads
Skill	Improvement relative to a reference model	SS_{SMAPE}	Skill based on an accuracy metric appropriate for non-Gaussian number sets
Discrimination	Data–range subset	Any of the above	Assesses what data ranges contribute most to metric scores
Reliability	Model–range subset	Any of the above	Assesses what model ranges contribute most to metric scores
Sliding threshold	Events at all threshold settings	STONE curve and other metrics	Identifies value intervals where model performs well

perhaps, equal increments in log-space). The sliding threshold technique is still an excellent method for determining intervals of the number sets where the model works well.

> **Quick and Easy for Section 12.1.2**
>
> Non-Gaussian distributions, or highly variable sets that span more than two orders of magnitude, are best assessed with metrics that use relative errors or logarithm operators.

12.1.3 Metrics for Event Detection Assessment

For event detection, the values themselves do not matter after the thresholds are set, so there is no split in whether the distributions are Gaussian or not. Table 12.3 lists some "best" metrics for event detection for each category. Most of the categories have more than one entry; let's go through the options. For accuracy, PC is good because it includes all cells from the contingency table and provides a single measure of the quality of the model at sorting the observations into events and nonevents. The difference between PC and F_1 is the inclusion of the correct negatives cell count, C. If C is big and yet the focus is on hits, then removing C from the accuracy formula is desirable. For precision, the two metrics assess different qualities of the relationship. The first (precision) is focused on modeled events, quantifying the split of these modeled events between observed events and nonevents. The second (recall) is the converse assessment, quantifying the split of the observed events between

Table 12.3 Summary of the best event detection metrics.

Category	Feature assessed	Best choice	Notes
Accuracy	Overall similarity	PC	Uses the full table
		F_1	Omits C from the calculation
Bias	Centroids	FB	If event symmetry is desired
Precision	Spreads	Precision	Hits within modeled events
		Recall	Hits within observed events
Association	Trends	ORSS	Assesses balance of the table
		MCC	Based on χ^2 score of the table
Extremes	Outliers	SEDS	Balanced version of EDS
Skill	Improvement relative to a reference model (random chance)	HSS	Balanced and widely used
		GSS	Minimizes influence of C count
		PSS	Discrimination-focused skill
		CSS	Reliability-focused skill
Discrimination	Data–range subset	POD	Hits within observed events
		POFD	False alarms in observed nonevents
Reliability	Model–range subset	PPV	Hits within modeled events
		MR	Misses within modeled nonevents
Sliding model threshold	Optimizing the model threshold	ROC curve and other metrics	Finds model threshold that best suits the usage need
Sliding data threshold	Data range where model is good	Alt-ROC curve and other metrics	Finds data range where model is of high quality for one threshold
Sliding both thresholds	Events at all threshold settings	STONE curve and other metrics	Finds model threshold that best suits the usage need

modeled events and modeled nonevents. It is good to calculate both of these metrics. For association, the two metrics listed are similar but derived in completed different ways. ORSS focuses on the balance of the good cells relative to the balance among the error cells, while MCC, a normalized version of the χ^2 significance test, compares the cell counts against an expected random table with the same row and column totals. Both have their merits. For skill, four different options are listed. All four of these options are based on random chance as the reference model, and the choice of which is best depends on the assessment being conducted. For a measure that symmetrically includes the full contingency table, HSS is the appropriate metric. If it is desirable to minimize the influence of C in the assessment, then GSS is most appropriate. If the focus of the analysis is on observed events, the PSS is really good, while if the focus is on modeled events, then CSS is best. Take your pick based on your usage needs. For discrimination and reliability, the two listed are those used for the ROC and

Alt-ROC curves. It is convenient to choose these instead of their complement formulas because all of these vary from one down to zero as the event identification threshold is swept from low to high. Note, however, that the interpretation is different among them, with POD and PPV having a perfect score of one and POFD and MR having a perfect score of zero.

If the data are precategorized into event and nonevent status, then sliding the model event identification threshold is the only threshold sweep that can be conducted. This is very useful to do because it reveals the threshold settings that optimize the score for each event detection metric. This culminates in the calculation of a ROC curve, which is interpreted as an assessment of the overall quality of the model to capture the physics behind the quantity being observed.

If the observed values are a set of continuous real numbers, then sliding the observed event identification threshold is possible. If there are specific thresholds of key interest, then setting the model threshold to this value and sweeping the data threshold allow for the identification of ranges of the data for which the model produces a high-quality sorting of the observations. This technique of sliding only the data threshold yields the Alt-ROC curve, which can be construed as an assessment of the overall quality of the model to sort observed events at a particular model threshold setting.

Similarly, if both the observed and modeled sets are real-valued numbers, then both thresholds can be swept from low to high values simultaneously. This reveals optimized metrics scores for which the model is good at reproducing observed events with both number sets using the same threshold. A special plot that can be made from this assessment is the STONE curve, which could have nonmonotonic features. Left-to-right wiggles in the STONE curve reveal threshold intervals where there is a cluster of points in the false alarms quadrant—that is, a cluster of overestimated values. Up-to-down ripples in the STONE curve identify threshold intervals where there is a cluster of points in the misses quadrant—that is, a cluster of underestimated values.

> **Quick and Easy for Section 12.1.3**
>
> For event detection data-model comparison situations, there is a somewhat longer list of "best" metrics, depending on the emphasis of the evaluation.

12.2 Combining Metrics for Robust Data-Model Comparisons

It is good to state, as so often is the case in number set comparisons, that all of these choices of "best metric" listed in the tables in Section 12.1 are only guidelines. You are free to choose whatever metric you prefer for any of the categories, mixing and matching as you see fit to most appropriately address the assessment task you are conducting. These tables provide sets of metrics that would produce a robust comparison. The important point is to use several metrics, because when used in combination, they provide a much richer analysis of the data-model relationship. The next few subsections explore some of these combinations that work very well together.

12.2.1 The Accuracy–Bias–Precision Trifecta

Taken together, accuracy, bias, and precision form a complementary and powerful triad of metrics that allow for additional interpretation of the data-model comparison. Specifically, while the accuracy measure gives an overall goodness-of-fit indication, the metrics of bias and precision parse it into specific aspects of the number sets that contribute to the quality of the accuracy score. While it is not a simple relationship of bias plus precision equals accuracy, the first two inform about why an accuracy metric has the score it has.

12 Applications of Metrics and Uncertainty: Final Advice and Introductions to Advanced Topics | 375

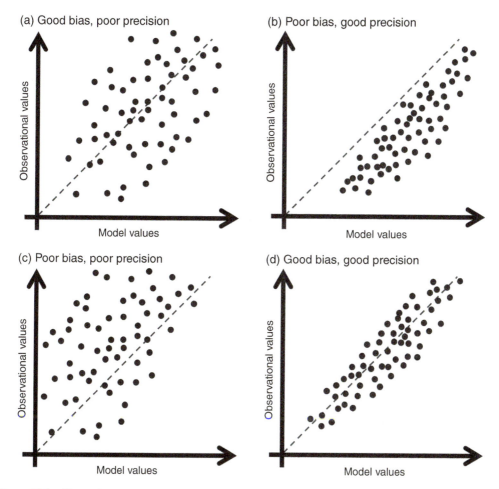

Figure 12.2 Illustrative data-model comparison scatterplots that reveal a relationship with good bias but poor precision, good precision but poor bias, poor bias and poor precision, and good precision and good bias (and therefore, also, good accuracy).

Figure 12.2 presents four examples of data-model comparison scatterplots for which the use of all three metrics categories together reveals additional information about the relationship. The top two panels would have nearly identical accuracy scores; the use of bias and precision metrics along with an accuracy metric would reveal the clear difference between the two distributions. For one, the imperfect accuracy score is mostly due to the spread of the values away from the ideal diagonal line, while for the other, the distribution has low spread but a significant offset, which would dominate the accuracy score. The lower left panel of Figure 12.2 would have a slightly worse accuracy score than the two in the top panels; the use of precision and bias along with it would reveal that this is due to a combination of poor spread and significant offset. The final panel shows an example with low scores for both precision and bias, and therefore the accuracy score would also be low and not need additional interpretation.

From accuracy alone, it can be learned that a model has some high or low quality at reproducing the observations. If it is not perfect, though, then the reason for this nonzero accuracy metric score can be deciphered from bias and precision. Context can be placed on the accuracy score with these two other metrics. Specifically, if the bias is rather small while the precision difference is relatively large, then this indicates that most of the accuracy metric score is from the differences in the spreads of the two number sets rather than the model having a large, systematic offset from the observed values. This understanding of the reason for a less-than-perfect accuracy score cannot be obtained from the accuracy metric itself but rather only from the use of several metrics in combination.

> **Quick and Easy for Section 12.2.1**
>
> The accuracy metric score is placed into context with the additional use of bias and precision metric scores.

12.2.2 The Accuracy–Association Connection

Metrics in the accuracy category are also complemented very nicely by metrics assessing association. The most common pairing of these two metrics is RMSE with R, and indeed many studies use only these two metrics. It is a good combination. The strengths of these two metrics blend well together.

The accuracy category includes metrics that focus on the overall closeness of the data and the model values. The association category encompasses the metrics measuring the similarity of the up–down trends within the two number sets. To be more specific, accuracy tests the closeness to the unity-slope line, while association tests the closeness to any straight line, regardless of the slope or y intercept. It is possible to have a good score in association, but not in accuracy. To illustrate this, Figure 12.3 shows two scatterplots, both with good association but one with poor accuracy and the other with good accuracy. In the left panel, the points clearly follow a linear relationship between the two parameters, but the line is not the dashed unity-slope line. Its association score should be quite good, but its accuracy score would not be as good. The right panel shows a set of points that are fairly tightly scattered around the unity-slope line, giving this relationship a high score in both accuracy and association.

If you only had the accuracy metric score, then you might conclude that the data-model relationship in the right-hand panel of Figure 12.3 is much better than the one in the left-hand panel. The association score for the left panel, however, might actually be higher than that for the right panel, because the points

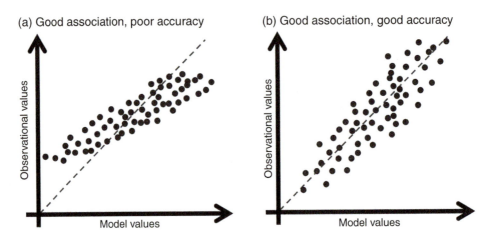

Figure 12.3 Illustrative data-model comparison scatterplots that reveal a relationship with good association but poor accuracy and another with both good association and good accuracy.

are "tighter" around the linear relationship. This difference with the unity-slope line is a systematic offset that can be corrected. Calculating the linear fit for the left panel provides a formula for adjusting the model values and obtaining a high-quality fit between the model and the observations.

> **Quick and Easy for Section 12.2.2**
>
> Association is a more general version of accuracy that does not require a perfect unity-slope fit. Pairing these two metrics together provides additional context to the data-model relationship, and even a fix for systematic bias.

12.2.3 The Association–Extremes Linkage

The metric category of extremes is sometimes hard to interpret. One way to do this is to use it in conjunction with a metric from the association category. Association focuses on the full-number-set linear connection, which is most likely dominated by the values in the center of the distribution. A comparison between the association metric score, and the linear relationship that it is based on, and the extremes metric scores can reveal additional information about the data-model relationship.

Figure 12.4 shows an example scatterplot with three lines overlaid on it. The first is the light-purple dashed line, showing the unity-slope perfect connection between the observations and the model output. The second is the green line, which is the linear fit determined from regression analysis. This comparison has a Pearson correlation coefficient of 0.82, a good metric score. Note that the green line has a positive y intercept and a slope that is less than one. The third red solid but segmented line shows the quantile–quantile values, with points centered in each 5% bin from lowest to highest. The lowest value in the red line is at 2.5%, located at the (x,y) coordinates of (0.54, 1.37), and the highest point along the red

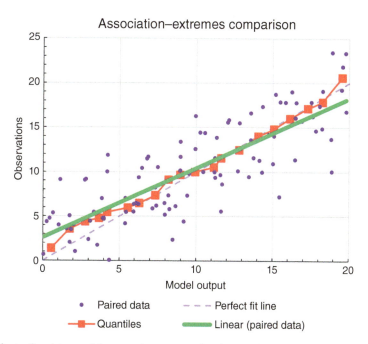

Figure 12.4 Illustrative data-model comparison scatterplot that emphasizes the connection between association and extremes.

line is at 97.5%, which has coordinates of (19.56, 20.62). These yield extreme metrics scores of $CPD_{\Delta,\varepsilon} = -0.83$ and $CPD_{\Delta,1-\varepsilon} = -1.06$. These two scores are quite good compared to the number set standard deviations, which are 5.4 and 5.7 for the data and model values, respectively.

It is seen in Figure 12.4 that the red and green lines largely overlap in location for most of the range. They are only different at the ends, when the red values diverge, below the green line at the low end and above the green line at the high end. At the low end, both the green and red lines are above the dashed unity-slope line, while at the high end they are on opposite sides of the dashed unity-slope line.

At the low end, it is seen that the overall trend (the green line) is that the model values are smaller than the observed values; there is a systematic underestimation at the lower portion of the range. This is not actually the case, though, because the red curve shows that the smallest quantile bin has values within one of each other and therefore much closer to the unity-slope line than the overall trend predicts.

At the high end, the overall trend of the green line is that the model values are larger than the observed values. This systematic overestimation at the higher portion of the range is not real, though. The upper-end extremes metric score is negative, indicating that the model is underpredicting the observed values in this region.

This example shows that metrics from association and extremes pair well together. The association metrics are often dominated by the central bulk of the distribution, while the extremes metrics largely (or completely) ignore that part of the distribution. From association you get the overall trend between the observation and model number sets, while the information from the extremes metrics can be quite different. Association metrics should not be trusted to inform you about the relationship in the lowermost or uppermost ends of the two distributions. They should not be expected to do this; they were not designed for assessing that aspect of the data-model relationship. There is a different metrics category for that, with its own set of equations.

> **Quick and Easy for Section 12.2.3**
>
> Association captures overall trends, while extremes focus on the outliers. They work well together, each providing context to the other.

12.2.4 Expanding Our Understanding of Skill

Skill is the derivative metric that uses another primary metric value and places historical context on it. Its standard form, the skill score, is a certain equation that includes metric scores from the current data-model comparison, the score from a reference model, and the perfect score for that metric. It is arranged such that a perfect skill score value is one and any value larger than zero indicates improvement of the current model over the reference model.

Because of the dominance of skill scores such as prediction efficiency and the Heidke skill score, skill could be perceived to be only a measure of accuracy (remember that PE is based on MSE and HSS is based on PC). Furthermore, PE uses the data set variance as its reference model, and HSS uses random chance as its reference model. Neither of these reference models are actual models, a fact that, again, could limit the perceived scope of metrics in the skill category to the thought that we must use one of these two options as our reference.

This does not have to be the case. *Any* metric can be used in the skill score formula, and *any* reference model can be used for the comparison conducted by this formula. If your particular assessment requires an emphasis on a different category, then developing a skill score based on a metric from that category is almost certainly more meaningful than simply calculating PE. Similarly, the reference model does not have to be based on the data or a random chance number set. If you want to compare a new model against an old model, then it is useful to create a skill score that does this directly, rather than requiring two metric scores to be calculated and then compared. It is good to remember this and expand our understanding of skill.

> **Quick and Easy for Section 12.2.4**
>
> Skill is the process of normalizing a metric score against the perfect score for that metric and the metric score for a reference model. Choosing the right metric and the right reference model will yield a skill score that is useful for your particular assessment.

12.2.5 Using Discrimination and Reliability Together

The combination of the two subsetting approaches of discrimination and reliability is a useful extra step to the data-model comparison analysis. This is highlighted in the example shown in Figure 12.5. The left panel shows discrimination subsetting into three data-value groupings, while the right panel shows reliability subsetting with three model value groupings. Both panels contain the same scatterplot of illustrative data-model pairs. Qualitatively, the discrimination plot shows that the middle subset has the highest-quality match between data and model values, but the reliability plot shows that the first subset has the highest-quality data-model comparison.

This information is useful, depending on your perspective on the comparison. Given a value from the observational set, if it is in the range of that center subset, then you know that the corresponding model value is most likely pretty good. If it is in one of the other two subset ranges, then the corresponding model value could be far from the observed value. That is, the model is best in this middle range of the observed values. If you have an observation in this middle subset range, then you can probably trust the model to correctly interpret the physics of the situation under examination.

Conversely, given a model value, if it is in the lowest subset range, then you know that it is most likely a good representation of the corresponding observation. If it is in either of the other two subset ranges, then the observations could be essentially anything across the full range of observational possibilities. So, if you are running the model with the hope of predicting the observations, then you can trust the

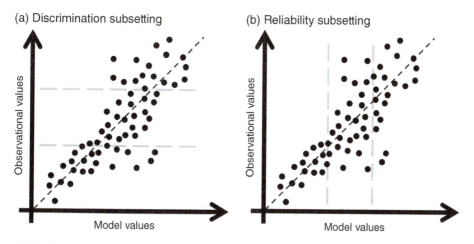

Figure 12.5 Discrimination and reliability subsetting used on the same data set to reveal portions of each range where the model has a high- and low-quality fit.

model output to be a good forecast only when the model values are in that first subset range. For all other model output values, it is not a particularly good model for prediction.

It will not always be the case that there will be this difference in the "best" subset from discrimination and reliability. Sometimes, though, this difference arises, as in the illustrative example shown in Figure 12.5. Therefore, it is useful to conduct the analysis for both subsetting options; the extra work can sometimes be very revealing about the relationship of the model output to the observed values.

> **Quick and Easy for Section 12.2.5**
>
> Sometimes, discrimination and reliability reveal a single part of the scatterplot where the model works well; other times, the two subsetting approaches can reveal different ranges of the data and model values where the model is good.

12.3 Uncertainty on Metrics

All of the metrics presented in this book should be interpreted not only against the uncertainty in the observed and modeled number sets themselves—like what is done for nRMSE (see Section 9.3.3, Chapter 9), for example—but also with respect to the uncertainty in the calculation of the metric itself. This is a worthwhile extra analysis step that provides a more thorough understanding of whether the calculated metric is similar to its perfect score. That is, a metric score that is, at first glance, far from that metric's perfect score might be initially diagnosed as poor, but if the uncertainty on this score is large, such that the error bar around the score includes the perfect score, then you need to adjust your understanding of the score.

The process of finding an uncertainty on the calculated metric score can be done analytically or numerically. The analytical approach is to replace every M and O value in the formulas for each metric with $M + \delta M$ and $O + \delta O$ and, using the techniques presented in Chapter 3, deriving equations for the uncertainties of these metrics. There are several formulas for which this is already derived and readily available. These formulas are often based on the assumption that both the model output and the observed values have a Gaussian distribution. If this is not the case, then analytically deriving an uncertainty can be a rather difficult mathematical problem.

It is usually a good choice to find these uncertainties numerically. There are several common approaches here. In no particular preferential order, the first to be discussed here is the application of the uncertainties of the observed and modeled values as a random change to the values themselves. That is, if a δM or δO uncertainty is known for all of the values, then a distribution can be assumed and value set of uncertainties can be determined. It does not matter whether the same uncertainty (a universal δM or δO value) or value-specific uncertainties (a unique δM_i or δO_i) are used. If the latter, then different distributions will be needed for each M_i or O_i value, otherwise the same distribution can be used for all M_i or O_i values. A random number can be used to select a perturbation quantity to add or subtract from M_i or O_i, done separately for all i values, and then a new set of metrics can be calculated. For event detection, there is the extra step of comparing these perturbed M_i and O_i values against the event identification thresholds to get cell counts for H, F, M, and C. The result is the same—an array of metric values. This should then be done many times to build up a distribution for the metrics, at least dozens if not hundreds or even thousands of times. From these distributions of values for each metric, a spread can be determined. If the distribution of the metric values looks Gaussian, then a standard deviation is appropriate. If not, then a spread based on quantile values is good, such as interquartile range or perhaps a plus and minus value that extends a similar 34% from the median, as does the standard deviation for a Gaussian distribution.

If the "model" is another observed quantity thought to be a driving parameter, then uncertainty propagation and calculation according to the rules in Chapter 3 can be applied. It is often the case, though, that output from a numerical model has no obvious uncertainty associated with it. A method for calculating uncertainties on this number set is to run the model many times with random uncertainties applied to its input parameters. Depending on the complexity of the model, this "ensemble calculational approach" could be quick or, more likely, rather time-consuming and resource-intensive. The thought is that you only have to do it once for a typical set of input observations, and then the same uncertainties can be used for comparing with many data sets. For a model input value set with a known uncertainty that is assumed to be a Gaussian spread around the actual value, a random number generator can be used to create many small deviations to that input parameter. This can be done for all inputs, if uncertainties are known. The model can then be run with these deviations applied as a plus or minus correction to the parameters, resulting in a different model value. This should be done several, if not many, times, to obtain a spread of model values. For each of these new model values, either the original data or, if its uncertainty is known, then a deviation of the data values according to its Gaussian spread can be used to calculate a new set of metric scores. From all of these new model and data-value combinations, a distribution of scores for each metric is found, from which an uncertainty on the metric score can be calculated.

Another good numerical approach is to conduct a bootstrap analysis, as was discussed in Chapter 6 regarding correlation coefficient uncertainty when comparing two data sets. The bootstrap method involves resampling the data-model pairs, with replacement, to create a "new" data-model number set, on which all of the metrics can be performed. This should then be done hundreds or even thousands of the times to produce a distribution of values for each metric. A histogram of this distribution should be plotted and tests conducted to determine if it is nearly Gaussian or not. An uncertainty can be calculated from this distribution of the metric values for all of the bootstrap resamplings, either the standard deviation of the distribution is Gaussian or an analogous value based on the quantiles, if it is not Gaussian.

These uncertainties could then be used in the interpretation of the metrics scores. In particular, these uncertainties can be used within the formulas for t tests for comparing metric scores from several models against each other. You could ask yourself, when I made this change to a physical process within the numerical model, did the metric score improve by a statistically significant amount? Without uncertainties on the metric scores, you are simply guessing about whether the improvement is large or small. Calculating uncertainty values on the metrics allows for the application of a Welch's t test (to pick a good one from Section 6.1, Chapter 6) that will place a probability on the potential similarity of the two metric scores.

> **Quick and Easy for Section 12.3**
>
> There are several good methods for determining uncertainties on metrics, such as perturbing the observed and modeled number set values according to their uncertainties or conducting a bootstrap analysis.

12.4 Uncertainty on Fit Performance Metrics for the Dst Running Example

For our running example of the O'Brien model for Dst introduced back in Chapter 9, a few bootstrap analyses were conducted to determine uncertainties on the fit performance metrics scores. First, a bootstrap with 500 resampling sets was done, and the resulting distributions of the scores are shown in Figure 12.6. One metric from each of the main

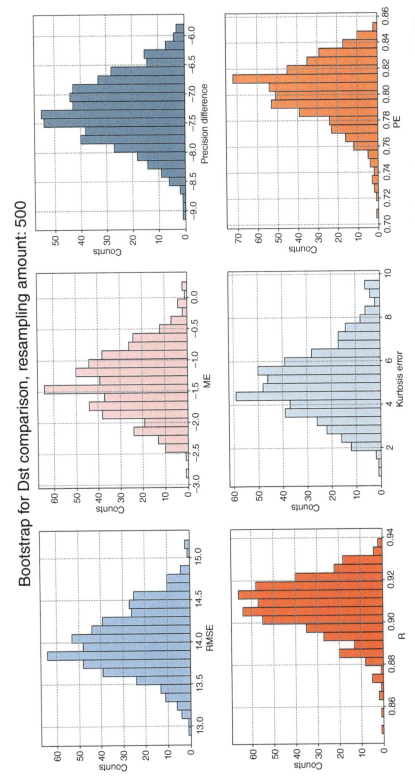

Figure 12.6 Histograms of some of the metric scores from a bootstrap analysis of the O'Brien Dst prediction model. These histograms were made with 500 resamplings of the original number sets.

Table 12.4 Metrics scores from the original comparison as well as from two versions of a bootstrap analysis, with 50 and 500 resampling sets.

Metric	Original score	Bootstrap $\bar{x}$ (M = 500)	Bootstrap σ (M = 500)	Bootstrap $\bar{x}$ (M = 50)	Bootstrap σ (M = 50)	Reported σ_x
RMSE	13.98 nT	13.96 nT	0.37 nT	13.98 nT	0.39 nT	0.4 nT
ME	−1.40 nT	−1.42 nT	0.53 nT	−1.44 nT	0.58 nT	0.5 nT
$P_{\sigma,\text{diff}}$	−7.24 nT	−7.25 nT	0.53 nT	−7.24 nT	0.56 nT	0.5 nT
R	0.909	0.907	0.013	0.910	0.012	0.013
k_Δ	4.79	5.07	1.55	4.67	1.54	1.6
PE	0.805	0.802	0.023	0.806	0.022	0.02

categories was selected for this example analysis, presenting RMSE, ME, $P_{\sigma,\text{diff}}$, R, k_Δ, and PE in the six panels, respectively. All of these histograms are well-formed Gaussians.

A comparison of the mean and standard deviation of these histograms is listed in Table 12.4, along with the original score. Extra digits are reported in order to reveal the convergence of the bootstrap mean with the original metric score. It is seen that the bootstrap mean values are very close to the original scores and well within the bootstrap standard deviation scores. The largest relative uncertainty is found to be on the kurtosis error, k_Δ. This is to be expected, as this metric is highly sensitive to the inclusion of even a few outliers in the number sets. As the resampling process either duplicates or omits the extreme values, the kurtosis of each number set could vary by a substantial amount. This results in a rather large spread of kurtosis values and a correspondingly large standard deviation for the kurtosis error metric.

To further demonstrate the convergence on a well-defined uncertainty value for each metric from the bootstrap method, a second resampling was conducted, but this time with M set to only 50. The resulting histograms of the metric score distributions are shown in Figure 12.7. These distributions are not quite as well formed into Gaussian shapes as when ten times more resampling sets were used, but most of them are not that far off from a normal curve. A full calculation to test the normality of these distributions could be conducted, but that is not the point of this illustration, which is to show convergence to the original metric scores. The mean values and standard deviations for these $M = 50$ bootstrap distributions are listed in the fifth and sixth columns of Table 12.4. It is seen that the mean values are still quite close to the original scores, in some cases actually closer to the original that for the $M = 500$ bootstrap analysis. Most of the standard deviations are slightly larger for the smaller-M bootstrap analysis, but not by much.

The final column in Table 12.4 shows the final version of the uncertainty that should be reported with each of the original metric scores. In this column, the values are truncated to the proper number of significant digits, taking into account our rules about this process from Chapter 1. Applying these to the base values, we have RMSE reported to three significant digits (14.0 nT), while ME, $P_{\sigma,\text{diff}}$, and PE reported to only two digits (−1.4 nT, −7.2 nT, and 0.81, respectively). Because the uncertainties for the other two metrics lead with a one in the first significant digit, the exception to the basic rule allows us to report two digits for these uncertainties, and therefore an extra digit on the base values (0.909 for R and 4.8 for k_Δ).

In Chapter 9, a second model for numerically calculating the Dst index was introduced at the end, the Burton model. This is an older

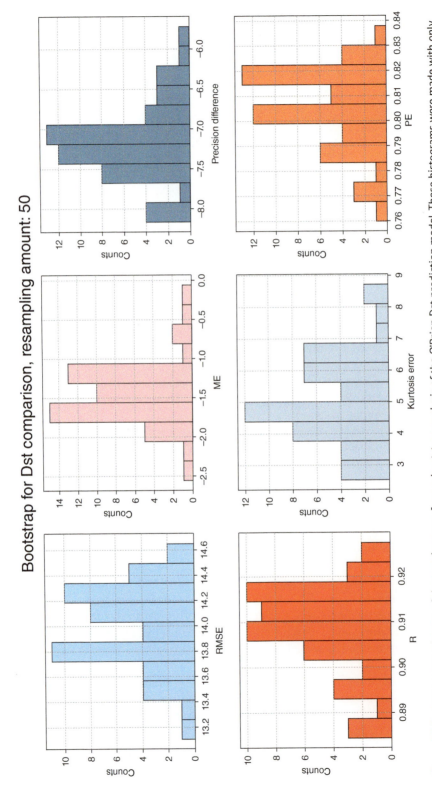

Figure 12.7 Histograms of some of the metric scores from a bootstrap analysis of the O'Brien Dst prediction model. These histograms were made with only 50 resamplings of the original number sets.

and simpler version but uses the same basic approach of applying a differential equation for the time advancement of Dst as a function of itself as well as solar-wind-driving parameters. In Chapter 9, the only calculation with the Burton model was for SS_{MSE}, using the Burton model output as the reference model. For the month of March 2015, the Burton model has an RMSE score of 19.5 nT. Is the O'Brien model RMSE score of 14.0 nT significantly better than the Burton model RMSE? Without uncertainties, we do not know.

A bootstrap analysis with $M = 500$ iterations yields an uncertainty of 0.7 nT on the Burton model RMSE. Now we have two scores, each with uncertainties, and equal numbers in the two sets ($N = 744$). In this case, a Welch's t test can be conducted (defined in Section 6.1.3, Chapter 6). The combined standard deviation $\sigma_{\bar{\Delta}}$ is found to be 0.029 and therefore $t = 190$. The probability for this is well below the 1% highly statistically significant value and so we can say, quantitatively, that the O'Brien model is significantly more accurate than the Burton model (at least against this March 2015 data set).

> **Quick and Easy for Section 12.4**
>
> A bootstrap analysis is applied to the running example of Dst prediction, finding uncertainties on the metric scores and using these in a t test for a hypothesis assessment.

12.5 A Recipe for Robust Comparisons

As stated at the beginning of this chapter, it is always good to start with a plot. At the very least, make line plots of the two number sets, against time, space, or whatever is the independent variable. An even better initial examination is to create three plots: an overlaid line plot; a scatter plot of the data values against the model output; and an overlaid histogram of the two distributions. Making even more of the full variety of plots available, as discussed in Chapter 2, is useful to explore, qualitatively, how these two number sets look relative to each other. Getting a general feeling for how the two number sets align is the best first step to robust data-model comparisons.

As part of this plotting, uncertainties should be included on the figures. If the value being plotted is derived from other measured quantities, each with their own uncertainties, then this involves the propagation of uncertainty to get a number appropriate for the derived quantity. Methods for doing this were discussed in Chapter 3.

Next, on each data set, conduct the single-data-set processing described in Chapter 4. Get to know the parameters that quantitatively reveal features of each number set. The most basic set is to calculate arithmetic mean and standard deviation, but an even better assessment of each number set can be achieved by considering median, mode, interquartile range, skew, kurtosis, and the others. These parameters for the data and model number sets can be compared, still at the qualitative level. This continues to build the overall impression of how the two number sets relate to each other.

This should be followed with an assessment of the type of distribution of each number set. Chapter 5 outlines a process for conducting this evaluation. The steps listed use many of the parameters calculated for each number set, but also include a few new values and tests.

Another step in the process is to do some or all of the two-data-set correlation processing given in Chapter 6. A few methods for finding an uncertainty on the correlation coefficient were given there. If the model is attempting to exactly reproduce the observations, then tests of the similarity of the number set means can be conducted as well.

If applicable, then doing the linear or nonlinear fitting calculation described in Chapter 7 can be very useful. This is the first quantitative data-model comparison being done, with a calculation that involves both of the number sets.

If it is a model that is attempting to exactly reproduce a continuous-valued data set, then the ideal fit should be a straight line with zero offset and unity slope. This might not be the case, though, as all models are approximations to the full set of physical properties of the system being observed. It could be, therefore, that the model does not match the data with a nice straight line, but rather with some kind of nonlinear best-fit curve through the data-model scatter plot.

All of these steps are simply a preamble to the use of metrics to assess the data-model relationship. At the very least, an accuracy calculation should be performed, yielding a quantitative yet greatly distilled understanding of how well the model reproduces the data. With the plots and calculations done earlier, this one metric can provide you with a pretty good grasp of how well the model is doing toward this goal. Adding in metrics from the bias, precision, and association categories further aids in interpreting the goodness of the data-model comparison.

The end of Chapter 8 included three key design elements for a good data-model comparison (Section 8.8). The first was selecting the right spatial or temporal interval for the comparison. The interval should include a span of observed values that will appropriately test the model for the features of interest. The second and related aspect is the spatial or temporal resolution. Data sets and model output are obtained at a particular cadence that may or may not be appropriate for the assessment that needs to be conducted. Think carefully about averaging or clustering in order to focus the comparison on the important features. Third and finally, another choice that needs to be considered is the nature of the quantity being compared. If the assessment is meant to focus on extreme events, then the comparison should be designed to highlight such events. Too often, a large data set is used and the statistics are dominated by quiet times. This can be overcome with the proper choice of metrics.

If your interest in using the model is focused on the outliers and one or both ends of the value range, then additional metrics will come in handy. Metrics from the extremes category specialize in highlighting these outliers and assessing the ability of the model to capture the frequency of these outliers. To further explore this aspect of the data-model relationship, the subsetting categories of discrimination and reliability are very useful because they split up the full set of data-model pairs into a smaller chunk of particular interest.

When choosing metrics to use and the approach of combining metric scores into an overall assessment, always keep in mind the reason for conducting the data-model comparison. In addition, it is good to keep in mind the level of the assessment within the application usability level (AUL) stages discussed in Chapter 8. The assessment conducted at the early AULs could be different than those used later in the process at higher AULs. This is because, at the higher AULs, the specific usage of the model is better clarified and focus can be given to those metrics that best assess that aspect of the data-model relationship.

A key element to interpreting any of the metrics is a calculation of the uncertainty in that value. Not only do the individual data or model values have an uncertainty, but the value obtained from a data-model comparison calculation also has an uncertainty associated with it. Knowing the uncertainty of a metric is critical to understanding whether the model is useful for a particular science investigation or operational application. It can also be very helpful when examining the metric score of one model relative to another model—overlapping uncertainties is an initial indication that the two models are similar.

This brings us to another best practice process for data-model comparisons. Take full advantage of the availability of the classic statistical analysis tools of t tests, χ^2 tests, and the F statistic. The t tests can be used to quantitatively assess the closeness of metric values from two models, helping you determine whether you can reasonably declare that one model is better than another (well, at least for that data-model comparison aspect being examined by that metric). The χ^2 tests are quantitative comparisons of two distributions—replacing the "expected value" with "model value" (or reversing the observational and model value usages in this formula) yields a measure of the closeness of the distributions. The F statistic is a quantitative assessment of the overall quality of the data-model comparison, essentially an accuracy-based skill parameter, but with its own unique formula, so not quite a skill score. The resulting t, χ^2, and F values from these tests can all be converted into a probability, often interpreted as the chance that the model and the data are from the same underlying population. This is desirable for a data-model comparison in which you want the model to exactly reproduce the data. The statistics profession is moving away from assigning significance to such probabilities, but they are still very informative to the researcher or operational user in evaluating the goodness of the model fit for their purposes.

These steps are summarized in Table 12.5. This provides a high-level strategy for conducting a robust comparison between two number sets. Also, note that these steps are not simply a linear progression. As the comparison is designed, the number sets could change, resulting in the need to make new initial plots and single-data-set processing. It might even result in a need for recalculation of the uncertainty propagation.

Table 12.5 Steps to a robust comparison between two number sets.

Step	Chapter or section	Notes
Propagate uncertainties	Chapter 3, Section 4.3	So there are error bars on all values
Make initial plots	Chapter 2, Section 8.3	Line plots, scatterplots, histograms
Single-data-set processing	Chapters 4 and 5	Assess normality, pick centroids and spreads
Design a good comparison	Section 8.8	Consider the important features and choose interval, resolution, and nature of comparison
Initial correlation	Chapter 6	Assess overall relationship
Line or curve fitting	Chapter 7	As appropriate and needed
Select additional metrics	Chapters 9, 10, and 11; Sections 12.1 and 12.2	Depends on the feature of importance for the comparison
Uncertainties on metrics	Section 12.4	For interpretative context

> **Quick and Easy for Section 12.5**
>
> When approaching any comparison of two number sets, this book provides resources for a thorough analysis of the similarities and differences between them. This section provides a quick-reference guide for applying all of these techniques.

12.6 Metrics and Decision-Making

Making decisions based on a data-model comparison assessment often involves the paradigm of **choice combination statistics**.

This is the process of using probabilities that certain outcomes could occur in order to make a judgment about the significance of a certain outcome. In the following text, the concept of choice combination statistics is introduced, followed by a specific example of making a decision on whether to adopt a new predictive model for operational space weather usage.

> **Choice combination statistics:** The process of using probabilities for events to occur in order to assess the significance of a result.

12.6.1 Choice Combination Statistics

The most basic example of choice combination statistics is flipping a coin; the result is either heads or tails and there is an equal probability of each result occurring. The chance of getting heads on three successive coin flips is the multiplication of the probability for each of the flips: 0.5 × 0.5 × 0.5 = 0.125, or 1/8. Another well-known example is rolling a die. A six-sided die has equal probability of each value occurring on any particular throw. The chance of throwing a six on three straight throws is a similar calculation: (1/6) × (1/6) × (1/6) = 1/216, or 0.00463, which is less than 1%. Conversely, the chance of getting no sixes in those three throws is this: (5/6) × (5/6) × (5/6) = 125/216, or 0.579, over half of the time there will be no sixes in three throws. These examples illustrate the math of probabilities. We will not delve deeply into this math, but rather focus on the applications of probability theory for the analysis of a data set.

For a meteorological context, let's make the parameter under consideration to be the chance of snowfall during the winter months in Ann Arbor, Michigan. This is a binomial problem because there are just two outcomes, it either snows or it does not during the 24-hour period of a day. The probability of snowfall can be any value between zero and one, it does not have to be 1/2 as in the coin-flip case. But let's make this a climatological average, though, so that we can assume that each day has the same probability of having snowfall. In this case, each day can be considered as a trial of our binomial probability parameter.

Let's define a few terms, first. For consistency with the earlier nomenclature, the number of days, or trials, is N. The probability of success will be denoted by p, a value between zero and one. The probability of failure will be called q, with $q = 1 - p$. The probability of ν successes in N trials is given by the **binomial distribution**, $B_{N,p}(\nu)$, written like this:

Binomial distribution: A special function that yields a probability of a certain number of events occurring out of a total number of trials.

$$B_{N,p}(\nu) = \binom{N}{\nu} p^\nu q^{N-\nu} \quad (12.1)$$

The last two terms in Equation (12.1) are integer powers of a value between zero and one, but what is the strange "N over ν" term? This is the binomial coefficient, written like this:

$$\binom{N}{\nu} = \frac{N(N-1)(N-2)\ldots(N-\nu-1)}{1 \cdot 2 \cdot 3 \cdot \ldots \cdot \nu} \quad (12.2)$$

We can use the factorial symbol ($a!$) to condense the product series in Equation (12.2) to write it as this:

$$\binom{N}{\nu} = \frac{N!}{\nu!(N-\nu)!} \quad (12.3)$$

Equation (12.1) can be used with any binomial problem where there is a recurring chance of success. For the coin flip, $p = q = 0.5$, and the math becomes easier because the two power terms combine to become p^N. In this case, the binomial distribution only depends on the changing value of ν in denominator of the binomial coefficient.

As with the Poisson distribution introduced back in Chapter 4, the binomial distribution also has a peak at some value of ν and decreases monotonically away from this ν value. The binomial distribution is truncated at both ends, though, because ν cannot be larger than N.

For the average and standard deviation of the binomial distribution, we will skip the derivation and go straight to the resulting formulas. The mean is given by this:

$$\bar{\nu} = \sum_{\nu=0}^{N} B_{N,p}(\nu) \quad (12.4)$$

The math of this summation works out to be a rather simple formula:

$$\bar{\nu} = Np \quad (12.5)$$

It is seen that, if the probability of success is less than the reciprocal of the number of the trials (i.e., if $p < 1/N$), then the average number of successes will be less than one. It is possible to have the distribution so truncated on the left side that the peak is missing from the range of possible ν values (which must be between 0 and N). In this case, the distribution has a maximum for $\nu = 0$ and decreases as ν increases.

The standard deviation has this form:

$$\sigma_\nu = \sqrt{Npq} \qquad (12.6)$$

On rewriting in terms of only p instead of p and q, we get:

$$\sigma_\nu = \sqrt{Np(1-p)} \qquad (12.7)$$

The equation is not quite the same as the Poisson counting uncertainty, which was the square root of the mean value, but rather contains an additional term, q, inside the square root operator. This is because of the truncation of the high-end tail, ν values above N.

A feature of the binomial distribution is that it is symmetric about the average value. This is clearly seen in the binomial coefficient, as the two terms in the denominator of Equation (12.3) are reversible. Similarly, because $q = 1 - p$, the polynomial terms are reversible once past the peak value. In fact, when both $\bar{\nu}$ and N are much greater than one, the binomial distribution approaches a Gaussian distribution. Therefore, given the formulas for the mean and standard deviation in Equations (12.5) and (12.7), the probability from the Gaussian distribution can be used instead of the probability from the binomial distribution. For a single ν value with a fairly small N, this is not much help. However, when N is large, then the factorial in the numerator of Equation (12.3) becomes impossible for most calculators to determine. Furthermore, when addressing a "this or more" question, then a z score can be used instead of a summation over many ν values using the binomial distribution. As with the Poisson distribution, this greatly simplifies the math for conducting hypothesis testing.

Continuing with the snowfall days example, there are, on average, 53 days with snowfall each year in Ann Arbor, Michigan. This occurs over the "snowy period" from mid-November to mid-April, a period covering 150 days. Of course, the chance of snowfall is not equal across this 5-month window, but for this example we will treat it that way. Let's say that a particular winter season had 40 snowfall days. Is this cause for concern, or is this within the yearly margin of error from binomial statistics?

For this case, we have parameters of $N = 150$, $p = 53/150 = 0.35$, and $q = 0.65$. We know that the average is 53 (we used Equation (12.5) to calculate p), and we can now find the standard deviation, $\sigma_\nu = \sqrt{(150)(0.35)(0.65)} = 5.84 \approx 6$. Because both 53 and 150 are much larger than one, the Gaussian distribution could be used instead of the binomial distribution. Actually, we cannot use a binomial distribution because $150!$ is too large for most calculators. We will use the Gaussian approximation to the binomial distribution, a good substitution in this case because the numbers are far from zero. If we interpret the question as asking the probability of getting this specific number of snow days, the binomial distribution looks this:

$$G_{53,6}(40) = \frac{1}{6\sqrt{2\pi}} e^{-\frac{(40-53)^2}{2(6)^2}} \qquad (12.8)$$

This yields $G_{53,6}(40) = 0.0063$. This probability is small, but it is only for this specific number of snowfall days. It cannot be compared against the significance thresholds. The test that should be considered is whether a 13-day discrepancy is "large" and therefore of concern. So, a way to address this question is to find the probability of having a discrepancy of 13 or larger. This is done with a z score and a two-sided probability:

$$z = \frac{|40 - 53|}{6} \qquad (12.9)$$

This results in $z = 2.2$. The two-sided probability for this z score is $P = 0.028$, which is below the 5% threshold for declaring it similar. It means that this is an unusually small snowfall year, at least according to this test and threshold. Is it really that unusual? That is a subjective call. Ann Arbor should expect 3 years every century with the number of snowfall days being this far from the average. If this is the first time in many years that it was this far off, then no, there is probably no need for concern. If this is the third time in a decade instead of a century, then yes, this is a sign that the climate is changing rapidly.

> **Quick and Easy for Section 12.6.1**
>
> The uncertainty on combinations of discrete choice options follows the binomial theorem, which also has an uncertainty proportional to the square root of N.

12.6.2 Example: Spacecraft-Charging Model

A common way that the binomial distribution is used in the natural sciences is when asking the question, "is my new model better than my old one?" This is perfectly handled by the binomial distribution and a comparison of the resulting probability against a significance threshold. A space weather example of this is with respect to spacecraft surface charging.

Earth's magnetic field is not only a decent shield against energetic particles from the Sun and deep space but also a fantastic particle accelerator. The magnetosphere, the bubble of space around Earth dominated by the Earth's internally generated magnetic field, regularly gets stretched by forces from the solar wind and interplanetary magnetic field sweeping past the planet. When it snaps back, energy is imparted into the electrically charged particles on those field lines. This is also the place where humans like to fly satellites. At geosynchronous orbit, 6.62 Earth radii from the center of the planet, there is a relatively crowded lineup of communications and military satellites. A little bit further inward, there are several constellations of global navigation satellite systems, including the American GPS satellites and European Galileo fleet. All of these satellites are dowsed in high-energy charged particles. When there is a sudden release of magnetic tension in geospace, a wave of freshly accelerated particles floods near-Earth space, and spacecraft can build up a static electricity charge on their surface.

To mitigate this issue, models are developed of how the electric charge will accumulate on a satellite surface. Other models are developed of the high-energy electron population in near-Earth space. These two numerical tools can, together, help satellite designers and operators prevent anomalies due to excessive surface charging.

Let's say that you are at a company that operates satellites and a research group within the firm develops a new algorithm for predicting spacecraft surface charging. Your company already has a model in operational usage for this purpose, but the research group claims that the new one is better. There is a cost to making the new code operational, so the management does not want to implement this change unless the new algorithm represents a significant improvement over the old one. You are asked to assess this significance.

In the AUL structure of evaluating a new model for operational decisions (introduced back in Chapter 8), this would be at level 6 (the end of Phase 2). The model works and has been verified and validated in a research and development context. Now it is going head-to-head with the previous model for possible advancement to Phase 3 and implementation for everyday use. One assessment is to conduct a data-model comparison against space environment observations. This might not be realistic or even necessary in this context, though. The fundamental question is whether it is better at predicting charging events than the existing model.

This issue can be addressed with an application of the binomial distribution. Let's assume that you have access to charging event episodes on your company's satellites. For simplicity, let's say that there are 10 satellites. Both codes are run for the same period of time, and you count how many of the surface-charging events were correctly predicted by each model. You then make a conclusion about which of the two models was better for each of the 10 satellites in the fleet.

The hypothesis to test is this: how many times does the new model have to "win" this contest in order for it to be declared statistically significantly better than the old model?

If the new model wins the head-to-head comparison for 5 of the 10 satellites, then the new model is equal to the old model and the null hypothesis—that the models are statistically the same—is correct. How many wins will it take to reach significance? Is it 6? Or 8? Or even all 10? The binomial distribution can help.

What is the probability of the new model being better for all 10 of the satellites? If the null hypothesis is that the models are equal, then we can set $p = 0.5$, and therefore $q = 0.5$. For our case with 10 satellites, we have $N = 10$, and we can use the binomial distribution to obtain a probability of the $\nu = 10$ outcome:

$$B_{10,0.5}(10) = \binom{10}{10}(0.5)^{10} \quad (12.10)$$

Equation (12.10) expands to be:

$$B_{10,0.5}(10) = \left(\frac{10!}{10! \cdot 0!}\right)\left(\frac{1}{2^{10}}\right) \quad (12.11)$$

This yields $B_{10,0.5}(10) \cong 0.001$, well below both the 5% and 1% thresholds for significance, and so if the new model is better for all 10 satellites, then you can recommend implementation for operational usage.

For 9 or 10 satellites, the calculation gets a bit longer, because we have to add the binomial distribution probabilities for $\nu = 9$ and $\nu = 10$. This is a key concept regarding the use of the binomial distribution for probabilities at a given ν value or greater—the probabilities for each ν value above the threshold must be added. The resulting probability is found like this:

$$P_{\nu \geq 9} = B_{10,0.5}(9) + B_{10,0.5}(10) \quad (12.12)$$

Expanding out these terms yields:

$$P_{\nu \geq 9} = \binom{10}{9}\left(\frac{1}{2^{10}}\right) + \left(\frac{1}{2^{10}}\right)$$
$$= \left(\frac{10!}{9! \cdot 1!} + 1\right)\left(\frac{1}{2^{10}}\right) \quad (12.13)$$

A bit more math changes it to:

$$P_{\nu \geq 9} = \binom{10}{9}\left(\frac{1}{2^{10}}\right) + \left(\frac{1}{2^{10}}\right)$$
$$= \left(\frac{10!}{9! \cdot 1!} + 1\right)\left(\frac{1}{2^{10}}\right) \quad (12.14)$$

The resulting value is $P_{\nu \geq 9} \cong 0.011$. This also is below the threshold for statistical significance (0.05), but it is now above the threshold to be declared very significantly different (0.01) from the null hypothesis. You could still confidently recommend to management to spend the money to bring the new model into operational usage.

Let's add in the next value, $\nu = 8$, to the probability calculation:

$$P_{\nu \geq 8} = \sum_{\nu=8}^{10} B_{10,0.5}(\nu) \quad (12.15)$$

Inserting the binomial distribution equation and conducting the summation give us this:

$$P_{\nu \geq 8} = \frac{56}{2^{10}} \quad (12.16)$$

This probability, $P_{\nu \geq 8} \cong 0.055$, is now above the 5% significance level. Tradition would have you declare that, with the new model being better for 8 of the 10 satellites, it is not statistically different from being equal to the old model. Whether or not management decides to switch in this case is not a simple no answer, though. The cost of switching still needs to be included, as does the cost of a satellite anomaly due to spacecraft surface charging. If surface-charging effects are a small concern for your satellite fleet, then the decision would probably be no. If surface-charging effects had caused serious problems for even one of the company's satellites, though, the cost of implementation might be worth it. The point is that the traditional hypothesis testing threshold of 5% or 1% is not the full story and decision-specific context is important.

What if the timelines for the 10 satellites were split into "early satellite life" and "later satellite life" epochs? This is a reasonable action because, as satellites age, systems can become susceptible to lower levels of surface charging. The algorithm for predicting surface charging should take this system degradation into account, and one way to ensure that it is emphasized in the assessment is to make these time intervals their own trials in the set. In this case, these two intervals could be treated as different contributions to the test suite, doubling the number of trials. With $N = 20$, is the same "win proportion" of $\nu = 16$ on the cusp of significance? This would require a summation of all probabilities from ν values of 16 and greater. While this is still a tractable calculation with the binomial distribution, it is easier with the Gaussian distribution. The binomial distribution properties mentioned in Section 12.6.1 can be used to obtain the input parameters to the Gaussian, with Equation (12.5) providing the formula for the mean, which works out to be 10. The binomial distribution also gives us a spread, using Equation (12.7) for the standard deviation, we get $\sigma = 2.2$. Moreover, we could now use the z score to quickly find the probability for obtaining a win rate equal to or higher than some value. For this case, we have:

$$z = \frac{|16-10|}{2.2} \quad (12.17)$$

This z score, $z = 2.73$, has a one-sided probability of $P = 0.003$. This is below the 1% level, so the new model can be declared to be highly significantly different from the old model.

As a final extension to this example, let's change the situation so that the number of satellites in the test suite is now 100 rather than 10. This is now too many for use directly with the binomial distribution, but the binomial distribution can be used to get an average and standard deviation value for use with the Gaussian distribution. Specifically, we have $\bar{\nu} = 50$ and $\sigma_\nu = 5$. To keep the win proportion the same, let's say that the new model is better for 80 of the 100 satellites in the set. The z score calculation looks like this:

$$z = \frac{|80-50|}{5} \quad (12.18)$$

This gives $z = 6.0$, which is a very large z score and has a one-sided probability of $P < 10^{-8}$, so the improvement is highly significant.

As with Poisson counting statistics, the absolute uncertainty increases with increasing number of trials, but the relative uncertainty decreases. This can work to your advantage if trials are relatively inexpensive to conduct.

> **Quick and Easy for Section 12.6.2**
>
> Choice combination statistics is particularly useful for making decisions about adopting new methodologies.

12.7 Additional Advanced Topics

It has provided the basic set that allows for comprehensive evaluations, but there are plenty of other methods in existence that were not covered. There are so many additional topics of data analysis, including many just for comparing two number sets, that a section like this will certainly be woefully inadequate. That said, here are a few good words to know because they are common enough that someone conducting natural science research will inevitably come across them. These descriptions have been kept intentionally short; they are not meant to fully describe the concept but rather introduce it, so that you at least know a few key terms. Many example studies that use these advanced topics are given in "Further Reading" (Section 12.9) at the end of the chapter.

12.7.1 Periodicity Analysis

Many phenomena in nature have cycles. It is useful to identify the periods of these cycles and the relative power in certain periods compared with others, or against a background noise floor of the data set. The primary method for determining cycles in a number set is the **Fourier transform**. This is a process of converting the number set from a spatial or temporal domain into a frequency domain. The transformed number set is no longer a function of space or time but rather frequency or period, with the wave power amplitude as the new quantity, rather than the data set values themselves. An inverse Fourier transform is needed to convert this frequency-domain information back into the original units as a function of space or time coordinates.

A more complicated but arguably more robust method is **wavelet analysis**. While Fourier analysis uses a single interval for all wavelengths, wavelet analysis uses only a subset of the full number set for each wavelength, determining the power of oscillations at that particular wavelength, separately from all other wavelengths. The process sweeps through all wavelengths of interest, each with its own subset interval. The issue this addresses is that a Fourier transform requires enough values to resolve the longest wavelength, and therefore many cycles of the smallest wavelengths are included in the number set undergoing the transformation. If the short-wavelength signal is only present for part of the interval, then the power of this wavelength will be the average for the full interval and not capture the intensity of the wave during its short-lived "burst." Wavelet analysis uses a different, and more appropriate, interval for each wavelength under consideration, therefore resolving these small-scale bursts of short-wavelength periodicities in the number set.

Another way to conduct periodicity analysis is through the correlation coefficient. When a number set is plotted against itself, it is called an **autocorrelation**. At first glance, this seems like a meaningless task; the values will, by definition, exactly match and will yield $R = 1$. The power of the autocorrelation is when one of the (identical) number sets are shifted relative to the other, a technique called **time-lagged autocorrelation**. If the values are similar to their immediate neighbors (in space or time), then the correlation coefficient will still be large. As the shift increases, this similarity will fade and the R score will decrease. If there is a natural periodicity within the number set, then at the half-period shift, the correlation should maximize in the negative direction, and at the full-period shift, it should have a relative maximum in the positive direction. These peaks reveal the periodicity and their amplitude reveals the intensity of this particular signal within the number set. Examining the amplitudes of these peaks several period shifts away shows the persistence of this signal across several wavelengths.

12.7.2 Time-Lagged Analysis

A related topic to periodicity analysis is conducting the two-number-set analysis with a shift between the number sets. This shift is usually a function of time but could be a

---

Fourier transform: The process of converting values from a space or time domain into wave power within a frequency domain.

Wavelet analysis: The process of converting values from a space or time domain into wave power within a frequency domain, using a wavelength-specific subset of the values.

Autocorrelation: The process of correlating a data set with itself.

Time-lagged autocorrelation: The process of correlating a data set with itself, using a range of shifts, to detect periodicities in the number set.

spatial shift. This **time-lagged analysis** is a powerful tool for assessing leading and trailing factors of a particular phenomenon. It was the technique used in Chapter 1 for the pair of papers examining solar wind driving of geospace storm events; one study showed a strong connection to solar wind density, while a second, larger-database study did not find this correlation.

Creating a time-shifted array is straightforward. For a non-shifted paired set, we have (x_i, y_i). To shift, you simply pair the y_i value with a different index of the x number set, offset by j, so that you have (x_{i+j}, y_i). The shift can be in either direction; the shift index j can be positive or negative. Usually, a negative shift is desired, in order to determine causality of the x parameter leading the response in the y number set. This process can be done with the number sets flipped, though, so it really does not matter. Just be sure to pay attention to which variable's indices are being shifted, so that you understand which parameter leads and which lags. There is a small issue at the end of the arrays, where the shifted $i + j$ index value might be before the first element of after the last element. These y_i values have to be excluded from the analysis. If $j < 0$, for example, then the omitted y_i values are at the beginning of the y array, with a corresponding number of omitted x values at the end of the array. It is often useful to create an array of j values, resulting in many new paired number sets. If you want to maximize the usage of the values, then each of these will have different lengths of $N - |j|$. If $N >> j$, then this does not influence the comparison between the results.

All of this can be done with a data-model pair. Instead of (x_i, y_i), we have (M_i, O_i). Everything else is the same.

Once you have these new paired number sets, then any of the tools discussed in this book can be applied. Arguably the most common calculation is to determine R for each of these sets and plot R versus j. This is called a **lagged correlation** and reveals the shift at which the peak correlation between the two original number sets occurs. Any analysis technique from the earlier text can be used, though.

12.7.3 Additional Tests

There are many existing tests that have not been mentioned yet. Some of these could be used in place of the various t tests for hypothesis assessment, such as the likelihood ratio test or the Wald test. Others assess the distribution function in a similar manner as the χ^2 and K–S tests, such as the Cramér–von Mises test which has a quadratic assessment of the cumulative probability distributions, or the Jarque–Bera test that creates a t score based on skewness and kurtosis. There are also tests on the rank-order distribution, such as Kendall's tau assessment, related to R_S but with a slightly different formulation. Another fit performance extremes category metric is the exclusion parameter, which quantifies a percentage of the model values that have a relative difference of more than some cutoff threshold away from the corresponding observed value. Many other tests exist beyond the set provided in this book.

12.7.4 Multidimensional Data Analysis

If you have two spatial variables, or one space variable but the data are also a function of time, then you have a multidimensional data set. Many of the techniques discussed in this

---

Time-lagged analysis: Comparing two number sets from different times – the driving parameter ahead of the response – to determine time lags in causal relationships.

Lagged correlation: Calculating correlation coefficients between two number sets with one systematically shifted relative to the other, revealing the peak correlation as a function of shift.

book are applicable to multidimensional sets with little or no modification. One option is to flatten the array, either by concatenating all rows into a single, very long column, or collapsing one dimension by summing or averaging the values within each row. In this case, you now have a 1D array and the processing techniques from the early chapters apply.

Curve fitting is possible in multiple dimensions. In this case, you are actually creating a best-fit surface instead of a line. To expand linear regression into a two-dimensional planar fit, instead of $y = A + Bx$, the initial formula now becomes $z = A + Bx + Cy$. The procedure of determining the normal equations is exactly the same as that discussed in Chapter 7 for a single linear regression. Similar analogous formulas can be written for higher-order polynomials or other functions and the normal equations for many such functional forms have been worked out.

When a parameter is a function of multiple variables, then there is a maximum somewhere in that 2D array of values (or higher dimensionality). This allows for a new style of comparisons to be made. Using all but one of the variables, the location of the peak in the number set at each time can be determined, creating an array of peak location versus time. Note that the one variable remaining is often time—and has been called time here—but could be a spatial dimension. This is now a number set that can be compared against a similar value from a different multidimensional array. The value of the maximum could also be similarly extracted to become a new array, or the width of the maximum out to some absolute or relative value threshold (say, 10% or 50% of the peak value, whatever makes sense for your investigation). All of these new number sets can be extracted from a driver and response number set pair and the analysis tools from this book can be used to assess the relationship.

There exists an entirely separate set of analysis tools for multidimensional number sets. There is one in particular that should be mentioned here—**principal component analysis (PCA)**—which is a process of distilling average patterns in the multidimensional number set. This technique collapses the full number set in one of the dimensions, usually time, creating a median or average pattern in the other remaining dimensions. The baseline pattern is then subtracted from all of the individual patterns and the averaging process is conducted a second time. This reveals another "secondary" pattern, which can also be subtracted from all of the individual patterns. These patterns are referred to as **empirical orthogonal functions (EOFs)**, and, given enough of them, they can be used to recreate any of the individual patterns from the multidimensional number set. The EOF patterns usually correspond to typical physical states of the system being observed. When one of the EOF patterns dominates at a particular time, then this reveals the processes governing the system at that time.

There is a very real issue with multidimensional data sets called the **curse of dimensionality**. This is the multiplicative effect that makes large-scale data sets difficult to process and interpret. For example, consider 100 data points along a straight line. If instead we

---

Principal component analysis: A method of determining average patterns in a multidimensional data set, known as empirical orthogonal functions, which can then be used to reconstruct specific patterns as a weighted sum of the EOFs.

Empirical orthogonal functions: Average patterns derived from a collection of multidimensional data which can be used to reconstruct any specific pattern.

Curse of dimensionality: The problem that adding dimensions to parameter space of a data set (or model) quickly expands the number of values by orders or magnitude.

wanted a grid of values, we now have 10 000 data points—100 × 100—in order to maintain the same resolution. In three dimensions, this becomes one million data points. The level of information quick rose by four orders of magnitude just by adding two spatial dimensions. This can be quantified as a formula: for $1/N$ resolution in each of M dimensions, N^M points are required for full coverage. The data set can quickly become intractable, unless resolution is sacrificed.

12.7.5 Multidimensional Data-Model Comparisons

When the data and the model output are both multidimensional, there is some ambiguity about how to apply and calculate the metrics discussed in this book. One approach is to simply concatenate the variables, treating the values in each multidimensional number set as a very long single-dimensional array. In this case, all of the metrics from this book can be calculated as listed. This does not take into account the multidimensional nature of the information, though, and additional techniques could be applied.

Other assessments can focus on aspects of the values within the array. For example, the location of the peak is a quantity that can be compared between the data and the model. Similarly, an area could be determined around each relative maxima, either above some absolute threshold or a relative threshold that is a percentage of the peak value, and these areas could then be analyzed and compared. Note that "area" does not have to be a region in two-dimensional space, but could be a volume or other region within multidimensional space. In fact, one of the dimensions could be time, creating a space–time integral of values above the specified threshold.

The lagged analysis discussed in Section 12.7.2 is sometimes useful for multidimensional data-model comparisons. It could be the case that the model is pretty good, but the output is shifted from the observed values, either in space or in time. A lagged correlation analysis can be used to identify these shifts, or confirm that there is no shift, between the model output and the observed values.

In addition, there are special techniques for identifying clusters in a multidimensional data set. **Cluster analysis** can be done with two one-dimensional number sets or with any higher-level dimensionality. This typically involves the calculation of a distance parameter between all points in the multidimensional scatterplot and then using certain techniques to identify groupings of points. There are two main methodologies, hierarchical clustering and nonhierarchical clustering. Hierarchical clustering methods form small groups of points and then merge these together to form a nested list of identified clusters. The most popular nonhierarchical clustering method is K-means processing, which allows for the reassignment of small groups later in the process.

Another multidimensional comparison method is **canonical correlation analysis**. This is the process of designing new collections of variables that best represent the connection between the two number sets. It involves multivariable correlation coefficients to build a new formulaic relationship between a set of possible driving parameters and another set of response parameters. It returns new system-state input and response variables that combine the drivers into single number set as well as combine the response variables into a single output set. Canonical correlation analysis is useful if the main objective of the assessment is to

---

Cluster analysis: The process of identifying clusters of points within a multidimensional number set.

Canonical correlation analysis: The process of distilling a collection of input and output number sets into a single, combined input and output set that represents the state of the full system.

determine if this collection of input values are the most important ones that govern the response in the collection of output values. This tool helps you determine the system-level connection between the full set of inputs to the full set of output quantities.

12.7.6 Uncertainty Quantification

A method mentioned earlier in Section 12.3 is running the model with a randomized set of input conditions that represents the uncertainty of the boundary conditions. This variability of inputs is not only limited to external boundary conditions but could also be applied to internal model factors for which there is uncertainty in the specific setting. Take, for instance, collision cross sections in an atmospheric general circulation model, or scattering coefficients in a radiative transfer code. This process of introducing random variations to model settings, either internal or external, is generally referred to as **uncertainty quantification**.

Uncertainty quantification is more than simply code verification and validation. **Verification** refers to a check that the code is solving the equations correctly, often done by applying simple input functions for which there are analytical solutions. Verification still involves a data-model comparison, but the "data" in a verification check are often this analytical solution or output from a previously developed and well-trusted code. That is, verification tests your numerical approach. All of the data set analysis tools and data-model comparison metrics can be (and should be) used in verification, and the tables in Section 12.1 can be used as a guideline for how to construct a robust verification methodology.

Validation is the process of assessing whether the code is solving the correct set of equations for the observation set to which it is being compared. That is, it requires real data from the environment that the model is seeking to describe. Data-model differences reveal inadequate physical processes—or perhaps inadequate numerical implementation—in the model. Declaring "good" validation is often a use-specific judgment call. Adding or subtracting physical processes in the model during a validation exercise allows for the assessment of the importance of these processes at governing the quantities being measured. Determining which processes to include or exclude for a particular assessment is numerical experiment design and should be tailored to the specific hypothesis being tested. Validation is often a part of the scientific process, systematically changing key factors within the model to determine which one dominates the signal seen in the observations. Validation can also be completely separate from the scientific process, especially when used as part of the AUL framework to assess a model's ability to perform at a quality level that justifies its use for operational decision-making.

Uncertainty quantification does not involve a data-model comparison but rather is the creation of a spread of output values from the model. This number set of output values can be analyzed (as done in Chapter 4) to yield uncertainties on the model output, which can then be incorporated into later data-model comparisons. The only data involved could be the initial or boundary conditions for the simulations.

Ideally, uncertainty quantification for a model should be done between the verification and validation steps. This allows for these

---

Uncertainty quantification: The process of determining uncertainty values on model output.

Verification: The process of determining if the model is correctly and accurately solving the selected set of equations.

Validation: The process of determining if the model is solving the correct set of equations for the data set to which it is being compared.

model output uncertainties to be taken into account when calculating metrics and interpreting their scores. Sometimes it is not recognized as a required element until after an initial data-model comparison. Models often solve for many variables across large spatial domains spanning long time intervals, and it can be daunting to consider an uncertainty quantification process that attempts to determine standard deviations everywhere in the domain. Selecting specific data for comparison helps identify what aspects of the model require uncertainty values, and this can focus the uncertainty quantification process on what is really needed. In any case, all of the tools from Chapters 4 and 5 on analyzing a single data set can be used to determine the proper spread to use as the uncertainty on a given model output variable at a particular location or time.

12.7.7 Design of Experiments

It was just mentioned that numerical models could be run several times, adding and subtracting physical processes in the code configuration, to assess the importance of these terms within a relevant measured data set. Conversely, additional data could be taken in different situations, places, times, and driving conditions. If resources are unlimited, then obtaining observations or conducting simulations for every possible combination of input and output connection can be done. Sometimes, though, this additional experimentation—either observational or numerical—is expensive and there is a limit to the volume of numbers that can be generated for a particular analysis. In this case, care should be taken to set up the tests in a way that efficiently uses the available resources.

This is a field of optimization theory called **design of experiments**. Methods have been created to combine the variation of several parameters and then extract out the signal from particular elements within that parameter set. When you are at the stage of creating the experiment to test a newly developed hypothesis about a strange feature seen in a data set, applying these sampling optimization techniques can greatly save time and effort. This is particularly important for multidimensional systems. When there are many parameters that drive or govern a particular phenomenon, an unsophisticated approach to exploring this parameter space—varying each one separately and in full—could be rather time-consuming. It is useful to take advantage of existing methods that optimize the selection of settings to maximize the information returned in the most efficient manner.

12.7.8 Geographical Information System (GIS) Analysis

If you are a researcher in a natural science discipline, then there is a high chance that you will someday come in contact with **geographical information systems** (GIS). This is a method of transforming data from one projection to another, specifically plotting geophysical data on the Earth using topographical or political boundaries. It is a way to parse the data into discrete bins of human interest. These boundaries in GIS projections are often far more complex than a simple (x,y) Cartesian grid, with complicated polygons or intricate curve sets defining each region on the map. GIS software does this projection and mapping for you. Once in this format, different number sets can be compared.

Design of experiments: The process of determining the minimum set and configuration of experiments needed to explore a range of input parameters.

Geographical information system: A method of transforming data from one projection to another, especially with respect to geophysical or political boundaries.

12.7.9 Machine Learning

Machine learning refers to the advanced statistical techniques for computers to determine—that is, "learn"—relationships between number sets. For the natural sciences, it is often thought of in terms of optimizing the predictive correlation between an input and an output number set. Concepts such as a neural network, random forest, decision tree, supervised learning, logistic regression, reinforcement learning, deep learning, gradient descent, back propagation, and support vector machines all fit into the general category of machine-learning methods.

A machine-learning algorithm is essentially the iterative curve-fitting approach discussed in Chapter 7. Some will even include any curve-fitting procedure, all the way down to linear regression, within the umbrella of machine learning. Usually, though, machine learning refers to certain specifications on the type of formulaic connection between the input and output, the metric being optimized, and the process of forward-feeding or back feeding information within the optimization routine.

Some machine-learning techniques are "supervised" which have a human-created list of pre-learned examples. This might be event identification, where you have found events manually for a small portion of a large database and want to use a machine-learning technique to identify similar events in the remainder of the database. There is also "unsupervised" machine learning, in which no pre-defined events are given and the technique searches for patterns and clusters within the database. This might find new categorizations that you did not even know existed.

A big category of machine learning are artificial neural networks. A "simple" neural network directly connects the input layer to the output layer by optimizing the linear combinations of the input array elements for each output array element, minimizing a chosen loss function. A "deep learning" neural network is one in which one or more layers of connective nodes are inserted between the input and output sets, allowing for a more complex interweaving of associations (and often a better match to the output).

There is a strong connection between machine-learning techniques and the data-model comparison metrics discussed in Chapters 9 and 10. A metric must be chosen to be optimized during the learning process. Because of its similarity to variance and its recognition as a measure of overall accuracy, a metric that is very often used in these situations is MSE. This is not the only choice that can or even should be made. Other metrics, or even better normalized combinations of metrics, could be used for determining the optimal coefficients.

Once "learned," these coefficients can then be applied to other, similar input to predict an outcome. This makes machine-learning techniques powerful tools for forecasting. In addition, though, they can be used to investigate physical relationships. The coefficients can be analyzed to reveal the dominant connections between input and output layers. Sometimes the identified relationships surprise you, highlighting an input that you thought might have an influence but had not been appreciated in previous statistical compilations or physical modeling.

Machine learning: The collection of methods for which the code automatically iterates to an optimized set of coefficients connecting an input and an output number set.

Quick and Easy for Section 12.7
It is good to know about these additional topics in data analysis and modeling metrics. This book provides a solid background for diving into these other subjects.

12.8 Uncertainty and the Scientist

Uncertainty, a running theme of this book, not only pertains to scientific analysis and how scientists conduct research but also to the scientists themselves. Making a living out of applying the scientific method means that you are going to be wrong with your guess. You will be wrong a lot, and you might often feel stupid and inadequate during this work. Impostor syndrome is very real among scientists because, by the very definition of "doing science," you are doing something hard that no one else has done before. You are pushing the envelope of what humanity knows about a topic, expanding our collective understanding with your new findings. Scientists want to get to step 6 (publish) with a successful test of their hypothesis, but the need to go back to an earlier step in the method is a very common reality facing scientists throughout their careers.

A usual step in becoming a practicing scientist is going through graduate school toward a PhD degree. This is where you are taught to be an expert in a particular field and trained to conduct original research. Many students entering graduate school are the best of their undergraduate science cohort. Yet, depression is common among graduate students. Why? Because this is usually the first time in their lives that these smart people are wrong without a clue about how to make it right. This is sometimes the first time that they cannot get the answer. This is often the first time that they are faced with a challenging problem that, by the nature of "original research," has not been done before.

It can often be the case that a "strange thing" has to be set aside for a while, perhaps years, because you cannot make the right guess at the process that is causing it. In addition, sometimes the published hypothesis of a scientist is shown to be wrong by others, when new observations are made or theoretical concepts are devised. As a professional researcher, you will have to be accepting of the notion of changing your mind. You have to be comfortable with uncertainty.

Figure 12.8   Chicken Little is lost. Sometimes you might feel lost, too, while conducting scientific investigations. Artwork by Asher/Anya Hurst.

As you are going through a difficult time of rejection or burnout, a feeling of inadequacy can grow. You might feel mentally or emotionally lost; you might even feel like a lost chicken, as is the case with Chicken Little being physically lost as seen in Figure 12.8. Please know that you are most definitely not alone in having these thoughts and feelings. Moreover, you do not have to go through this alone. Seek help. At the very least, talk to a colleague or a trusted friend about your frustrations. Perhaps even more appropriately, get in contact with a professional therapist. Much has been learned about how minds work and there are numerous strategies for getting through difficult times.

Conversely, there will be times of great joy and excitement in conducting scientific research. People do not choose a scientific research career for the money; the pay is often good but you will most likely not become spectacularly rich investigating nature for a living. People choose these career paths because they feel a great sense of awe toward the topic they are studying. It is good to hold on to that sense of wonder that brought you into the natural sciences and to actively look for it and appreciate it in your everyday work life. That feeling could arise during every step of the scientific

The scientific method: annotated

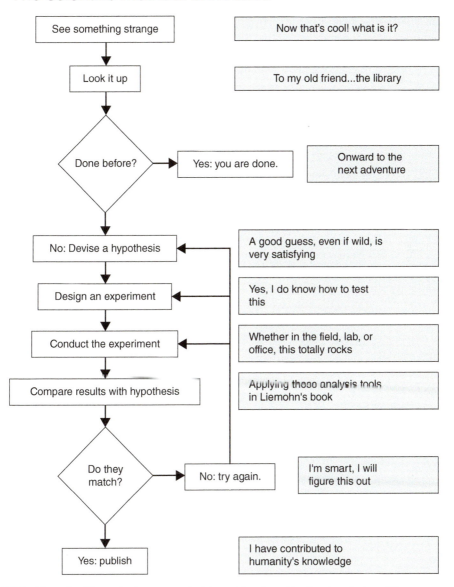

Figure 12.9 The scientific method, annotated with positive thoughts about each step of the process.

process, as illustrated in the annotated version of the scientific method in Figure 12.9. We should not save that sense of accomplishment for only the final publication, but rather celebrate each step along the journey.

Conveying this excitement and enjoyment toward your chosen career, as well as the uncertainty you regularly face within it, will hopefully be recognized by those around you. This will hopefully result in others vicariously experiencing a bit of your awe and wonder at the natural world and will increase transparency about the process of discovering new knowledge. It will hopefully lead to stamping out the specter of anti-intellectualism that occasionally waxes through society. It will hopefully lead to better trust in science and better understanding of the humanity of scientists.

Therefore, at every stage in your scientific career, embrace the uncertainty. Taking the path of scientific research means that self-doubt is going to be part of your work life; knowing that it is going to arise means that you can prepare yourself to get through it. Other researchers will argue with you and show evidence that your ideas are wrong. Remember, though, that you are not your thoughts; you are not your work. You are a smart person taking on a difficult challenge, and you are aware that obstacles will arise along your path. Acknowledge the problem and muster the courage to not let it overcome you. When you or others show that your guess was wrong, remember that it was only a guess. Embracing the uncertainty that is part of scientific research will free you from the bondage that it can otherwise place on your state of mind.

Go out into the world and conquer new frontiers. Science awaits.

Quick and Easy for Section 12.8

The process of doing science is hard and you could experience self-doubt, rejection, or burnout along the way. Please know that you are not alone in having these feelings and that help is available to get you through the difficult times.

12.9 Further Reading

I again refer readers to Murphy's 1991 paper as a helpful treatise on reasons for adopting a robust approach to data-model comparisons:

Murphy, A. H. (1991). Forecast verification: Its complexity and dimensionality. *Monthly Weather Review, 119*(7), 1590–1601. https://doi.org/10.1175/1520-0493(1991)119%3C1590:FVICAD%3E2.0.CO;2

More discussion on developing a robust metrics assessment is given here:

Liemohn, M. W., Shane, A. D., Azari, A. R., Petersen, A. K., Swiger, B. M., & Mukhopadhyay, A. (2021). RMSE is not enough: Guidelines to robust data-model comparisons for magnetospheric physics. *Journal of Atmospheric and Solar-Terrestrial Physics, 218*, 105624. https://doi.org/10.1016/j.jastp.2021.105624

Here is a good paper on log accuracy methods for highly variable data sets:

Morley, S. K., Brito, T. V., & Welling, D. T. (2018). Measures of model performance based on the log accuracy ratio. *Space Weather, 16*, 69–88. https://doi.org/10.1002/2017SW001669

The NOAA GOES data are available at the National Geophysical Data Center: https://satdat.ngdc.noaa.gov/sem/goes/data/full/

There are many great resources for the additional topics listed in Section 12.7. To keep this further reading list tractable, only a few good references (sometimes just one) per subsection are listed.

Here is one of the numerous books that includes a lengthy presentation on periodicity analysis:

Oppenheim, A. V., Schafer, R. W., & Buck, J. R. (1999). *Discrete-Time Signal Processing* (2nd edition). Prentice Hall, Upper Saddle River, N.J. ISBN 0-13-754920-2

A recent study that describes Fourier transforms and power spectral densities in a space physics context is here:

Di Matteo, S., Viall, N. M., & Kepko, L. (2021). Power spectral density background estimate and signal detection via the multitaper method. *Journal of Geophysical Research: Space Physics, 126*, e2020JA028748. https://doi.org/10.1029/2020JA028748

One on wavelet analysis in assessing hydrological models, including a piecewise linear fit in one of the plots, is:

Bittner, D., Engel, M., Wohlmuth, B., Labat, D., & Chiogna, G. (2021). Temporal scale-dependent sensitivity analysis for hydrological model parameters using the discrete wavelet transform and active subspaces. *Water Resources Research*, 57, e2020WR028511. https://doi.org/10.1029/2020WR028511

Time-lagged analysis is incorporated into this study on the rate of ocean mammals (whales and dolphins) stranding themselves on beaches:

Zellar, R., Pulkkinen, A., Moore, K., Rousseaux, C. S., & Reeb, D. (2021). Oceanic and atmospheric correlations to cetacean mass stranding events in Cape Cod, Massachusetts, USA. *Geophysical Research Letters*, 48, e2021GL093697. https://doi.org/10.1029/2021GL093697

A study on the intensity of geomagnetic activity that includes time-lagged analysis:

Hajra, R., Marques de Souza Franco, A., Echer, E., & Bolzan, M. J. A. (2021). Long-term variations of the geomagnetic activity: A comparison between the strong and weak solar activity cycles and implications for the space climate. *Journal of Geophysical Research: Space Physics*, 126, e2020JA028695. https://doi.org/10.1029/2020JA028695

Regarding the many additional tests that could be conducted when comparing two number sets, a good collection is included throughout this book by Alan Chave:

Chave, Alan D. (2017). *Computational Statistics in the Earth Sciences*. Cambridge University Press, Cambridge. https://doi.org/10.1017/9781316156100

A usage of the exclusion parameter for space physics can be found here:

Mukhopadhyay, A., Jia, X., Welling, D. T., & Liemohn, M. W. (2021). Global magnetohydrodynamic simulations: Performance quantification of magnetopause distances and convection potential prediction. *Frontiers in Astronomy and Space Science*, 8, 637197. https://doi.org/10.3389/fspas.2021.637197

For more information on principal component analysis, canonical correlation analysis, and cluster analysis, a thorough reference is part 3 of the book by Daniel Wilks:

Wilks, D. S. (2019). *Statistical Methods in the Atmospheric Sciences* (4th edition). Academic Press, Oxford. https://www.elsevier.com/books/statistical-methods-in-the-atmospheric-sciences/wilks/978-0-12-815823-4

A recent study that uses this technique is this one on how croplands influence atmospheric carbon fluxes:

Sun, W., Fang, Y., Luo, X., Shiga, Y. P., Zhang, Y., Andrews, A. E., et al. (2021). Midwest US croplands determine model divergence in North American carbon fluxes. *AGU Advances*, 2, e2020AV000310. https://doi.org/10.1029/2020AV000310

Here is another that used PCA to detect subsurface oceans on outer planet moons:

Cochrane, C. J., Persinger, R. R., Vance, S. D., Midkiff, E. L., Castillo-Rogez, J., Luspay-Kuti, A., et al. (2022). Single- and multi-pass magnetometric subsurface ocean detection and characterization in icy worlds using Principal Component Analysis (PCA): Application to Triton. *Earth and Space Science*, 9, e2021EA002034. https://doi.org/10.1029/2021EA002034

A use of canonical correlation analysis for precipitation data interpretation is this one:

Wang, G., Zhuang, Y., Fu, R., Zhao, S., & Wang, H. (2021). Improving seasonal prediction of California winter precipitation using canonical correlation analysis. *Journal of Geophysical Research: Atmospheres*, 126, e2021JD034848. https://doi.org/10.1029/2021JD034848

CCA has also been used in space weather applications, like here:

Borovsky, J. E. (2014), Canonical correlation analysis of the combined solar wind and geomagnetic index data sets. *Journal of Geophysical Research: Space Physics*, 119, 5364–5381, doi:10.1002/2013JA019607.

A good review of uncertainty quantification, with application toward the Earth sciences, is given here:

Scheidt, C., Li, L., & Caers, J. (2018). Data science for uncertainty quantification. In C. Scheidt, L. Li, & J. Caers (Eds.), *Quantifying Uncertainty in Subsurface Systems*. https://doi.org/10.1002/9781119325888.ch3

A good paper with model uncertainty quantification that is focused on weather and climate is here:

Slingo, J., & Palmer, T. (2011). Uncertainty in weather and climate prediction. *Philosophical Transactions of the Royal Society A, 369*, 4751–4767. http://doi.org/10.1098/rsta.2011.0161

A seismic usage of uncertainty quantification with model predictions is here:

Zou, C., Zhao, L., Xu, M., Chen, Y., & Geng, J. (2021). Porosity prediction with uncertainty quantification from multiple seismic attributes using Random Forest. *Journal of Geophysical Research: Solid Earth, 126*, e2021JB021826. https://doi.org/10.1029/2021JB021826

An excellent reference on the methods of optimally designing experiments is given in Part 2 of this book by Mason and coauthors:

Mason, R. L., Gunst, R. F., & Hess, J. L (1989). *Statistical Design and Analysis of Experiments, with Applications to Engineering and Science*. John Wiley & Sons, New York.

A geoscience example that uses Latin Hypercube Sampling is this one on groundwater allocation planning:

Siade, A. J., Cui, T., Karelse, R. N., & Hampton, C. (2020). Reduced-dimensional Gaussian process machine learning for groundwater allocation planning using swarm theory. *Water Resources Research, 56*, e2019WR026061. https://doi.org/10.1029/2019WR026061

This group used LHS for a method of tuning machine-learning parameters during a real-time run of the code:

Balogh, B., Saint-Martin, D., & Ribes, A. (2022). How to calibrate a dynamical system with neural network based physics? *Geophysical Research Letters, 49*, e2022GL097872. https://doi.org/10.1029/2022GL097872

For additional details about geographical information systems, here is a classic book on the topic:

Burroughs, P. A., McDonnell, R. A., & Lloyd, C. D. (1998). *Principles of Geographical Information Systems* (2nd edition). Oxford University Press. ISBN 978-0-19-874274-5.

There are many papers that include GIS graphics within their analysis. Here is one on tree cover changes:

Wang, J. A., Randerson, J. T., Goulden, M. L., Knight, C. A., & Battles, J. J. (2022). Losses of tree cover in California driven by increasing fire disturbance and climate stress. *AGU Advances, 3*, e2021AV000654. https://doi.org/10.1029/2021AV000654

Here is another using GIS analysis to examine seismic activity and earthquake early warning systems:

Brooks, B. A., Protti, M., Ericksen, T., Bunn, J., Vega, F., Cochran, E. S., et al. (2021). Robust earthquake early warning at a fraction of the cost: ASTUTI Costa Rica. *AGU Advances, 2*, e2021AV000407. https://doi.org/10.1029/2021AV000407

For machine learning, this is a nice (and short!) write-up in *Eos* of applying machine learning to geophysical problems, in particular solid Earth physics:

Curtis, A., O'Malley, D., Beroza, G. C., Johnson, P. A., & Li, E. (2020), Tackling 21st century geoscience problems with machine learning, *Eos, 101*. https://doi.org/10.1029/2020EO150184

And here is a follow-up article, Published on 07 October 2020. also in *Eos*, on the same topic:

Bortnik, J., & Camporeale, E. (2021). Ten ways to apply machine learning in Earth and space sciences, *Eos, 102*. https://doi.org/10.1029/2021EO160257. Published on 29 June 2021.

In addition, here are two books that I have found useful for coming up to speed on the topic, with application to Python code implementation methods:

Müller, A. C., & Guido, S. (2016). *Introduction to Machine Learning with Python*. O'Reilly Media, Inc. ISBN 9781449369415. https://www.oreilly.com/library/view/introduction-to-machine/9781449369880/

Igual, L., & Segui, S. (2017). *An Introduction to Data Science: A Python Approach to Concepts, Techniques and Applications.* Springer. https://doi.org/10.1007/978-3-319-50017-1

Here is the special collection of research articles on machine learning in solid Earth physics related to the *Eos* article above:
https://agupubs.onlinelibrary.wiley.com/doi/toc/10.1002/(ISSN)2169-9356.MACHLRN1

And here is another special collection of research articles on machine learning in space physics:
https://www.frontiersin.org/research-topics/10384/machine-learning-in-heliophysics#articles

And a third special collection from the field of hydrology:
https://agupubs.onlinelibrary.wiley.com/doi/toc/10.1002/(ISSN)1944-7973.MACHINELEARN

There are numerous *studies* that incorporate machine-learning techniques in geoscience research, but here are three from different disciplines. Specifically, climate change:

Pastick, N. J., Wylie, B. K., Rigge, M. B., Dahal, D., Boyte, S. P., Jones, M. O., et al. (2021). Rapid monitoring of the abundance and spread of exotic annual grasses in the western United States using remote sensing and machine learning. *AGU Advances*, 2, e2020AV000298. https://doi.org/10.1029/2020AV000298

Solar wind categorization:

Li, H., Wang, C., Cui, T., & Xu, F. (2020). Machine learning approach for solar wind categorization. *Earth and Space Science*, 7, e2019EA000997. https://doi.org/10.1029/2019EA000997

And detecting deep ocean subduction zones:

He, B., Wei, M., Watts, D. R., & Shen, Y. (2020). Detecting slow slip events from seafloor pressure data using machine learning. *Geophysical Research Letters*, 47, e2020GL087579. https://doi.org/10.1029/2020GL087579

This is an excellent—and short—essay on the feeling of inadequacy in STEM research, how hard it is to do research, and the necessity of embracing "productive stupidity" in research life:

Schwartz, M. A. (2008). The importance of stupidity in scientific research. *Journal of Cell Science*, 121, 1771. https://doi.org/10.1242/jcs.033340

Here is a much longer article with substantial references to additional scholarly work on rejection, impostor syndrome, and burnout:

Jaremka, L. M., et al. (2020). Common academic experiences no one talks about: Repeated rejection, impostor syndrome, and burnout. *Perspectives on Psychological Science*, 15, 519–543. https://doi.org/10.1177/1745691619898848

Regarding the feelings of inadequacy on the part of the mentor, here is a nice article on the topic with some practical advice:

Stalcup, A.M. (2016). Some ruminations on graduate students. *Analytical and Bioanalytical Chemistry*, 408, 6239–6243. https://doi.org/10.1007/s00216-016-9755-x

Here is a recent research article investigating the concept of seeing wonder in nature and letting that drive a passion for natural science discoveries:

Gottlieb, S., Keltner, D., & Lombrozo, T. (2018). Awe as a scientific emotion. *Cognitive Science*, 42, 2081–2094. https://doi.org/10.1111/cogs.12648

And here is another on scientists feeling awe toward their work:

Cuzzolino, M. P. (2021). *"The Awe is In the Process"*: The nature and impact of professional scientists' experiences of awe. *Science Education*, 105, 681–706. https://doi.org/10.1002/sce.21625

Philip Abelson's editorial in *Science* magazine on anti-intellectualism is here:

Abelson, P. H. (1976). Enough of pessimism. *Science*, 191, 29. https://doi.org/10.1126/science.191.4222.29

The famous book by Hofstadter on anti-intellectualism in America (yes, from 60 years ago, this is not a new development):

Hofstadter, R. (1963). *Anti-Intellectualism in American Life. Alfred A. Knopf, New York.*

Finally, here is an article with additional thoughts by me on this topic:

Liemohn, M. W. (2023). Conducting space physics research: Wind the frog, work hard, and be nice. *Perspectives of Earth and Space Scientists*, 4, e2022CN000197. https://doi.org/10.1029/2022CN000197

12.10 Exercises in Geoscience

1. In Section 12.6.2, the case of 20 trials was briefly presented and investigated for the case of 16 successes for the new model. For 20 trials, determine the ν cutoff number of successes to meet probabilities of 20%, 10%, 5%, and 1%. Remember to add probabilities for all ν values at and above the one of interest.

2. Using the ASOS weather data and HRRR model output from the Chapter 10 homework:
   A Choose several of the metrics from those exercises and conduct a 50-element bootstrap analysis. A good set could be F_1, FB, and ORSS.
   B Calculate the metric scores for each of the bootstrap resamplings.
   C Make a histogram of each of the metric score arrays.
   D Calculate the mean and standard deviation from these metric score arrays.
   E Write a paragraph discussing the validity of using the calculated standard deviations from the bootstrap analysis as uncertainties on the original metric scores.

3. Using the Vostok paleoclimate air temperature and CO_2 concentration data sets from the Chapter 8 homework:
   A Conduct a time-lagged analysis by shifting which temperature value is paired with which CO_2 concentration value, changing the index of the temperature value by up to 30 (± 3000 years) in each direction.
   B With these shifted paired arrays, calculate R as a function of the time shift.
   C Create a line plot of this metric score versus time shift.
   D Write a paragraph commenting on the features of this line plot of time-lagged R.

Index

Tables are indicated by bold page references. Figures are indicated by italicized page references.

a

absolute uncertainty   4–5, 6
   stepwise uncertainty
      propagation   85
accuracy
   association   *376*, 376–377
   bias and precision
      374–376, *375*
   event detection
      metrics   310–311,
      311, **373**
   fit performance metrics
      250, 250–262, *251*, *253*
      centroid   263
      degrees of freedom
         253–255
      Gaussian
         distribution   **368**
      MAD   251–252, 258
      MAE   252–253, *253*, 255,
         257, 263
      MSE   251–252, *253*,
         254–255, 278
      non-Gaussian
         distributions   **372**
      normalization factor
         256–257, **257**
      RMSE   252, *253*, 255,
         257, **257**
   model metrics   230

addition
   Gaussian distribution,
      weighted average
      146–148
   two number sets
      linear fit   184
      polynomial fitting   196
      uncertainty propagation
         77–78
      formula   92
   uncertainty quantification
      397
alternative ROC curve (Alt-ROC),
   event detection metrics
   event identification
      threshold sliding
      346–347, *347*, 359–362,
      361, **373**
analysis of variance (ANOVA)
   188–191, **190**
annotations, plots   30, 32
ANOVA. *See* analysis of
   variance
application usability level
   (AUL)   386
   choice combination
      statistics   390
   model metrics   236–237, *237*
   uncertainty quantification
      397

area under the curve (AUC)
   345–346, 359–362, **361**
arithmetic mean   96–97
   fit performance metrics,
      reliability   285
   histogram   99–100
   standard deviation   102, 111
   two number sets
      covariance   158
      ozone and temperature
      170
artificial neural
   networks   399
association
   accuracy   *376*, 376–377
   event detection metrics
      314–317, **373**
   extremes   *377*, 377–378
   fit performance metrics
      268–272, *269*
   Gaussian
      distribution   **368**
   non-Gaussian
      distributions   **372**
   Pearson linear correlation
      coefficient   269–270
   Spearman rank-order
      correlation coefficient
      270–272, *271*
   model metrics   231

Data Analysis for the Geosciences: Essentials of Uncertainty, Comparison, and Visualization, Advanced Textbook 5, First Edition.
Michael W. Liemohn.
© 2024 American Geophysical Union. Published 2024 by John Wiley & Sons, Inc.
Companion website: www.wiley.com/go/liemohn/uncertaintyingeosciences

Index

asymmetric uncertainty 142–144
atmospheric chemistry 169
AUC. *See* area under the curve
AUC skill score 346
AUL. *See* application usability level
autocorrelation 393

b

bar chart. *See* histogram
basic rule of significant digits 4
bias 264
  accuracy and precision 374–376, *375*
  event detection metrics 311–313, *312*, **373**
  fit performance metrics 262–265
    Gaussian distribution **368**
    non-Gaussian distributions **372**
  model metrics 230
Binder, H. 56–57, *57*, *58*
binomial distribution, choice combination statistics 388–392
bootstrap method
  Dst index 383
  two number sets
    ozone and temperature *175*, 175–177, *176*
    uncertainty 173–175
  uncertainty 381
box-and-whisker scatterplot *43*, 43–44, *44*
box plot
  Gaussian distribution 124
  quantiles 42, 43, 105
Burton equation 279–280

c

canonical correlation analysis 396–397
carbon dioxide and temperature *48*, 48–50, *50*
causation, correlation analysis 177

CCD. *See* charge-coupled device
central limit theorem, Gaussian distribution 123
  asymmetric uncertainty 143
centroid 95–100, *98*, *99*. *See also* mean; median; mode
  event detection metrics **373**
  fit performance metrics
    accuracy 251
    ME 263
    Gaussian distribution 13, 14, 126, *126*
      kurtosis 132
      lake pH 141, 142, 149
      skewness coefficient 128
      weighted average 146–148
      z test and 15, 18
  running average 45
  selection of 111–112
  two number sets
    covariance 158–159
    linear fit 183
CFCs. *See* chlorofluorocarbons
change vector 205
Chapman layer 200–201
charge-coupled device (CCD) 116
Chauvenet's criterion, Gaussian distribution
  lake pH 148–149
  outliers 144–145
Chicken Little 2, *2*, *24*, 80, 81, 400, *400*
Chimborazo Map (Naturegemälde) 51
chi-squared coefficient 132–137
chi-squared test
  Gaussian distribution 132–137
    degrees of freedom 134, **134**, **135**
    histogram 133–134, **134**
    monotonicity 136
    probability **134**, 134–135
    z score 136, **136**
  two number sets 154–155, 386–387

chlorofluorocarbons (CFCs)
  carbon dioxide and temperature 48
  ozone hole 201–203
choice combination statistics 387–392
clarity, plots 30–31
Clayton skill score (CSS), event detection metrics 323–324, **373**
Cleveland, W. S. *59*, 59–60
cluster analysis 396
coefficient normalization 206–207
coefficient of determination 162, 163
colorblindness 63, *63*
color scales 63–64, *64*
The Commercial and Political Atlas (Playfair) 51, *52*
contingency tables, event detection metrics 299–301, *300*, 302
  EDS 319
  event identification threshold sliding 349–352, *351*, *352*
  HSS 321
  odds ratio *315*, 315–316
  precipitation *307*
  precision 314
  SEDS 321
  significance of 307–309, *308*
convergence
  bootstrap method 383
  two number sets
    bootstrap method 174
    iterative curve fitting 207
    jackknife method 173
  uncertainty 393
correct negatives 299
correlation analysis
  canonical 396–397
  causation 177
  two number sets 157–168, *159*, **162**, **165**, *166*, *168*
correlation coefficient of the logs 167–167, *168*
cosine curve 12, *12*
counting statistics 113–116
  GCRs 116–119, *118*

covariance   158–160, *159*
CPD. *See* cumulative probability distribution
critical success index (CSI),
   event detection metrics   310–311, *311*
   event identification threshold sliding   337
   HSS   322
critical value
   of Kolmogorov-Smirnov test   137
   two number sets, linear fit   183
CSI. *See* critical success index
CSS. *See* Clayton skill score
cumulative probability distribution (CPD)
   association and extremes   378
   fit performance metrics, extremes   272–276, *274*, **275**, 277
   Gaussian distribution   14
   Kolmogorov-Smirnov test   137–139, *138*
   lake pH   142
   outliers   144
   z test and   15, 16
   model metrics   222–223, *223*
   two number sets, Kolmogorov-Smirnov test   154
curse of dimensionality   395–396
curve fitting. *See also* linear fit
   iterative
      machine learning   399
      two number sets   203–208, *204–206*
   multidimensional data analysis   395
   running average   47
   two number sets   181–209, **387**
      gradient descent curve fitting   208
      iterative   203–208
      Spearman rank-order correlation coefficient   166

d
decile   103
degrees of freedom
   fit performance metrics
      accuracy   253–255
   Gaussian distribution
      chi-squared test   134, **134**, **135**
      spread   108
      two number sets
         linear fit   185, 189
         Student's *t* test   156
         Welch's *t* test   157
delete-*d* jackknife   173
dependent variable
   exponential fitting   197
   uncertainty propagation   69
design of experiments   398
deuteranopia   63, *63*
dichotomous metrics   295
dimensionless parameters   216
discrepancy
   fit performance metrics
      accuracy   250, *250*, 260
      Dst index   248
   spread
      Gaussian distribution   127
      mean   101
      two number sets, polynomial fitting   195
   uncertainty   2
discrimination
   event detection metrics   325–326, **373**
   CSS   324
   fit performance metrics   281, 281–283, **283**
   Gaussian distribution   368
   non-Gaussian distributions   372
   model metrics   234
   reliability   379–380
distributions. *See also specific types*
   uncertainty   12, *12*, 380
disturbance storm time index (Dst index)
   event detection metrics
      event identification threshold sliding   349–362, *350–353*, **354**, **356**, *356*, **357**, **358**, **359**, **360**, **361**
   fit performance metrics   245–250, *247*, *248*, **249**, 258–259
      accuracy   251, 252, 253, 255
      association   270
      bias   263–264
      discrimination   282, *283*
      extremes   275–277
      precision   267, 268
      reliability   **284**, 284–285
      skill   278–280, *279*
      uncertainty   381–385, *382*, **383**, *384*
division
   arithmetic mean   96
   fit performance metrics, extremes   276
   uncertainty propagation   80–81
   formula   92
   independent variables   81–82
Dst index. *See* disturbance storm time index
d value
   degrees of freedom   255
   linear fit   190
   RMSE   255
   two number sets, Welch's *t* test   157

e
Earth's atmosphere
   atmospheric chemistry   169
   carbon dioxide and temperature   48, 48–50, *50*
   density, uncertainty propagation   72–75
   ozone and temperature   168–171, *169–171*, **170**
   ozone hole   200–203
Earth's magnetic field. *See also* disturbance storm time index
   choice combinations statistics   390–392
   plots   33–34, *34*

EDS. *See* extreme dependency score
empirical model (reanalysis model)   214, **215**
empirical orthogonal functions (EOFs)   395
ephemeris data (metadata)   28–29
error bar   27
event detection metrics   243
  accuracy   310–311, *311*
  assessment   372–374, *373*
  association   314–317
  bias   311–313, *312*
  contingency tables 299–301, *300*, 302
    EDS   319
    HSS   321
    odds ratio   *315*, 315–316
    precipitation   307
    precision   314
    SEDS   321
    significance of 307–309, *308*
  CSI   322
    event identification threshold sliding   337
  CSS   **373**
  data-model comparisons *301*, 301–303
  defining event   296–299, *297*
  dichotomous metrics   295
  discrimination   325–326
  EDS   *318*, 318–321
    event identification threshold sliding 340, *340*
  event identification threshold 297–298, *298*
    precipitation   306–307
    uncertainty   380
  event identification threshold sliding   333–363, *335*, *336*
    Alt-ROC   346–347, *347*, 359–362, **361**, **373**
    AUC   345–346, 359–362, **361**
    contingency tables 349–352, *351*, 352

DST index   349–362, *350–353*, **354**, **356**, *356*, **357**, *358*, **359**, *360*, **361**
F_1 score   352, 353, *353*, 354, **354**, *356*, *357*, 358–359
FAR   335
FB   352, 353, *353*, 354, **354**, *356*, *357*, 358–359
HSS   352, 353, *353*, 354, **354**, *356*, *357*, 358–359
MR   335, 348
NPV   335, 340
ORSS   343, 352, 353, *353*, 354, **354**, *357*, 358–359
POD   334, 345, 347–349, *348*
POFD   334, 336–337, 344–345, 347–349, *348*
power of   362–365
PPV   335, 348–349
RMSE   349
ROC curve   344–346, *345*, 359–362, *360*, **361**, **373**
scatterplots   *336*, *337*, 349, *350*
SEDS   340
STONE curve   347–349, *348*, 359–362, **361**, **373**
sweeping data threshold 340, 340–341
sweeping modeled and data thresholds simultaneously 342, 342–343
sweeping modeled threshold   337–340, *339*, 340
extremes   317–321, *318*, *320*
F_1 score   **373**
FAR   327
  event identification threshold sliding   335
FB   **373**
fit performance metrics 295–296, 301, 313

GSS   **373**
histogram
  HRRR   *305*
  precipitation   *305*, 306, 307
HRRR   303–307, *304*, *305*
HSS   **373**
  event identification threshold sliding 352, 353, *353*, 354, **354**, *356*, *357*, 358–359
MCC   **373**
MR   326–327, **373**
  event identification threshold sliding 335, 348
NPV   327
  event identification threshold sliding 335, 340
ORSS   316–317, **373**
  event identification threshold sliding 343, 352, 353, *353*, 354, **354**
PC   310–311, **373**
  HSS   321
POD   325–326, **373**
  event identification threshold sliding 334, 345, 347–349, *348*
POFD   325–327, **373**
  event identification threshold sliding   334, 336–337, 344–345, 347–349, *348*
  zero   336–337, 346
PPV   326, **373**
  event identification threshold sliding 335, 348–349
  precipitation   303–307, *304*, *305*, *307*, 327–328
  precision   313–314
PSS   **373**
recall   313–314
reliability   326–327, 334
  subsets   327, 334, 343

Index   411

RMSE, event identification
    threshold
    sliding   349
ROC curve, event
    identification
    threshold sliding
    344–346, *345*
  scatterplots   *297*, 297–299
  event identification
    threshold sliding
    336, *337*, *349*, *350*
  HRRR   305
  precipitation   305
SEDS   319–321, *320*, **373**
  contingency tables   321
  event identification
    threshold sliding
    340, *340*
  square-root rule   306
  STONE curve, event
    identification
    threshold sliding
    347–349, *348*
  subsets, reliability   327,
    334, 343
  summary   328, *329*
  uncertainty   301
event identification threshold,
    event detection
    metrics
    297–298, *298*
  precipitation   306–307
  sliding   333–363, *335–337*
    Alt-ROC   346–347, *347*
    AUC   345–346
    contingency tables
      349–352, *351*, *352*
    DST index   349–362,
      350–353, **354**, **356**,
      356, **357**, *358*, *359*,
      360, **361**
    F_1 score   352, 353, *353*,
      354, **354**, *356*
    FB   352, 353, *353*, 354,
      354, *356*
    HSS   352, 353, *353*, 354,
      354, *356*
    MR   348
    ORSS   352, 353, *353*,
      354, **354**
    POD   347–349, *348*

POFD   347–349, *348*
PPV   348–349
RMSE   349
ROC curve   344–346, *345*
scatterplots   *336*, *337*,
    349, *350*
STONE curve   347–349,
    348
sweeping data threshold
    340, 340–341
sweeping modeled and
    data thresholds
    simultaneously
    342, 342–343
sweeping modeled
    threshold   337–340,
    339, *340*
uncertainty   380
exception to the basic rule of
    significant digits   5–6
exclusive percentiles
    103–104
explanatory graphics
    31–33, *32*
exploratory graphics
    31–33, *32*
exponential fitting   197–198
exponential functions
  geometric mean   96
  uncertainty propagation   83
  formula   92
extreme dependency score
    (EDS), event detection
    metrics   *318*, 318–321
  event identification threshold
    sliding   340, *340*
extremes
  association   377, 377–378
  event detection metrics
    317–321, *318*, *320*, **373**
  fit performance metrics
    272–277
  Gaussian distribution
    368
  non-Gaussian distributions
    372

f

F_1 score, event detection
    metrics   310–311,
    311, **373**

event identification threshold
    sliding   352, 353,
    353, 354, **354**, *356*,
    357, 358–359
  precision and recall   314
faculae   226, 228
false alarm ratio (FAR), event
    detection metrics
    327
  event identification threshold
    sliding   335
false alarms, event detection
    metrics   299, 314
false discovery rate (FDR)   327
false negative rate (FNR),
    event detection
    metrics   326
  event identification threshold
    sliding   334
FAR. *See* false alarm ratio
FB. *See* frequency bias
FDR. *See* false discovery rate
feedback loop   49
fit performance metrics
    243–287
  accuracy   250, 250–262,
    251, *253*
  centroid   263
  degrees of freedom
    253–255
  Gaussian distribution
    368
  MAD   251–252, 258
  MAE   252–253, *253*, 255,
    257, 263
  MAPE   258–260
  ME   263
  MSA   260–261
  MSE   251–252, *253*,
    254–255, 278
  natural logarithm
    operator   260, 261
  non-Gaussian
    distributions   **372**
  normalization factor
    256–257, **257**
  PE   278
  percentage   257–261
  RMSE   252, *253*, 255,
    257, **257**
  SMAPE   259–260

fit performance metrics (*cont'd*)
  association 268–272, *269*
    Gaussian distribution **368**
    non-Gaussian distributions **372**
    Pearson linear correlation coefficient 269–270
    Spearman rank-order correlation coefficient 270–272, *271*
  bias
    Gaussian distribution 264, **368**
    geometric mean error 263
    ME 262–264
    non-Gaussian distributions **372**
    SSPB 265
  defined *244*, 244–245
  discrimination *281*, 281–283, **283**
    Gaussian distribution **368**
    non-Gaussian distributions **372**
  Dst index 245–250, *247*, *248*, **249**, 258–259
    association 270
    bias 263–264
    discrimination 282, *283*
    extremes 275–277
    precision 267, 268
    reliability **284**, 284–285
    skill 278–280, *279*
    uncertainty 381–385, *382*, **383**, *384*
  event detection metrics 295–296, 301, 313
  extremes
    CPD 272–276, *274*, **275**
    Gaussian distribution **368**
    HDE *273*, 273–275
    LDE 272–275, *273*, **275**
    non-Gaussian distributions **372**
    skew and kurtosis 276–277

Gaussian distribution 368–369
  bias 264, **368**
  histogram 244, *244*
    Dst index 383, *384*
  MAE
    discrimination 281, 282
    Gaussian distribution 369
    reliability 285
  ME, Dst index 383, **383**
  MSA, SSPB 265
  MSE, skill score 280
  non-Gaussian distributions 369–372, *371*, **372**
  normalization factor, Gaussian distribution 369
  PE
    Dst index 383, **383**
    Gaussian distribution 369
  precision 266–268, *267*
    Gaussian distribution **368**
    logarithmic modeling yield 267
    non-Gaussian distributions **372**
    standard deviation 268
    YI 266–267
  reliability 283–286, **284**, *284*
    Gaussian distribution **368**
    non-Gaussian distributions **372**
    Pearson correlation coefficient 286
  RMSE
    Dst index 383, **383**
    Gaussian distribution 369
    skill 278–281, *279*
    Gaussian distribution **368**
    non-Gaussian distributions **372**
  SMAPE 259–260
    discrimination 282
    reliability 285

standard deviation, Gaussian distribution 369
  subsets
    discrimination 282, 283
    reliability 284, *284*, 286
    summary 287, *288–290*
  fitted model 214, 215, **215**
  FNR. *See* false negative rate
  forecast ratio (FR), event detection metrics 327
  formula *92*
  Fourier transform 393
  FR. *See* forecast ratio
  fractional uncertainty 4
    uncertainty propagation 75–76
      division 80–81
      formula *92*
      multiplication 81
      stepwise 85
  frequency bias (FB), event detection metrics 311–313, *312*, **373**
    event identification threshold sliding 352, 353, *353*, 354, **354**, *356*, *357*, 358–359
  F statistic, two number sets 386–387
    linear fit 188–191, **190**, **191**
  full width at half maximum (FWHM)
    Gaussian distribution 127
    spread *106*, 106–107
    two number sets, bootstrap method 174

g
galactic cosmic rays (GCRs) 116–119, *118*
Galileo 9, 51, 227
Gantt charts *59*
Gaussian distribution *13*, 13–14, *15*
  assessment 123–149
  asymmetric uncertainty 142–144
  box plot 124

Index | 413

central limit theorem   123
centroid   126, *126*
   kurtosis   132
   skewness coefficient   128
chi-squared test   132–137
   degrees of freedom   134, **134**, **135**
   histogram   133–134, **134**
   monotonicity   136
   probability   **134**, 134–135
   z score   136, **136**
fit performance metrics   368–369
   accuracy   252
      extremes   275
      precision   267
   bias   264, **368**
Dst index   248–249
FWHM   107, 127
histogram   *124*, 124–125, *125*
   chi-squared test   133–134
   exponential fitting   198
   Kolmogorov-Smirnov test   **137**, 137–139, **138**, *138*
   kurtosis   130–132, *131*
   lake pH   *139–141*, 139–142, **141**, **142**, 148–149
   skewness coefficient   *127*, **128**, 128–130
HWHM   107
IQR   105
lake pH   141
maximum likelihood   209, 216
outliers   144–145
Poisson distribution   114–115
   chi-squared test   133, 136
population, chi-squared test   133
probability distribution   36, 183
quantile   *103*, 103–104
quantiles   111
random number   27
random uncertainty   111
spread   126–127, **127**
   uncertainty   381

standard deviation   102
   choice combination statistics   389, 392
   spread   127
   uncertainty   380
two number sets
   bootstrap method   174
   exponential fitting   197–198
   iterative curve fitting   207
   linear fit   181–182, *182*, 185, 187–188
   ozone and temperature   170, 171
   Pearson linear correlation   163
   polynomial fitting   195
   Spearman rank-order correlation coefficient   164, 166
   uncertainty propagation   74
   visualization   123
   weighted average   146–148
   z score   124
   z test and   15–18, *16*, **17**
      comparing two numbers   18–19
GCRs. *See* galactic cosmic rays
generalized linear coefficient fitting, two number sets   196–197
geomagnetic field   33
geometric mean   96–97
   fit performance metrics
      ME   263
      reliability   285
geometric mean error   263
Geophysical Research Letters (GRL)   53–57, *54*, *55*, *57*, *58*
geospace   245
geospace storm (magnetic storm)   246
Geostationary Operational Environmental Satellite (GOES)   370
Gilbert skill score (GSS)   324–325, **373**
GIS. *See* graphical information systems

GOES. *See* Geostationary Operational Environmental Satellite
goodness of fit
   accuracy, bias, and precision   374–376, *375*
   model metrics   216, 235–236
   gradient descent curve fitting   208
graphical information systems (GIS)   398
grayscale coloring   *59*
greenhouse gas   48
   planetary equilibrium temperature   88–89
GRL. See Geophysical Research Letters
GSS. *See* Gilbert skill score

h

half width at half maximum (HWHM)
   Gaussian distribution, asymmetric uncertainty   143
   spread   106–107
harmonic mean   96–97
Haugland, S. M.   54, *54*
HDE. *See* high distribution error
Heidke skill score (HSS)
   event detection metrics   321–323, **373**
   event identification threshold sliding   352, 353, *353*, 354, **354**, *356*, *357*, 358–359
   skill   378
HFCs. *See* hydrofluorocarbons
high distribution error (HDE)   *273*, 273–275
high-resolution rapid refresh (HRRR)   327–328
   event detection metrics   303–307, *304*, *305*
   event identification threshold sliding   349, 354, 355, 362

histogram. *See also* overlaid histogram
  arithmetic mean 99–100
  for distribution of number sets 12, *12*
  event detection metrics
    HRRR 305
    precipitation 305, 306, 307
  fit performance metrics 244, *244*
    accuracy 250, 250–251, *251*
    Dst index 248, **248**, 249, 383, *384*
  FWHM 106
  Gaussian distribution *124*, 124–125, *125*
    chi-squared test 133–134, **134**
    exponential fitting 198
    Kolmogorov-Smirnov test **137**, 137–139, **138**, *138*
    kurtosis 130–132, *131*
    lake pH *139–141*, 139–142, **141**, *142*, 148–149
    skewness coefficient *127*, **128**, 128–130
  levels of abstraction 59
  maximum likelihood 99
  mode 99, *99*, 111
  model metrics 219–221, *220*
  overlaid 39–40, *40*
  plots 35–39, *38*
  probability distributions 35–39, *38*
  for Saturn's moons 10, *10*
  scatterplot 41, *42*
  two number sets
    chi-squared test 154
    linear fit 182, 185–186, *186*
    ozone and temperature 170, *170*
  hits, event detection metrics contingency tables 299

hourly temperatures, uncertainty propagation 90, 90–91, *91*
HRRR. *See* high-resolution rapid refresh
HSS. *See* Heidke skill score
Hubble Space Telescope (HST) 116
hue *59*, 60
Humboldt, Alexander von 51
HWHM. *See* half width at half maximum
hydraulic fracturing 191–194, *192*, **193**, *193*
hydrofluorocarbons (HFCs) 48
hydrostatic equilibrium 72–73
hypothesis. *See also* null hypothesis
  scientific method 23
  two number sets, chi-squared test 154
hypothesis testing
  choice combination statistics 391
  Gaussian distribution 123
  two number sets, Kolmogorov-Smirnov test 154–155

i
Ice Age 48
ideal gas law 73
inclusive percentiles 104
independent variables
  two number sets, covariance 159
  uncertainty propagation 69–72, 91
    Earth's atmospheric density 72–75
    multiplication and division 81–82
interglacial periods 48
interplanetary magnetic field, GCRs 116–117
interquartile range (IQR) box plot 42–43

fit performance metrics
  Dst index 248, **248**, 249
  MAE 264
  MSA 261
  Gaussian distribution 105
  quantiles 105
  running average 45
  two number sets
    bootstrap method 174
    ozone and temperature **170**
interval for comparison, model metrics 237–239
IQR. *See* interquartile range
iterative curve fitting
  machine learning 399
  two number sets 203–208, *204–206*

j
jackknife method, two number sets
  ozone and temperature *175*, 175–177
  uncertainty 172–173

k
K-means processing 396
know the aspects of the plot on which you want audience to focus 29, 30
know the audience for whom it is being made 29, 30
know the main point being conveyed 29–30
Kolmogorov-Smirnov statistic 137
  Gaussian distribution, lake pH 141–142
Kolmogorov-Smirnov test
  Gaussian distribution histogram 137–139, **137–139**, *138*
  two number sets 154–155
kurtosis
  fit performance metrics
    accuracy 252, 253
    extremes 276–277
  Gaussian distribution histogram 130–132, *131*

Index | 415

two number sets, jackknife method 172
kurtosis coefficient
  fit performance metrics
    Dst index 248, **248**
    reliability 285
  Gaussian distribution, lake pH 141
  two number sets, ozone and temperature **170**, 171
kurtosis error 276–277

l

lagged correlation 394
lake pH, Gaussian distribution
  histogram *139–141*, 139–142, **141**, **142**, 148–149
LDE. *See* low distribution error
levels of abstraction 58–60, *59*
lightness *59*
likelihood ratio test (Wald test) 394
linear fit
  fit performance metrics
    accuracy 254
  two number sets
    ANOVA 188–191, **190**
    curve fitting 181–194, *182*, *186*, **190**, **191**, *192*, **193**, *193*
    F statistic 188–191, **190**, **191**
    hydraulic fracturing 191–194, *192*, **193**, *193*
    ozone hole 200–203
    weighted average 187–188
linear interpolation 104
linearity 157–168, *159*, **162**, **165**, *166*, *168*
line plots *54*, 54–55
logarithmic functions. *See also* correlation coefficient of the logs; natural logarithm operator
  geometric mean 96

uncertainty propagation 84
  formula 92
logarithmic modeling yield 267
low distribution error (LDE) 272–275, *273*, **275**
L-p norm 107–108
LU decomposition 195

m

machine learning 399
MAD. *See* mean absolute difference
MAE. *See* mean absolute error
magnetic field. *See also* disturbance storm time index of Earth
  choice combinations statistics 390–392
  plots 33–34, *34*
  fit performance metrics 245–250, *247*, **248**, **249**
  interplanetary, GCRs 116–117
  solar, TSI 226
magnetic storm (geospace storm) 246
magnetometer 33, 245–246
MAPE. *See* mean absolute percentage error
Matthews correlation coefficient (MCC) 317, **373**
maximum likelihood
  Gaussian distribution 209, 216
  histogram 99
  polynomial fitting 195
  probability distribution 146
  two number sets
    linear fit 183
    polynomial fitting 195
Maxwell's equations 33
MCC. *See* Matthews correlation coefficient
McGill, R. *59*, 59–60
McPherron, R. L. *21*, 246
ME. *See* mean error

mean. *See also* arithmetic mean; geometric mean; standard deviation of the mean
  discrepancy, spread 101
  Dst index, fit performance metrics 248, **248**
  Gaussian distribution
    lake pH 140, 141
    weighted average 147
  running average 45
  standard deviation 112–113
  two number sets
    bootstrap method 174
    covariance 158, 159, 160
    jackknife method 172
    linear fit 182
    Student's *t* test 155
    versions 96–97
mean absolute difference (MAD)
  fit performance metrics
    accuracy 251–252, 258
    Dst index **248**
    spread 101–102
    L-p norm 107–108
mean absolute error (MAE), fit performance metrics
  accuracy 252–253, *253*, 255, 257, 263–264
  discrimination 281, 282
  Gaussian distribution 369
  ME 263–264
  reliability 285
mean absolute percentage error (MAPE) 258–260
mean error (ME), fit performance metrics 262–264
  discrimination 281
  Dst index 383, **383**
  reliability 285
  skill score 280
mean square for error (MSE)
  fit performance metrics
    accuracy 251–252, *253*, 254–255, 278
    PE 278
    skill score 280
  machine learning 399

mean square for error (MSE) (cont'd)
  skill   378
  two number sets linear fit 189–191
mean square for regression (MSR)   189–191
measurement technique uncertainty   7–8
median   98, 98–99
  box plot   42
  Dst index, fit performance metrics   248, **248**
  Gaussian distribution, lake pH   141
  running average   45
  spread   111
  two number sets, ozone and temperature   **170**
median symmetric accuracy (MSA)
  fit performance metrics 260–261
  SSPB   265
metadata. *See* ephemeris data
metrics. *See* event detection metrics; fit performance metrics; model metrics
Milankovitch cycles   49
misses, event detection metrics contingency tables   299
miss ratio (MR), event detection metrics   326–327, **373**
  event identification threshold sliding   335, 348
mode   98, 98–99
  Gaussian distribution, asymmetric uncertainty   143
  histogram   99, *99*, 111
modeling yield (YI)   266–267
model metrics, 213–239. *See also* event detection metrics; fit performance metrics
  accuracy   230
  association   231
  AUL   236–237, *237*

bias   230
CPD   222–223, *223*
defined   **214**, 214–217
dimensionless parameters 216
discrimination   234
goodness of fit   216, 235–236
histogram   219–221, *220*
interval for comparison 237–239
nature of the quantity being compared   238
overlaid histogram   *221*, 221–222, *222*
precision   231
quantile-quantile plot 224–226, **225**, *225*
reliability   234
resolution for comparison 238
scatterplots   217–219, *218*, *219*
skill   231–234, *233*
subsets   234–235
TSI   226–229, *227–229*
two number sets visualization 217–226, *218–222*, *224*, **225**, *225*
monotonicity
  choice combination statistics   388
  ephemeris data   29
  event detection metrics, Dst index   351
FWHM   106
Gaussian distribution chi-squared test   136
Spearman rank-order correlation coefficient 164, 166
  two number sets, Spearman rank-order correlation coefficient   166
Montreal Protocol   202
MR. *See* miss ratio
MSA. *See* median symmetric accuracy
MSE. *See* mean square for error
MSR. *See* mean square for regression

multidimensional data analysis   394–396
multidimensional data-model comparisons 396–397
multidimensional iterative curve fitting 206–207
multiplication
  event detection metrics, PSS   323
  two number sets
    covariance   158
    linear fit   183, 184–185
    Pearson linear correlation 161
    polynomial fitting   196
    uncertainty propagation 78–81
      formula   *92*
      independent variables 81–82

n
Naked Statistics (Wheelan)   19
National Centers for Environmental Prediction (NCEP)   303
National Oceanic and Atmospheric Administration (NOAA)   303
National Weather Service (NWS)   303
natural exponential operator
  Gaussian distribution   13
  two number sets, linear fit   183
  uncertainty propagation 73, 83, 84
natural logarithm operator
  arithmetic mean   96
  event detection metrics, EDS   319
  fit performance metrics
    accuracy   260, 261
    bias   265
  uncertainty propagation   84
Naturegemälde (Chimborazo Map)   51

Index

nature of the quantity being compared 238
Navier-Stokes momentum equation 72
NCEP. *See* National Centers for Environmental Prediction
near-ultraviolet light 200
negative predictive value (NPV), event detection metrics 327
  event identification threshold sliding 335, 340
Nightingale, Florence 51, *53*
nitrogen oxides (Nox) 168–171, *169–171*, **170**
NOAA. *See* National Oceanic and Atmospheric Administration
nonlinear fit 194–200, *198*
normal distribution. *See* Gaussian distribution
normal equations
  multidimensional data analysis 395
  two number sets
    linear fit 184
    polynomial fitting 195–196
normalization factor
  fit performance metrics
    accuracy 256–257, **257**
    Gaussian distribution 369
    Gaussian distribution 14
Nox. *See* nitrogen oxides
NPV. *See* negative predictive value
null hypothesis
  choice combination statistics 391
  Gaussian distribution, lake pH 142
numerical model 214, **215**
NWS. *See* National Weather Service

o

O'Brien, T. P. O. *21*, 246. *See also* disturbance storm time index

odds ratio, event detection metrics *315*, 315–316
odds ratio skill score (ORSS), event detection metrics 316–317, **373**
event identification threshold sliding 343, 352, 353, *353*, 354, **354**, *357*, 358–359
one-dimensional data 27–33, *28, 31, 32*
one-dimensional iterative curve fitting 203–205
one-sided z test *16*, 16–17, **17**, *17*
ORSS. *See* odds ratio skill score
outliers
  event detection metrics **373**
  fit performance metrics
    extremes 276
    skill 278
  Gaussian distribution 144–145
    kurtosis 131
  two number sets, Spearman rank-order correlation coefficient 164
overlaid histogram 39–40, *40*
  model metrics *221*, 221–222, *222*
ozone and temperature, two number sets 168–171, *169–171*, **170**
  bootstrap method *175*, 175–177, *176*
  jackknife method *175*, 175–177
ozone hole 200–203

p

parameter variability uncertainty 7–8
PC. *See* proportion correct
PCA. *See* principal component analysis
PDFs. *See* probability distribution functions

PE. *See* performance efficiency
Pearson linear correlation coefficient
  association and extremes 377
  fit performance metrics
    association 269–270
    PE 278
    reliability 286
    TSI 227–228
  two number sets 161–163, **162**
  ozone and temperature 171
Peirce skill score (PSS) 323, **373**
percentage uncertainty 4–6, **6**, **7**
  uncertainty propagation 75–76
  formula *92*
percentile 103
performance efficiency (PE)
  fit performance metrics
    Dst index 383, **383**
    Gaussian distribution 369
    skill 278–279
    skill score 280
  skill 378
periodicity analysis 392–393
perturbation
  Earth's magnetic field 34
  Gaussian distribution 13
  random number 380
  random uncertainty 111
  two number sets, iterative curve fitting 207
  uncertainty propagation 70
piecewise linear fitting 198–199, *199*
pie charts, levels of abstraction *59*
Piet, H. 53, **54**, 56
planetary equilibrium temperature, uncertainty propagation 87–90
Playfair, William 51, *52*

Index

plots 27–64. *See also specific types*
annotations 30, 32
carbon dioxide and temperature 48, 48–50, *50*
clarity 30–31
Earth's magnetic field 33–34, *34*
ephemeris data 28–29
error bar 27
explanatory graphics 31–33, *32*
exploratory graphics 31–33, *32*
know the audience for whom it is being made 29, 30
know the main point being conveyed 29–30
one-dimensional data 27–33, *28*, *31*, *32*
probability distributions 35–39, *38*
running average 44–48, *46*, *47*
two data sets against each other 39–47, *40–44*, *46*, *47*
POD. *See* probability of detection
POFD. *See* probability of false detection
Poisson distribution 113–116, *114*
choice combination statistics 388–289, 392
Gaussian distribution, chi-squared test 133, 136
GCRs 118–119
Poisson uncertainty 113–116
polynomial fitting, two number sets 194–196
population
Gaussian distribution
chi-squared test 133
weighted average 147–148
spread 108–109
two number sets
chi-squared test 154
Student's *t* test 155
Welch's *t* test 157
z test 15
positive feedback loop 49
positive predictive value (PPV), event detection metrics 326, **373**
event identification threshold sliding 335
STONE curve 348, 349
power density, planetary equilibrium temperature 88
power laws, uncertainty propagation 82
formula 92
PPV. *See* positive predictive value
precipitation, event detection metrics 303–307, *304*, *305*, *307*, 327–328
precision
accuracy and bias 374–376, *375*
event detection metrics 313–314, **373**
fit performance metrics 266–268, *267*
Gaussian distribution **368**
non-Gaussian distributions **372**
model metrics 231
precision difference
bias 376
fit performance metrics 268
precision ratio 268
principal component analysis (PCA) 395
probability
choice combination statistics 388–390
fit performance metrics accuracy, precision 267
Gaussian distribution 14
chi-squared test **134**, 134–135
outliers 145
weighted average 146–147
z test 17
GCRs 118–119
Poisson distribution 114
two number sets
linear fit 183, 184
ozone and temperature 170, 171
Pearson linear correlation 161
polynomial fitting 195
Welch's *t* test 157
probability distribution. *See also* cumulative probability distribution; histogram
Gaussian distribution 183
maximum likelihood 146
plots 35–39, *38*
probability distribution functions (PDFs)
Gaussian distribution 14
two number sets, linear fit 182, 185
probability of detection (POD)
event detection metrics 325–326, **373**
event identification threshold sliding 334, 345
event detection metrics event identification threshold sliding, STONE curve 347–349, *348*
probability of false detection (POFD)
event detection metrics 325–327, **373**
event identification threshold sliding 334, 336–337, 344–345, 347–349, *348*
STONE curve 347–349, *348*
zero 336–337, 346

Program for Research on
Oxidants,
Photochemistry,
Emissions and
Transport
(PROPHET)   169
proportion correct (PC),
event detection
metrics
310–311, **373**
HSS   321
protanopia   63, *63*
PSS. *See* Peirce skill score
Pythagorean theorem   71

q
quadrature
Gaussian distribution,
asymmetric
uncertainty   143
two number sets, Welch's *t*
test   157
uncertainty propagation
71, 110
quantile
box plot   42, 43, 105
fit performance metrics,
extremes   276
Gaussian distribution
103, 103–104, 111
lake pH   141, **141**
IQR   105
running average   45
spread   102–105, *103*,
104, 111
two number sets, ozone
and temperature
170
quantile-quantile plot
association and
extremes   377
model metrics   224–226,
225, *225*
quartile   103. *See also*
interquartile
range
box-and-whisker plot   43
Gaussian distribution,
asymmetric
uncertainty
143

r
random number (generator)
convergence   173
Gaussian distribution   27
multidimensional iterative
curve fitting
206–207
perturbation   380
Spearman rank-order
correlation
coefficient   164
two number sets, jackknife
method   173
random uncertainty
8, 109–111
rank order. *See also* Spearman
rank-order
correlation
coefficient
box plot   42
reanalysis model (empirical
model)   214
recall, event detection
metrics   313–314
relative operating
characteristic curve
(ROC curve), event
detection metrics
event identification
threshold
sliding   344–346,
345, 359–362, *360*,
361, 373
STONE curve   347–348
reliability
discrimination   379–380
event detection metrics
326–327, 334, **373**
subsets   327, 334, 343
fit performance metrics
283–286, **284**, *284*
Gaussian
distribution   **368**
non-Gaussian
distributions   **372**
model metrics   234
resampling, two number sets
bootstrap method   173–175
jackknife method   172–173
resolution for comparison
238

Rice's rule   36–37
RMS. *See* root mean square
RMSE. *See* root mean square
error
ROC curve. *See* relative
operating
characteristic curve
ROC score   346
root mean square (RMS)   101
root mean square error
(RMSE)
event detection metrics event
identification
threshold sliding   349
fit performance metrics
accuracy   252, *253*, 255,
257, **257**
Dst index   383, **383**
Gaussian distribution   369
rounding
exception to the basic rule of
significant digits
5, 6
quantiles   104
running average
fit performance metrics,
discrimination   282
plots   44–48, *46*, *47*

s
sample set. *See also* population
centroids   112
correlation   177
cosine curve   12, *12*
delete-*d*-jackknife   173
jackknife method   173
Schwarz inequality   160
two number sets, linear
fit   185
weighted average   146
sample size
histogram   36
standard deviation of the
mean   113
Student's *t* test   155, 156
two number sets, Student's *t*
test   156
uncertainty   22
saturation   *59*, 60
Saturn's moons, uncertainty
with   9, 9–11, *10*

Index

scale height   73–74
scatterplots   40–41, *40–42*
   accuracy and association   376, 376–377
   association and extremes   377, 377–378
   box-and-whisker   43, 43–44, *44*
   carbon dioxide and temperature   49–50, *59*
   discrimination and reliability   379–380
   event detection metrics   297, 297–299
      contingency tables   299–301, *300*
      event identification threshold sliding   *336, 337, 349, 350*
      HRRR   305
      precipitation   305
   fit performance metrics   245
      accuracy   *250*, 250–251
      association   271, *271*
      discrimination   281, *281*
      reliability   284, *284*
   levels of abstraction   *59*
   model metrics   217–219, *218, 219*
   two number sets
      bootstrap method   174
      chi-squared test   154
      correlation coefficient of the logs   167–168, *168*
      covariance   *159*, 159–160
      linear fit   *182*, 185–186, *186*
      ozone and temperature   171, *171*
      Pearson linear correlation   161
      piecewise linear fitting   198–199, *199*
      Spearman rank-order correlation coefficient   164–166, **165**, *166*
Schwarz inequality   160

science, technology, engineering, and mathematics (STEM)   1
scientific method, uncertainty   22–24, *24*, *400*, 400–402
SEDS. *See* symmetric extreme dependency score
semi-log correlation coefficient   167
SEPs. *See* solar energetic particles
signal-to-noise ratio
   GCRs   119
   Poisson distribution   115–116
   Saturn's moons   *10*
significantly different
   chi-squared test   135, *136*
   choice combination statistics   391
   event detection metrics contingency tables   309
   Kolmogorov-Smirnov test   138
   z test   17–18
significant numbers/digits   3–7, **6**, **7**
   exception to the basic rule of significant digits   5–6
skew
   fit performance metrics
      Dst index   248
      extremes   276–277
   Gaussian distribution   128–130
      asymmetric uncertainty   143
      weighted average   147–148
   two number sets
      bootstrap method   174
      jackknife method   172
skew error   276–277
skewness coefficient
   fit performance metrics
      Dst index   **248**
      extremes   276
      reliability   285

Gaussian distribution histogram   *127*, **128**, 128–130
   lake pH   141
   two number sets, ozone and temperature   **170**, 171
skill   378–379
   event detection metrics   373
   fit performance metrics   278–281, *279*
      Gaussian distribution   **368**
      non-Gaussian distributions   **372**
   model metrics   231–234, *233*
skill score
   event detection metrics
      AUC skill score   346
      CSS   323–324
      GSS   324–325
      HSS   321–323, 352, 353, *353*, 356
      ORSS   316–317, 343
      PSS   323
      ROC score   346
   fit performance metrics   279–281
   model metrics   231–234, *233*
sliding threshold of observation for numeric evaluation curve (STONE curve)   347–349, *348*, 359–362, **361**, 373
SMAPE. *See* symmetric mean absolute percentage error
Smith, J. P.   20, 22
Snow, K.   55, 55–56
solar energetic particles (SEPs)   369–370, *371*
solar magnetic field   226
solar wind density
   Dst index, fit performance metrics   247
   uncertainty   20–22, *21*

Index

space weather   20–22, *21*
Spearman rank-order correlation coefficient
    fit performance metrics, association   270–272, *271*
    two number sets   163–166, **165**, *166*
specificity, event detection metrics   326
spread   100–109
    event detection metrics   313, 314, **373**
    fit performance metrics
        accuracy   251
        bias   264
        Dst index   248
        reliability   285–286
    FWHM   *106*, 106–107
    Gaussian distribution   13, 14, 126–127, **127**
        asymmetric uncertainty   143
        lake pH   141, **141**, 142, 149
        outliers   145
        uncertainty   381
        uncertainty propagation   74
        weighted average   146–148
        z test and   15
    HWHM   106–107
    L-p norm   107–108
    population   108–109
    quantiles   102–105, *103*, *104*
    running average   45
    selection of   111–112
    standard deviation, population   109
    two number sets
        bootstrap method   174
        chi-squared test   154
        covariance   158–159
        exponential fitting   197–198
        linear fit   183, 185
        polynomial fitting   195
        Spearman rank-order correlation coefficient   164, 166
        Welch's *t* test   157
        uncertainty   109, 111–112
square-root rule
    event detection metrics   306
    probability distributions   36–37
SSE. *See* sum of squares for error
SSN. *See* sunspot number
SSPB. *See* symmetric signed percentage bias
SST. *See* total sum of squares
standard deviation
    arithmetic mean   102, 111
    fit performance metrics
        accuracy   251, 252, 275
        bias   264
        Dst index   **248**
        extremes   275, 276
        Gaussian distribution   369
        precision   268
        reliability   285
    Gaussian distribution   102
        chi-square test   132–137
        choice combination statistics   389, 392
        fit performance metrics   369
        lake pH   140, 142, 149
        spread   127
        uncertainty   380
        weighted average   146
    Poisson distribution   115
    running average   45
    spread   100, 102
    population   109
    two number sets
        bootstrap method   174
        jackknife method   172
        linear fit   185
        ozone and temperature   170, **170**
        Student's *t* test   155
        Welch's *t* test   157
standard deviation of the mean (standard error)   112–113
Gaussian distribution
    lake pH   141
    weighted average   147
    two number sets, jackknife method   172, 173
statistical significance
    choice combination statistics   391
    linear fit   190
    Pearson linear correlation coefficient   270
    Poisson distribution   116
    two number sets, Pearson linear correlation   161
Stefan-Boltzmann constant   88
STEM. *See* science, technology, engineering, and mathematics
stepwise uncertainty propagation   85–87
STONE curve. *See* sliding threshold of observation for numeric evaluation curve
stratosphere   201
Student's *t* test   155–156, **156**
Sturges' formula   36
subsets
    discrimination and reliability   380
    event detection metrics   **373**
        reliability   327, 334, 343
    fit performance metrics
        discrimination   282, 283
        reliability   284, *284*, 286
    model metrics   234–235
subtraction
    fit performance metrics
        accuracy   *250*
        extremes   276
    Gaussian distribution
        chi-squared test   135–136
    uncertainty propagation   77–78
    formula   *92*
    uncertainty quantification   397

Index

sum of squares for error (SSE) 188–190, **190**, 199, 203, 207, 208
   fit performance metrics, PE 278
sunspot number (SSN) *227*, 227–228, *228*
sunspots 226
superposed epoch analysis *54*, 54–55
symmetric extreme dependency score (SEDS), event detection metrics 319–321, *320*, **373**
   event identification threshold sliding 340, *340*
symmetric mean absolute percentage error (SMAPE), fit performance metrics 259–260
   discrimination 282
   reliability 285
symmetric signed percentage bias (SSPB) 265
   fit performance metrics, reliability 285
systematic error 109–111, *110*
systematic uncertainty 8

t

temperature
   carbon dioxide and *48*, 48–50, *50*
   hourly, uncertainty propagation *90*, 90–91, *91*
   ozone and
      bootstrap method *175*, 175–177, *176*
      jackknife method *175*, 175–177
      two number sets 168–171, *169–171*, **170**
   theoretical model 214, **215**
tightness 314
time-lagged analysis 393–394
time-lagged correlation 393

total solar irradiance (TSI) 226–229, *227–229*
total sum of squares (SST)
   fit performance metrics, PE 278
two number sets linear fit 189–191
trigonometric functions, uncertainty propagation 84–85
tritanopia 63, *63*
TSI. *See* total solar irradiance
two number sets 153–177
   bootstrap method
      ozone and temperature *175*, 175–177, *176*
      uncertainty 173–175
   chi-squared test 154–155, 386–387
   correlation analysis 157–168, *159*, **162**, **165**, *166*, *168*
   covariance 158–160, *159*
   curve fitting 181–209, **387**
      exponential 197–198
      generalized linear coefficient 196–197
      gradient descent 208
      iterative 203–208, *204–206*
      nonlinear 194–200, *198*
      piecewise 198–199, *199*
      polynomial 194–196
   F statistic 386–387
   Gaussian distribution, bootstrap method 174
   jackknife method
      ozone and temperature *175*, 175–177
      uncertainty 172–173
   Kolmogorov-Smirnov test 154–155
   linear fit 181–194, *182*, *186*, *192*
      ANOVA 188–191, **190**
      F statistic 188–191, **190**, **191**
      hydraulic fracturing 191–194, *192*, **193**, *193*
      ozone hole 200–203
      testing 188–191, **190**, **191**
      weighted average 187–188

linearity 157–168, *159*, **162**, **165**, *166*, *168*
   ozone and temperature 168–171, *169–171*, **170**
   Pearson linear correlation coefficient 161–163, **162**
   ozone and temperature 171
   robust comparison of 385–387, **387**
   Spearman rank-order correlation coefficient 163–166, **165**, *166*
Student's *t* test 155–156, **156**
time-lagged analysis 393–394
uncertainty 172–177, *175*, *176*, 386, **387**
uncertainty propagation **387**
visualization, model metrics 217–226, *218–222*, *224*, **225**, *225*
Welch's *t* test 156–157
two-sided z test *16*, 16–17, **17**, *17*

u

uncertainty 1–24
   absolute 4–5, 6
      stepwise uncertainty propagation 85
   asymmetric 142–144
   bootstrap method 381
   Chicken Little 2, *2*, 24, 80, *81*, 400, *400*
   comparing two numbers 11–19, *12*, *13*, *15*, *16*, **17**, *17*
   z test 18–19
   convergence 393
   discrepancy 2
   distributions 12, *12*, 380
   event detection metrics 301
   event identification threshold 380
   fit performance metrics
      accuracy 255
      Dst index 248, 381–385, *382*, **383**, *384*

Index

fractional 4
  uncertainty
    propagation 75–76, 80–81, 85, *92*
  Gaussian distribution 13, 13–14, *15*
    lake pH 148–149
    spread 381
    standard deviation 380
    weighted average 147
    z test and 15–18, *16*, **17**
  GCRs 119
  MAD 102
  meaning to values from 3
  measurement technique 7–8
  parameter variability 7–8
  percentage 4–6, **6**, **7**
    uncertainty propagation 75–76, *92*
  Poisson distribution 115
  random 8, 109–111
  sample size 22
  Saturn's moons 9, 9–11, *10*
  scientific method 22–24, *24*, *400*, 400–402
  significant numbers 3–7, **6**, **7**
  solar wind density 20–22, *21*
  space weather 20–22, *21*
  spread 109, 111–112
    Gaussian distribution 381
  standard deviation of the mean 113
  systematic 8
  two number sets 172–177, *175*, *176*, 386, **387**
    bootstrap method 173–175
    exponential fitting 197–198
    linear fit 185–186, *186*
  types 7–8
  use and misuse of statistics 19–20
uncertainty propagation 69–93
  addition 77–78
    formula *92*
  dependent variable 69

division 80–81
  formula *92*
  independent variables 81–82
  Earth's atmospheric density 72–75
exponential functions 83
  formula *92*
formula guide *92*
fractional uncertainty 75–76
  division 80–81
  formula *92*
  multiplication 81
  stepwise 85
hourly temperatures 90, 90–91, *91*
independent variables 69–72
  Earth's atmospheric density 72–75
  multiplication and division 81–82
logarithmic functions 84
  formula *92*
multiplication 78–81
  formula *92*
  independent variables 81–82
percentage uncertainty 75–76
  formula *92*
planetary equilibrium temperature 87–90
power laws 82
  formula *92*
quadrature 71, 110
special cases 77–82
stepwise 85–87
subtraction 77–78
  formula *92*
trigonometric functions 84–85
two number sets **387**
uncertainty quantification 397–398
universal time (UT) 34, *34*

v

validation, uncertainty quantification 397–398

variance 102
  Dst index, fit performance metrics **248**
  machine learning 399
  two number sets
    Pearson linear correlation 162, 163
    Student's *t* test 155
    Welch's *t* test 157
verification, uncertainty quantification 397–398
visualization. *See also* plots
  best practices 58–64, *59*, *60*, *62–64*
  colorblindness 63, *63*
  color scales 63–64, *64*
  Gaussian distribution 123
  history 52, *52*, *53*
  levels of abstraction 58–60, *59*
  modern-day example 53–57, *54*, *55*, *57*, *58*
  two number sets, model metrics 217–226, *218–222*, *224*, **225**, *225*
volatile organic compounds (VOCs), ozone and temperature 168–171, *169–171*, **170**

w

Wald test (likelihood ratio test) 394
wavelet analysis 393
weighted average
  Gaussian distribution 146–148
  lake pH 149
  two number sets
    linear fit 187–188
    Student's *t* test 156
  Welch's *t* test 156–157
Wheelan, Charles 19

y

YI. *See* modeling yield
Yule's Q score 316–317

Z

zero
  event detection metrics
    EDS 319
    HSS 322
    ORSS 316
    POFD 336–337, 346
  fit performance metrics
    accuracy 250, 258–259
    bias 265
    Dst index 248
    MAPE 258–259
    skill score 280
  Gaussian distribution 13
    kurtosis 132
    lake pH 142
    skewness coefficient 128, 129–130
  weighted average 147–148
  model metric 232
  Poisson distribution 114
  two number sets
    covariance 158, 159
    linear fit 183, 184, 186–187
    Pearson linear correlation 161, 162
    polynomial fitting 196
  uncertainty 3, 5
  uncertainty propagation
    trigonometric functions 85

z score 15
  chi-squared test
    Gaussian distribution 136, **136**
    two number sets 154
  choice combination statistics 389
  Gaussian distribution 124
    chi-squared test 136, **136**
    lake pH 148
    outliers 144
  GCRs 118–119
  two number sets, chi-squared test 154

z test 14–18, *16*, **17**, *17*
  comparing two numbers 18–19